Chemometrics

Chemometrics

Data Analysis for the Laboratory and Chemical Plant

Richard G. Brereton

University of Bristol, UK

WILEY

Other Wiley Editorial Offices

John Wiley & Sons Inc., 111 River Street, Hoboken, NJ 07030, USA

Jossey-Bass, 989 Market Street, San Francisco, CA 94103-1741, USA

Wiley-VCH Verlag GmbH, Boschstr. 12, D-69469 Weinheim, Germany

John Wiley & Sons Australia Ltd, 33 Park Road, Milton, Queensland 4064, Australia

John Wiley & Sons (Asia) Pte Ltd, 2 Clementi Loop #02-01, Jin Xing Distripark, Singapore 129809

John Wiley & Sons Canada Ltd, 22 Worcester Road, Etobicoke, Ontario, Canada M9W 1L1

Wiley also publishes its books in a variety of electronic formats. Some content that appears
in print may not be available in electronic books.

Library of Congress Cataloging-in-Publication Data

Brereton, Richard G.
   Chemometrics : data analysis for the laboratory and chemical plant / Richard Brereton.
    p. cm.
   Includes bibliographical references and index.
    ISBN 0-471-48977-8 (hardback : alk. paper) – ISBN 0-470-84911-8 (pbk. : alk. paper)
    1. Chemistry, Analytic–Statistical methods–Data processing. 2. Chemical
processes–Statistical methods–Data processing.   I. Title.
QD75.4.S8 B74 2002
543′.007′27–dc21   2002027212

British Library Cataloguing in Publication Data

A catalogue record for this book is available from the British Library

ISBN 0-471-48977-8 (Hardback)
ISBN 0-471-48978-6 (Paperback)

Typeset in 10/12pt Times by Laserwords Private Limited, Chennai, India
Printed and bound in Great Britain by Antony Rowe Ltd, Chippenham, Wiltshire
This book is printed on acid-free paper responsibly manufactured from sustainable forestry
in which at least two trees are planted for each one used for paper production.

Contents

Preface

This text is a product of several years activities from myself. First and foremost, the task of educating students in my research group from a wide variety of backgrounds over the past 10 years has been a significant formative experience, and this has allowed me to develop a large series of problems which we set every 3 weeks and present answers in seminars. From my experience, this is the best way to learn chemometrics! In addition, I have had the privilege to organise international quality courses mainly for industrialists with the participation as tutors of many representatives of the best organisations and institutes around the world, and I have learnt from them. Different approaches are normally taken when teaching industrialists who may be encountering chemometrics for the first time in mid-career and have a limited period of a few days to attend a condensed course, and university students who have several months or even years to practice and improve. However, it is hoped that this book represents a symbiosis of both needs.

In addition, it has been a great inspiration for me to write a regular fortnightly column for Chemweb (available to all registered users on *www.chemweb.com*) and some of the material in this book is based on articles first available in this format. Chemweb brings a large reader base to chemometrics, and feedback via e-mails or even travels around the world have helped me formulate my ideas. There is a very wide interest in this subject but it is somewhat fragmented. For example, there is a strong group of near-infrared spectroscopists, primarily in the USA, that has led to the application of advanced ideas in process monitoring, who see chemometrics as a quite technical industrially oriented subject. There are other groups of mainstream chemists who see chemometrics as applicable to almost all branches of research, ranging from kinetics to titrations to synthesis optimisation. Satisfying all these diverse people is not an easy task.

This book relies heavily on numerical examples: many in the body of the text come from my favourite research interests, which are primarily in analytical chromatography and spectroscopy; to have expanded the text more would have produced a huge book of twice the size, so I ask the indulgence of readers whose area of application may differ. Certain chapters, such as that on calibration, could be approached from widely different viewpoints, but the methodological principles are the most important and if you understand how the ideas can be applied in one area you will be able to translate to your own favourite application. In the problems at the end of each chapter I cover a wider range of applications to illustrate the broad basis of these methods. The emphasis of this book is on understanding ideas, which can then be applied to a wide variety of problems in chemistry, chemical engineering and allied disciplines.

It was difficult to select what material to include in this book without making it too long. Every expert to whom I have shown this book has made suggestions for new material. Some I have taken into account and I am most grateful for every proposal, others I have mentioned briefly or not at all, mainly for reasons of length and also to ensure that this text sees the light of day rather than constantly expands without end.

There are many outstanding specialist books for the enthusiast. It is my experience, though, that if you understand the main principles (which are quite few in number), and constantly apply them to a variety of problems, you will soon pick up the more advanced techniques, so it is the building blocks that are most important.

In a book of this nature it is very difficult to decide on what detail is required for the various algorithms: some readers will have no real interest in the algorithms, whereas others will feel the text is incomplete without comprehensive descriptions. The main algorithms for common chemometric methods are presented in Appendix A.2. Step-by-step descriptions of methods, rather than algorithms, are presented in the text. A few approaches that will interest some readers, such as cross-validation in PLS, are described in the problems at the end of appropriate chapters which supplement the text. It is expected that readers will approach this book with different levels of knowledge and expectations, so it is possible to gain a great deal without having an in-depth appreciation of computational algorithms, but for interested readers the information is nevertheless available. People rarely read texts in a linear fashion, they often dip in and out of parts of it according to their background and aspirations, and chemometrics is a subject which people approach with very different types of previous knowledge and skills, so it is possible to gain from this book without covering every topic in full. Many readers will simply use Add-ins or Matlab commands and be able to produce all the results in this text.

Chemometrics uses a very large variety of software. In this book we recommend two main environments, Excel and Matlab; the examples have been tried using both environments, and you should be able to get the same answers in both cases. Users of this book will vary from people who simply want to plug the data into existing packages to those that are curious and want to reproduce the methods in their own favourite language such as Matlab, VBA or even C. In some cases instructors may use the information available with this book to tailor examples for problem classes. Extra software supplements are available via the publisher's *www. SpectroscopyNOW.com* Website, together with all the datasets and solutions associated with this book.

The problems at the end of each chapter form an important part of the text, the examples being a mixture of simulations (which have an important role in chemometrics) and real case studies from a wide variety of sources. For each problem the relevant sections of the text that provide further information are referenced. However, a few problems build on the existing material and take the reader further: a good chemometrician should be able to use the basic building blocks to understand and use new methods. The problems are of various types, so not every reader will want to solve all the problems. Also, instructors can use the datasets to construct workshops or course material that go further than the book.

I am very grateful for the tremendous support I have had from many people when asking for information and help with datasets, and permission where required. Chemweb is thanked for agreement to present material modified from articles originally published in their e-zine, *The Alchemist*, and the Royal Society of Chemistry for permission to base the text of Chapter 5 on material originally published in *The Analyst* [**125**, 2125–2154 (2000)]. A full list of acknowledgements for the datasets used in this text is presented after this preface.

Tom Thurston and Les Erskine are thanked for a superb job on the Excel add-in, and Hailin Shen for outstanding help with Matlab. Numerous people have tested out the answers to the problems. Special mention should be given to Christian Airiau, Kostas

Zissis, Tom Thurston, Conrad Bessant and Cevdet Demir for access to a comprehensive set of answers on disc for a large number of exercises so I can check mine. In addition, several people have read chapters and made detailed comments, particularly checking numerical examples. In particular, I thank Hailin Shen for suggestions about improving Chapter 6 and Mohammed Wasim for careful checking of errors. In some ways the best critics are the students and postdocs working with me, because they are the people that have to read and understand a book of this nature, and it gives me great confidence that my co-workers in Bristol have found this approach useful and have been able to learn from the examples.

Finally I thank the publishers for taking a germ of an idea and making valuable suggestions as to how this could be expanded and improved to produce what I hope is a successful textbook, and having faith and patience over a protracted period.

Bristol, February 2002   Richard Brereton

Supplementary Information

Supplementary information is available on the publisher's spectroscopyNOW website.

To access this information, go to www.spectroscopynow.com and select the 'Chemometrics' channel. A website for the book is available – you should be able to access this either via the "Features" on the opening page or the left-hand side "Education" menu. If in doubt, use the search facility to find the book, or send an e-mail to chemometrics@wiley.co.uk.

The website contains the following.

1. Extensive worked solutions to all problems in the book.
2. All the datasets both in the problems and the main text, organised as tables in Word, available as a single downloadable zip file. These are freely available to all readers of the book, but you are asked to acknowledge their source in any publication or report, for example via citations.
3. VBA code for PCA and labelling points as described in Section A.4.6.1. These are freely available.
4. Excel macros for MLR, PCA, PCR and PLS as described in Section A.4.6.2, written by Tom Thurston, based on original material by Les Erskine. These are freely available for private and academic educational uses, but if used for profit making activities such as consultancy or industrial research, or profit making courses, you must contact bris-chemom@bris.ac.uk for terms of agreement.
5. Matlab procedures corresponding to the main methods in the book, cross-referenced to specific sections, written by Hailin Shen. These are freely available for private and academic educational uses, but if used for profit making activities such as consultancy or industrial research, or profit making courses, you must contact bris-chemom@bris.ac.uk for terms of agreement.

A password is required for the Excel macros and Matlab procedures, as outlined in the website; this is available to all readers of the book. This corresponds to a specific word on a given line of a given page of the book. The password may change but there will always be current details available on-line. If there are problems, contact chemometrics@wiley.co.uk.

Acknowledgements

The following have provided me with sources of data for this text. All other case studies are simulations.

Dataset	Source
Problem 2.2	A. Nordin, L. Eriksson and M. Öhman, *Fuel*, **74**, 128–135 (1995)
Problem 2.6	G. Drava, University of Genova
Problem 2.7	I. B. Rubin, T. J. Mitchell and G. Goldstein, *Anal. Chem.*, **43**, 717–721 (1971)
Problem 2.10	G. Drava, University of Genova
Problem 2.11	Y. Yifeng, S. Dianpeng, H. Xuebing and W. Shulan, *Bull. Chem. Soc. Jpn.*, **68**, 1115–1118 (1995)
Problem 2.12	D. V. McCalley, University of West of England, Bristol
Problem 2.15	D. Vojnovic, B. Campisi, A. Mattei and L. Favreto, *Chemom. Intell. Lab. Syst.*, **27**, 205–219 (1995)
Problem 2.16	L. E. Garcia-Ayuso and M. D. Luque de Castro, *Anal. Chim. Acta*, **382**, 309–316 (1999)
Problem 3.8	K. D. Zissis, University of Bristol
Problem 3.9	C. Airiau, University of Bristol
Table 4.1	S. Dunkerley, University of Bristol
Table 4.2	D. V. McCalley, University of West of England, Bristol
Tables 4.14, 4.15	D. V. McCalley, University of West of England, Bristol
Problem 4.1	A. Javey, Chemometrics On-line
Problem 4.5	D. Duewer, National Institute of Standards and Technology, USA
Problem 4.7	S. Wold, University of Umeå (based on R. Cole and K. Phelps, *J. Sci. Food Agric.*, **30**, 669–676 (1979))
Problem 4.8	P. Bruno, M. Caselli, M. L. Curri, A. Genga, R. Striccoli and A. Traini, *Anal. Chim. Acta*, **410**, 193–202 (2000)
Problem 4.10	R. Vendrame, R. S. Braga, Y. Takahata and D. S. Galvão, *J. Chem. Inf. Comput. Sci.*, **39**, 1094–1104 (1999)
Problem 4.12	S. Dunkerley, University of Bristol
Table 5.1	S. D. Wilkes, University of Bristol
Table 5.20	S. D. Wilkes, University of Bristol
Problem 5.1	M. C. Pietrogrande, F. Dondi, P. A. Borea and C. Bighi, *Chemom. Intell. Lab. Syst.*, **5**, 257–262 (1989)
Problem 5.3	H. Martens and M. Martens, *Multivariate Analysis of Quality*, Wiley, Chichester, 2001, p. 14
Problem 5.6	P. M. Vacas, University of Bristol
Problem 5.9	K. D. Zissis, University of Bristol

Problem 6.1   S. Dunkerley, University of Bristol
Problem 6.3   S. Dunkerley, University of Bristol
Problem 6.5   R. Tauler, University of Barcelona (results published in
                     R. Gargallo, R. Tauler and A. Izquierdo-Ridorsa, *Quim.*
                     Anali., **18**, 117–120 (1999))
Problem 6.6   S. P. Gurden, University of Bristol

1 Introduction

1.1 Points of View

There are numerous groups of people interested in chemometrics. One of the problems over the past two decades is that each group has felt it is dominant or unique in the world. This is because scientists tend to be rather insular. An analytical chemist will publish in analytical chemistry journals and work in an analytical chemistry department, a statistician or chemical engineer or organic chemist will tend to gravitate towards their own colleagues. There are a few brave souls who try to cross disciplines but on the whole this is difficult. However, many of the latest advances in theoretical statistics are often too advanced for routine chemometrics applications, whereas many of the problems encountered by the practising analytical chemist such as calibrating pipettes and checking balances are often too mundane to the statistician. Cross-citation analysis of different groups of journals, where one looks at which journal cites which other journal, provides fascinating insights into the gap between the theoretical statistics and chemometrics literature and the applied analytical chemistry journals. The potential for chemometrics is huge, ranging from physical chemistry such as kinetics and equilibrium studies, to organic chemistry such as reaction optimisation and QSAR, theoretical chemistry, most areas of chromatography and spectroscopy on to applications as varied as environmental monitoring, scientific archaeology, biology, forensic science, industrial process monitoring, geochemistry, etc., but on the whole there is no focus, the ideas being dissipated in each discipline separately. The specialist chemometrics community tends to be mainly interested in industrial process control and monitoring plus certain aspects of analytical chemistry, mainly near-infrared spectroscopy, probably because these are areas where there is significant funding for pure chemometrics research. A small number of tutorial papers, reviews and books are known by the wider community, but on the whole there is quite a gap, especially between computer based statisticians and practising analytical chemists.

This division between disciplines spills over into industrial research. There are often quite separate data analysis and experimental sections in many organisations. A mass spectrometrist interested in principal components analysis is unlikely to be given time by his or her manager to spend a couple of days a week mastering the various ins and outs of modern chemometric techniques. If the problem is simple, that is fine; if more sophisticated, the statistician or specialist data analyst will muscle in, and try to take over the project. But the statistician may have no feeling for the experimental difficulties of mass spectrometry, and may not understand when it is most effective to continue with the interpretation and processing of data, or when to suggest changing some mass spectrometric parameters.

All these people have some interest in data analysis or chemometrics, but approach the subject in radically different ways. Writing a text that is supposed to appeal to a broad church of scientists must take this into account. The average statistician likes to build on concepts such as significance tests, matrix least squares and so on. A statistician is unlikely to be satisfied if he or she cannot understand a method in algebraic terms. Most texts, even most introductory texts, aimed at statisticians contain a fair amount of algebra. Chemical engineers, whilst not always so keen to learn about distributions and significance tests, are often very keen on matrix algebra, and a chemometrics course taught by a chemical engineer will often start with matrix least squares and linear algebra.

Practical chemists, on the other hand, often think quite differently. Many laboratory based chemists are doing what they are doing precisely because at an early phase in their career they were put off by mathematicians. This is especially so with organic chemists. They do not like ideas expressed in terms of formal maths, and equations are 'turn offs'. So a lecture course aimed at organic chemists would contain a minimum of maths. Yet some of these people recognise later in their career that they do need data analytical tools, even if these are to design simple experiments or for linear regression, or in QSAR. They will not, however, be attracted to chemometrics if they are told they are required first to go on a course on advanced statistical significance testing and distributions, just to be able to perform a simple optimisation in the laboratory. I was told once by a very advanced mathematical student that it was necessary to understand Gallois field theory in order to perform multilevel calibration designs, and that everyone in chemometrics should know what Krilov space is. Coming from a discipline close to computing and physics, this may be true. In fact, the theoretical basis of some of the methods can be best understood by these means. However, tell this to an experimentalist in the laboratory that this understanding is required prior to performing these experiments and he or she, even if convinced that chemometrics has an important role, will shy away. In this book we do not try to introduce the concepts of Gallois field theory or Krilov space, although I would suspect not many readers would be disappointed by such omissions.

Analytical chemists are major users of chemometrics, but their approach to the subject often causes big dilemmas. Many analytical chemists are attracted to the discipline because they are good at instrumentation and practical laboratory work. The majority spend their days recording spectra or chromatograms. They know what to do if a chromatographic column needs changing, or if a mass spectrum is not fragmenting as expected. Few have opted to work in this area specifically because of their mathematical background, yet many are confronted with huge quantities of data. The majority of analytical chemists accept the need for statistics and a typical education would involve some small level of statistics, such as comparison of means and of errors and a little on significance tests, but the majority of analytical texts approach these subjects with a minimum of maths. A number then try to move on to more advanced data analysis methods, mainly chemometrics, but often do not recognise that a different knowledge base and skills are required. The majority of practising analytical chemists are not mathematicians, and find equations difficult; however, it is important to have some understanding of the background to the methods they use. Quite correctly, it is not necessary to understand the statistical theory of principal components analysis or singular value decomposition or even to write a program to perform this (although it is in fact very easy!). However, it is necessary to have a feel for methods for data scaling, variable selection and interpretation of the principal components, and if one has such knowledge it probably is not too difficult to expand one's understanding to the algorithms themselves. In fact, the algorithms are a relatively small part of the data

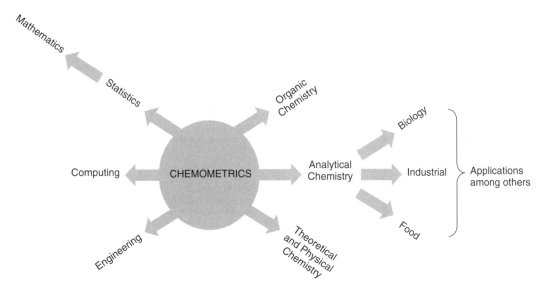

Figure 1.1
How chemometrics relates to other disciplines

analysis, and even in a commercial chemometric software package PCA or PLS (two popular approaches) may involve between 1 and 5 % of the code.

The relationship of chemometrics to different disciplines is indicated in Figure 1.1. On the left are the enabling sciences, mainly quite mathematical and not laboratory based. Statistics, of course, plays a major role in chemometrics, and many applied statisticians will be readers of this book. Statistical approaches are based on mathematical theory, so statistics falls between mathematics and chemometrics. Computing is important as much of chemometrics relies on software. However, chemometrics is not really computer science, and this book will not describe approaches such as neural networks or genetic programming, despite their potential importance in helping solve many complex problems in chemistry. Engineers, especially chemical and process engineers, have an important need for chemometric methods in many areas of their work, and have a quite different perspective from the mainstream chemist.

On the right are the main disciplines of chemistry that benefit from chemometrics. Analytical chemistry is probably the most significant area, although some analytical chemists make the mistake of claiming chemometrics uniquely as their own. Chemometrics has a major role to play and had many of its origins within analytical chemistry, but is not exclusively within this domain. Environmental chemists, biologists, food chemists as well as geochemists, chemical archaeologists, forensic scientists and so on depend on good analytical chemistry measurements and many routinely use multivariate approaches especially for pattern recognition, and so need chemometrics to help interpret their data. These scientists tend to identify with analytical chemists. The organic chemist has a somewhat different need for chemometrics, primarily in the areas of experimental design (e.g. optimising reaction conditions) and QSAR (quantitative structure–analysis relationships) for drug design. Finally, physical chemists such as spectroscopists, kineticists and materials scientists often come across methods for signal deconvolution and multivariate data analysis.

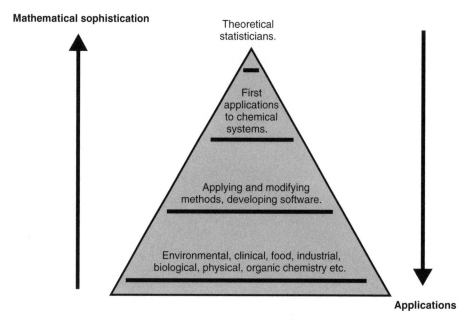

Figure 1.2
People interested in chemometrics

Different types of people will be interested in chemometrics, as illustrated in Figure 1.2. The largest numbers are application scientists. Many of these will not have a very strong mathematical background, and their main interest is to define the need for data analysis, to design experiments and to interpret results. This group may consist of some tens of thousands of people worldwide, and is quite large. A smaller number of people will apply methods in new ways, some of them developing software. These may well be consultants that interface with the users: many specialist academic research groups are at this level. They are not doing anything astoundingly novel as far as theoretical statisticians are concerned, but they will take problems that are too tough and complex for an applications scientist and produce new solutions, often tinkering with the existing methods. Industrial data analysis sections and dedicated software houses usually fit into this category too. There will be a few thousand people in such categories worldwide, often organised into diverse disciplines. A rather smaller number of people will be involved in implementing the first applications of computational and statistical methods to chemometrics. There is a huge theoretical statistical and computational literature of which only a small portion will eventually be useful to chemists. In-vogue approaches such as multimode data analysis, Bayesian statistics, and wavelet transforms are as yet not in common currency in mainstream chemistry, but fascinate the more theoretical chemometrician and over the years some will make their way into the chemists' toolbox. Perhaps in this group there are a few hundred or so people around the world, often organised into very tightly knit communities. At the top of the heap are a very small number of theoreticians. Not much of chemical data analysis is truly original from the point of view of the mathematician – many of the 'new' methods might have been reported in the mathematical literature 10, 20 or even 50 years ago; maybe the number of mathematically truly original chemometricians is 10 or less. However, mathematical novelty is not the only sign of innovation. In fact, much of science involves connecting ideas. A good chemometrician may have the mathematical ability to understand the ideas of the theoretician and then translate these into potential applications. He or she needs to be a good listener and to be able to link the various levels of the triangle. Chemical data analysis differs from more unitary disciplines such as organic chemistry, where most scientists have a similar training base, and above a certain professional level the difference is mainly in the knowledge base.

Readers of this book are likely to be of two kinds, as illustrated in Figure 1.3. The first are those who wish to ascend the triangle, either from outside or from a low level. Many of these might be analytical chemists, for example an NIR spectroscopist who has seen the need to process his or her data and may wish some further insight into the methods being used. Or an organic chemist might wish to have the skills to optimise a synthesis, or a food chemist may wish to be able to interpret the tools for relating the results of a taste panel to chemical constituents. Possibly you have read a paper, attended a conference or a course or seen some software demonstrated. Or perhaps in the next-door laboratory, someone is already doing some chemometrics, perhaps you have heard about experimental design or principal components analysis and need some insight into the methods. Maybe you have some results but have little idea how to interpret them and perhaps by changing parameters using a commercial package you are deluged with graphs and not really certain whether they are meaningful. Some readers might be MSc or PhD students wishing to delve a little deeper into chemometrics.

The second group already has some mathematical background but wishes to enter the triangle from the side. Some readers of this book will be applied statisticians, often

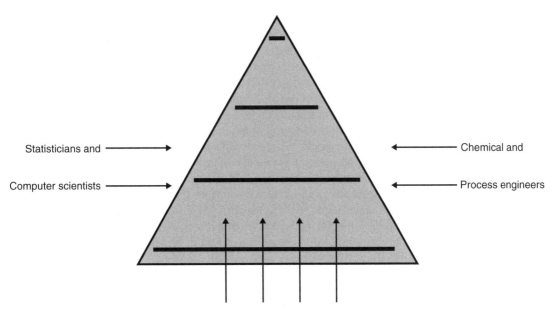

Figure 1.3
Groups of people with potential interest in this text

working in industry. Matrix algebra, significance tests and distributions are well known, but what is needed is to brush up on techniques as applied specifically to chemical problems. In some organisations there are specific data processing sections and this book is aimed as a particularly useful reference for professionals working in such an environment. Because there are not a large number of intensive courses in chemical data analysis, especially leading to degrees, someone with a general background in statistical data analysis who has moved job or is taking on extra responsibilities will find this book a valuable reference. Chemical engineers have a special interest in chemometrics and many are encountering the ideas when used to monitor processes.

1.2   Software and Calculations

The key to chemometrics is to understand how to perform meaningful calculations on data. In most cases these calculations are too complex to do by hand or using a calculator, so it is necessary to use some software.

The approach taken in this text, which differs from many books on chemometrics, is to understand the methods using numerical examples. Some excellent texts and reviews are more descriptive, listing the methods available together with literature references and possibly some examples. Others have a big emphasis on equations and output from packages. This book, however, is based primarily on how I personally learn and understand new methods, and how I have found it most effective to help students working with me. Data analysis is not really a knowledge based subject, but more a skill based subject. A good organic chemist may have an encyclopaedic knowledge of reactions in their own area. The best supervisor will be able to list to his or her students thousands of reactions, or papers or conditions that will aid their students, and with experience this knowledge base grows. In chemometrics, although there are quite a number of named methods, the key is not to learn hundreds of equations by heart, but to understand a few basic principles. These ideas, such as multiple linear regression, occur again and again but in different contexts. To become skilled in chemometric data analysis, what is required is practice in manipulating numbers, not an enormous knowledge base. Although equations are necessary for the formal description of methods, and cannot easily be avoided, it is easiest to understand the methods in this book by looking at numbers. So the methods described in this text are illustrated using numerical examples which are available for the reader to reproduce. The datasets employed in this book are available on the publisher's Website. In addition to the main text there are extensive problems at the end of each main chapter. All numerical examples are fairly small, designed so that you can check all the numbers yourselves. Some are reduced versions of larger datasets, such as spectra recorded at 5 nm rather than 1 nm intervals. Many real examples, especially in chromatography and spectroscopy, simply differ in size to those in this book. Also, the examples are chosen so that they are feasible to analyse fairly simply.

One of the difficulties is to decide what software to employ in order to analyse the data. This book is not restrictive and you can use any approach you like. Some readers like to program their own methods, for example in C or Visual Basic. Others may like to use a statistical packages such as SAS or SPSS. Some groups use ready packaged chemometrics software such as Pirouette, Simca, Unscrambler and several others on the market. One problem with using packages is that they are often very focused in their facilities. What they do they do excellently, but if they cannot do what you want you may be stuck, even for relatively simple calculations. If you have an excellent multivariate package but want to use a Kalman filter, where do you turn? Perhaps you have the budget to buy another package, but if you just want to explore the method, the simplest implementation takes only an hour or less for an experienced Matlab programmer to implement. In addition, there are no universally agreed definitions, so a 'factor' or 'eigenvector' might denote something quite different according to the software used. Some software has limitations, making it unsuitable for many applications of chemometrics, a very simple example being the automatic use of column centring in PCA in most general statistical packages, whereas many chemometric methods involve using uncentred PCA.

Nevertheless, many of the results from the examples in this book can successfully be obtained using commercial packages, but be aware of the limitations, and also understand the output of any software you use. It is important to recognise that the definitions used in this book may differ from those employed by any specific package. Because there are a huge number of often incompatible definitions available, even for fairly common parameters, in order not to confuse the reader we have had to adopt one single definition for each parameter, so it is important to check carefully with your favourite package or book or paper if the results appear to differ from those presented in this book. It is not the aim of this text to replace an international committee that defines chemometrics terms. Indeed, it is unlikely that such a committee would be formed because of the very diverse backgrounds of those interested in chemical data analysis.

However, in this text we recommend that readers use one of two main environments.

The first is Excel. Almost everyone has some familiarity with Excel, and in Appendix A.4 specific features that might be useful for chemometrics are described. Most calculations can be performed quite simply using normal spreadsheet functions. The exception is principal components analysis (PCA), for which a small program must be written. For instructors and users of VBA (a programming language associated with Excel), a small macro that can be edited is available, downloadable from the publisher's Website. However, some calculations such as cross-validation and partial least squares (PLS), whilst possible to set up using Excel, can be tedious. It is strongly recommended that readers do reproduce these methods step by step when first encountered, but after a few times, one does not learn much from setting up the spreadsheet each time. Hence we also provide a package that contains Excel add-ins for VBA to perform PCA, PLS, MLR (multiple linear regression) and PCR (principal components regression), that can be installed on PCs which have at least Office 97, Windows 98 and 64 Mbyte memory. The software also contains facilities for validation. Readers of this book should choose what approach they wish to take.

A second environment, that many chemical engineers and statisticians enjoy, is Matlab, described in Appendix A.5. Historically the first significant libraries of programs in chemometrics first became available in the late 1980s. Quantum chemistry, originating in the 1960s, is still very much Fortran based because this was the major scientific programming environment of the time, and over the years large libraries have been developed and maintained, so a modern quantum chemist will probably learn Fortran. The vintage of chemometrics is such that a more recent environment to scientific programming has been adopted by the majority, and many chemometricians swap software using Matlab. The advantage is that Matlab is very matrix oriented and it is most convenient to think in terms of matrices, especially since most data are multivariate. Also, there are special facilities for performing singular value decomposition (or PCA) and the pseudoinverse used in regression, meaning that it is not necessary to program these basic functions. The user interface of Matlab is not quite as user-friendly as Excel and is more suited to the programmer or statistician rather than the laboratory based chemist. However, there have been a number of recent enhancements, including links to Excel, that allow easy interchange of data, which enables simple programs to be written that transfer data to and from Excel. Also, there is no doubt at all that matrix manipulation, especially for complex algorithms, is quite hard in VBA and Excel. Matlab is an excellent environment for learning the nuts and bolts of chemometrics. A slight conceptual problem with Matlab is that it is possible to avoid looking at the raw numbers, whereas most users of Excel will be forced to look at the raw numerical data in detail, and I have come across experienced Matlab users who are otherwise very good at chemometrics but who sometimes miss quite basic information because they are not constantly examining the numbers – so if you are a dedicated Matlab user, look at the numerical information from time to time!

An ideal situation would probably involve using both Excel and Matlab simultaneously. Excel provides a good interface and allows flexible examination of the data, whereas Matlab is best for developing matrix based algorithms. The problems in this book have been tested both in Matlab and Excel and identical answers obtained. Where there are quirks of either package, the reader is guided. If you are approaching the triangle of Figure 1.3 from the sides you will probably prefer Matlab, whereas if you approach it from the bottom, it is more likely that Excel will be your choice.

Two final words of caution are needed. The first is that some answers in this book have been rounded to a few significant figures. Where intermediate results of a calculation have been presented, putting these intermediate results back may not necessarily result in exactly the same numerical results as retaining them to higher accuracy and continuing the calculations. A second issue that often perplexes new users of multivariate methods is that it is impossible to control the sign of a principal component (see Chapter 4 for a description of PCA). This is because PCs involve calculating square roots which may give negative as well as positive answers. Therefore, using different packages, or even the same package but with different starting points, can result in reflected graphs, with scores and loadings that are opposite in sign. It is therefore unlikely to be a mistake if you obtain PCs that are opposite in sign to those in this book.

1.3   Further Reading

There have been a large number of texts and review articles covering differing aspects of chemometrics, often aimed at a variety of audiences. This chapter summarises some of the most widespread. In most cases these texts will allow the reader to delve further into the methods introduced within this book. In each category only a few main books will be mentioned, but most have extensive bibliographies allowing the reader to access information especially from the primary literature.

1.3.1   General

The largest text in chemometrics is published by Massart and co-workers, part of two volumes [1,2]. These volumes provide an in-depth summary of many modern chemometric methods, involving a wide range of techniques, and many references to the literature. The first volume, though, is strongly oriented towards analytical chemists, but contains an excellent grounding in basic statistics for measurement science. The books are especially useful as springboards for the primary literature. This text is a complete rewrite of the original book published in 1988 [3], which is still cited as a classic in the analytical chemistry literature.

Otto's book on chemometrics [4] is a welcome recent text, that covers quite a range of topics but at a fairly introductory level. The book looks at computing in general in analytical chemistry including databases, and instrumental data acquisition. It does not deal with the multivariate or experimental design aspects in a great deal of detail but is a very clearly written introduction for the analytical chemist, by an outstanding educator.

Beebe and co-workers at Dow Chemical have recently produced a book [5] which is useful for many practitioners, and contains very clear descriptions especially of multivariate calibration in spectroscopy. However there is a strong 'American School' originating in part from the pioneering work of Kowalski in NIR spectroscopy and process control, and whilst covering the techniques required in this area in an out-standing way, and is well recommended as a next step for readers of this text working in this application area, it lacks a little in generality, probably because of the very close association between NIR and chemometrics in the minds of some. Kramer has produced a somewhat more elementary book [6]. He is well known for his consultancy company and highly regarded courses, and his approach is less mathematical. This will suit some people very well, but may not be presented in a way that suits statisticians and chemical engineers.

One of the first ever texts in the area of chemometrics was co-authored by Kowal-ski [7]. The book is somewhat mathematical and condensed, but provided a good manual for the mathematically minded chemometrician of the mid-1980s, and is a use-ful reference. Kowalski also edited a number of symposium volumes in the early days of the subject. An important meeting, the NATO Advanced Study School in Cosenza, Italy, in 1983, brought together many of the best international workers in this area and the edited volume from this is a good snapshot of the state-of-the-art of the time [8], although probably the interest is more for the historians of chemometrics.

The present author published a book on chemometrics about a decade ago [9], which has an emphasis on signal resolution and minimises matrix algebra, and is an introductory tutorial book especially for the laboratory based chemist. The jour-nal *Chemometrics and Intelligent Laboratory Systems* published regular tutorial review articles over its first decade or more of existence. Some of the earlier articles are good introductions to general subjects such as principal components analysis, Fourier transforms and Matlab. They are collected together as two volumes [10,11]. They also contain some valuable articles on expert systems.

Meloun and co-workers published a two volume text in the early 1990s [12,13]. These are very thorough texts aimed primarily at the analytical chemist. The first volume contains detailed descriptions of a large number of graphical methods for handling analytical data, and a good discussion of error analysis, and the second volume is a very detailed discussion of linear and polynomial regression.

Martens and Martens produced a recent text which gives quite a detailed discussion on how multivariate methods can be used in quality control [14], but covers sev-eral aspects of modern chemometrics, and so should be classed as a general text on chemometrics.

1.3.2 Specific Areas

There are a large number of texts and review articles dealing with specific aspects of chemometrics, interesting as a next step up from this book, and for a comprehensive chemometrics library.

1.3.2.1 Experimental Design

In the area of experimental design there are innumerable texts, many written by statisticians. Specifically aimed at chemists, Deming and Morgan produced a highly regarded book [15] which is well recommended as a next step after this text. Bayne and Rubin have written a clear and thorough text [16]. An introductory book discussing mainly factorial designs was written by Morgan as part of the Analytical Chemistry by Open Learning Series [17]. For mixture designs, involving compositional data, the classical statistical text by Cornell is much cited and recommended [18], but is quite mathematical.

1.3.2.2 Pattern Recognition

There are several books on pattern recognition and multivariate analysis. An introduction to several of the main techniques is provided in an edited book [19]. For more statistical in-depth descriptions of principal components analysis, books by Joliffe [20] and Mardia and co-authors [21] should be read. An early but still valuable book by Massart and Kaufmann covers more than just its title theme 'cluster analysis' [22] and provides clear introductory material.

1.3.2.3 Multivariate Curve Resolution

Multivariate curve resolution is the main topic of Malinowski's book [23]. The author is a physical chemist and so the book is oriented towards that particular audience, and especially relates to the spectroscopy of mixtures. It is well known because the first edition (in 1980) was one of the first major texts in chemometrics to contain formal descriptions of many common algorithms such as principal components analysis.

1.3.2.4 Multivariate Calibration

Multivariate calibration is a very popular area, and the much reprinted classic by Martens and Næs [24] is possibly the most cited book in chemometrics. Much of the text is based around NIR spectroscopy which was one of the major success stories in applied chemometrics in the 1980s and 1990s, but the clear mathematical descriptions of algorithms are particularly useful for a wider audience.

1.3.2.5 Statistical Methods

There are numerous books on general statistical methods in chemistry, mainly oriented towards analytical and physical chemists. Miller and Miller wrote a good introduction [25] that takes the reader through many of the basic significance tests, distributions, etc. There is a small amount on chemometrics in the final chapter. The Royal Society of Chemistry publish a nice introductory tutorial text by Gardiner [26]. Caulcutt and Boddy's book [27] is also a much reprinted and useful reference. There are several other competing texts, most of which are very thorough, for example, in describing applications of the t-test, F-test and analysis of variance (ANOVA) but which do not progress much into modern chemometrics. If you are a physical chemist, Gans' viewpoint on deconvolution and curve fitting may suit you more [28], covering many regression methods, but remember that physical chemists like equations more than analytical chemists and so approach the topic in a different manner.

1.3.2.6 Digital Signal Processing and Time Series

There are numerous books on digital signal processing (DSP) and Fourier transforms. Unfortunately, many of the chemically based books are fairly technical in nature and oriented towards specific techniques such as NMR; however, books written primarily by and for engineers and statisticians are often quite understandable. A recommended reference to DSP contains many of the main principles [29], but there are several similar books available. For nonlinear deconvolution, Jansson's book is well known [30]. Methods for time series analysis are described in more depth in an outstanding and much reprinted book written by Chatfield [31].

1.3.2.7 Articles

We will not make very great reference to the primary literature in this text. Many of the authors of well regarded texts first published material in the form of research, review and tutorial articles, which then evolved into books. However, it is worth mentioning a very small number of exceptionally well regarded tutorial papers. A tutorial by Wold and co-workers on principal components analysis [32] in the 1980s is a citation classic in the annals of chemometrics. Geladi and Kowalski's tutorial [33] on partial least squares is also highly cited and a good introduction. In the area of moving average filters, Savitsky and Golay's paper [34] is an important original source.

1.3.3 Internet Resources

Another important source of information is via the Internet. Because the Internet changes very rapidly, it is not practicable in this text to produce a very comprehensive list of Websites; however, some of the best resources provide regularly updated links to other sites, and are likely to be maintained over many years.

A good proportion of the material in this book is based on an expanded version of articles originally presented in ChemWeb's e-zine the *Alchemist*. Registration is free [35] and past articles are in the chemometrics archive. There are several topics that are not covered in this book. Interested readers are also referred to an article which provides a more comprehensive list of Web resources [36].

Wiley's *Chemometrics World* is a comprehensive source of information freely available to registered users via their SpectroscopyNOW Website [37], and the datasets and software from this book are available via this Website.

There are one or two excellent on-line textbooks, mainly oriented towards statisticians. Statsoft have a very comprehensive textbook [38] that would allow readers to delve into certain topics introduced in this text in more detail. Hyperstat also produce an on-line statistics textbook, mainly dealing with traditional statistical methods, but their Website also provides references to other electronic tutorial material [39], including Stockburger's book on multivariate statistics [40].

1.4   References

1. D. L. Massart, B. G. M. Vandeginste, L. M. C. Buydens, S. De Jong, P. J. Lewi and J. Smeyers-Verbeke, *Handbook of Chemometrics and Qualimetrics Part A*, Elsevier, Amsterdam, 1997.
2. B. M. G. Vandeginste, D. L. Massart, L. M. C. Buydens, S. de Jong, P. J. Lewi and J. Smeyers-Verbeke, *Handbook of Chemometrics and Qualimetrics Part B*, Elsevier, Amsterdam, 1998.
3. D. L. Massart, B. G. M. Vandeginste, S. N. Deming, Y. Michotte, and L. Kaufman, *Chemometrics: a Textbook*, Elsevier, Amsterdam, 1988.
4. M. Otto, *Chemometrics: Statistics and Computer Applications in Analytical Chemistry*, Wiley-VCH, Weinheim, 1998.
5. K. R. Beebe, R. J. Pell and M. B. Seasholtz, *Chemometrics: a Practical Guide*, Wiley, New York, 1998.
6. R. Kramer, *Chemometrics Techniques for Quantitative Analysis*, Marcel Dekker, New York, 1998.
7. M. A. Sharaf, D. L. Illman and B. R. Kowalski, *Chemometrics*, Wiley, New York, 1996.
8. B. R. Kowalski (Editor), *Chemometrics: Mathematics and Statistics in Chemistry*, Reidel, Dordrecht, 1984.
9. R. G. Brereton, *Chemometrics: Applications of Mathematics and Statistics to Laboratory Systems*, Ellis Horwood, Chichester, 1990.
10. D. L. Massart, R. G. Brereton, R. E. Dessy, P. K. Hopke, C. H. Spiegelman and W. Wegscheider (Editors), *Chemometrics Tutorials*, Elsevier, Amsterdam, 1990.
11. R. G. Brereton, D. R. Scott, D. L. Massart, R. E. Dessy, P. K. Hopke, C. H. Spiegelman and W. Wegscheider (Editors), *Chemometrics Tutorials II*, Elsevier, Amsterdam, 1992.
12. M. Meloun, J. Militky and M. Forina, *Chemometrics for Analytical Chemistry*, Vol. 1, Ellis Horwood, Chichester, 1992.
13. M. Meloun, J. Militky and M. Forina, *Chemometrics for Analytical Chemistry*, Vol. 2, Ellis Horwood, Chichester, 1994.
14. H. Martens and M. Martens, *Multivariate Analysis of Quality*, Wiley, Chichester, 2000.
15. S. N. Deming and S. L. Morgan, *Experimental Design: a Chemometric Approach*, Elsevier, Amsterdam, 1994.
16. C. K. Bayne and I. B. Rubin, *Practical Experimental Designs and Optimisation Methods for Chemists*, VCH, Deerfield Beach, FL, 1986.
17. E. Morgan, *Chemometrics: Experimental Design*, Wiley, Chichester, 1995.
18. J. A. Cornell, *Experiments with Mixtures: Design, Models, and the Analysis of Mixture Data*, Wiley, New York, 2nd edn, 1990.
19. R. G. Brereton (Editor), *Multivariate Pattern Recognition in Chemometrics, Illustrated by Case Studies*, Elsevier, Amsterdam, 1992.
20. I. T. Joliffe, *Principal Components Analysis*, Springer-Verlag, New York, 1987.
21. K. V. Mardia, J. T. Kent and J. M. Bibby, *Multivariate Analysis*, Academic Press, London, 1979.
22. D. L. Massart and L. Kaufmann, *The Interpretation of Analytical Chemical Data by the Use of Cluster Analysis*, Wiley, New York, 1983.
23. E. R. Malinowski, *Factor Analysis in Chemistry*, Wiley, New York, 2nd edn, 1991.
24. H. Martens and T. Næs, *Multivariate Calibration*, Wiley, Chichester, 1989.
25. J. N. Miller and J. Miller, *Statistics for Analytical Chemistry*, Prentice-Hall, Hemel Hempstead, 1993.
26. W. P. Gardiner, *Statistical Analysis Methods for Chemists: a Software-based Approach*, Royal Society of Chemistry, Cambridge, 1997.
27. R. Caulcutt and R. Boddy, *Statistics for Analytical Chemists*, Chapman and Hall, London, 1983.

28. P. Gans, *Data Fitting in the Chemical Sciences: by the Method of Least Squares*, Wiley, Chichester, 1992.
29. P. A. Lynn and W. Fuerst, *Introductory Digital Signal Processing with Computer Applications*, Wiley, Chichester, 2nd edn, 1998.
30. P. A. Jansson (Editor), *Deconvolution: with Applications in Spectroscopy*, Academic Press, New York, 1984.
31. C. Chatfield, *Analysis of Time Series: an Introduction*, Chapman and Hall, London, 1989.
32. S. Wold, K. Esbensen and P. Geladi, *Chemom. Intell. Lab. Syst.*, **2**, 37 (1987).
33. P. Geladi and B. R. Kowalski, *Anal. Chim. Acta*, **185**, 1 (1986).
34. A. Savitsky and M. J. E. Golay, *Anal. Chem.*, **36**, 1627 (1964).
35. *www.chemweb.com.*
36. R. G. Brereton, Chemometrics on the Net, *Alchemist*, 2 April 2001 *(www.chemweb.com).*
37. *www.spectroscopynow.com.*
38. *www.statsoft.com/textbook/stathome.html.*
39. *http://davidmlane.com/hyperstat/.*
40. *www.psychstat.smsu.edu/MultiBook/mlt00.htm.*

2 Experimental Design

2.1 Introduction

Although all chemists acknowledge the need to be able to design laboratory based experiments, formal statistical (or chemometric) rules are rarely taught as part of mainstream chemistry. In contrast, a biologist or psychologist will often spend weeks carefully constructing a formal statistical design prior to investing what could be months or years in time-consuming and often unrepeatable experiments and surveys. The simplest of experiments in chemistry are relatively quick and can be repeated, if necessary under slightly different conditions, so not all chemists see the need for formalised experimental design early in their career. For example, there is little point spending a week constructing a set of experiments that take a few hours to perform. This lack of expertise in formal design permeates all levels from management to professors and students. However, most real world experiments are expensive; for example, optimising conditions for a synthesis, testing compounds in a QSAR study, or improving the chromatographic separation of isomers can take days or months of people's time, and it is essential under such circumstances to have a good appreciation of the fundamentals of design.

There are several key reasons why the chemist can be more productive if he or she understands the basis of design, including the following four main areas.

1. *Screening*. These types of experiments involve seeing which factors are important for the success of a process. An example may be the study of a chemical reaction, dependent on proportion of solvent, catalyst concentration, temperature, pH, stirring rate, etc. Typically 10 or more factors might be relevant. Which can be eliminated, and which should be studied in detail? Approaches such as factorial or Plackett–Burman designs (Sections 2.3.1–2.3.3) are useful in this context.
2. *Optimisation*. This is one of the commonest applications in chemistry. How to improve a synthetic yield or a chromatographic separation? Systematic methods can result in a better optimum, found more rapidly. Simplex is a classical method for optimisation (Section 2.6), although several designs such as mixture designs (Section 2.5) and central composite designs (Section 2.4) can also be employed to find optima.
3. *Saving time*. In industry, this is possibly the major motivation for experimental design. There are obvious examples in optimisation and screening, but even more radical cases, such as in the area of quantitative structure–property relationships. From structural data, of existing molecules, it is possible to predict a small number of compounds for further testing, representative of a larger set of molecules. This allows enormous savings in time. Fractional factorial, Taguchi and Plackett–Burman designs (Sections 2.3.2 and 2.3.3) are good examples, although almost all experimental designs have this aspect in mind.
4. *Quantitative modelling*. Almost all experiments, ranging from simple linear calibration in analytical chemistry to complex physical processes, where a series of

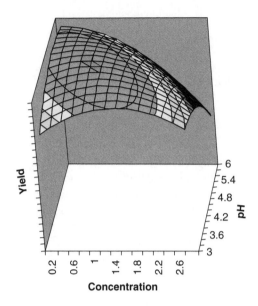

Figure 2.1
Yield of a reaction as a function of pH and catalyst concentration

observations are required to obtain a mathematical model of the system, benefit from good experimental design. Many such designs are based around the central composite design (Section 2.4), although calibration designs (Section 2.3.4) are also useful.

An example of where systematic experimental design is valuable is the optimisation of the yield of a reaction as a function of reagent concentration. A true representation is given in Figure 2.1. In reality this contour plot is unknown in advance, but the experimenter wishes to determine the pH and concentration (in mM) that provides the best reaction conditions. To within 0.2 of a pH and concentration unit, this optimum happens to be pH 4.4 and 1.0 mM. Many experimentalists will start by guessing one of the factors, say concentration, then finding the best pH at that concentration.

Consider an experimenter who chooses to start the experiment at 2 mM and wants to find the best pH. Figure 2.2 shows the yield at 2.0 mM. The best pH is undoubtedly a low one, in fact pH 3.4. So the next stage is to perform the experiments at pH 3.4 and improve on the concentration, as shown in Figure 2.3. The best concentration is 1.4 mM. These answers, pH 3.4 and 1.4 mM, are far from the true values.

The reason for this problem is that the influences of pH and temperature are not independent. In chemometric terms, they 'interact'. In many cases, interactions are commonsense. The best pH in one solvent may be different to that in another solvent. Chemistry is complex, but how to find the true optimum, by a quick and efficient manner, and be confident in the result? Experimental design provides the chemist with a series of rules to guide the optimisation process which will be explored later.

A rather different example relates to choosing compounds for biological tests. Consider the case where it is important to determine whether a group of compounds is harmful, often involving biological experiments. Say there are 50 potential compounds in the group. Running comprehensive and expensive tests on each compound

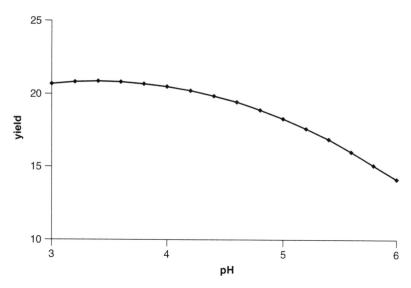

Figure 2.2
Cross-section through surface of Figure 2.1 at 2 mM

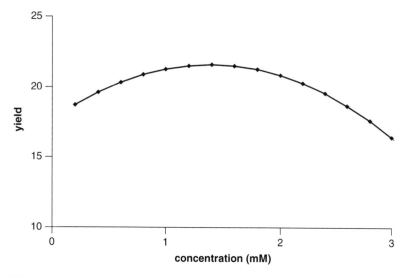

Figure 2.3
Cross-section through surface of Figure 2.1 at pH 3.4

is prohibitive. However, it is likely that certain structural features will relate to toxicity. The trick of experimental design is to choose a selection of the compounds and then decide to perform tests only this subset.

Chemometrics can be employed to develop a mathematical relationship between chemical property descriptors (e.g. bond lengths, polarity, steric properties, reactivities, functionalities) and biological functions, via a computational model such as principal components analysis. The question asked is whether it is really necessary to test all

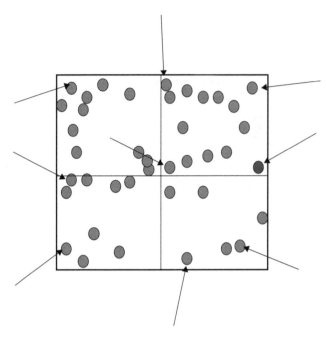

Figure 2.4
Choice of nine molecules based on two properties

50 compounds for this model? The answer is no. Choosing a set of 8 or 16 compounds may provide adequate information to predict not only the influence of the remaining compounds (and this can be tested), but any unknown in the group.

Figure 2.4 illustrates a simple example. An experimenter is interested in studying the influence of hydrophobicity and dipoles on a set of candidate compounds, for example, in chromatography. He or she finds out these values simply by reading the literature and plots them in a simple graph. Each circle in the figure represents a compound. How to narrow down the test compounds? One simple design involves selecting nine candidates, those at the edges, corners and centre of the square, indicated by arrows in the diagram. These candidates are then tested experimentally, and represent a typical range of compounds. In reality there are vastly more chemical descriptors, but similar approaches can be employed, using, instead of straight properties, statistical functions of these to reduce the number of axes, often to about three, and then choose a good and manageable selection of compounds.

The potential uses of rational experimental design throughout chemistry are large, and some of the most popular designs will be described below. Only certain selective, and generic, classes of design are discussed in this chapter, but it is important to recognise that the huge number of methods reported in the literature are based on a small number of fundamental principles. Most important is to appreciate the motivations of experimental design rather than any particular named method. The material in this chapter should permit the generation of a variety of common designs. If very specialist designs are employed there must be correspondingly specialist reasons for such choice, so the techniques described in this chapter should be applicable to most common situations. Applying a design without appreciating the motivation is dangerous.

For introductory purposes multiple linear regression (MLR) is used to relate the experimental response to the conditions, as is common to most texts in this area, but it is important to realise that other regression methods such as partial least squares (PLS) are applicable in many cases, as discussed in Chapter 5. Certain designs, such as those of Section 2.3.4, have direct relevance to multivariate calibration. In some cases multivariate methods such as PLS can be modified by inclusion of squared and interaction terms as described below for MLR. It is important to remember, however, that in many areas of chemistry a lot of information is available about a dataset, and conceptually simple approaches based on MLR are often adequate.

2.2 Basic Principles

2.2.1 Degrees of Freedom

Fundamental to the understanding of experimental designs is the idea of degrees of freedom. An important outcome of many experiments is the measurement of errors. This can tell us how confidently a phenomenon can be predicted; for example, are we really sure that we can estimate the activity of an unknown compound from its molecular descriptors, or are we happy with the accuracy with which a concentration can be determined using spectroscopy? In addition, what is the weak link in a series of experiments? Is it the performance of a spectrometer or the quality of the volumetric flasks? Each experiment involves making a series of observations, which allow us to try to answer some of these questions, the number of degrees of freedom relating to the amount of information available for each answer. Of course, the greater the number of degrees of freedom, the more certain we can be of our answers, but the more the effort and work are required. If we have only a limited amount of time available, it is important to provide some information to allow us to answer all the desired questions.

Most experiments result in some sort of *model*, which is a mathematical way of relating an experimental *response* to the value or state of a number of *factors*. An example of a response is the yield of a synthetic reaction; the factors may be the pH, temperature and catalyst concentration. An experimenter wishes to run a reaction under a given set of conditions and predict the yield. How many experiments should be performed in order to provide confident predictions of the yield at any combination of the three factors? Five, ten, or twenty? Obviously, the more experiments, the better are the predictions, but the greater the time, effort and expense. So there is a balance, and experimental design helps to guide the chemist as to how many and what type of experiments should be performed.

Consider a linear calibration experiment, for example measuring the peak height in electronic absorption spectroscopy as a function of concentration, at five different concentrations, illustrated in Figure 2.5. A chemist may wish to fit a straight line model to the experiment of the form

$$y = b_0 + b_1 x$$

where y is the response (in this case the peak height), x is the value of the factor (in this case concentration) and b_0 and b_1 are the *coefficients* of the model. There are two coefficients in this equation, but five experiments have been performed. More than enough experiments have been performed to give an equation for a straight line, and the

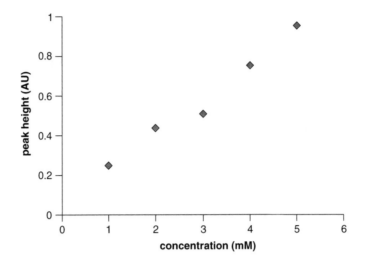

Figure 2.5
Graph of spectroscopic peak height against concentration at five concentrations

remaining experiments help answer the question 'how well is the linear relationship obeyed?' This could be important to the experimenter. For example, there may be unknown interferents, or the instrument might be very irreproducible, or there may be nonlinearities at high concentrations. Hence not only must the experiments be used to determine the equation relating peak height to concentration but also to answer whether the relationship is truly linear and reproducible.

The ability to determine how well the data fit a linear model depends on the number of degrees of freedom which is given, in this case, by

$$D = N - P$$

where N is the number of experiments and P the number of coefficients in the model. In this example

- $N = 5$
- $P = 2$ (the number of coefficients in the equation $y = b_0 + b_1 x$)

so that

- $D = 3$

There are three degrees of freedom allowing us to determine the ability of predict the model, often referred to as the *lack-of-fit*.

From this we can obtain a value which relates to how well the experiment obeys a linear model, often referred to as an *error*, or by some statisticians as a *variance*. However, this error is a simple number, which in the case discussed will probably be expressed in absorbance units (AU). Physical interpretation is not so easy. Consider an error that is reported as 100 mAU: this looks large, but then express it as AU and it becomes 0.1. Is it now a large error? The absolute value of the error must be compared with something, and here the importance of *replication* comes into play. It is useful to repeat the experiment a few times under identical conditions: this gives an idea of the reproducibility of the experimental sometimes called the *analytical* or *experimental* error. The larger the error, the harder it is to make good predictions. Figure 2.6 is of a linear calibration experiment with large errors: these may be due to many reasons, for example, instrumental performance, quality of volumetric flasks, accuracy of weighings and so on. It is hard to see visually whether the results can be adequately described by a linear equation or not. The reading that results in the experiment at the top right hand corner of the graph might be a 'rogue' experiment, often called an *outlier*. Consider a similar experiment, but with lower experimental error (Figure 2.7). Now it looks as

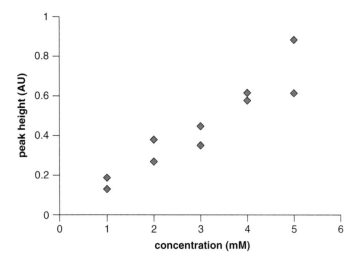

Figure 2.6
Experiment with high instrumental errors

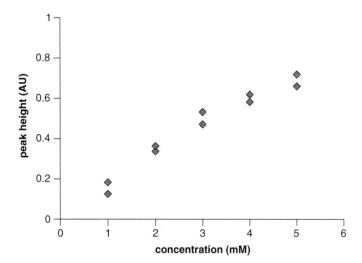

Figure 2.7
Experiment with low instrumental errors

if a linear model is unlikely to be suitable, but only because the experimental error is small compared with the deviation from linearity. In Figures 2.6 and 2.7, an extra five degrees of freedom (the five replicates) have been added to provide information on experimental error. The degrees of freedom available to test for lack-of-fit to a linear model are now given by

$$D = N - P - R$$

where R equals the number of replicates, so that

$$D = 10 - 2 - 5 = 3$$

Although this number remains the same as in Figure 2.5, five extra experiments have been performed to give an idea of the experimental error.

In many designs it is important to balance the number of unique experiments against the number of replicates. Each replicate provides a degree of freedom towards measuring experimental error. Some investigators use a degree of freedom tree which represents this information; a simplified version is illustrated in Figure 2.8. A good rule of thumb is that the number of replicates (R) should be similar to the number of degrees of freedom for the lack-of-fit (D), unless there is an overriding reason for studying one aspect of the system in preference to another. Consider three experimental designs in Table 2.1. The aim is to produce a linear model of the form

$$y = b_0 + b_1 x_1 + b_2 x_2$$

The response y may represent the absorbance in a spectrum and the two xs the concentrations of two compounds. The value of P is equal to 3 in all cases.

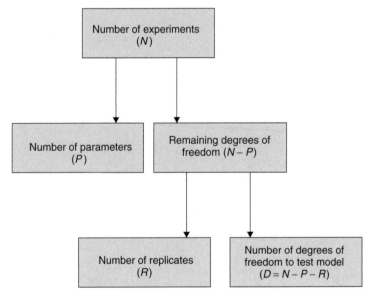

Figure 2.8
Degree of freedom tree

Table 2.1 Three experimental designs.

Experiment No.	Design 1		Design 2		Design 3	
	A	B	A	B	A	B
1	1	1	1	2	1	3
2	2	1	2	1	1	1
3	3	1	2	2	3	3
4	1	2	2	3	3	1
5	2	2	3	2	1	3
6	3	2	2	2	1	1
7	1	3	2	2	3	3
8	2	3	2	2	3	1
9	3	3				

- *Design 1.* This has a value of R equal to 0 and $D = 6$. There is no information about experimental error and all effort has gone into determining the model. If it is known with certainty, in advance, that the response is linear (or this information is not of interest) this experiment may be a good one, but otherwise relatively too little effort is placed in measuring replicates. Although this design may appear to provide an even distribution over the experimental domain, the lack of replication information could, in some cases, lose crucial information.
- *Design 2.* This has a value of R equal to 3 and $D = 2$. There is a reasonable balance between taking replicates and examining the model. If nothing much is known of certainty about the system, this is a good design taking into account the need to economise on experiments.
- *Design 3.* This has a value of R equal to 4 and $D = 1$. The number of replicates is rather large compared with the number of unique experiments. However, if the main aim is simply to investigate experimental reproducibility over a range of concentrations, this approach might be useful.

It is always possible to break down a set of planned experiments in this manner, and is a recommended first step prior to experimentation.

2.2.2 Analysis of Variance and Comparison of Errors

A key aim of experimentation is to ask how significant a factor is. In Section 2.2.1 we discussed how to design an experiment that allows sufficient degrees of freedom to determine the significance of a given factor; below we will introduce an important way of providing numerical information about this significance.

There are many situations in where this information is useful, some examples being listed.

- In an enzyme catalysed extraction, there are many possible factors that could have an influence over the extraction efficiency, such as incubation temperature, extraction time, extraction pH, stirring rates and so on. Often 10 or more possible factors can be identified. Which are significant and should be studied or optimised further?
- In linear calibration, is the baseline important? Are there curved terms; is the concentration too high so that the Beer–Lambert law is no longer obeyed?

- In the study of a simple reaction dependent on temperature, pH, reaction time and catalyst concentration, are the interactions between these factors important? In particular, are higher order interactions (between more than two factors) significant?

A conventional approach is to set up a mathematical model linking the response to coefficients of the various factors. Consider the simple linear calibration experiment, of Section 2.2.1 where the response and concentration are linked by the equation

$$y = b_0 + b_1 x$$

The term b_0 represents an intercept term, which might be influenced by the baseline of the spectrometer, the nature of a reference sample (for a double beam instrument) or the solvent absorption. Is this term significant? Extra terms in an equation will *always* improve the fit to a straight line, so simply determining how well a straight line is fitted to the data does not provide the full picture.

The way to study this is to determine a model of the form

$$y = b_1 x$$

and ask how much worse the fit is to the data. If it is not much worse, then the extra (intercept) term is not very important. The overall lack-of-fit to the model excluding the intercept term can be compared with the replicate error. Often these errors are called variances, hence the statistical term *analysis of variance*, abbreviated to *ANOVA*. If the lack-of-fit is much larger than the replicate error, it is significant, hence the intercept term must be taken into account (and the experimenter may wish to check carefully how the baseline, solvent background and reference sample influence the measurements).

Above, we discussed how an experiment is divided up into different types of degrees of freedom, and we need to use this information in order to obtain a measure of significance.

Two datasets, A and B, are illustrated in Figures 2.9 and 2.10: the question asked is whether there is a significant intercept term; the numerical data are given in Table 2.2. These provide an indication as to how serious a baseline error is in a series of instrumental measurements. The first step is to determine the number of degrees of freedom. For each experiment

- N (the total number of experiments) equals 10;
- R (the number of replicates) equals 4, measured at concentrations 1, 3, 4 and 6 mM.

Two models can be determined, the first without an intercept of the form $y = bx$ and the second with an intercept of the form $y = b_0 + b_1 x$. In the former case

$$D = N - R - 1 = 5$$

and in the latter case

$$D = N - R - 2 = 4$$

The tricky part comes in determining the size of the errors.

- The total replicate error can be obtained by observing the difference between the responses under identical experimental concentrations. For the data in Table 2.2, the

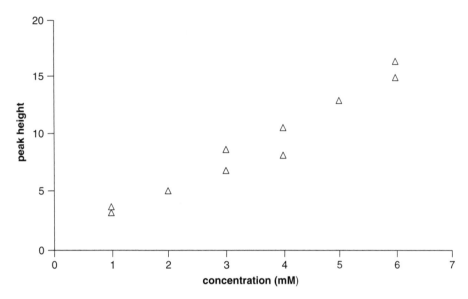

Figure 2.9
Graph of peak height against concentration for ANOVA example, dataset A

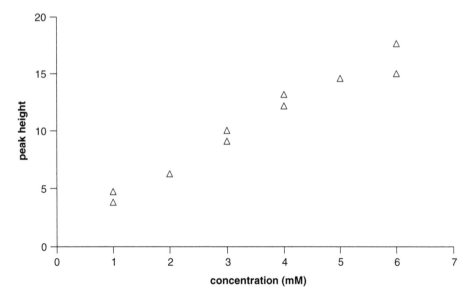

Figure 2.10
Graph of peak height against concentration for ANOVA example, dataset B

replication is performed at 1, 3, 4 and 6 mM. A simple way of determining this error is as follows.

1. Take the average reading at each replicated level or concentration.
2. Determine the differences between this average and the true reading for each replicated measurement.

Table 2.2 Numerical information for data-sets A and B.

Concentration	A	B
1	3.803	4.797
1	3.276	3.878
2	5.181	6.342
3	6.948	9.186
3	8.762	10.136
4	10.672	12.257
4	8.266	13.252
5	13.032	14.656
6	15.021	17.681
6	16.426	15.071

3. Then calculate the sum of squares of these differences (note that the straight sum will always be zero).

This procedure is illustrated in Table 2.3(a) for the dataset A and it can be seen that the replicate sum of squares equals 5.665 in this case.

Algebraically this sum of squares is defined as

$$S_{rep} = \sum_{i=1}^{I} (\bar{y}_i - y_i)^2$$

where $\bar{y}_i$ is the mean response at each unique experimental condition: if, for example, only one experiment is performed at a given concentration it equals the response, whereas if three replicated experiments are performed under identical conditions, it is the average of these replicates. There are R degrees of freedom associated with this parameter.

• The total residual error sum of squares is simply the sum of square difference between the observed readings and those predicted using a best fit model (for example obtained using standard regression procedures in Excel). How to determine the best fit model using multiple linear regression will be described in more detail in Section 2.4. For a model with an intercept, $y = b_0 + b_1 x$, the calculation is presented in Table 2.3(b), where the predicted model is of the form $y = 0.6113 + 2.4364x$, giving a residual sum of square error of $S_{resid} = 8.370$.

Algebraically, this can be defined by

$$S_{resid} = \sum_{i=1}^{I} (y_i - \hat{y}_i)^2$$

and has $(N - P)$ degrees of freedom associated with it.

It is also related to the difference between the total sum of squares for the raw dataset given by

$$S_{total} = \sum_{i=1}^{I} y_i^2 = 1024.587$$

Table 2.3 Calculation of errors for dataset A, model including intercept.

(a) Replicate error

Concentration	Replicate		Difference	Squared difference
	Absorbance	Average		
1	3.803		0.263	0.069
1	3.276	3.540	−0.263	0.069
2	5.181			
3	6.948		−0.907	0.822
3	8.762	7.855	0.907	0.822
4	10.672		1.203	1.448
4	8.266	9.469	−1.203	1.448
5	13.032			
6	15.021		−0.702	0.493
6	16.426	15.724	0.702	0.493
Sum of square replicate error				**5.665**

(b) Overall error (data fitted using univariate calibration)

Concentration	Absorbance	Fitted data	Difference	Squared difference
1	3.803	3.048	0.755	0.570
1	3.276	3.048	0.229	0.052
2	5.181	5.484	−0.304	0.092
3	6.948	7.921	−0.972	0.945
3	8.762	7.921	0.841	0.708
4	10.672	10.357	0.315	0.100
4	8.266	10.357	−2.091	4.372
5	13.032	12.793	0.238	0.057
6	15.021	15.230	−0.209	0.044
6	16.426	15.230	1.196	1.431
Total squared error				**8.370**

and the sum of squares for the predicted data:

$$S_{reg} = \sum_{i=1}^{I} \hat{y}_i^2 = 1016.207$$

so that

$$S_{resid} = S_{total} - S_{reg} = 1024.587 - 1016.207 = 8.370$$

- The lack-of-fit sum of square error is simply the difference between these two numbers or 2.705, and may be defined by

$$S_{lof} = S_{resid} - S_{rep} = 8.370 - 5.665$$

or

$$S_{lof} = \sum_{i=1}^{I} (\bar{y}_i - \hat{y}_i)^2 = S_{mean} - S_{reg}$$

Table 2.4 Error analysis for datasets A and B.

		A	B
Model without intercept		$y = 2.576x$	$y = 2.948x$
Total error sum of squares	S_{resid}	9.115	15.469
Replicate error sum of squares $(d.f. = 4)$	S_{rep}	5.665 (mean = 1.416)	4.776 (mean = 1.194)
Difference between sum of squares $(d.f. = 5)$: lack-of-fit	S_{lof}	3.450 (mean = 0.690)	10.693 (mean = 2.139)
Model with intercept		$y = 0.611 + 2.436x$	$y = 2.032 + 2.484x$
Total error sum of squares	S_{resid}	8.370	7.240
Replicate error sum of squares $(d.f. = 4)$	S_{rep}	5.665 (mean = 1.416)	4.776 (mean = 1.194)
Difference between sum of squares $(d.f. = 4)$: lack-of-fit	S_{lof}	2.705 (mean = 0.676)	2.464 (mean = 0.616)

where

$$S_{mean} = \sum_{i=1}^{I} \bar{y}_i^2$$

and has $(N - P - R)$ degrees of freedom associated with it.

Note that there are several equivalent ways of calculating these errors.

There are, of course, two ways in which a straight line can be fitted, one with and one without the intercept. Each generates different error sum of squares according to the model. The values of the coefficients and the errors are given in Table 2.4 for both datasets. Note that although the size of the term for the intercept for dataset B is larger than dataset A, this does not in itself indicate significance, unless the replicate error is taken into account.

Errors are often presented either as mean square or root mean square errors. The root mean square error is given by

$$s = \sqrt{(S/d)}$$

where d is the number of degrees of freedom associated with a particular sum of squares. Note that the calculation of residual error for the overall dataset differs according to the authors. Strictly this sum of squares should be divided by $(N - P)$ or, for the example with the intercept, 8 ($=10 - 2$). The reason for this is that if there are no degrees of freedom for determining the residual error, the apparent error will be equal to exactly 0, but this does not mean too much. Hence the root mean square residual error for dataset A using the model with the intercept is strictly equal to $\sqrt{(8.370/8)}$ or 1.0228. This error can also be converted to a percentage of the mean reading for the entire dataset (which is 9.139), resulting in a mean residual of 11.19 % by this criterion. However, it is also possible, provided that the number of parameters is significantly less than the number of experiments, simply to divide by N for the residual error, giving a percentage of 10.01 % in this example. In many areas of experimentation, such as principal components analysis and partial least squares regression (see Chapter 5, Section 5.5), it is not always easy to analyse the degrees of freedom in a straightforward manner, and sometimes acceptable, if, for example, there are 40 objects in a dataset, simply to divide by the mean residual error by the number of objects. Many mathematicians debate the meaning of probabilities and errors: is there an inherent physical (or natural) significance to an error, in which case the difference between 10 and 11 % could mean something or do errors primarily provide general guidance as to how good and useful a set of results is? For chemists, it is more important to get a ballpark figure for an error rather than debate the ultimate meaning of the number numbers. The degrees of freedom would have to take into account the number of principal components in the model, as well as data preprocessing such as normalisation and standardisation as discussed in Chapter 4. In this book we adopt the convention of dividing by the total number of degrees of freedom to get a root mean square residual error, unless there are specific difficulties determining this number.

Several conclusions can be drawn from Table 2.4.

- The replicate sum of squares is obviously the same no matter which model is employed for a given experiment, but differs for each experiment. The two experiments result in roughly similar replicate errors, suggesting that the experimental procedure (e.g. dilutions, instrumental method) is similar in both cases. Only four degrees of freedom are used to measure this error, so it is unlikely that these two measured replicate errors will be exactly equal. Measurements can be regarded as samples from a larger population, and it is necessary to have a large sample size to obtain very close agreement to the overall population variance. Obtaining a high degree of agreement may involve several hundred repeated measurements, which is clearly overkill for such a comparatively straightforward series of experiments.
- The total error reduces when an intercept term is added in both cases. This is inevitable and does not necessarily imply that the intercept is significant.
- The difference between the total error and the replicate error relates to the lack-of-fit. The bigger this is, the worse is the model.
- The lack-of-fit error is slightly smaller than the replicate error, in all cases except when the intercept is removed from the model for the dataset B, where it is large, 10.693. This suggests that adding the intercept term to the second dataset makes a big difference to the quality of the model and so the intercept is significant.

Conventionally these numbers are often compared using ANOVA. In order for this to be meaningful, the sum of squares should be divided by the number of degrees of freedom to give the 'mean' sum of squares in Table 2.4. The reason for this is that the larger the number of measurements, the greater the underlying sum of squares will be. These mean squares can are often called variances, and it is simply necessary to compare their sizes, by taking ratios. The larger the ratio to the mean replicate error, the greater is the significance. It can be seen that in all cases apart from the model

Table 2.5 ANOVA table: two parameter model, dataset B.

Source of variation	Sum of squares	Degrees of freedom	Mean sum of squares	Variance ratio
Total	1345.755	10	134.576	
Regression	1338.515	2	669.258	
Residual	7.240	8	0.905	
Replicate	4.776	4	1.194	
Lack-of-fit	2.464	4	0.616	0.516

without the intercept arising from dataset B, the mean lack-of-fit error is considerably less than the mean replicate error. Often the results are presented in tabular form; a typical example for the two parameter model of dataset B is given in Table 2.5, the five sums of squares S_{total}, S_{reg}, S_{resid}, S_{rep} and S_{lof}, together with the relevant degrees of freedom, mean square and variance ratio, being presented. The number 0.516 is the key to assess how well the model describes the data and is often called the F-ratio between the mean lack-of-fit error and the mean replicate error, which will be discussed in more detail in Section 2.2.4.4. Suffice it to say that the higher this number, the more significant is an error. A lack-of-fit that is much less than the replicate error is not significant, within the constraints of the experiment.

Most statistical packages produce ANOVA tables if required, and it is not always necessary to determine these errors manually, although it is important to appreciate the principles behind such calculations. However, for simple examples a manual calculation is often quite and a good alternative to the interpretation of the output of complex statistical packages.

The use of ANOVA is widespread and is based on these simple ideas. Normally two mean errors are compared, for example, one due to replication and the other due to lack-of-fit, although any two errors or variances may be compared. As an example, if there are 10 possible factors that might have an influence over the yield in a synthetic reaction, try modelling the reaction removing one factor at a time, and see how much the lack-of-fit error increases: if not much relative to the replicates, the factor is probably not significant. It is important to recognise that reproducibility of the reaction has an influence over apparent significance also. If there is a large replicate error, then some significant factors might be missed out.

2.2.3 Design Matrices and Modelling

The design matrix is a key concept. A design may consist of a series of experiments performed under different conditions, e.g. a reaction at differing pHs, temperatures, and concentrations. Table 2.6 illustrates a typical experimental set-up, together with an experimental response, e.g. the rate constant of a reaction. Note the replicates in the final five experiments: in Section 2.4 we will discuss such an experimental design commonly called a *central composite* design.

2.2.3.1 Models

It is normal to describe experimental data by forming a mathematical relationship between the factors or independent variables such as temperature and a response or dependent variable such as a synthetic yield, a reaction time or a percentage impurity. A typical equation for three factors might be of the form

$\hat{y} =$ (response)

$\quad b_0 +$ (an intercept or average)

$\quad b_1 x_1 + b_2 x_2 + b_3 x_3 +$ (linear terms depending on each of the three factors)

$\quad b_{11} x_1^2 + b_{22} x_2^2 + b_{33} x_3^2 +$ (quadratic terms depending on each of the three factors)

$\quad b_{12} x_1 x_2 + b_{13} x_1 x_3 + b_{23} x_2 x_3$ (interaction terms between the factors).

Notice the 'hat' on top of the y; this is because the equation estimates its value, and is unlikely to give an exact value that agrees experimentally because of error.

Table 2.6 Typical experimental design.

pH	Temperature (°C)	Concentration (mM)	Response (y)
6	60	4	34.841
6	60	2	16.567
6	20	4	45.396
6	20	2	27.939
4	60	4	19.825
4	60	2	1.444
4	20	4	37.673
4	20	2	23.131
6	40	3	23.088
4	40	3	12.325
5	60	3	16.461
5	20	3	33.489
5	40	4	26.189
5	40	2	8.337
5	40	3	19.192
5	40	3	16.579
5	40	3	17.794
5	40	3	16.650
5	40	3	16.799
5	40	3	16.635

The justification for these terms is as follows.

- The intercept is an average in certain circumstances. It is an important term because the average response is not normally achieved when the factors are at their average values. Only in certain circumstances (e.g. spectroscopy if it is known there are no baseline problems or interferents) can this term be ignored.
- The linear terms allow for a direct relationship between the response and a given factor. For some experimental data, there are only linear terms. If the pH increases, does the yield increase or decrease and, if so, by how much?
- In many situations, quadratic terms are important. This allows curvature, and is one way of obtaining a maximum or minimum. Most chemical reactions have an optimum performance at a particular pH, for example. Almost all enzymic reactions work in this way. Quadratic terms balance out the linear terms.
- Earlier in Section 2.1, we discussed the need for interaction terms. These arise because the influence of two factors on the response is rarely independent. For example, the optimum pH at one temperature may differ from that at a different temperature.

Some of these terms may not be very significant or relevant, but it is up to the experimenter to check this using approaches such as ANOVA (Section 2.2.2) and significance tests (Section 2.2.4). In advance of experimentation it is often hard to predict which factors are important.

2.2.3.2 Matrices

There are 10 terms or parameters in the equation above. Many chemometricians find it convenient to work using matrices. Although a significant proportion of traditional

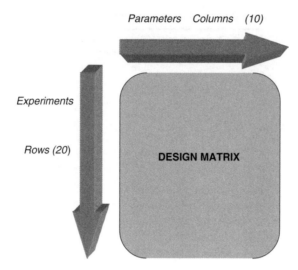

Parameters   Columns   (10)

Experiments

Rows (20)

DESIGN MATRIX

Figure 2.11
Design matrix

texts often shy away from matrix based notation, with modern computer packages and spreadsheets it is easy and rational to employ matrices. The design matrix is simply one in which

- the *rows* refer to experiments and
- the *columns* refer to individual parameters in the mathematical model or equation linking the response to the values of the individual factors.

In the case described, the design matrix consists of

- 20 rows as there are 20 experiments and
- 10 columns as there are 10 parameters in the model, as is illustrated symbolically in Figure 2.11.

For the experiment discussed above, the design matrix is given in Table 2.7. Note the first column, of 1s: this corresponds to the intercept term, b_0, which can be regarded as multiplied by the number 1 in the equation. The figures in the table can be checked numerically. For example, the interaction term between pH and temperature for the first experiment is 360, which equals 6×60, and appears in the eighth column of the first row corresponding the term b_{12}.

There are two considerations required when computing a design matrix, namely

- the number and arrangement of the experiments, including replication and
- the mathematical model to be tested.

It is easy to see that

- the 20 responses form a vector with 20 rows and 1 column, called y;
- the design matrix has 10 columns and 20 rows, as illustrated in Table 2.7, and is called D; and
- the 10 coefficients of the model form a vector with 10 rows and 1 column, called b.

Table 2.7 Design matrix for the experiment in Table 2.6.

Intercept	Linear terms			Quadratic terms			Interaction terms		
b_0	b_1	b_2	b_3	b_{11}	b_{22}	b_{33}	b_{12}	b_{13}	b_{23}
Intercept	pH	Temp	Conc	pH^2	Temp^2	Conc^2	pH × temp	pH × conc	Temp × conc
1	6	60	4	36	3600	16	360	24	240
1	6	60	2	36	3600	4	360	12	120
1	6	20	4	36	400	16	120	24	80
1	6	20	2	36	400	4	120	12	40
1	4	60	4	16	3600	16	240	16	240
1	4	60	2	16	3600	4	240	8	120
1	4	20	4	16	400	16	80	16	80
1	4	20	2	16	400	4	80	8	40
1	6	40	3	36	1600	9	240	18	120
1	4	40	3	16	1600	9	160	12	120
1	5	60	3	25	3600	9	300	15	180
1	5	20	3	25	400	9	100	15	60
1	5	40	4	25	1600	16	200	20	160
1	5	40	2	25	1600	4	200	10	80
1	5	40	3	25	1600	9	200	15	120
1	5	40	3	25	1600	9	200	15	120
1	5	40	3	25	1600	9	200	15	120
1	5	40	3	25	1600	9	200	15	120
1	5	40	3	25	1600	9	200	15	120
1	5	40	3	25	1600	9	200	15	120

2.2.3.3 Determining the Model

The relationship between the response, the coefficients and the experimental conditions can be expressed in matrix form by

$$\hat{y} = D.b$$

as illustrated in Figure 2.12. It is simple to show that this is the matrix equivalent to the equation introduced in Section 2.2.3.1. It is surprisingly easy to calculate b (or the coefficients in the model) knowing D and y using MLR (multiple linear regression). This approach will be discussed in greater detail in Chapter 5, together with other potential ways such as PCR and PLS.

- If D is a square matrix, then there are exactly the same number of experiments as coefficients in the model and

$$b = D^{-1}.y$$

- If D is not a square matrix (as in the case in this section), then use the pseudo-inverse, an easy calculation in Excel, Matlab and almost all matrix based software, as follows:

$$b = (D'.D)^{-1}.D'.y$$

The idea of the pseudo-inverse is used in several places in this text, for example, see Chapter 5, Sections 5.2 and 5.3, for a general treatment of regression. A simple

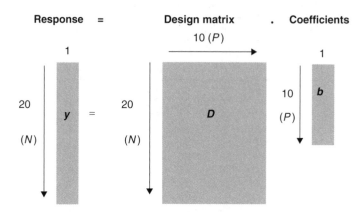

Figure 2.12
Relationship between response, design matrix and coefficients

derivation is as follows:

$$y \approx D.b \text{ so } D'.y \approx D'.D.b \text{ or } (D'.D)^{-1}.D'.y \approx (D'.D)^{-1}.(D'.D).b \approx b$$

In fact we obtain estimates of b from regression, so strictly there should be a hat on top of the b, but in order to simplify the notation we ignore the hat and so the approximation sign becomes an equals sign.

It is important to recognise that for some designs there are several alternative methods for calculating these regression coefficients, which will be described in the relevant sections, but the method of regression described above will *always* work provided that the experiments are designed appropriately. A limitation prior to the computer age was the inability to determine matrix inverses easily, so classical statisticians often got around this by devising methods often for summing functions of the response, and in some cases designed experiments specifically to overcome the difficulty of inverses and for ease of calculation. The dimensions of the square matrix $(D'.D)$ equal the number of parameters in a model, and so if there are 10 parameters it would not be easy to compute the relevant inverse manually, although this is a simple operation using modern computer based packages.

There are a number of important consequences.

- If the matrix D is a square matrix, the estimated values of $\hat{y}$ are identical with the observed values y. The model provides an exact fit to the data, and there are no degrees of freedom remaining to determine the lack-of-fit. Under such circumstances there will not be any replicate information but, nevertheless, the values of b can provide valuable information about the size of different effects. Such a situation might occur, for example, in factorial designs (Section 2.3). The residual error between the observed and fitted data will be zero. This does not imply that the predicted model exactly represents the underlying data, simply that the number of degrees of freedom is insufficient for determination of prediction errors. In all other circumstances there is likely to be an error as the predicted and observed response will differ.

- The matrix D – or $D'.D$ (if the number of experiments is more than the number of parameters) – must have an inverse. If it does not, it is impossible to calculate the coefficients b. This is a consequence of poor design, and may occur if two terms or factors are correlated to each other. For well designed experiments this problem will not occur. Note that a design in which the number of experiments is less than the number of parameters has no meaning.

2.2.3.4 Predictions

Once b has been determined, it is then possible to predict y and so calculate the sums of squares and other statistics as outlined in Sections 2.2.2 and 2.2.4. For the data in Table 2.6, the results are provided in Table 2.8, using the pseudo-inverse to obtain b and then predict $\hat{y}$. Note that the *size* of the parameters does not necessarily indicate significance, in this example. It is a common misconception that the larger the parameter the more important it is. For example, it may appear that the b_{22} parameter is small (0.020) relative to the b_{11} parameter (0.598), but this depends on the physical measurement units:

- the pH range is between 4 and 6, so the square of pH varies between 16 and 36 or by 20 units overall;
- the temperature range is between 20 and 60 °C, the squared range varying between 400 and 3600 or by 3200 units overall, which is a 160-fold difference in range compared with pH;
- therefore, to be of equal importance b_{22} would need to be 160 times smaller than b_{11};
- since the ratio $b_{11}:b_{22}$ is 29.95, in fact b_{22} is considerably more significant than b_{11}.

Table 2.8 The vectors b and $\hat{y}$ for data in Table 2.6.

Parameter		Predicted y
b_0	58.807	35.106
b_1	−6.092	15.938
b_2	−2.603	45.238
b_3	4.808	28.399
b_{11}	0.598	19.315
b_{22}	0.020	1.552
b_{33}	0.154	38.251
b_{12}	0.110	22.816
b_{13}	0.351	23.150
b_{23}	0.029	12.463
		17.226
		32.924
		26.013
		8.712
		17.208
		17.208
		17.208
		17.208
		17.208
		17.208

In Section 2.2.4 we discuss in more detail how to tell whether a given parameter is significant, but it is very dangerous indeed to rely on visual inspection of tables of regression parameters and make deductions from these without understanding carefully how the data are scaled.

If carefully calculated, three types of information can come from the model.

- The size of the coefficients can inform the experimenter how significant the coefficient is. For example, does pH significantly improve the yield of a reaction? Or is the interaction between pH and temperature significant? In other words, does the temperature at which the reaction has a maximum yield differ at pH 5 and at pH 7?
- The coefficients can be used to construct a model of the response, for example the yield of a reaction as a function of pH and temperature, and so establish the optimum conditions for obtaining the best yield. In this case, the experimenter is not so interested in the precise equation for the yield but is very interested in the best pH and temperature.
- Finally, a quantitative model may be interesting. Predicting the concentration of a compound from the absorption in a spectrum requires an accurate knowledge of the relationship between the variables. Under such circumstances the precise value of the coefficients is important. In some cases it is known that there is a certain kind of model, and the task is mainly to obtain a regression or calibration equation.

Although the emphasis in this chapter is on multiple linear regression techniques, it is important to recognise that the analysis of design experiments is not restricted to such approaches, and it is legitimate to employ multivariate methods such as principal components regression and partial least squares as described in detail in Chapter 5.

2.2.4 Assessment of Significance

In many traditional books on statistics and analytical chemistry, large sections are devoted to significance testing. Indeed, an entire and very long book could easily be written about the use of significance tests in chemistry. However, much of the work on significance testing goes back nearly 100 years, to the work of Student, and slightly later to R. A. Fisher. Whereas their methods based primarily on the t-test and F-test have had a huge influence in applied statistics, they were developed prior to the modern computer age. A typical statistical calculation, using pen and paper and perhaps a book of logarithm or statistical tables, might take several days, compared with a few seconds on a modern microcomputer. Ingenious and elaborate approaches were developed, including special types of graph papers and named methods for calculating the significance of various effects.

These early methods were developed primarily for use by specialised statisticians, mainly trained as mathematicians, in an environment where user-friendly graphics and easy analysis of data were inconceivable. A mathematical statistician will have a good feeling for the data, and so is unlikely to perform calculations or compute statistics from a dataset unless satisfied that the quality of data is appropriate. In the modern age everyone can have access to these tools without a great deal of mathematical expertise but, correspondingly, it is possible to misuse methods in an inappropriate manner. The practising chemist needs to have a numerical and graphical feel for the significance of his or her data, and traditional statistical tests are only one of a battery of approaches used to determine the significance of a factor or effect in an experiment.

Table 2.9 Coding of data.

Variable	Units	−1	+1
pH	−Log[H$^+$]	4	6
Temperature	°C	20	60
Concentration	mM	2	4

This section provides an introduction to a variety of approaches for assessing significance. For historical reasons, some methods such as cross-validation and independent testing of models are best described in the chapters on multivariate methods (see Chapters 4 and 5), although the chemometrician should have a broad appreciation of all such approaches and not be restricted to any one set of methods.

2.2.4.1 Coding

In Section 2.2.3, we introduced an example of a three factor design, given in Table 2.6, described by 10 regression coefficients. Our comment was that the significance of the coefficients cannot easily be assessed by inspection because the physical scale for each variable is different. In order to have a better idea of the significance it is useful to put each variable on a comparable scale. It is common to *code* experimental data. Each variable is placed on a common scale, often with the highest coded value of each variable equal to +1 and the lowest to −1. Table 2.9 represents a possible way to scale the data, so for factor 1 (pH) a coded value (or level) or −1 corresponds to a true pH of 4. Note that coding does not need to be linear: in fact pH is actually measured on a logarithmic scale.

The design matrix simplifies considerably, and together with the corresponding regression coefficients is presented in Table 2.10. Now the coefficients are approximately on the same scale, and it appears that there are radical differences between these new numbers and the coefficients in Table 2.8. Some of the differences and their interpretation are listed below.

- The coefficient b_0 is very different. In the current calculation it represents the predicted response in the centre of the design, where the coded levels of the three factors are $(0, 0, 0)$. In the calculation in Section 2.2.3 it represents the predicted response at 0 pH units, 0 °C and 0 mM, conditions that cannot be reached experimentally. Note also that this approximates to the mean of the entire dataset (21.518) and is close to the average over the six replicates in the central point (17.275). For a perfect fit, with no error, it will equal the mean of the entire dataset, as it will for designs centred on the point $(0, 0, 0)$ in which the number of experiments equals the number of parameters in the model such as a factorial designs discussed in Section 2.6.
- The relative size of the coefficients b_{11} and b_{22} changes dramatically compared with Table 2.8, the latter increasing hugely in apparent size when the coded dataset is employed. Provided that the experimenter chooses appropriate physical conditions, it is the coded values that are most helpful for interpretation of significance. A change in pH of 1 unit is more important than a change in temperature of 1 °C. A temperature range of 40 °C is quite small, whereas a pH range of 40 units would be almost unconceivable. Therefore, it is important to be able to compare directly the size of parameters in the coded scale.

Table 2.10 Coded design matrix together with values of coded coefficients.

x_0	x_1	x_2	x_3	x_1^2	x_2^2	x_3^2	x_1x_2	x_1x_3	x_2x_3
1	1	1	1	1	1	1	1	1	1
1	1	1	−1	1	1	1	1	−1	−1
1	1	−1	1	1	1	1	−1	1	−1
1	1	−1	−1	1	1	1	−1	−1	1
1	−1	1	1	1	1	1	−1	−1	1
1	−1	1	−1	1	1	1	−1	1	−1
1	−1	−1	1	1	1	1	1	−1	−1
1	−1	−1	−1	1	1	1	1	1	1
1	1	0	0	1	0	0	0	0	0
1	−1	0	0	1	0	0	0	0	0
1	0	1	0	0	1	0	0	0	0
1	0	−1	0	0	1	0	0	0	0
1	0	0	1	0	0	1	0	0	0
1	0	0	−1	0	0	1	0	0	0
1	0	0	0	0	0	0	0	0	0
1	0	0	0	0	0	0	0	0	0
1	0	0	0	0	0	0	0	0	0
1	0	0	0	0	0	0	0	0	0
1	0	0	0	0	0	0	0	0	0
1	0	0	0	0	0	0	0	0	0

Values

b_0	b_1	b_2	b_3	b_{11}	b_{22}	b_{33}	b_{12}	b_{13}	b_{23}
17.208	5.343	−7.849	8.651	0.598	7.867	0.154	2.201	0.351	0.582

- Another very important observation is that the *sign* of significant parameters can also change as the coding of the data is changed. For example, the sign of the parameter for b_1 is negative (−6.092) in Table 2.8 but positive (+5.343) in Table 2.10, yet the size and sign of the b_{11} term do not change. The difference between the highest and lowest true pH (2 units) is the same as the difference between the highest and lowest coded values of pH, also 2 units. In Tables 2.8 and 2.10 the value of b_1 is approximately 10 times greater in magnitude than b_{11} and so might appear much more significant. Furthermore, it is one of the largest terms apart from the intercept. What has gone wrong with the calculation? Does the value of y increase with increasing pH or does it decrease? There can be only one physical answer. The clue to change of sign comes from the mathematical transformation. Consider a simple equation of the form

$$y = 10 + 50x - 5x^2$$

and a new transformation from a range of raw values between 9 and 11 to coded values between −1 and +1, so that

$$c = x - 10$$

where c is the coded value. Then

$$y = 10 + 50(c + 10) - 5(c + 10)^2$$
$$= 10 + 50c + 500 - 5c^2 - 100c - 500$$
$$= 10 - 50c - 5c^2$$

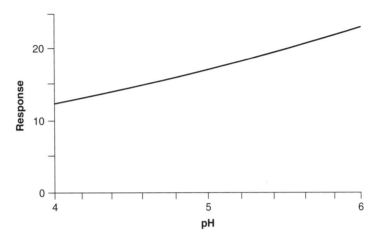

Figure 2.13
Graph of estimated response versus pH at the central temperature of the design in Table 2.6

an apparent change in sign. Using raw data, we might conclude that the response increases with increasing x, whereas with the coded data, the opposite conclusion might be drawn. Which is correct? Returning to our example, although the graph of the response depends on interaction effects, and so the relationship between y and pH is different at each temperature and concentration, but at the central point of the design, it is given in Figure 2.13, increasing monotonically over the experimental region. Indeed, the average value of the response when the pH is equal to 6 is higher than the average value when it is equal to 4. Hence it is correct to conclude that the response increases with pH, and the negative coefficient of Table 2.8 is misleading. Using coded data provides correct conclusions about the trends whereas the coefficients for the raw data may lead to incorrect deductions.

Therefore, without taking great care, misleading conclusions can be obtained about the significance and influence of the different factors. It is essential that the user of simple chemometric software is fully aware of this, and always interprets numbers in terms of physical meaning.

2.2.4.2 Size of Coefficients

The simplest approach to determining significance is simply to look at the magnitude of the coefficients. Provided that the data are coded correctly, the larger the coefficient, the greater is its significance. This depends on each coded factor varying over approximately the same range (between $+1$ and -1 in this case). Clearly, small differences in range are not important, often the aim is to say whether a particular factor has a significant influence or not rather than a detailed interpretation of the size of the coefficients. A value of 5.343 for b_1 implies that on average the response is higher by 5.343 if the value of b_1 is increased by one coded pH unit. This is easy to verify, and provides an alternative, classical, approach to the calculation of the coefficients:

1. consider the 10 experiments at which b_1 is at a coded level of either $+1$ or -1, namely the first 10 experiments;

2. then group these in five pairs, each of which the levels of the other two main factors are identical; these pairs are {1, 5}, {2, 6}, {3, 7}, {4, 8} and {9, 10};
3. take the difference between the responses at the levels and average them:

$$[(34.841 - 19.825) + (16.567 - 1.444) + (45.396 - 37.673)$$

$$+ (27.939 - 23.131) + (23.088 - 12.325)]/5$$

which gives an answer of 10.687 representing the average change in value of the response when the pH is increased from a coded value of -1 to one of $+1$, half of which equals the coefficient 5.343.

It is useful to make practical deductions from the data which will guide the experimenter.

- The response varies over a range of 43.953 units between the lowest and highest observation in the experimental range.
- Hence the linear effect of pH, on average, is to increase the response by twice the coded coefficient or 10.687 units over this range, approximately 25 % of the variation, probably quite significant. The effect of the interaction between pH and concentration (b_{13}), however, is only 0.702 units or a very small contribution, rather less than the replicate error, so this factor is unlikely to be useful.
- The squared terms must be interpreted slightly differently. The lowest possible coded value for the squared terms is 0, not -1, so we do not double these values to obtain an indication of significance, the range of variation of the squared terms being between 0 and $+1$, or half that of the other terms.

It is not necessary, of course, to have replicates to perform this type of analysis. If the yield of a reaction varies between 50 and 90 % over a range of experimental conditions, then a factor that contributes, on average, only 1 % of this increase is unlikely to be too important. However, it is vital in all senses that the factors are coded for meaningful comparison. In addition, certain important properties of the design (namely orthogonality) which will be discussed in detail in later sections are equally important.

Provided that the factors are coded correctly, it is fairly easy to make qualitative comparisons of significance simply by examining the size of the coefficients either numerically and graphically. In some cases the range of variation of each individual factor might differ slightly (for example squared and linear terms above), but provided that this is not dramatic, for rough indications the sizes of the factors can be legitimately compared. In the case of two level factorial designs (described in Sections 2.6–2.8), each factor is normally scaled between -1 and $+1$, so all coefficients are on the same scale.

2.2.4.3 Student's t-Test

An alternative, statistical indicator, based on Student's t-test, can be used, provided that more experiments are performed than there are parameters in the model. Whereas this and related statistical indicators have a long and venerated history, it is always important to back up the statistics by simple graphs and considerations about the data. There are many diverse applications of a t-test, but in the context of analysing the significance of factors on designed experiments, the following the main steps are used

Table 2.11 Calculation of t-statistic.

(a) Matrix $(\boldsymbol{D'.D})^{-1}$

	b_0	b_1	b_2	b_3	b_{11}	b_{22}	b_{33}	b_{12}	b_{13}	b_{23}
b_0	**0.118**	0.000	0.000	0.000	−0.045	−0.045	−0.045	0.000	0.000	0.000
b_1	0.000	**0.100**	0.000	0.000	0.000	0.000	0.000	0.000	0.000	0.000
b_2	0.000	0.000	**0.100**	0.000	0.000	0.000	0.000	0.000	0.000	0.000
b_3	0.000	0.000	0.000	**0.100**	0.000	0.000	0.000	0.000	0.000	0.000
b_{11}	−0.045	0.000	0.000	0.000	**0.364**	−0.136	−0.136	0.000	0.000	0.000
b_{22}	−0.045	0.000	0.000	0.000	−0.136	**0.364**	−0.136	0.000	0.000	0.000
b_{33}	−0.045	0.000	0.000	0.000	−0.136	−0.136	**0.364**	0.000	0.000	0.000
b_{12}	0.000	0.000	0.000	0.000	0.000	0.000	0.000	**0.125**	0.000	0.000
b_{13}	0.000	0.000	0.000	0.000	0.000	0.000	0.000	0.000	**0.125**	0.000
b_{23}	0.000	0.000	0.000	0.000	0.000	0.000	0.000	0.000	0.000	0.125

(b) Values of t and significance

s	v	$\sqrt{sv}$	b	t	% Probability
b_0	0.118	0.307	17.208	56.01	>99.9
b_1	0.100	0.283	5.343	18.91	>99.9
b_2	0.100	0.283	−7.849	−27.77	>99.9
b_3	0.100	0.283	8.651	30.61	>99.9
b_{11}	0.364	0.539	0.598	1.11	70.7
b_{22}	0.364	0.539	7.867	14.60	>99.9
b_{33}	0.364	0.539	0.154	0.29	22.2
b_{12}	0.125	0.316	2.201	6.97	>99.9
b_{13}	0.125	0.316	0.351	1.11	70.7
b_{23}	0.125	0.316	0.582	1.84	90.4

and are illustrated in Table 2.11 for the example described above using the coded values of Table 2.9.

1. Calculate the matrix $(\boldsymbol{D'D})^{-1}$. This will be a square matrix with dimensions equal to the number of parameters in the model.
2. Calculate the error sum of squares between the predicted and observed data (compare the actual response in Table 2.6 with the predictions of Table 2.8):

$$S_{resid} = \sum_{i=1}^{I} (y_i - \hat{y}_i)^2 = 7.987$$

3. Take the mean the error sum of squares (divided by the number of degrees of freedom available for testing for regression):

$$s = S_{resid}/(N - P) = 7.987/(20 - 10) = 0.799$$

Note that the t-test is not applicable to data where the number of experiments equals the number of parameters, such as full factorial designs discussed in Section 2.3.1, where all possible terms are included in the model.

4. For each of the P parameters (=10 in this case), take the appropriate number from the diagonal of the matrix of Table 2.11(a) obtained in step 1 above. This

is called the variance for each parameter, so that, for example, $v_{11} = 0.364$ (the variance of b_{11}).

5. For each coefficient, b, calculate $t = b/\sqrt{sv}$. The higher this ratio, the more significant is the coefficient. This ratio is used for the t-test.

6. The statistical significance can then be obtained from a two-tailed t-distribution (this is described in detail in Appendix A.3.4), or most packages such as Excel have simple functions for the t-test. Take the absolute value of the ratio calculated above. If you use a table, along the left-hand column of a t-distribution table are tabulated degrees of freedom, which equal the number available to test for regression, or $N - P$ or 10 in this case. Along the columns, locate the percentage probability (often the higher the significance the smaller is the percentage, so simply subtract from 1). The higher this probability, the greater is the confidence that the factor is significant. So, using Table A.4 we see that a critical value of 4.1437 indicates 99.9 % certainty that a parameter is significant for 10 degrees of freedom, hence any value above this is highly significant. 95 % significance results in a value of 1.8125, so b_{23} is just above this level. In fact, the numbers in Table 2.11 were calculated using the Excel function TDIST, which gives provides probabilities for any value of t and any number of degrees of freedom. Normally, fairly high probabilities are expected if a factor is significant, often in excess of 95 %.

2.2.4.4 F-test

The F-test is another alternative. A common use of the F-test is together with ANOVA, and asks how significant one variance (or mean sum of squares) is relative to another one; typically, how significant the lack-of-fit is compared with the replicate error. Simply determine the mean square lack-of-fit to replicate errors (e.g. see Table 2.4) and check the size of this number. F-distribution tables are commonly presented at various probability levels. We use a one-tailed F-test in this case as the aim is to see whether one variance is significantly bigger than another, not whether it differs significantly; this differs from the t-test, which is two tailed in the application described in Section 2.2.4.3. The columns correspond to the number of degrees of freedom for S_{lof} and the rows to S_{rep} (in the case discussed in here). The table allows one to ask how significant is the error (or variance) represented along the columns relative to that represented along the rows. Consider the proposed models for datasets A and B both excluding the intercept. Locate the relevant number [for a 95 % confidence that the lack-of-fit is significant, five degrees of freedom for the lack-of-fit and four degrees of freedom for the replicate error, this number is 6.26, see Table A.3 (given by a distribution often called $F_{(5,4)}$), hence an F-ratio must be greater than this value for this level of confidence]. Returning to Table 2.4, it is possible to show that the chances of the lack-of-fit to a model without an intercept are not very high for the data in Figure 2.9 (ratio = 0.49), but there is some doubt about the data arising from Figure 2.10 (ratio = 1.79); using the FDIST function in Excel we can see that the probability is 70.4 %, below the 95 % confidence that the intercept is significant, but still high enough to give us some doubts. Nevertheless, the evidence is not entirely conclusive. A reason is that the intercept term (2.032) is of approximately the same order of magnitude as the replicate error (1.194), and for this level of experimental variability it will never be possible to predict and model the presence of an intercept of this size with a high degree of confidence.

Table 2.12 F-ratio for experiment with low experimental error.

Concentration	Absorbance		Model with intercept	Model without intercept
1	3.500	b_0	0.854	n/a
1	3.398	b_1	2.611	2.807
2	6.055			
3	8.691	S_{reg}	0.0307	1.4847
3	8.721	S_{rep}	0.0201	0.0201
4	11.249	S_{lof}	0.0107	1.4646
4	11.389			
5	13.978			
6	16.431			
6	16.527	F-ratio	0.531	58.409

The solution is perform new experiments, perhaps on a different instrument, in which the reproducibility is much greater. Table 2.12 is an example of such a dataset, with essential statistics indicated. Now the F-ratio for the lack-of-fit without the intercept becomes 58.409, which is significant at the $>99\%$ level (critical value from Table A.2) whereas the lack-of-fit with the intercept included is less than the experimental error.

2.2.4.5 Normal Probability Plots

For designs where there are no replicates (essential for most uses of the F-test) and also where there no degrees of freedom available to assess the lack-of-fit to the data (essential for a t-test), other approaches can be employed to examine the significance of coefficients.

As discussed in Section 2.3, two-level factorial designs are common, and provided that the data are appropriately coded, the size of the coefficients relates directly to their significance. Normally several coefficients are calculated, and an aim of experimentation is to determine which have significance, the next step possibly then being to perform another more detailed design for quantitative modelling of the significant effects. Often it is convenient to present the coefficients graphically, and a classical approach is to plot them on normal probability paper. Prior to the computer age, a large number of different types of statistical graph paper were available, assisting data analysis. However, in the age of computers, it is easy to obtain relevant graphs using simple computer packages.

The principle of normal probability plots is that if a series of numbers is randomly selected, they will often form a normal distribution (see Appendix A.3.2). For example, if I choose seven numbers randomly, I would expect, in the absence of systematic effects, that these numbers would be approximately normally distributed. Hence if we look at the size of seven effects, e.g. as assessed by their values of b (provided that the data are properly coded and the experiment is well designed, of course), and the effects are simply random, on average we would expect the size of each effect to occur evenly over a normal distribution curve. In Figure 2.14, seven lines are indicated on the normal distribution curve (the horizontal axis representing standard deviations from the mean) so that the areas between each line equal one-seventh of the total area (the areas at the extremes adding up to 1/7 in total). If, however, an effect is very large, it will fall at a very high or low value, so large that it is unlikely to be arise from

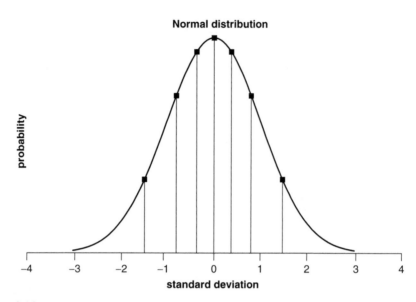

Figure 2.14
Seven lines, equally spaced in area, dividing the normal distribution into eight regions, the six central regions and the sum of the two extreme regions having equal areas

Table 2.13   Normal probability calculation.

Effect	Coefficient	$(p - 0.5)/7$	Standard deviation
b_1	−6.34	0.0714	−1.465
b_{12}	−0.97	0.2143	−0.792
b_{13}	0.6	0.3571	−0.366
b_{123}	1.36	0.5	0
b_3	2.28	0.6429	0.366
b_{12}	5.89	0.7858	0.792
b_2	13.2	0.9286	1.465

random processes, and is significant. Normal probability plots can be used to rank the coefficients in size (the most negative being the lowest, the most positive the highest), from the rank determine the likely position in the normal probability plot and then produce a graph of the coefficients against this likely position. The insignificant effects should form approximately on a straight line in the centre of the graph, significant effects will deviate from this line.

Table 2.13 illustrates the calculation.

1. Seven possible coefficients are to be assessed for significance. Note that the b_0 coefficient cannot be analysed in this way.
2. They are ranked from 1 to 7 where p is the rank.
3. Then the values of $(p - 0.5)/7$ are calculated. This indicates where in the normal distribution each effect is likely to fall. For example, the value for the fourth coefficient is 0.5, meaning that the coefficient might be expected in the centre of the distribution, corresponding to a standard deviation from the mean of 0, as illustrated in Figure 2.14.

4. Then work out how many standard deviations corresponding to the area under the normal curve calculated in step 3, using normal distribution tables or standard functions in most data analysis packages. For example, a probability of 0.9286 (coefficient b_2) falls at 1.465 standard deviations. See Table A.1 in which a 1.46 standard deviations correspond to a probability of 0.927 85 or use the NORMINV function in Excel.

5. Finally, plot the size of the effects against the value obtained in step 4, to give, for the case discussed, the graph in Figure 2.15. The four central values fall roughly on a straight line, suggesting that only coefficients b_1, b_2 and b_{12}, which deviate from the straight line, are significant.

Like many classical methods of data analysis, the normal probability plot has limitations. It is only useful if there are several factors, and clearly will not be much use in the case of two or three factors. It also assumes that a large number of the factors are not significant, and will not give good results if there are too many significant effects. However, in certain cases it can provide useful preliminary graphical information, although probably not much used in modern computer based chemometrics.

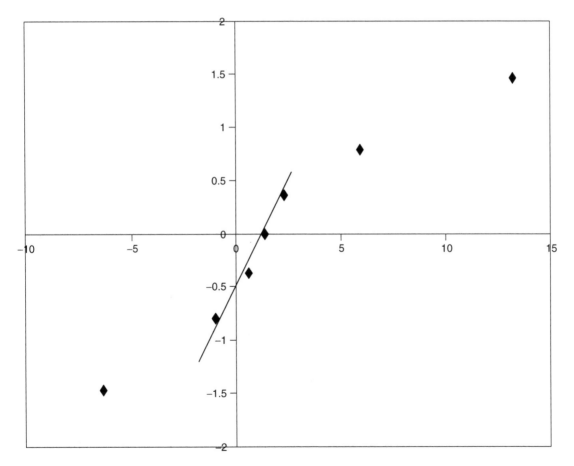

Figure 2.15
Normal probability plot

2.2.4.6 Dummy Factors

Another very simple approach is to include one or more dummy factors. These can be built into a design, and might, for example, be the colour of shoes worn by the experimenter, some factor that is not likely to have a real effect on the experiment; level −1 might correspond to black shoes and level +1 to brown shoes. Mathematical models can be built including this factor, and effects smaller than this factor ignored (remembering as ever to ensure that the scaling of the data is sensible).

2.2.4.7 Limitations of Statistical Tests

Whereas many traditionalists often enjoy the security that statistical significance tests give, it is important to recognise that these tests do depend on assumptions about the underlying data that may not be correct, and a chemist should be very wary of making decisions based only on a probability obtained from a computerised statistical software package without looking at the data, often graphically. Some typical drawbacks are as follows.

- Most statistical tests assume that the underlying samples and experimental errors fall on a normal distribution. In some cases this is not so; for example, when analysing some analytical signals it is unlikely that the noise distribution will be normal: it is often determined by electronics and sometimes even data preprocessing such as the common logarithmic transform used in electronic absorption and infrared spectroscopy.
- The tests assume that the measurements arise from the same underlying population. Often this is not the case, and systematic factors will come into play. A typical example involves calibration curves. It is well known that the performance of an instrument can vary from day to day. Hence an absorption coefficient measured on Monday morning is not necessarily the same as the coefficient measured on Tuesday morning, yet all the coefficients measured on Monday morning might fall into the same class. If a calibration experiment is performed over several days or even hours, the performance of the instrument may vary and the only real solution is to make a very large number of measurements over a long time-scale, which may be impractical.
- The precision of an instrument must be considered. Many typical measurements, for example, in atomic spectroscopy, are recorded to only two significant figures. Consider a dataset in which about 95 % of the readings were recorded between 0.10 and 0.30 absorbance units, yet a statistically designed experiment tries to estimate 64 effects. The t-test provides information on the significance of each effect. However, statistical tests assume that the data are recorded to indefinite accuracy, and will not take this lack of numerical precision into account. For the obvious effects, chemometrics will not be necessary, but for less obvious effects, the statistical conclusions will be invalidated because of the low numerical accuracy in the raw data.

Often it is sufficient simply to look at the size of factors, the significance of the lack-of-fit statistics, perform simple ANOVA or produce a few graphs, to make valid scientific deductions. In most cases, significance testing is used primarily for a preliminary modelling of the data and detailed experimentation should be performed after eliminating those factors that are deemed unimportant. It is not necessary to have a very detailed theoretical understanding of statistical significance tests prior to the design and analysis of chemical experiments, although a conceptual appreciation of, for example, the importance of coding is essential.

2.2.5 Leverage and Confidence in Models

An important experimental question relates to how well quantitative information can be predicted after a series of experiments has been carried out. For example, if observations have been made between 40 and 80 °C, what can we say about the experiment at 90 °C? It is traditional to cut off the model sharply outside the experimental region, so that the model is used to predict only within the experimental limits. However, this approaches misses much information. The ability to make a prediction often reduces smoothly from the centre of the experiments, being best at 60 °C and worse the further away from the centre in the example above. This does not imply that it is impossible to make any statement about the response at 90 °C, simply that there is less confidence in the prediction than at 80 °C, which, in turn, is predicted less well than at 60 °C. It is important to be able to visualise how the ability to predict a response (e.g. a synthetic yield or a concentration) varies as the independent factors (e.g. pH, temperature) are changed.

When only one factor is involved in the experiment, the predictive ability is often visualised by confidence bands. The 'size' of these confidence bands depends on the magnitude of the experimental error. The 'shape', however, depends on the experimental design, and can be obtained from the design matrix (Section 2.2.3) and is influenced by the arrangement of experiments, replication procedure and mathematical model. The concept of *leverage* is used as a measure of such confidence. The mathematical definition is given by

$$H = D.(D'.D)^{-1}.D'$$

where D is the design matrix. This new matrix is sometimes called the hat matrix and is a square matrix with the number of rows and columns equal to the number of experiments. Each of n experimental points has a value of leverage h_n (the diagonal element of the hat matrix) associated with it. Alternatively, the value of leverage can be calculated as follows:

$$h_n = d_n.(D'.D)^{-1}.d'_n$$

where d_n is the row of the design matrix corresponding to an individual experiment. The steps in determining the values of leverage for a simple experiment are illustrated in Table 2.14.

1. Set up design matrix.
2. Calculate $(D'D)^{-1}$. Note that this matrix is also used in the t-test, as discussed in Section 2.2.4.3.
3. Calculate the hat matrix and determine the diagonal values.
4. These diagonal values are the values of leverage for each experiment.

This numerical value of leverage has certain properties.

Table 2.14 Calculation of leverage.

Step 1

1	1	−1	1	1	−1
1	1	1	1	1	1
1	−1	−1	1	1	1
1	−1	1	1	1	−1
1	1.5	0	2.25	0	0
1	−1.5	0	2.25	0	0
1	0	1.5	0	2.25	0
1	0	−1.5	0	2.25	0
1	0	0	0	0	0
1	0	0	0	0	0
1	0	0	0	0	0
1	0	0	0	0	0

$\underbrace{}_{X}$

$$\underbrace{}_{D}$$

Step 2

0.248	0	0	−0.116	−0.116	0
0	0.118	0	0	0	0
0	0	0.118	0	0	0
−0.116	0	0	0.132	0.033	0
−0.116	0	0	0.033	0.132	0
0	0	0	0	0	0.25

$$(D'.D)^{-1}$$

Step 3

0.597	−0.139	−0.139	0.126	0.300	−0.053	−0.053	0.300	0.015	0.015	0.015	0.015
−0.139	0.597	0.126	−0.139	0.300	−0.053	0.300	−0.053	0.015	0.015	0.015	0.015
−0.139	0.126	0.597	−0.139	−0.053	0.300	−0.053	0.300	0.015	0.015	0.015	0.015
0.126	−0.139	−0.139	0.597	−0.053	0.300	0.300	−0.053	0.015	0.015	0.015	0.015
0.300	0.300	−0.053	−0.053	0.655	0.126	−0.110	−0.110	−0.014	−0.014	−0.014	−0.014
−0.053	−0.053	0.300	0.300	0.126	0.655	−0.110	−0.110	−0.014	−0.014	−0.014	−0.014
−0.053	0.300	−0.053	0.300	−0.110	−0.110	0.655	0.126	−0.014	−0.014	−0.014	−0.014
0.300	−0.053	0.300	−0.053	−0.110	−0.110	0.126	0.655	−0.014	−0.014	−0.014	−0.014
0.015	0.015	0.015	0.015	−0.014	−0.014	−0.014	−0.014	0.248	0.248	0.248	0.248
0.015	0.015	0.015	0.015	−0.014	−0.014	−0.014	−0.014	0.248	0.248	0.248	0.248
0.015	0.015	0.015	0.015	−0.014	−0.014	−0.014	−0.014	0.248	0.248	0.248	0.248
0.015	0.015	0.015	0.015	−0.014	−0.014	−0.014	−0.014	0.248	0.248	0.248	0.248

$$D.(D'.D)^{-1}.D'$$

Step 4

		h_n
1	−1	0.597
1	1	0.597
−1	−1	0.597
−1	1	0.597
1.5	0	0.655
−1.5	0	0.655
0	1.5	0.655
0	−1.5	0.655
0	0	0.248
0	0	0.248
0	0	0.248
0	0	0.248

- The value is always greater than 0.
- The lower the value, the higher is the confidence in the prediction. A value of 1 indicates very poor prediction. A value of 0 indicates perfect prediction and will not be achieved.
- If there are P coefficients in the model, the sum of the values for leverage at each experimental point adds up to P. Hence the sum of the values of leverage for the 12 experiments in Table 2.14 is equal to 6.

In the design in Table 2.14, the leverage is lowest at the centre, as expected. However, the value of leverage for the first four points in slightly lower than that for the second four points. As discussed in Section 2.4, this design is a form of central composite design, with the points 1–4 corresponding to a factorial design and points 5–8 to a star design. Because the leverage, or confidence, in the model differs, the design is said to be nonrotatable, which means that the confidence is not solely a function of distance from the centre of experimentation. How to determine rotatability of designs is discussed in Section 2.4.3.

Leverage can also be converted to equation form simply by substituting the algebraic expression for the coefficients in the equation

$$h = d.(D'.D)^{-1}.d'$$

where, in the case of Table 2.14,

$$d = (1 \quad x_1 \quad x_2 \quad x_1^2 \quad x_2^2 \quad x_1 x_2)$$

to give an equation, in this example, of the form

$$h = 0.248 - 0.116(x_1^2 + x_2^2) + 0.132(x_1^4 + x_2^4) + 0.316x_1^2 x_2^2$$

The equation can be obtained by summing the appropriate terms in the matrix $(D'.D)^{-1}$. This is illustrated graphically in Figure 2.16. Label each row and column by the corresponding terms in the model, and then find the combinations of terms in the matrix

	1	x_1	x_2	x_1^2	x_2^2	$x_1 x_2$
1	0.248	0	0	−0.116	−0.116	0
x_1	0	0.118	0	0	0	0
x_2	0	0	0.118	0	0	0
x_1^2	−0.116	0	0	0.132	0.033	0
x_2^2	−0.116	0	0	0.033	0.132	0
$x_1 x_2$	0	0	0	0	0	0.25

Figure 2.16
Method of calculating equation for leverage term $x_1^2 x_2^2$

that result in the coefficients of the leverage equation, for $x_1{}^2 x_2{}^2$ there are three such combinations so that the term $0.316 = 0.250 + 0.033 + 0.033$.

This equation can also be visualised graphically and used to predict the confidence of any point, not just where experiments were performed. Leverage can also be used to predict the confidence in the prediction under any conditions, which is given by

$$y_{\pm} = s\sqrt{[F_{(1,N-P)}(1 + h)]}$$

where s is the root mean square residual error given by $\sqrt{S_{resid}/(N - P)}$ as determined in Section 2.2.2, the F-statistic as introduced in Section 2.2.4.4, which can be obtained at any desired level of confidence but most usually at 95 % limits and is one-sided. Note that this equation technically refers to the confidence in the individual prediction and there is a slightly different equation for the mean response after replicates have been averaged given by

$$y_{\pm} = s\sqrt{\left[F_{(1,N-P)}\left(\frac{1}{M} + h\right)\right]}$$

See Section 2.2.1 for definitions of N and P; M is the number of replicates taken at a given point, for example if we repeat the experiment at 10 mM five times, $M = 5$. If repeated only once, then this equation is the same as the first one.

Although the details of these equations may seem esoteric, there are two important considerations:

- the shape of the confidence bands depends entirely on leverage;
- the size depends on experimental error.

These and most other equations developed by statisticians assume that the experimental error is the same over the entire response surface: there is no satisfactory agreement for how to incorporate heteroscedastic errors. Note that there are several different equations in the literature according to the specific aims of the confidence interval calculations, but for brevity we introduce only two which can be generally applied to most situations.

Table 2.15 Leverage for three possible single variable designs using a two parameter linear model.

Concentration			Leverage		
Design A	Design B	Design C	Design A	Design B	Design C
1	1	1	0.234	0.291	0.180
1	1	1	0.234	0.291	0.180
1	2	1	0.234	0.141	0.180
2	2	1	0.127	0.141	0.180
2	3	2	0.127	0.091	0.095
3	3	2	0.091	0.091	0.095
4	3	2	0.127	0.091	0.095
4	4	3	0.127	0.141	0.120
5	4	3	0.234	0.141	0.120
5	5	4	0.234	0.291	0.255
5	5	5	0.234	0.291	0.500

To show how leverage can help, consider the example of univariate calibration; three designs A–C (Table 2.15) will be analysed. Each experiment involves performing 11 experiments at five different concentration levels, the only difference being the arrangement of the replicates. The aim is simply to perform linear calibration to produce a model of the form $y = b_0 + b_1 x$, where x is the concentration, and to compare how each design predicts confidence. The leverage can be calculated using the design matrix $\mathbf{D}$, which consists of 11 rows (corresponding to each experiment) and two columns (corresponding to each coefficient). The hat matrix consists of 11 rows and 11 columns, the numbers on the diagonal being the values of leverage for each experimental point. The leverage for each experimental point is given in Table 2.15. It is also possible to obtain a graphical representation of the equation as shown in Figure 2.17 for designs A–C.

What does this tell us?

- Design A contains more replicates at the periphery of the experimentation than design B, and so results in a flatter graph. This design will provide predictions that are fairly even throughout the area of interest.
- Design C shows how replication can result in a major change in the shape of the curve for leverage. The asymmetric graph is a result of replication regime. In fact, the best predictions are no longer in the centre of experimentation.

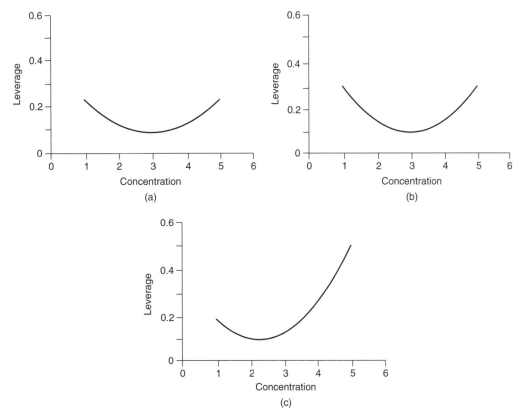

Figure 2.17
Graph of leverage for designs in Table 2.15.

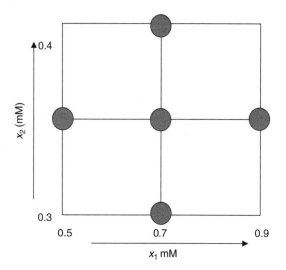

Figure 2.18
Two factor design consisting of five experiments

This approach can used for univariate calibration experiments more generally. How many experiments are necessary to produce a given degree of confidence in the prediction? How many replicates are sensible? How good is the prediction outside the region of experimentation? How do different experimental arrangements relate? In order to obtain an absolute value of the confidence of predictions, it is also necessary, of course, to determine the experimental error, but this together, with the leverage, which is a direct consequence of the design and model, is sufficient information. Note that leverage will change if the model changes.

Leverage is most powerful as a tool when several factors are to be studied. There is no general agreement as to the definition of an experimental space under such circumstances. Consider the simple design of Figure 2.18, consisting of five experiments. Where does the experimental boundary stop? The range of concentrations for the first compound is 0.5–0.9 mM and for the second compound 0.2–0.4 mM. Does this mean we can predict the response well when the concentrations of the two compounds are at 0.9 and 0.4 mM, respectively? Probably not, as some people would argue that the experimental region is a circle, not a square. For this nice symmetric arrangement of experiments it is possible to envisage an experimental region, but imagine telling the laboratory worker that if the concentration of the second compound is 0.34 mM then if the concentration of the first is 0.77 mM the experiment is within the region, whereas if it is 0.80 mM it is outside. There will be confusion as to where the model starts and stops. For some supposedly simply designs such as a full factorial design the definition of the experimental region is even harder to conceive.

The best solution is to produce a simple graph to show how confidence in the prediction varies over the experimental region. Consider the two designs in Figure 2.19. Using a very simple linear model, of the form $y = b_1x_1 + b_2x_2$, the leverage for both designs is as given Figure 2.20. The consequence of the different experimental arrangements is now obvious, and the result in the second design on the confidence in predictions can be seen. Although a two factor example is fairly straightforward, for multifactor designs

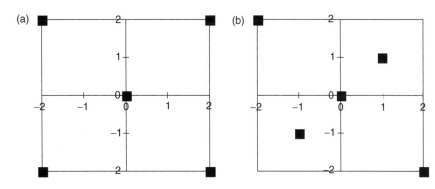

Figure 2.19
Two experimental arrangements

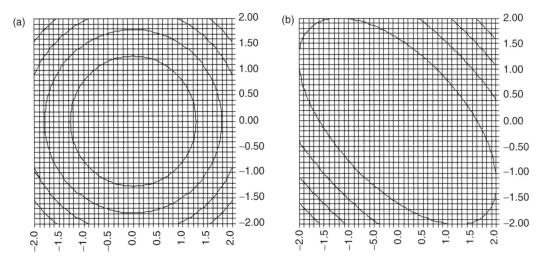

Figure 2.20
Graph of leverage for experimental arrangements in Figure 2.19

(e.g. mixtures of several compounds) it is hard to produce an arrangement of samples in which there is symmetric confidence in the results over the experimental domain.

Leverage can show the effect of changing an experimental parameter such as the number of replicates, or, in the case of central composite design, the position of the axial points (see Section 2.4). Some interesting features emerge from this analysis. For example, confidence is not always highest in the centre of experimentation, depending on the number of replicates. The method in this section is an important tool for visualising how changing design relates to the ability to make quantitative predictions.

2.3 Factorial Designs

In this and the remaining sections of this chapter we will introduce a number of possible designs, which can be understood using the building blocks introduced in Section 2.2.

Factorial designs are some of the simplest, often used for screening or when there are a large number of possible factors. As will be seen, they have limitations, but are the easiest to understand. Many designs are presented as a set of rules which provide the experimenter with a list of conditions, and below we will present the rules for many of the most common methods.

2.3.1 Full Factorial Designs

Full factorial designs at two levels are mainly used for screening, that is, to determine the influence of a number of effects on a response, and to eliminate those that are not significant, the next stage being to undertake a more detailed study. Sometimes, where detailed predictions are not required, the information from factorial designs is adequate, at least in situations where the aim is fairly qualitative (e.g. to improve the yield of a reaction rather than obtain a highly accurate rate dependence that is then interpreted in fundamental molecular terms).

Consider a chemical reaction, the performance of which is known to depend on pH and temperature, including their interaction. A set of experiments can be proposed to study these two factors, each at two levels, using a two level, two factor experimental design. The number of experiments is given by $N = l^k$, where l is the number of levels (=2), and k the number of factors (=2), so in this case $N = 4$. For three factors, the number of experiments will equal 8, and so on, provided that the design is performed at two levels only. The following stages are used to construct the design and interpret the results.

1. The first step is to choose a high and low level for each factor, for example, $30°$ and $60°$, and pH 4 and 6.
2. The next step is to use a standard design. The value of each factor is usually coded (see Section 2.2.4.1) as − (low) or + (high). Note that some authors use −1 and +1 or even 1 and 2 for low and high. When reading different texts, do not get confused: always first understand what notation has been employed. There is no universally agreed convention for coding; however, design matrices that are symmetric around 0 are almost always easier to handle computationally. There are four possible unique sets of experimental conditions which can be represented as a table analogous to four binary numbers, 00 (−−), 01 (−+), 10 (+−) and 11 (++), which relate to a set of physical conditions.
3. Next, perform the experiments and obtain the response. Table 2.16 illustrates the coded and true set of experimental conditions plus the response, which might, for example be the percentage of a by-product, the lower the better. Something immediately appears strange from these results. Although it is obvious that the higher

Table 2.16 Coding of a simple two factor, two level design and the response.

Experiment No.	Factor 1	Factor 2	Temperature	pH	Response
1	−	−	30	4	12
2	−	+	30	6	10
3	+	−	60	4	24
4	+	+	60	6	25

Table 2.17 Design matrix.

Intercept	Temperature	pH	Temp. * pH	b_0	b_1	b_2	b_{12}
1	30	4	120	+	−	−	−
1	30	6	180	+	−	+	−
1	60	4	240	+	+	−	+
1	60	6	360	+	+	+	+

the temperature, the higher is the percentage by-product, there does not seem to be any consistent trend as far as pH is concerned. Provided that the experimental results were recorded correctly, this suggests that there must be an interaction between temperature and pH. At a lower temperature, the percentage decreases with increase in pH, but the opposite is observed at a higher temperature. How can we interpret this?

4. The next step, of course, is to analyse the data, by setting up a design matrix (see Section 2.2.3). We know that an interaction term must be taken into account, and set up a design matrix as given in Table 2.17 based on a model of the form $y = b_0 + b_1 x_1 + b_2 x_2 + b_{11} x_1 x_2$. This can be expressed either as a function of the true or coded concentrations, but, as discussed in Section 2.2.4.1, is probably best as coded values. Note that four possible coefficients that can be obtained from the four experiments. Note also that each of the columns in Table 2.17 is different. This is an important and crucial property and allows each of the four possible terms to be distinguished uniquely from one another, and is called *orthogonality*. Observe, also, that squared terms are impossible because four experiments can be used to obtain only a maximum of four terms, and also the experiments are performed at only two levels; ways of introducing such terms will be described in Section 2.4.

5. Calculate the coefficients. It is not necessary to employ specialist statistical software for this. In matrix terms, the response can be given by $y = D.b$, where b is a vector of the four coefficients and D is presented in Table 2.17. Simply use the matrix inverse so that $b = D^{-1}.y$. Note that there are no replicates and the model will exactly fit the data. The parameters are listed below.

   - For raw values:
     intercept $= 10$
     temperature coefficient $= 0.2$
     pH coefficient $= -2.5$
     interaction term $= 0.05$
   - For coded values:
     intercept $= 17.5$
     temperature coefficient $= 6.75$
     pH coefficient $= -0.25$
     interaction term $= 0.75$

6. Finally, interpret the coefficients. Note that for the raw values, it appears that pH is much more influential than temperature, and also that the interaction is very small. In addition, the intercept term is not the average of the four readings. The reason why this happens is that the intercept is the predicted response at pH 0 and $0\,°C$, conditions unlikely to be reached experimentally. The interaction term appears very small, because units used for temperature correspond to a range of $30\,°C$ as opposed to a pH range of 2. A better measure of significance comes from

the coded coefficients. The effect of temperature is overwhelming. Changing pH has a very small influence, which is less than the interaction between the two factors, explaining why the response is higher at pH 4 when the reaction is studied at 30 °C, but the opposite is true at 60 °C.

Two level full factorial designs (also sometimes called saturated factorial designs), as presented in this section, take into account all linear terms, and all possible k way interactions. The number of different types of terms can be predicted by the binomial theorem [given by $k!/(k - m)!m!$ for mth-order interactions and k factors, e.g. there are six two factor ($=m$) interactions for four factors ($=k$)]. Hence for a four factor, two level design, there will be 16 experiments, the response being described by an equation with a total of 16 terms, of which

- there is one interaction term;
- four linear terms such as b_1;
- six two factor interaction terms such as b_1b_2;
- four three factor interactions terms such as $b_1b_2b_3$;
- and one four factor interaction term $b_1b_2b_3b_4$.

The coded experimental conditions are given in Table 2.18(a) and the corresponding design matrix in Table 2.18(b). In common with the generally accepted conventions, a + symbol is employed for a high level and a − for a low level. The values of the interactions are obtained simply by multiplying the levels of the individual factors together. For example, the value of $x_1x_2x_4$ for the second experiment is + as it is a product of − × − × +. Several important features should be noted.

- Every column (apart from the intercept) contains exactly eight high and eight low levels. This property is called balance.
- Apart from the first column, each of the other possible pairs of columns have the property that each for each experiment at level + for one column, there are equal number of experiments for the other columns at levels + and −. Figure 2.21 shows a graph of the level of any one column (apart from the first) plotted against the level of any other column. For any combination of columns 2–16, this graph will be identical, and is a key feature of the design. It relates to the concept of *orthogonality*, which is used in other contexts throughout this book. Some chemometricians regard each column as a vector in space, so that any two vectors are at right angles to each other. Algebraically, the *correlation coefficient* between each pair of columns equals 0. Why is this so important? Consider a case in which the values of two factors (or indeed any two columns) are related as in Table 2.19. In this case, every time the first factor is at a high level, the second is at a low level and vice versa. Thus, for example, every time a reaction is performed at pH 4, it is also performed at 60 °C, and every time it is performed at pH 6, it is also performed at 30 °C, so how can an effect due to increase in temperature be distinguished from an effect due to decrease in pH? It is impossible. The two factors are correlated. The only way to be completely sure that the influence of each effect is independent is to ensure that the columns are orthogonal, that is, not correlated.
- The other remarkable property is that the inverse of the design matrix is related to the transpose by

$$\boldsymbol{D}^{-1} = (1/N)\boldsymbol{D}'$$

Table 2.18 Four factor, two level full factorial design.

(a) Coded experimental conditions

Experiment No.	Factor 1	Factor 2	Factor 3	Factor 4
1	−	−	−	−
2	−	−	−	+
3	−	−	+	−
4	−	−	+	+
5	−	+	−	−
6	−	+	−	+
7	−	+	+	−
8	−	+	+	+
9	+	−	−	−
10	+	−	−	+
11	+	−	+	−
12	+	−	+	+
13	+	+	−	−
14	+	+	−	+
15	+	+	+	−
16	+	+	+	+

(b) Design matrix

x_0	x_1	x_2	x_3	x_4	x_1x_2	x_1x_3	x_1x_4	x_2x_3	x_2x_4	x_3x_4	$x_1x_2x_3$	$x_1x_2x_4$	$x_1x_3x_4$	$x_2x_3x_4$	$x_1x_2x_3x_4$
+	−	−	−	−	+	+	+	+	+	+	−	−	−	−	+
+	−	−	−	+	+	+	−	+	−	−	−	+	+	+	−
+	−	−	+	−	+	−	+	−	+	−	+	−	+	+	−
+	−	−	+	+	+	−	−	−	−	+	+	+	−	−	+
+	−	+	−	−	−	+	+	−	−	+	+	+	−	+	−
+	−	+	−	+	−	+	−	−	+	−	+	−	+	−	+
+	−	+	+	−	−	−	+	+	−	−	−	+	+	−	+
+	−	+	+	+	−	−	−	+	+	+	−	−	−	+	−
+	+	−	−	−	−	−	−	+	+	+	+	+	+	−	−
+	+	−	−	+	−	−	+	+	−	−	+	−	−	+	+
+	+	−	+	−	−	+	−	−	+	−	−	+	−	+	+
+	+	−	+	+	−	+	+	−	−	+	−	−	+	−	−
+	+	+	−	−	+	−	−	−	−	+	−	−	+	+	+
+	+	+	−	+	+	−	+	−	+	−	−	+	−	−	−
+	+	+	+	−	+	+	−	+	−	−	+	−	−	−	−
+	+	+	+	+	+	+	+	+	+	+	+	+	+	+	+

where there are N experiments. This a general feature of all saturated two level designs, and relates to an interesting classical approach for determining the size of each effect. Using modern matrix notation, the simplest method is simply to calculate

$$b = D^{-1}.y$$

but many classical texts use an algorithm involving multiplying the response by each column (of coded coefficients) and dividing by the number of experiments to obtain the value of the size of each factor. For the example in Table 2.16, the value of the effect due to temperature can be given by

$$b_1 = (-1 \times 12 - 1 \times 10 + 1 \times 24 + 1 \times 25)/4 = 6.75$$

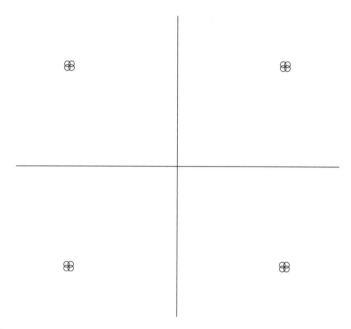

Figure 2.21
Graph of levels of one term against another in the design in Table 2.18

Table 2.19   Correlated factors.

−	+
+	−
−	+
+	−
−	+
+	−
−	+
+	−
−	+
+	−
−	+
+	−
−	+
+	−
−	+
+	−

identical with that obtained using simply matrix manipulations. Such method for determining the size of the effects was extremely useful to classical statisticians prior to matrix oriented software, and still widely used, but is limited only to certain very specific designs. It is also important to recognise that some texts divide the expression above by $N/2$ rather than N, making the classical numerical value of the effects equal to twice those obtained by regression. So long as all the effects are on the same scale, it does not matter which method is employed when comparing the size of each factor.

An important advantage of two level factorial designs is that some factors can be 'categorical' in nature, that is, they do not need to refer to a quantitative parameter. One factor may be whether a reaction mixture is stirred (+ level) or not (− level), and another whether it is carried out under nitrogen or not. Thus these designs can be used to ask qualitative questions. The values of the b parameters relate directly to the significance or importance of these factors and their interactions.

Two level factorial designs can be used very effectively for screening, but also have pitfalls.

- They only provide an approximation within the experimental range. Note that for the model above it is possible to obtain nonsensical predictions of negative percentage yields outside the experimental region.
- They cannot take quadratic terms into account, as the experiments are performed only at two levels.
- There is no replicate information.
- If all possible interaction terms are taken into account no error can be estimated, the F-test and t-test not being applicable. However, if it is known that some interactions are unlikely or irrelevant, it is possible to model only the most important factors. For example, in the case of the design in Table 2.18, it might be decided to model only the intercept, four single factor and six two factor interaction terms, making 11 terms in total, and to ignore the higher order interactions. Hence,
  - $N = 16$;
  - $P = 11$;
  - $(N - P) = 5$ terms remain to determine the fit to the model.

Some valuable information about the importance of each term can be obtained under such circumstances. Note, however, the design matrix is no longer square, and it is not possible to use the simple approaches above to calculate the effects, regression using the pseudo-inverse being necessary.

However, two level factorial designs remain popular largely because they are extremely easy to set up and understand; also, calculation of the coefficients is very straightforward. One of the problems is that once there are a significant number of factors involved, it is necessary to perform a large number of experiments: for six factors, 64 experiments are required. The 'extra experiments' really only provide information about the higher order interactions. It is debatable whether a six factor, or even four factor, interaction is really relevant or even observable. For example, if an extraction procedure is to be studied as a function of (a) whether an enzyme is present or not, (b) incubation time, (c) incubation temperature, (d) type of filter, (e) pH and (f) concentration, what meaning will be attached to the higher order interactions, and even if they are present can they be measured with any form of confidence? And is it economically practicable or sensible to spend such a huge effort studying these interactions? Information such as squared terms is not available, so detailed models of the extraction behaviour are not available either, nor is any replicate information being gathered. Two improvements are as follows. If it is desired to reduce the number experiments by neglecting some of the higher order terms, use designs discussed in Sections 2.3.2 and 2.3.3. If it is desired to study squared or higher order terms whilst reducing the number of experiments, use the designs discussed in Section 2.4.

Sometimes it is not sufficient to study an experiment at two levels. For example, is it really sufficient to use only two temperatures? A more detailed model will be

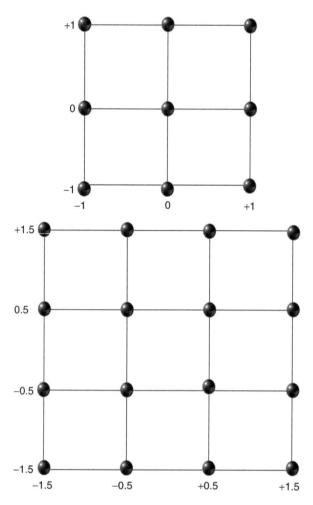

Figure 2.22
Three and four level full factorial designs

obtained using three temperatures; in addition, such designs either will allow the use of squared terms or, if only linear terms used in the model, some degrees of freedom will be available to assess goodness-of-fit. Three and four level designs for two factors are presented in Figure 2.22 with the values of the coded experimental conditions in Table 2.20. Note that the levels are coded to be symmetrical around 0, and so that each level differs by one from the next. These designs are called multilevel factorial designs. The number of experiments can become very large if there are several factors, for example, a five factor design at three levels involves 3^5 or 243 experiments. In Section 2.3.4 we will discuss how to reduce the size safely and in a systematic manner.

2.3.2   Fractional Factorial Designs

A weakness of full factorial designs is the large number of experiments that must be performed. For example, for a 10 factor design at two levels, 1024 experiments are

Table 2.20 Full factorial designs corresponding to Figure 2.22.

(a) Three levels

−1	−1
−1	0
−1	+1
0	−1
0	0
0	+1
+1	−1
+1	0
+1	+1

(b) Four levels

−1.5	−1.5
−1.5	−0.5
−1.5	+0.5
−1.5	+1.5
−0.5	−1.5
−0.5	−0.5
−0.5	+0.5
−0.5	+1.5
+0.5	−1.5
+0.5	−0.5
+0.5	+0.5
+0.5	+1.5
+1.5	−1.5
+1.5	−0.5
+1.5	+0.5
+1.5	+1.5

required, which may be impracticable. These extra experiments do not always result in useful or interesting extra information and so are wasteful of time and resources. Especially in the case of screening, where a large number of factors may be of potential interest, it is inefficient to run so many experiments in the first instance. There are numerous tricks to reduce the number of experiments.

Consider a three factor, two level design. Eight experiments are listed in Table 2.21, the conditions being coded as usual. Figure 2.23 is a symbolic representation of the experiments, often presented on the corners of a cube, whose axes correspond to each factor. The design matrix for all the possible coefficients can be set up as is also illustrated in Table 2.21 and consists of eight possible columns, equal to the number of experiments. Some columns represent interactions, such as the three factor interaction, that are not very likely. At first screening we may primarily wish to say whether the three factors have any real influence on the response, not to study the model in detail. In a more complex situation, we may wish to screen 10 possible factors, and reducing the number of factors to be studied further to three or four makes the next stage of experimentation easier.

How can we reduce the number of experiments safely and systematically? Two level fractional factorial designs are used to reduce the number of experiments by 1/2, 1/4, 1/8 and so on. Can we halve the number of experiments? At first glance, a simple

Table 2.21 Full factorial design for three factors together with the design matrix.

Experiment No.	Factor 1	Factor 2	Factor 3	Design matrix							
				x_0	x_1	x_2	x_3	x_1x_2	x_1x_3	x_2x_3	$x_1x_2x_3$
1	+	+	+	+	+	+	+	+	+	+	+
2	+	+	−	+	+	+	−	+	−	−	−
3	+	−	+	+	+	−	+	−	+	−	−
4	+	−	−	+	+	−	−	−	−	+	+
5	−	+	+	+	−	+	+	−	−	+	−
6	−	+	−	+	−	+	−	−	+	−	+
7	−	−	+	+	−	−	+	+	−	−	+
8	−	−	−	+	−	−	−	+	+	+	−

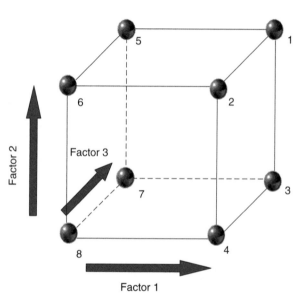

Figure 2.23
Representation of a three factor, two level design

approach might be to take the first four experiments of Table 2.21. However, these would leave the level of the first factor at +1 throughout. A problem is that we now no longer study the variation of this factor, so we do not obtain any information on how factor 1 influences the response, and are studying the wrong type of variation, in fact such a design would remove all four terms from the model that include the first factor, leaving the intercept, two single factor terms and the interaction between factors 2 and 3, not the hoped for information, unless we know that factor 1 and its interactions are insignificant.

Can a subset of four experiments be selected that allows us to study all three factors? Rules have been developed to produce these fractional factorial designs obtained by taking the correct subset of the original experiments. Table 2.22 illustrates a possible fractional factorial design that enables all factors to be studied. There are a number of important features:

Table 2.22 Fractional factorial design.

Experiment No.	Factor 1	Factor 2	Factor 3	Matrix of effects							
				x_0	x_1	x_2	x_3	x_1x_2	x_1x_3	x_2x_3	$x_1x_2x_3$
1	+	+	+	+	+	+	+	+	+	+	+
2	+	−	−	+	+	−	−	−	−	+	+
3	−	−	+	+	−	−	+	+	−	−	+
4	−	+	−	+	−	+	−	−	+	−	+

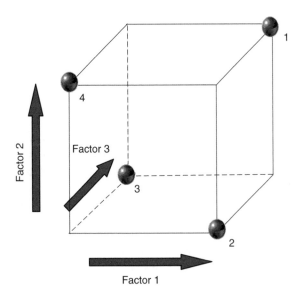

Figure 2.24
Fractional factorial design

- every column in the experimental matrix is different;
- in each column, there are an equal number of − and + levels;
- for each experiment at level + for factor 1, there are equal number of experiments for factors 2 and 3 which are at levels + and −, and the columns are orthogonal.

The properties of this design can be understood better by visualisation (Figure 2.24): half the experiments have been removed. For the remainder, each face of the cube now corresponds to two rather than four experiments, and every alternate corner corresponds to an experiment.

The matrix of effects in Table 2.22 is also interesting. Whereas the first four columns are all different, the last four each correspond to one of the first four columns. For example, the x_1x_2 column exactly equals the x_3 column. What does this imply in reality? As the number of experiments is reduced, the amount of information is correspondingly reduced. Since only four experiments are now performed, it is only possible to measure four unique factors. The interaction between factors 1 and 2 is said to be *confounded* with factor 3. This might mean, for example, that, using this design the interaction between temperature and pH is indistinguishable from the influence of concentration alone. However, not all interactions will be significant, and the purpose of a preliminary experiment is often simply to sort out which main factors should be studied in detail later. When calculating the effects, it is important to use only four unique columns in the design matrix, rather than all eight columns, as otherwise the matrix will not have an inverse.

Note that two level fractional factorial designs only exist when the number of experiments equals a power of 2. In order to determine the minimum number of experiments do as follows:

- determine how many terms are interesting;
- then construct a design whose size is the next greatest power of 2.

Setting up a fractional factorial design and determining which terms are confounded is relatively straightforward and will be illustrated with reference to five factors.

A half factorial design involves reducing the experiments from 2^k to 2^{k-1} or, in this case, from 32 to 16.

1. In most cases, the aim is to
   - confound k factor interactions with the intercept;
   - $(k-1)$ factor interactions with single factor interactions;
   - up to $(k-1)/2$ factor interactions with $(k-1)/2 + 1$ factor interactions if the number of factors is odd, or $k/2$ factor interactions with themselves if the number of factors is even.

   i.e. for five factors, confound 0 factor interactions (intercept) with 5, 1 factor interactions (pure variables) with 4, and 2 factor interactions with 3 factor interactions, and for six factors, confound 0 with 6, 1 with 5, 2 with 4 factor interactions, and 3 factor interactions with themselves.
2. Set up a $k-1$ factor design for the first $k-1$ factors, i.e. a 4 factor design consisting of 16 experiments.
3. Confound the kth (or final) factor with the product of the other factors by setting the final column as either $-$ or $+$ the product of the other factors. A simple notation is often used to analyse these designs, whereby the final column is given by $k = +1*2*\ldots*(k-1)$ or $k = -1*2*\ldots*(k-1)$. The case where $5 = +1*2*3*4$ is illustrated in Table 2.23. This means that a four factor interaction (most unlikely to have any physical meaning) is confounded with the fifth factor. There are, in fact, only two different types of design with the properties of step 1 above. Each design is denoted by how the intercept (**I**) is confounded, and it is easy to show that this design is of the type $\mathbf{I} = +1*2*3*4*5$, the other possible design being of type $\mathbf{I} = -1*2*3*4*5$. Table 2.23, therefore, is one possible half factorial design for five factors at two levels.
4. It is possible to work out which of the other terms are confounded with each other, either by multiplying the columns of the design together or from first principles, as follows. Every column multiplied by itself will result in a column of $+$ signs or **I** as the square of either -1 or $+1$ is always $+1$. Each term will be confounded with one other term in this particular design. To demonstrate which term $1*2*3$ is confounded with, simply multiply **5** by **4** since $5 = 1*2*3*4$, so $5*4 = 1*2*3*4*4 = 1*2*3$ since $4*4$ equals **I**. These interactions for the design of Table 2.23 are presented in Table 2.24.
5. In the case of negative numbers, ignore the negative sign. If two terms are correlated, it does not matter if the correlation coefficient is positive or negative, they cannot

Table 2.23 Confounding factor 5 with the product of factors 1–4.

Factor 1 **1**	Factor 2 **2**	Factor 3 **3**	Factor 4 **4**	Factor 5 **+1∗2∗3∗4**
−	−	−	−	+
−	−	−	+	−
−	−	+	−	−
−	−	+	+	+
−	+	−	−	−
−	+	−	+	+
−	+	+	−	+
−	+	+	+	−
+	−	−	−	−
+	−	−	+	+
+	−	+	−	+
+	−	+	+	−
+	+	−	−	+
+	+	−	+	−
+	+	+	−	−
+	+	+	+	+

Table 2.24 Confounding interaction terms in design of Table 2.23.

I	+1∗2∗3∗4∗5
1	+2∗3∗4∗5
2	+1∗3∗4∗5
3	+1∗2∗4∗5
4	+1∗2∗3∗5
5	+1∗2∗3∗4
1∗2	+3∗4∗5
1∗3	+2∗4∗5
1∗4	+2∗3∗5
1∗5	+2∗3∗4
2∗3	+1∗4∗5
2∗4	+1∗3∗5
2∗5	+1∗3∗4
3∗4	+1∗2∗5
3∗5	+1∗2∗4
4∗5	+1∗2∗3

be distinguished. In practical terms, this implies that if one term increases the other decreases.

Note that it is possible to obtain other types of half factorials, but these may involve, for example, confounding single factor terms with two factor interactions.

A smaller factorial design can be constructed as follows.

1. For a 2^{-f} fractional factorial, first set up the design consisting of 2^{k-f} experiments for the first $k - f$ factors, i.e. for a quarter ($f = 2$) of a five ($=k$) factorial experiment, set up a design consisting of eight experiments for the first three factors.
2. Determine the lowest order interaction that must be confounded. For a quarter of a five factorial design, second-order interactions must be confounded. Then, almost

Table 2.25 Quarter factorial design.

Factor 1 **1**	Factor 2 **2**	Factor 3 **3**	Factor 4 **−1∗2**	Factor 5 **1∗2∗3**
−	−	−	−	−
−	−	+	−	+
−	+	−	+	+
−	+	+	+	−
+	−	−	+	+
+	−	+	+	−
+	+	−	−	−
+	+	+	−	+

arbitrarily (unless there are good reasons for specific interactions to be confounded) set up the last two columns as products (times − or +) of combinations of the other columns, with the proviso that the products must include as least as many terms as the lowest order interaction to be confounded. Therefore, for our example, any two factor (or higher) interaction is entirely valid. In Table 2.25, a quarter factorial design where **4 = −1∗2** and **5 = 1∗2∗3** is presented.

3. Confounding can be analysed as above, but now each term will be confounded with three other terms for a quarter factorial design (or seven other terms for an eighth factorial design).

In more complex situations, such as 10 factor experiments, it is unlikely that there will be any physical meaning attached to higher order interactions, or at least that these interactions are not measurable. Therefore, it is possible to select specific interactions that are unlikely to be of interest, and consciously reduce the experiments in a systematic manner by confounding these with lower order interactions.

There are obvious advantages in two level fractional factorial designs, but these do have some drawbacks:

• there are no quadratic terms, as the experiments are performed only at two levels;
• there are no replicates;
• the number of experiments must be a power of two.

Nevertheless, this approach is very popular in many exploratory situations and has the additional advantage that the data are easy to analyse. It is important to recognise, however, that experimental design has a long history, and a major influence on the minds of early experimentalists and statisticians has always been ease of calculation. Sometimes extra experiments are performed simply to produce a design that could be readily analysed using pencil and paper. It cannot be over-stressed that inverse matrices were very difficult to calculate manually, but modern computers now remove this difficulty.

2.3.3 Plackett–Burman and Taguchi Designs

Where the number of factors is fairly large, the constraint that the number of experiments must equal a power of 2 can be rather restrictive. Since the number of experiments must always exceed the number of factors, this would mean that 32 experiments are required for the study of 19 factors, and 64 experiments for the study of 43 factors. In order

Table 2.26 A Plackett–Burman design.

		Factors										
		1	2	3	4	5	6	7	8	9	10	11
Experiments	1	−	−	−	−	−	−	−	−	−	−	−
	2	+	−	+	−	−	−	+	+	+	−	+
	3	+	+	−	+	−	−	−	+	+	+	−
	4	−	+	+	−	+	−	−	−	+	+	+
	5	+	−	+	+	−	+	−	−	−	+	+
	6	+	+	−	+	+	−	+	−	−	−	+
	7	+	+	+	−	+	+	−	+	−	−	−
	8	−	+	+	+	−	+	+	−	+	−	−
	9	−	−	+	+	+	−	+	+	−	+	−
	10	−	−	−	+	+	+	−	+	+	−	+
	11	+	−	−	−	+	+	+	−	+	+	−
	12	−	+	−	−	−	+	+	+	−	+	+

to overcome this problem and reduce the number of experiments, other approaches are needed.

Plackett and Burman published their classical paper in 1946, which has been much cited by chemists. Their work originated from the need for war-time testing of components in equipment manufacture. A large number of factors influenced the quality of these components and efficient procedures were required for screening. They proposed a number of two level factorial designs, where the number of experiments is a multiple of four. Hence designs exist for 4, 8, 12, 16, 20, 24, etc., experiments. The number of experiments exceeds the number of factors, k, by one.

One such design is given in Table 2.26 for 11 factors and 12 experiments and has various features.

- In the first row, all factors are at the same level.
- The first column from rows 2 to k is called a *generator*. The key to the design is that there are only certain allowed generators which can be obtained from tables. Note that the number of factors will always be an odd number equal to $k = 4m - 1$ (or 11 in this case), where m is any integer. If the first row consists of −, the generator will consist of $2m$ (=6 in this case) experiments at the + level and $2m - 1$ (=5 in this case) at the − level, the reverse being true if the first row is at the + level. In Table 2.26, the generator is $+ + - + + + + - - - + -$.
- The next $4m - 2$ (=10) columns are generated from the first column simply by shifting the down cells by one row. This is indicated by diagonal arrows in the table. Notice that experiment 1 is not included in this procedure.
- The level of factor j in experiment (or row) 2 equals to the level of this factor in row k for factor $j - 1$. For example, the level of factor 2 in experiment 2 equals the level of factor 1 in experiment 12.

There are as many high as low levels of each factor over the 12 experiments, as would be expected. The most important property of the design, however, is *orthogonality*. Consider the relationship between factors 1 and 2.

- There are six instances in which factor 1 is at a high level and six at a low level.
- For each of the six instances at which factor 1 is at a high level, in three cases factor 2 is at a high level, and in the other three cases it is at a low level. A similar

Table 2.27 Generators for Plackett–Burman design; first row is at – level.

Factors	Generator
7	$+++-+--$
11	$+ + - + + + - - - + -$
15	$+ + + + - + - + + - - + - - -$
19	$+ + - + + + + - + - + - - - - + + -$
23	$+ + + + + - + - + + - - + + - - + - + - - - -$

relationship exists where factor 1 is at a low level. This implies that the factors are orthogonal or uncorrelated, an important condition for a good design.

• Any combination of two factors is related in a similar way.

Only certain generators possess all these properties, so it is important to use only known generators.

Standard Plackett–Burman designs exist for 7, 11, 15, 19 and 23 factors; generators are given in Table 2.27. Note that for 7 and 15 factors it is also possible to use conventional fractional factorial designs as discussed in Section 2.3.2. However, in the old adage all roads lead to Rome, in fact fractional factorial and Plackett–Burman designs are equivalent, the difference simply being in the way the experiments and factors are organised in the data table. In reality, it should make no difference in which order the experiments are performed (in fact, it is best if the experiments are run in a randomised order) and the factors can be represented in any order along the rows. Table 2.28 shows that for 7 factors, a Plackett–Burman design is the same as a sixteenth factorial ($=2^{7-4} = 8$ experiments), after rearranging the rows, as indicated by the arrows. The confounding of the factorial terms is also indicated. It does not really matter which approach is employed.

If the number of experimental factors is less that of a standard design (a multiple of 4 minus 1), the final factors are *dummy* ones. Hence if there are only 10 real factors, use an 11 factor design, the final factor being a dummy one: this may be a variable that has no effect on the experiment, such as the technician that handed out the glassware or the colour of laboratory furniture.

If the intercept term is included, the design matrix is a square matrix, so the coefficients for each factor are given by

$$b = D^{-1}.y$$

Table 2.28 Equivalence of Plackett–Burman and fractional factorial design for seven factors.

Plackett–Burman design								Fractional factorial design						
								1	2	3	$4=-1*3$	$5=1*2*3$	$6=-1*2$	$7=-2*3$
–	–	–	–	–	–	–		–	–	–	–	–	–	–
+	+	+	–	+	–	–		–	–	+	+	+	–	+
–	+	+	+	–	+	–		–	+	–	–	+	+	+
–	–	+	+	+	–	+		–	+	+	+	–	+	–
+	–	–	+	+	+	–		+	–	–	+	+	+	–
–	+	–	–	+	+	+		+	–	+	–	–	+	+
+	–	+	–	–	+	+		+	+	–	+	–	–	+
+	+	–	+	–	–	+		+	+	+	–	+	–	–

Provided that coded values are used throughout, since there are no interaction or squared terms, the size of the coefficients is directly related to importance. An alternative method of calculation is to multiply the response by each column, dividing by the number of experiments:

$$b_j = \sum_{i=1}^{N} x_{ij} y_i / N$$

as in normal full factorial designs where x_{ij} is a number equal to $+1$ or -1 according to the value in the experimental matrix. If one or more dummy factor is included, it is easy to compare the size of the real factors with that of the dummy factor, and factors that are demonstrably larger in magnitude have significance.

An alternative approach comes from the work of Glenichi Taguchi. His method of quality control was much used by Japanese industry, and only fairly recently was it recognised that certain aspects of the theory are very similar to Western practices. His philosophy was that consumers desire products that have constant properties within narrow limits. For example, a consumer panel may taste the sweetness of a product, rating it from 1 to 10. A good marketable product may result in a taste panel score of 8: above this value the product is too sickly, and below it the consumer expects the product to be sweeter. There will be a huge number of factors in the manufacturing process that might cause deviations from the norm, including suppliers of raw materials, storage and preservation of the food and so on. Which factors are significant? Taguchi developed designs for screening large numbers of potential factors.

His designs are presented in the form of a table similar to that of Plackett and Burman, but with a 1 for a low and 2 for a high level. Superficially, Taguchi designs might appear different, but by changing the notation, and swapping rows and columns around, it is possible to show that both types of design are identical and, indeed, the simpler designs are the same as the well known partial factorial designs. There is a great deal of controversy surrounding Taguchi's work; while many statisticians feel that he has 'reinvented the wheel', he was an engineer, but his way of thinking had a major and positive effect on Japanese industrial productivity. Before globalisation and the Internet, there was less exchange of ideas between different cultures. His designs are part of a more comprehensive approach to quality control in industry.

Taguchi designs can be extended to three or more levels, but construction becomes fairly complicated. Some texts do provide tables of multilevel screening designs, and it is also possible to mix the number of levels, for example having one factor at two levels and another at three levels. This could be useful, for example, if there are three alternative sources of one raw material and two of another. Remember that the factors can fall into discrete categories and do not have to be numerical values such as temperature or concentrations. A large number of designs have been developed from Taguchi's work, but most are quite specialist, and it is not easy to generalise. The interested reader is advised to consult the source literature.

2.3.4 Partial Factorials at Several Levels: Calibration Designs

Two level designs are useful for exploratory purposes and can sometimes result in useful models, but in many areas of chemistry, such as calibration (see Chapter 5 for more details), it is desirable to have several levels, especially in the case of spectra of mixtures. Much of chemometrics is concerned primarily with linearly additive models of the form

$$X = C.S$$

where X is an observed matrix, such as a set of spectra, each row consisting of a spectrum, and each column of a wavelength, C is a matrix of, e.g., concentrations, each row consisting of the concentration of a number of compounds in a spectrum, and S could consist of the corresponding spectra of each compound. There are innumerable variations on this theme, in some cases where all the concentrations of all the components in a mixture are known, the aim being to develop a calibration model that predicts the concentrations from an unknown spectrum, to cases where the concentrations of only a few components in a mixture are known. In many situations, it is possible to control the experiments by mixing up components in the laboratory but in other cases this is not practicable, samples being taken from the field. A typical laboratory based experiment might involve recording a series of four component mixtures at five concentration levels.

A recommended strategy is as follows:

1. perform a calibration experiment, by producing a set of mixtures on a series of compounds of known concentrations to give a 'training set';
2. then test this model on an independent set of mixtures called a 'test set';
3. finally, use the model on real data to produce predictions.

More detail is described in Chapter 5, Section 5.6. Many brush aside the design of training sets, often employing empirical or random approaches. Some chemometricians recommend huge training sets of several hundred samples so as to get a representative distribution, especially if there are known to be half a dozen or more significant components in a mixture. In large industrial calibration models, such a procedure is often considered important for robust predictions. However, this approach is expensive in time and resources, and rarely possible in routine laboratory studies. More seriously, many instrumental calibration models are unstable, so calibration on Monday might vary significantly from calibration on Tuesday; hence if calibrations are to be repeated at regular intervals, the number of spectra in the training set must be limited. Finally, very ambitious calibrations can take months or even years, by which time the instruments and often the detection methods may have been replaced.

For the most effective calibration models, the nature of the training set must be carefully thought out using rational experimental design. Provided that the spectra are linearly additive, and there are no serious baseline problems or interactions there are standard designs that can be employed to obtain training sets. It is important to recognise that the majority of chemometric techniques for regression and calibration assume linear additivity. In the case where this may not be so, either the experimental conditions can be modified (for example, if the concentration of a compound is too high so that the absorbance does not obey the Beer–Lambert law, the solution is simply diluted) or various approaches for multilinear modelling are required. It is important to recognise that there is a big difference between the application of chemometrics to primarily analytical or physical chemistry, where it is usual to be able to attain conditions of linearity, and organic or biological chemistry (e.g. QSAR), where often this is not possible. The designs in this section are most applicable in the former case.

In calibration it is normal to use several concentration levels to form a model. Indeed, for information on lack-of-fit and so predictive ability, this is essential. Hence two level factorial designs are inadequate and typically four or five concentration levels are required for each compound. However, chemometric techniques are most useful for multicomponent mixtures. Consider an experiment carried out in a mixture of methanol and acetone. What happens if the concentrations of acetone and methanol in a training set are completely correlated? If the concentration of acetone increases, so does that of methanol, and similarly with a decrease. Such an experimental arrangement is shown in Figure 2.25. A more satisfactory design is given in Figure 2.26, in which the two

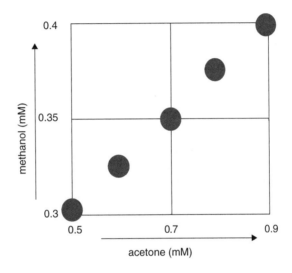

Figure 2.25
Poorly designed calibration experiment

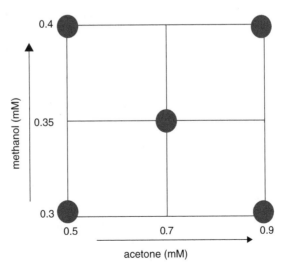

Figure 2.26
Well designed calibration experiment

concentrations are completely uncorrelated or orthogonal. In the former design there is no way of knowing whether a change in spectral characteristic results from change in concentration of acetone or of methanol. If this feature is consciously built into the training set and expected in all future samples, there is no problem, but if a future sample arises with a high acetone and low methanol concentration, calibration software will give a wrong answer for the concentration of each component. This is potentially very serious, especially when the result of chemometric analysis of spectral data is used to make decisions, such as the quality of a batch of pharmaceuticals, based on the concentration of each constituent as predicted by computational analysis of spectra. Some packages include elaborate diagnostics for so-called outliers, which may in many cases be perfectly good samples but ones whose correlation structure differs from that of the training set. In this chapter we will emphasize the importance of good design. In the absence of any certain knowledge (for example, that in all conceivable future samples the concentrations of acetone and methanol will be correlated), it is safest to design the calibration set so that the concentrations of as many compounds as possible in a calibration set are orthogonal.

A guideline to designing a series of multicomponent mixtures for calibration is described below.

1. Determine how many components in the mixture ($=k$) and the maximum and minimum concentration of each component. Remember that, if studied by spectroscopy or chromatography, the overall absorbance when each component is at a maximum should be within the Beer–Lambert limit (about 1.2 AU for safety).
2. Decide how many concentration levels are required each compound ($=l$), typically four or five. Mutually orthogonal designs are only possible if the number of concentration levels is a prime number or a power of a prime number, meaning that they are possible for 3, 4, 5, 7, 8 and 9 levels but not 6 or 10 levels.
3. Decide on how many mixtures to produce. Designs exist involving $N = ml^p$ mixtures, where l equals the number of concentration levels, p is an integer at least equal to 2, and m an integer at least equal to 1. Setting both m and p at their minimum values, at least 25 experiments are required to study a mixture (of more than one component) at five concentration levels, or l^2 at l levels.
4. The maximum number of mutually orthogonal compound concentrations in a mixture design where $m = 1$ is four for a three level design, five for a four level design and 12 for a five level design, so using five levels can dramatically increase the number of compounds that we can study using calibration designs. We will discuss how to extend the number of mutually orthogonal concentrations below. Hence choose the design and number of levels with the number of compounds of interest in mind.

The method for setting up a calibration design will be illustrated by a five level, eight compound, 25 experiment, mixture. The theory is rather complicated so the design will be presented as a series of steps.

1. The first step is to number the levels, typically from -2 (lowest) to $+2$ (highest), corresponding to coded concentrations, e.g. the level $-2 = 0.7$ mM and level $+2 = 1.1$ mM; note that the concentration levels can be coded differently for each component in a mixture.

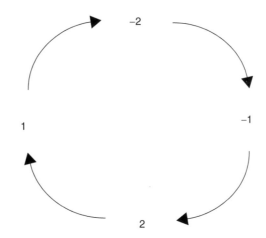

Figure 2.27
Cyclic permuter

2. Next, choose a *repeater* level, recommended to be the middle level, 0. For a 5 level design, and 7 to 12 factors (=components in a mixture), it is essential that this is 0. The first experiment is at this level for all factors.

3. Third, select a *cyclical permuter* for the remaining $(l - 1)$ levels. This relates each of these four levels as will be illustrated below; only certain cyclic generators can be used, namely $-2 \longrightarrow -1 \longrightarrow 2 \longrightarrow 1 \longrightarrow -2$ and $-2 \longrightarrow 1 \longrightarrow 2 \longrightarrow -1 \longrightarrow -2$ which have the property that factors j and $j + l + 1$ are orthogonal (these are listed in Table 2.30, as discussed below). For less than $l + 2$ (=7) factors, any permuter can be used so long as it includes all four levels. One such permuter is illustrated in Figure 2.27, and is used in the example below.

4. Finally, select a *difference vector*; this consists of $l - 1$ numbers from 0 to $l - 2$, arranged in a particular sequence (or four numbers from 0 to 3 in this example). Only a very restricted set of such vectors are acceptable of which {0 2 3 1} is one. The use of the difference vector will be described below.

5. Then generate the first column of the design consisting of l^2 (=25) levels in this case, each level corresponding to the concentration of the first compound in the mixture in each of 25 experiments.

   (a) The first experiment is at the repeater level for each factor.

   (b) The $l - 1$ (=4) experiments 2, 8, 14 and 20 are at the repeater level (=0 in this case). In general, the experiments 2, $2 + l + 1$, $2 + 2(l + 1)$ up to $2 + (l - 1) \times (l + 1)$ are at this level. These divide the columns into "blocks" of 5 (=l) experiments.

   (c) Now determine the levels for the first block, from experiments 3 to 7 (or in general, experiments 3 to $2 + l$). Experiment 3 can be at any level apart from the repeater. In the example below, we use level -2. The key to determining the levels for the next four experiments is the difference vector. The conditions for the fourth experiment are obtained from the difference vector and cyclic generator. The difference vector {0 2 3 1} implies that the second experiment of the block is zero cyclical differences away from the third experiment or

−2 using the cyclic permuter of Figure 2.27. The next number in the difference vector is 2, making the fifth experiment at level 2 which is two cyclic differences from −2. Continuing, the sixth experiment is three cyclic differences from the fifth experiment or at level −1, and the final experiment of the block is at level 2.

(d) For the second block (experiments 9 to 13), simply shift the first block by one cyclic difference using the permuter of Figure 2.27 and continue until the last (or fourth) block is generated.

6. Then generate the next column of the design as follows:
   (a) the concentration level for the first experiment is always at the repeater level;
   (b) the concentration for the second experiment is at the same level as the third experiment of the previous column, up to the 24th [or $(l^2 − 1)$th etc.] experiment;
   (c) the final experiment is at the same level as the second experiment for the previous column.

7. Finally, generate successive columns using the principle in step 6 above.

The development of the design is illustrated in Table 2.29. Note that a full five level factorial design for eight compounds would require 5^8 or 390 625 experiments, so there has been a dramatic reduction in the number of experiments required.

There are a number of important features to note about the design in Table 2.29.

• In each column there are an equal number of −2, −1, 0, +1 and +2 levels.

Table 2.29 Development of a multilevel partial factorial design.

Experiments		Factor 1	Factor 2	Factor 3	Factor 4	Factor 5	Factor 6	Factor 7	Factor 8
	1	0	0	0	0	0	0	0	0
Repeater →	2	0	−2	−2	2	−1	2	0	−1
	3	−2	−2	2	−1	2	0	−1	−1
	4	−2	2	−1	2	0	−1	−1	1
Block 1	5	2	−1	2	0	−1	−1	1	2
	6	−1	2	0	−1	−1	1	2	1
	7	2	0	−1	−1	1	2	1	0
Repeater →	8	0	−1	−1	1	2	1	0	2
	9	−1	−1	1	2	1	0	2	2
	10	−1	1	2	1	0	2	2	−2
Block 2	11	1	2	1	0	2	2	−2	1
	12	2	1	0	2	2	−2	1	−2
	13	1	0	2	2	−2	1	−2	0
Repeater →	14	0	2	2	−2	1	−2	0	1
	15	2	2	−2	1	−2	0	1	1
	16	2	−2	1	−2	0	1	1	−1
Block 3	17	−2	1	−2	0	1	1	−1	−2
	18	1	−2	0	1	1	−1	−2	−1
	19	−2	0	1	1	−1	−2	−1	0
Repeater →	20	0	1	1	−1	−2	−1	0	−2
	21	1	1	−1	−2	−1	0	−2	−2
	22	1	−1	−2	−1	0	−2	−2	2
Block 4	23	−1	−2	−1	0	−2	−2	2	−1
	24	−2	−1	0	−2	−2	2	−1	2
	25	−1	0	−2	−2	2	−1	2	0

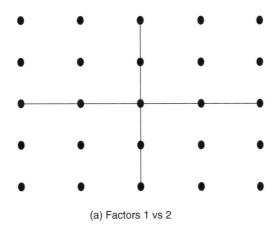

(a) Factors 1 vs 2

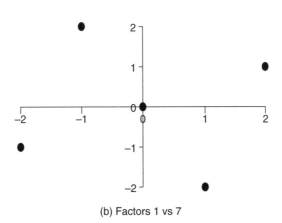

(b) Factors 1 vs 7

Figure 2.28
Graph of factor levels for the design in Table 2.29

- Each column is orthogonal to every other column, that is the correlation coefficient is 0.
- A graph of the levels of any two factors against each other is given in Figure 2.28(a) for each combination of factors except factors 1 and 7, and 2 and 8, for which a graph is given in Figure 2.28(b). It can be seen that in most cases the levels of any two factors are distributed exactly as they would be for a full factorial design, which would require almost half a million experiments. The nature of the difference vector is crucial to this important property. Some compromise is required between factors differing by $l + 1$ (or 6) columns, such as factors 1 and 7. This is unavoidable unless more experiments are performed.

Table 2.30 summarises information to generate some common designs, including the difference vectors and cyclic permuters, following the general rules above for different designs. Look for the five factor design and it can be seen that {0 2 3 1} is one possible difference vector, and also the permuter used above is one of two possibilities.

Table 2.30 Parameters for construction of a multilevel calibration design.

Levels	Experiments	Max. no. of orthogonal factors	Repeater	Difference vectors	Cyclic permuters
3	9	4	Any	{01}, {10}	
4	16	5	Any	{021}, {120}	
5	25	12	0	{0231}, {1320},	$-2 \to -1 \to 2 \to 1 \to -2,$
				{2013}, {3102}	$-2 \to 1 \to 2 \to -1 \to -2$
7	49	16	0	{241 035}, {514 302},	$-3 \to 2 \to 3 \to -1 \to 1 \to -2 \to -3,$
				{451 023}, {124 350},	$-3 \to 1 \to -1 \to 2 \to 3 \to -2 \to -3,$
				{530 142}, {203 415},	$-3 \to -2 \to 3 \to 2 \to -1 \to 1 \to -3,$
				{320 154}, {053 421}	$-3 \to -2 \to 1 \to -1 \to 3 \to 2 \to -3$

It is possible to expand the number of factors using a simple trick of matrix algebra. If a matrix A is orthogonal, then the matrix

$$\begin{pmatrix} A & A \\ A & -A \end{pmatrix}$$

is also orthogonal. Therefore, new matrices can be generated from the original orthogonal designs, to expand the number of compounds in the mixture.

2.4   Central Composite or Response Surface Designs

Two level factorial designs are primarily useful for exploratory purposes and calibration designs have special uses in areas such as multivariate calibration where we often expect an independent linear response from each component in a mixture. It is often important, though, to provide a more detailed model of a system. There are two prime reasons. The first is for optimisation – to find the conditions that result in a maximum or minimum as appropriate. An example is when improving the yield of synthetic reaction, or a chromatographic resolution. The second is to produce a detailed quantitative model: to predict mathematically how a response relates to the values of various factors. An example may be how the near-infrared spectrum of a manufactured product relates to the nature of the material and processing employed in manufacturing.

Most exploratory designs do not involve recording replicates, nor do they provide information on squared terms; some, such as Plackett–Burman and highly fractional factorials, do not even provide details of interactions. In the case of detailed modelling it is often desirable at a first stage to reduce the number of factors via exploratory designs as described in Section 2.3, to a small number of main factors (perhaps three or four) that are to be studied in detail, for which both squared and interaction terms in the model are of interest.

2.4.1   Setting Up the Design

Many designs for use in chemistry for modelling are based on the central composite design (sometimes called a response surface design), the main principles of which will be illustrated via a three factor example, in Figure 2.29 and Table 2.31. The first step,

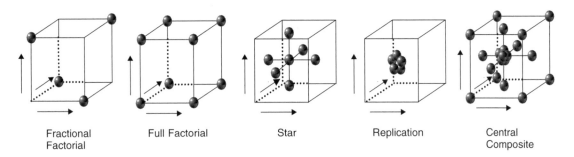

Fractional   Full Factorial   Star   Replication   Central
Factorial   Composite

Figure 2.29
Elements of a central composite design, each axis representing one factor

of course, is to code the factors, and it is always important to choose sensible physical values for each of the factors first. It is assumed that the central point for each factor is 0, and the design is symmetric around this. We will illustrate the design for three factors, which can be represented by points on a cube, each axis corresponding to a factor. A central composite design is constructed as several superimposed designs.

- The smallest possible *fractional factorial* three factor design consists of four experiments, used to estimate the three linear terms and the intercept. Such as design will not provide estimates of the interactions, replicates or squared terms.
- Extending this to eight experiments provides estimates of all interaction terms. When represented by a cube, these experiments are placed on the eight corners, and are consist of a *full factorial design*. All possible combinations of $+1$ and -1 for the three factors are observed.
- Another type of design, often designated a *star design*, can be employed to estimate the squared terms. In order to do this, at least three levels are required for each factor, often denoted by $+a$, 0 and $-a$, with level 0 being in the centre. The reason for this is that there must be at least three points to fit a quadratic. Points where one factor is at level $+a$ are called *axial* points. Each axial point involves setting one factor at level $\pm a$ and the remaining factors at level 0. One simple design sets a equal to 1, although, as we will see below, this value of the axial point is not always recommended. For three factors, a star design consists of the centre point, and six in the centre (or above) each of the six faces of the cube.
- Finally it is often useful to be able estimate the experimental error (as discussed in Section 2.2.2), and one method is to perform extra *replicates* (typically five) in the centre. Obviously other approaches to replication are possible, but it is usual to replicate in the centre and assume that the error is the same throughout the response surface. If there are any overriding reasons to assume that heteroscedasticity of errors has an important role, replication could be performed at the star or factorial points. However, much of experimental design is based on classical statistics where there is no real detailed information about error distributions over an experimental domain, or at least obtaining such information would be unnecessarily laborious.
- Performing a full factorial design, a star design and five replicates, results in 20 experiments. This design is a type of *central composite design*. When the axial or star points are situated at $a = \pm 1$, the design is sometimes called a *face centred cube design* (see Table 2.31).

Table 2.31 Construction of a
central composite design.

Fractional factorial		
1	1	1
1	−1	−1
−1	−1	1
−1	1	−1

Full factorial		
1	1	1
1	1	−1
1	−1	1
1	−1	−1
−1	1	1
−1	1	−1
−1	−1	1
−1	−1	−1

Star		
0	0	−1
0	0	1
0	1	0
0	−1	0
1	0	0
−1	0	0
0	0	0

Replication in centre		
0	0	0
0	0	0
0	0	0
0	0	0
0	0	0

Central composite		
1	1	1
1	1	−1
1	−1	1
1	−1	−1
−1	1	1
−1	1	−1
−1	−1	1
−1	−1	−1
0	0	−1
0	0	1
0	1	0
0	−1	0
1	0	0
−1	0	0
0	0	0
0	0	0
0	0	0
0	0	0
0	0	0
0	0	0

2.4.2 Degrees of Freedom

In this section we analyse in detail the features of such designs. In most cases, however many factors are used, only two factor interactions are computed so higher order interactions are ignored, although, of course, provided that sufficient degrees of freedom are available to estimate the lack-of-fit, higher interactions are conceivable.

- The first step is to set up a model. A full model including all two factor interactions consists of $1 + 2k + [k(k-1)]/2 = 1 + 6 + 3$ or 10 parameters in this case, consisting of
    - 1 intercept term (of the form b_0),
    - 3 (=k) linear terms (of the form b_1),
    - 3 (=k) squared terms (of the form b_{11}),
    - and 3 (=[k(k-1)]/2) interaction terms (of the form b_{12}),

    or in equation form

$$\hat{y} = b_0 + b_1x_1 + b_2x_2 + b_3x_3 + b_{11}x_1^2 + b_{22}x_2^2 + b_{33}x_3^2$$

$$+ b_{12}x_1x_2 + b_{13}x_1x_3 + b_{23}x_2x_3$$

- A degree of freedom tree can be drawn up as illustrated in Figure 2.30. We can see that
    - there are 20 (=N) experiments overall,
    - 10 (=P) parameters in the model,
    - 5 (=R) degrees of freedom to determine replication error,
    - and 5 (=N − P − R) degrees of freedom for the lack-of-fit.

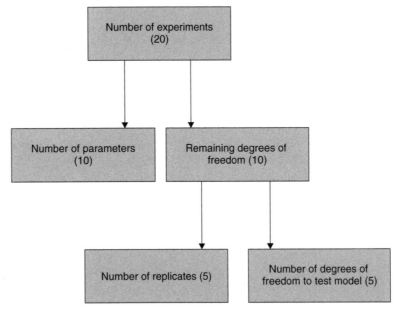

Figure 2.30
Degrees of freedom for central composite design of Table 2.31

Note that the number of degrees of freedom for the lack-of-fit equals that for repli-cation in this case, suggesting quite a good design.

The total number of experiments, N (=20), equals the sum of

- 2^k (=8) factorial points, often represented as the corners of the cube,
- $2k + 1$ (=7) star points, often represented as axial points on (or above) the faces of the cube plus one in the centre,
- and R (=5) replicate points, in the centre.

There are a large number of variations on this theme but each design can be defined by four parameters, namely

1. the number of factorial or cubic points (N_f);
2. the number of axial points (N_a), usually one less than the number of points in the star design;
3. the number of central points (N_c), usually one more than the number of replicates;
4. the position of the axial points a.

In most cases, it is best to use a full factorial design for the factorial points, but if the number of factors is large, it is legitimate to reduce this and use a partial factorial design. There are almost always $2k$ axial points.

The number of central points is often chosen according to the number of degrees of freedom required to assess errors via ANOVA and the F-test (see Sections 2.2.2 and 2.2.4.4), and should be approximately equal to the number of degrees of freedom for the lack-of-fit, with a minimum of about four unless there are special reasons for reducing this.

2.4.3 Axial Points

The choice of the position of the axial (or star) points and how this relates to the number of replicates in the centre is an interesting issue. Whereas many chemists use these designs fairly empirically, it is worth noting two statistical properties that influence the property of these designs. It is essential to recognise, though, that there is no single perfect design, indeed many of the desirable properties of a design are incompatible with each other.

1. *Rotatability* implies that the confidence in the predictions depends only on the distance from the centre of the design. For a two factor design, this means that all experimental points in a circle of a given radius will be predicted equally well. This has useful practical consequences, for example, if the two factors correspond to concentrations of acetone and methanol, we know that the further the concentrations are from the central point the lower is the confidence. Methods for visualising this were described in Section 2.2.5. Rotatability does not depend on the number of replicates in the centre, but only on the value of a, which should equal $\sqrt[4]{N_f}$, where N_f is the number of factorial points, equal to 2^k if a full factorial is used, for this property. Note that the position of the axial points will differ if a fractional factorial is used for the cubic part of the design.
2. *Orthogonality* implies that all the terms (linear, squared and two factor interactions) are orthogonal to each other in the design matrix, i.e. the correlation coefficient

between any two terms (apart from the zero order term where it is not defined) equals 0. For linear and interaction terms this will always be so, but squared terms are not so simple, and in the majority of central composite designs they are not orthogonal. The rather complicated condition is

$$a = \sqrt{\frac{\sqrt{N \times N_f} - N_f}{2}}$$

which depends on the number of replicates since a term for the overall number of experiments is included in the equation. A small lack of orthogonality in the squared terms can sometimes be tolerated, but it is often worth checking any particular design for this property.

Interestingly these two conditions are usually not compatible, resulting in considerable dilemmas to the theoretician, although in practical situations the differences of a for the two different properties are not so large, and in some cases it is not very meaningful experimentally to get too concerned about small differences in the acial points of the design. Table 2.32 analyses the properties of three two factor designs with a model of the form $y = b_0 + b_1x_1 + b_2x_2 + b_{11}x_1^2 + b_{22}x_{22}^2 + b_{12}x_1x_2$ ($P = 6$). Design A is rotatable, Design B is orthogonal and Design C has both properties. However, the third is extremely inefficient in that seven replicates are required in the centre, indeed half the design points are in the centre, which makes little practical sense. Table 2.33 lists the values of a for rotatability and orthogonality for different numbers of factors and replicates. For the five factor design a half factorial design is also tabulated, whereas in all other cases the factorial part is full. It is interesting that for a two factor design with one central point (i.e. no replication), the value of a for orthogonality is 1, making it identical with a two factor, three level design [see Table 2.20(a)], there being four factorial and five star points or 3^2 experiments in total.

Terminology varies according to authors, some calling only the rotatable designs true central composite designs. It is very important to recognise that the literature on statistics is very widespread throughout science, especially in experimental areas such as biology, medicine and chemistry, and to check carefully an author's precise usage of terminology. It is important not to get locked in a single textbook (even this one!), a single software package or a single course provider. In many cases, to simplify, a single terminology is employed. Because there are no universally accepted conventions, in which chemometrics differs from, for example, organic chemistry, and most historical attempts to set up committees have come to grief or been dominated by one specific strand of opinion, every major group has its own philosophy.

The real experimental conditions can be easily calculated from a coded design. For example, if coded levels $+1$, 0 and -1 for a rotatable design correspond to temperatures of 30, 40 and 50 °C for a two factor design, the axial points correspond to temperatures of 25.9 and 54.1 °C, whereas for a four factor design these points are 20 and 60 °C. Note that these designs are only practicable where factors can be numerically defined, and are not normally employed if some data are categorical, unlike factorial designs. However, it is sometimes possible to set the axial points at values such as ± 1 or ± 2 under some circumstance to allow for factors that can take discrete values, e.g. the number of cycles in an extraction procedure, although this does restrict the properties of the design.

Table 2.32 Three possible two factor central composite designs.

Design A

−1	−1	Rotatability	$\checkmark$
−1	1	Orthogonality	$\times$
1	−1	N_c	6
1	1	a	1.414
−1.414	0		
1.414	0	Lack-of-fit (d.f.)	3
0	−1.414	Replicates (d.f.)	5
0	1.414		
0	0		
0	0		
0	0		
0	0		
0	0		

Design B

−1	−1	Rotatability	$\times$
−1	1	Orthogonality	$\checkmark$
1	−1	N_c	6
1	1	a	1.320
−1.320	0		
1.320	0	Lack-of-fit (d.f.)	3
0	−1.320	Replicates (d.f.)	5
0	1.320		
0	0		
0	0		
0	0		
0	0		
0	0		

Design C

−1	−1	Rotatability	$\checkmark$
−1	1	Orthogonality	$\checkmark$
1	−1	N_c	8
1	1	a	1.414
−1.414	0		
1.414	0	Lack-of-fit (d.f.)	3
0	−1.414	Replicates (d.f.)	7
0	1.414		
0	0		
0	0		
0	0		
0	0		
0	0		
0	0		
0	0		

A rotatable four factor design consists of 30 experiments, namely

- 16 factorial points at all possible combinations of ±1;
- nine star points, including a central point of $(0, 0, 0, 0)$ and eight points of the form $(\pm 2, 0, 0, 0)$, etc.;

Table 2.33 Position of the axial points for rotatability and orthogonality for central composite designs with varying number of replicates in the centre.

k	Rotatability	Orthogonality N_c		
		4	5	6
2	1.414	1.210	1.267	1.320
3	1.682	1.428	1.486	1.541
4	2.000	1.607	1.664	1.719
5	2.378	1.764	1.820	1.873
5 (half factorial)	2.000	1.719	1.771	1.820

- typically five further replicates in the centre; note that a very large number of replicates (11) would be required to satisfy orthogonality with the axial points at two units, and this is probably overkill in many real experimental situations. Indeed, if resources are available for so many replicates, it might make sense to replicate different experimental points to check whether errors are even over the response surface.

2.4.4 Modelling

Once the design is performed it is then possible to calculate the values of the terms using regression and design matrices or almost any standard statistical software and assess the significance of each term using ANOVA, F-tests and t-tests if felt appropriate. It is important to recognise that these designs are mainly employed in order to produce a detailed model, and also to look at interactions and higher order (quadratic) terms. The number of experiments becomes excessive if the number of factors is large, and if more than about five significant factors are to be studied, it is best to narrow down the problem first using exploratory designs, although the possibility of using fractional factorials on the corners helps. Remember also that it is conventional (but not always essential) to ignore interaction terms above second order.

After the experiments have been performed, it is then possible to produce a detailed mathematical model of the response surface. If the purpose is optimisation, it might then be useful, for example, by using contour or 3D plots, to determine the position of the optimum. For relatively straightforward cases, partial derivatives can be employed to solve the equations, as illustrated in Problems 2.7 and 2.16, but if there are a large number of terms an analytical solution can be difficult and also there can be more than one optimum. It is recommended always to try to look at the system graphically, even if there are too many factors to visualise the whole of the experimental space at once. It is also important to realise that there may be other issues that affect an optimum, such as expense of raw materials, availability of key components, or even time. Sometimes a design can be used to model several responses, and each one can be analysed separately; perhaps one might be the yield of a reaction, another the cost of raw materials and another the level of impurities in a produce. Chemometricians should resist the temptation to insist on a single categorical 'correct' answer.

2.4.5   Statistical Factors

Another important use of central composite designs is to determine a good range of compounds for testing such as occurs in quantitative structure–activity relationships (QSARs). Consider the case of Figure 2.4. Rather than the axes being physical variables such as concentrations, they can be abstract mathematical or statistical variables such as principal components (see Chapter 4). These could come from molecular property descriptors, e.g. bond lengths and angles, hydrophobicity, dipole moments, etc. Consider, for example, a database of several hundred compounds. Perhaps a selection are interesting for biological tests. It may be very expensive to test all compounds, so a sensible strategy is to reduce the number of compounds. Taking the first two PCs as the factors, a selection of nine representative compounds can be obtained using a central composite design as follows.

1. Determine the principal components of the original dataset.
2. Scale each PC, for example, so that the highest score equals $+1$ and the lowest score equals -1.
3. Then choose those compounds whose scores are closest to the desired values. For example, in the case of Figure 2.4, choose a compound whose score is closest to $(-1, -1)$ for the bottom left-hand corner, and closest to $(0, 0)$ for the centre point.
4. Perform experimental tests on this subset of compounds and then use some form of modelling to relate the desired activity to structural data. Note that this modelling does not have to be multilinear modelling as discussed in this section, but could also be PLS (partial least squares) as introduced in Chapter 5.

2.5   Mixture Designs

Chemists and statisticians use the term 'mixture' in different ways. To a chemist, any combination of several substances is a mixture. In more formal statistical terms, however, a mixture involves a series of factors whose total is a constant sum; this property is often called 'closure' and will be discussed in completely different contexts in the area of scaling data prior to principal components analysis (Chapter 4, Section 4.3.6.5 and Chapter 6, Section 6.2.3.1). Hence in statistics (and chemometrics) a solvent system in HPLC or a blend of components in products such as paints, drugs or food is considered a mixture, as each component can be expressed as a proportion and the total adds up to 1 or 100 %. The response could be a chromatographic separation, the taste of a foodstuff or physical properties of a manufactured material. Often the aim of experimentation is to find an optimum blend of components that tastes best, or provide the best chromatographic separation, or the material that is most durable.

Compositional mixture experiments involve some specialist techniques and a whole range of considerations must be made before designing and analysing such experiments. The principal consideration is that the value of each factor is constrained. Take, for example, a three component mixture of acetone, methanol and water, which may be solvents used as the mobile phase for a chromatographic separation. If we know that there is 80 % water in the mixture, there can be no more than 20 % acetone or methanol in the mixture. If there is also 15 % acetone, the amount of methanol is fixed at 5 %. In fact, although there are three components in the mixtures, these translate into two independent factors.

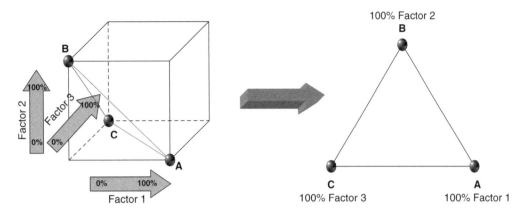

Figure 2.31
Three component mixture space

2.5.1 Mixture Space

Most chemists represent their experimental conditions in mixture space, which corresponds to all possible allowed proportions of components that add up to 100 %. A three component mixture can be represented by a triangle (Figure 2.31), which is a two-dimensional cross-section of a three-dimensional space, represented by a cube, showing the allowed region in which the proportions of the three components add up to 100 %. Points within this triangle or mixture space represent possible mixtures or blends:

- the three corners correspond to single components;
- points along the edges correspond to binary mixtures;
- points inside the triangle correspond to ternary mixtures;
- the centre of the triangle corresponds to an equal mixture of all three components;
- all points within the triangle are physically allowable blends.

As the number of components increases, so does the dimensionality of the mixture space. Physically meaningful mixtures can be represented as points in this space:

- for two components the mixture space is simply a straight line;
- for three components it is a triangle;
- for four components it is a tetrahedron.

Each object (pictured in Figure 2.32) is called a *simplex* – the simplest possible object in space of a given dimensionality: the dimensionality is one less than the number of factors or components in a mixture, so a tetrahedron (three dimensions) represents a four component mixture.

There are a number of common designs which can be envisaged as ways of determining a sensible number and arrangement of points within the simplex.

2.5.2 Simplex Centroid

2.5.2.1 Design

These designs are probably the most widespread. For k factors they involve performing $2^k - 1$ experiments, i.e. for four factors, 15 experiments are performed. It involves all

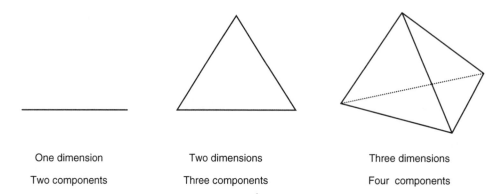

Figure 2.32
Simplex in one, two and three dimensions

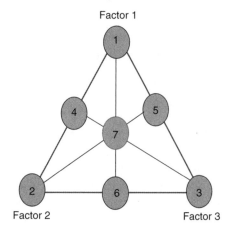

Figure 2.33
Three factor simplex centroid design

possible combinations of the proportions 1, 1/2 to $1/k$ and is best illustrated by an example. A three factor design consists of

- three single factor combinations;
- three binary combinations;
- one ternary combination.

These experiments are represented graphically in mixture space of Figure 2.33 and tabulated in Table 2.34.

2.5.2.2   Model

Just as previously, a model and design matrix can be obtained. However, the nature of the model requires some detailed thought. Consider trying to estimate model of the form

$$y = c_0 + c_1x_1 + c_2x_2 + c_3x_3 + c_{11}x_1^2 + c_{22}x_2^2 + c_{33}x_3^2 + c_{12}x_1x_2 + c_{13}x_1x_3 + c_{23}x_2x_3$$

Table 2.34 Three factor simplex centroid design.

Experiment		Factor 1	Factor 2	Factor 3
1		1	0	0
2	Single factor	0	1	0
3		0	0	1
4		1/2	1/2	0
5	Binary	1/2	0	1/2
6		0	1/2	1/2
7	Ternary	1/3	1/3	1/3

This model consists of 10 terms, impossible if only seven experiments are performed. How can the number of terms be reduced? Arbitrarily removing three terms such as the quadratic or interaction terms has little theoretical justification. A major problem with the equation above is that the value of x_3 depends on x_1 and x_2, since it equals $1 - x_1 - x_2$ so there are, in fact, only two independent factors. If a design matrix consisting of the first four terms of the equation above was set up, it would not have an inverse, and the calculation is impossible. The solution is to set up a reduced model. Consider, instead, a model consisting only of the first three terms:

$$y = a_0 + a_1 x_1 + a_2 x_2$$

This is, in effect, equivalent to a model containing just the three single factor terms without an intercept since

$$y = a_0(x_1 + x_2 + x_3) + a_1 x_1 + a_2 x_2 = (a_0 + a_1)x_1 + (a_0 + a_2)x_2 + a_0 x_3$$

$$= b_1 x_1 + b_2 x_2 + b_3 x_3$$

It is not possible to produce a model contain both the intercept and the three single factor terms. Closed datasets, such as occur in mixtures, have a whole series of interesting mathematical properties, but it is primarily important simply to watch for these anomalies.

The two common types of model, one with an intercept and one without an intercept term, are related. Models excluding the intercept are often referred to as Sheffé models and those with the intercept as Cox models. Normally a full Sheffé model includes all higher order interaction terms, and for this design is given by

$$y = b_1 x_1 + b_2 x_2 + b_3 x_3 + b_{12} x_1 x_2 + b_{13} x_1 x_3 + b_{23} x_2 x_3 + b_{123} x_1 x_2 x_3$$

Since seven experiments have been performed, all seven terms can be calculated, namely

- three one factor terms;
- three two factor interactions;
- one three factor interaction.

The design matrix is given in Table 2.35, and being a square matrix, the terms can easily be determined using the inverse. For interested readers, the relationship between the two types of models is explored in more detail in the Problems, but in most cases we recommend using a Sheffé model.

Table 2.35 Design matrix for a three factor simplex centroid design.

x_1	x_2	x_3	x_1x_2	x_1x_3	x_2x_3	$x_1x_2x_3$
1.000	0.000	0.000	0.000	0.000	0.000	0.000
0.000	1.000	0.000	0.000	0.000	0.000	0.000
0.000	0.000	1.000	0.000	0.000	0.000	0.000
0.500	0.500	0.000	0.250	0.000	0.000	0.000
0.500	0.000	0.500	0.000	0.250	0.000	0.000
0.000	0.500	0.500	0.000	0.000	0.250	0.000
0.333	0.333	0.333	0.111	0.111	0.111	0.037

2.5.2.3 Multifactor Designs

A general simplex centroid design for k factors consists of $2^k - 1$ experiments, of which there are

- k single blends;
- $k \times (k-1)/2$ binary blends, each component being present in a proportion of 1/2;
- $k!/[(k-m)!m!]$ blends containing m components (these can be predicted by the binomial coefficients), each component being present in a proportion of $1/m$;
- one blend consisting of all components, each component being present in a proportion of $1/k$.

Each type of blend yields an equivalent number of interaction terms in the Sheffé model. Hence for a five component mixture and three component blends, there will be $5!/[(5-3)!3!] = 10$ mixtures such as (1/3 1/3 1/3 0 0) containing all possible combinations, and 10 terms such as $b_1b_2b_3$.

It is normal to use all possible interaction terms in the mixture model, although this does not leave any degrees of freedom for determining lack-of-fit. Reducing the number of higher order interactions in the model but maintaining the full design is possible, but this must be carefully thought out, because each term can also be re-expressed, in part, as lower order interactions using the Cox model. This will, though, allow the calculation of some measure of confidence in predictions. It is important to recognise that the columns of the mixture design matrix are not orthogonal and can never be, because the proportion of each component depends on all others, so there will always be some correlation between the factors.

For multifactor mixtures, it is often impracticable to perform a full simplex centroid design and one approach is to simply to remove higher order terms, not only from the model but also the design. A five factor design containing up to second-order terms is presented in Table 2.36. Such designs can be denoted as $\{k, m\}$ simplex centroid designs, where k is the number of components in the mixture and m the highest order interaction. Note that at least binary interactions are required for squared terms (in the Cox model) and so for optimisation.

2.5.3 Simplex Lattice

Another class of designs called *simplex lattice* designs have been developed and are often preferable to the reduced simplex centroid design when it is required to reduce the number of interactions. They span the mixture space more evenly.

Table 2.36 A $\{5, 2\}$ simplex centroid design.

1	0	0	0	0
0	1	0	0	0
0	0	1	0	0
0	0	0	1	0
0	0	0	0	1
1/2	1/2	0	0	0
1/2	0	1/2	0	0
1/2	0	0	1/2	0
1/2	0	0	0	1/2
0	1/2	1/2	0	0
0	1/2	0	1/2	
0	1/2	0	0	1/2
0	0	1/2	1/2	0
0	0	1/2	0	1/2
0	0	0	1/2	1/2

A $\{k, m\}$ simplex lattice design consists of all possible combinations of 0, $1/m$, $2/m, \ldots, m/m$ or a total of

$$N = (k + m - 1)!/[(k - 1)!m!]$$

experiments where there are k factors. A $\{3, 3\}$ simplex lattice design can be set up analogous to the $\{3, 3\}$ simplex centroid design given in Table 2.34. There are

- three single factor experiments,
- six experiments where one factor is at 2/3 and the other at 1/3,
- and one experiment where all factors are at 1/3,

resulting in $5!/(2!3!) = 10$ experiments in total, as illustrated in Table 2.37 and Figure 2.34. Note that there are now more experiments than are required for a full Sheffé model, so some information about the significance of each parameter could be obtained; however, no replicates are measured. Generally, though, chemists mainly use mixture models for the purpose of optimisation or graphical presentation of results. Table 2.38 lists how many experiments are required for a variety of $\{k, m\}$ simplex centroid designs.

Table 2.37 Three factor simplex lattice design.

Experiment		Factor 1	Factor 2	Factor 3
1		1	0	0
2	Single factor	0	1	0
3		0	0	1
4		2/3	1/3	0
5		1/3	2/3	0
6		2/3	0	1/3
7	Binary	1/3	0	2/3
8		0	2/3	1/3
9		0	1/3	2/3
10	Ternary	1/3	1/3	1/3

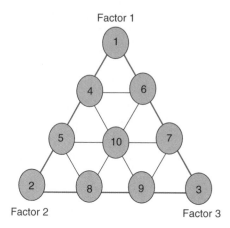

Figure 2.34
Three factor simplex lattice design

Table 2.38   Number of experiments required for various simplex lattice designs, with different numbers of factors and interactions.

Factors (k)	Interactions (m)				
	2	3	4	5	6
2	3				
3	6	10			
4	10	20	35		
5	15	35	70	126	
6	21	56	126	252	462

2.5.4   Constraints

In chemistry, there are frequently constraints on the proportions of each factor. For example, it might be of interest to study the effect of changing the proportion of ingredients in a cake. Sugar will be one ingredient, but there is no point baking a cake using 100 % sugar and 0 % of each other ingredient. A more sensible approach is to put a constraint on the amount of sugar, perhaps between 2 and 5 %, and look for solutions in this reduced mixture space. A good design will only test blends within the specified regions.

Constrained mixture designs are often difficult to set up, but there are four fundamental situations, exemplified in Figure 2.35, each of which requires a different strategy.

1. Only a *lower bound* for each factor is specified in advance.
   - The first step is to determine whether the proposed lower bounds are feasible. The sum of the lower bounds must be less than one. For three factors, lower bounds of 0.5, 0.1 and 0.2 are satisfactory, whereas lower bounds of 0.3, 0.4 and 0.5 are not.
   - The next step is to determine new upper bounds. For each factor these are 1 minus the sum of the lower bounds for all other factors. If the lower bounds for three factors are 0.5, 0.1 and 0.2, then the upper bound for the first factor is

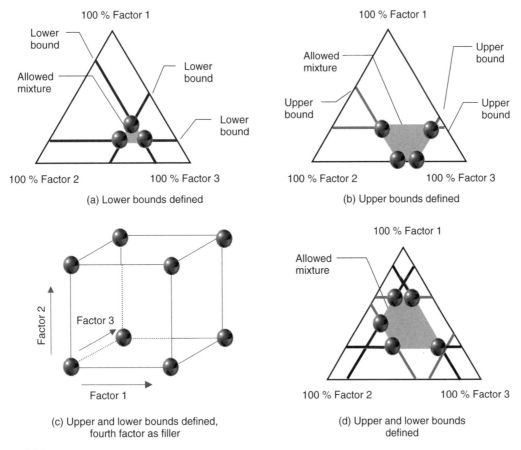

Figure 2.35
Four situations encountered in constrained mixture designs

$1 - 0.1 - 0.2 = 0.7$, so the upper bound of one factor plus the lower bounds of the other two must equal one.

- The third step is to take a standard design and the recalculate the conditions, as follows:

$$x_{new,f} = x_{old,f}(U_{,f} - L_f) + L_f$$

where L_f and U_f are the lower and upper bounds for factor f. This is illustrated in Table 2.39.

The experiments fall in exactly the same pattern as the original mixture space. Some authors call the vertices of the mixture space 'pseudo-components', so the first pseudo-component consists of 70 % of pure component 1, 10 % of pure component 2 and 20 % of pure component 3. Any standard design can now be employed. It is also possible to perform all the modelling on the pseudo-components and convert back to the true proportions at the end.

2. An *upper bound* is placed on each factor in advance. The constrained mixture space often becomes somewhat more complex dependent on the nature of the upper bounds. The trick is to find the extreme corners of a polygon in mixture space,

Table 2.39 Constrained mixture design with three lower bounds.

	Simple centroid design			Constrained design		
	Factor 1	Factor 2	Factor 3	Factor 1	Factor 2	Factor 3
	1.000	0.000	0.000	0.700	0.100	0.200
	0.000	1.000	0.000	0.500	0.300	0.200
	0.000	0.000	1.000	0.500	0.100	0.400
	0.500	0.500	0.000	0.600	0.200	0.200
	0.500	0.000	0.500	0.600	0.100	0.300
	0.000	0.500	0.500	0.500	0.200	0.300
	0.333	0.333	0.333	0.567	0.167	0.267
Lower	0.5	0.1	0.2			
Upper	0.7	0.3	0.4			

Table 2.40 Constrained mixture designs with upper bounds established in advance.

(a) Upper bounds	0.3	0.4	0.5	
1	0.3	0.4	0.3	Factors 1 and 2 high
2	0.3	0.2	0.5	Factors 1 and 3 high
3	0.1	0.4	0.5	Factors 2 and 3 high
4	0.3	0.3	0.4	Average of experiments 1 and 2
5	0.2	0.4	0.4	Average of experiments 1 and 3
6	0.2	0.3	0.5	Average of experiments 2 and 3
7	0.233	0.333	0.433	Average of experiments 1, 2 and 3

(b) Upper bounds	0.7	0.5	0.2	
1	0.7	0.1	0.2	Factors 1 and 3 high
2	0.3	0.5	0.2	Factors 2 and 3 high
3	0.7	0.3	0.0	Factor 1 high, factor 2 as high as possible
4	0.5	0.5	0.0	Factor 2 high, factor 1 as high as possible
5	0.7	0.2	0.1	Average of experiments 1 and 3
6	0.4	0.5	0.1	Average of experiments 2 and 4
7	0.5	0.3	0.2	Average of experiments 1 and 2
8	0.6	0.4	0.0	Average of experiments 3 and 4

perform experiments at these corners, midway along the edges and, if desired, in the centre of the design. There are no hard and fast rules as the theory behind these designs is complex. Recommended guidance is provided below for two situations. The methods are illustrated in Table 2.40 for a three factor design.

- If the sum of all $(k - 1)$ upper bounds is ≤ 1, then do as follows:
  - (a) set up k experiments where all but one factor is its upper bound [the first three in Table 2.40(a)]; these are the extreme vertices of the constrained mixture space;
  - (b) then set up binary intermediate experiments, simply the average of two of the k extremes;
  - (c) if desired, set up ternary experiments, and so on.

- If this condition is not met, the constrained mixture space will resemble an irregular polygon as in Figure 2.35(b). An example is illustrated in Table 2.40(b).
  (a) Find the extreme vertices for those combinations of $(k-1)$ factors that are less than one, of which there are two in this example.
  (b) Each missing vertex (one in this case) increases the number of new vertices by one. If, for example, it is impossible to simultaneously reach maxima for factors 1 and 2, create one new vertex with factor 1 at its highest level (U_1), factor 3 at 0 and factor 2 at $(1 - U_1)$, with another vertex for factor 2 at U_2, factor 3 at 0 and factor 1 at $(1 - U_2)$.
  (c) If there are v vertices, calculate extra experimental points between the vertices. Since the figure formed by the vertices in (b) has four sides, there will be four extra experiments, making eight in total. This is equivalent to performing one experiment on each corner of the mixture space in Figure 2.35(b), and one experiment on each edge.
  (d) Occasionally, one or more experiments are performed in the middle of the new mixture space, which is the average of the v vertices.
  Note that in some circumstances, a three factor constrained mixture space may be described by a hexagon, resulting in 12 experiments on the corners and edges. Provided that there are no more than four factors, the constrained mixture space is often best visualised graphically, and an even distribution of experimental points can be determined by geometric means.

3. Each factor has *an upper and lower bound and a $(k+1)$th factor is added* (the fourth in this example), so that the total comes to 100 %; this additional factor is called a filler. An example might be where the fourth factor is water, the others being solvents, buffer solutions, etc. This is common in chromatography, for example, if the main solvent is aqueous. Standard designs such as factorial designs can be employed for the three factors in Figure 2.35(c), with the proportion of the final factor computed from the remainder, given by $(1 - x_1 - x_2 - x_3)$. Of course, such designs will only be available if the upper bounds are low enough that their sum is no more than (often much less than) one. However, in some applications it is common to have some background filler, for example flour in baking of a cake, and active ingredients that are present in small amounts.

4. *Upper and lower bounds* defined in advance. In order to reach this condition, the sum of the upper bound for each factor plus the lower bounds for the remaining factors must not be greater than one, i.e. for three factors

$$U_1 + L_2 + L_3 \leq 1$$

and so on for factors 2 and 3. Note that the sum of all the upper bounds together must be at least equal to one. Another condition for three factors is that

$$L_1 + U_2 + U_3 \geq 1$$

otherwise the lower bound for factor 1 can never be achieved, similar conditions applying to the other factors. These equations can be extended to designs with more factors. Two examples are illustrated in Table 2.41, one feasible and the other not.

The rules for setting up the mixture design are, in fact, straightforward for three factors, provided that the conditions are met.

Table 2.41 Example of simultaneous constraints in mixture designs.

Impossible conditions			
Lower	0.1	0.5	0.4
Upper	0.6	0.7	0.8

Possible conditions			
Lower	0.1	0.0	0.2
Upper	0.4	0.6	0.7

1. Determine how many vertices; the maximum will be six for three factors. If the sum of the upper bound for one factor and the lower bounds for the remaining factors equal one, then the number of vertices is reduced by one. The number of vertices also reduces if the sum of the lower bound of one factor and the upper bounds of the remaining factors equals one. Call this number v. Normally one will not obtain conditions for three factors for which there are less than three vertices, if any less, the limits are too restrictive to show much variation.
2. Each vertex corresponds to the upper bound for one factor, the lower bound for another factor and the final factor is the remainder, after subtracting from 1.
3. Order the vertices so that the level of one factor remains constant between vertices.
4. Double the number of experiments, by taking the average between each successive vertex (and also the average between the first and last), to provide $2v$ experiments. These correspond to experiments on the edges of the mixture space.
5. Finally it is usual to perform an experiment in the centre, which is simply the average of all the vertices.

Table 2.42 illustrates two constrained mixture designs, one with six and the other with five vertices. The logic can be extended to several factors but can be complicated.

Table 2.42 Constrained mixture design where both upper and lower limits are known in advance.

(a) Six vertices			
Lower	0.1	0.2	0.3
Upper	0.4	0.5	0.6

Step 1
- $0.4 + 0.2 + 0.3 = 0.9$
- $0.1 + 0.5 + 0.3 = 0.9$
- $0.1 + 0.2 + 0.6 = 0.9$

so $v = 6$

Steps 2 and 3 Vertices

A	0.4	0.2	0.4
B	0.4	0.3	0.3
C	0.1	0.5	0.4
D	0.2	0.5	0.3
E	0.1	0.3	0.6
F	0.2	0.2	0.6

Table 2.42   (*continued*)

Steps 4 and 5 Design

1	A	0.4	0.2	0.4
2	Average A & B	0.4	0.25	0.35
3	B	0.4	0.3	0.3
4	Average B & C	0.25	0.4	0.35
5	C	0.1	0.5	0.4
6	Average C & D	0.15	0.5	0.35
7	D	0.2	0.5	0.3
8	Average D & E	0.15	0.4	0.45
9	E	0.1	0.3	0.6
10	Average E & F	0.15	0.25	0.6
11	F	0.2	0.2	0.6
12	Average F & A	0.3	0.2	0.5
13	Centre	0.2333	0.3333	0.4333

(b) Five vertices

Lower	0.1	0.3	0
Upper	0.7	0.6	0.4

Step 1

- $0.7 + 0.3 + 0.0 = 1.0$
- $0.1 + 0.6 + 0.0 = 0.7$
- $0.1 + 0.3 + 0.4 = 0.8$

so $v = 5$

Steps 2 and 3 Vertices

A	0.7	0.3	0.0
B	0.4	0.6	0.0
C	0.1	0.6	0.3
D	0.1	0.5	0.4
E	0.3	0.3	0.4

Steps 4 and 5 Design

1	A	0.7	0.3	0.0
2	Average A & B	0.55	0.45	0.0
3	B	0.4	0.6	0.0
4	Average B & C	0.25	0.6	0.15
5	C	0.1	0.6	0.3
6	Average C & D	0.1	0.55	0.35
7	D	0.1	0.5	0.4
8	Average D & E	0.2	0.4	0.4
9	E	0.3	0.3	0.4
10	Average E & A	0.5	0.3	0.2
11	Centre	0.32	0.46	0.22

If one is using a very large number of factors all with constraints, as can sometimes be the case, for example in fuel or food chemistry where there may be a lot of ingredients that influence the quality of the product, it is probably best to look at the original literature as designs for multifactor constrained mixtures are very complex: there is insufficient space in this introductory text to describe all the possibilities in detail. Sometimes constraints might be placed on one or two factors, or one factor could have an upper limit, another a lower limit, and so on. There are no hard and fast rules, but when the number of factors is sufficiently small it is important to try to visualise the design. The trick is to try to obtain a fairly even distribution of experimental points over the mixture space. Some techniques, which will indeed have feasible design points, do not have this property.

2.5.5 Process Variables

Finally, it is useful to mention briefly designs for which there are two types of variable, conventional (often called process) variables, such as pH and temperature, and mixture variables, such as solvents. A typical experimental design is represented in Figure 2.36, in the case of two process variables and three mixture variables consisting of 28 experiments. Such designs are relatively straightforward to set up, using the principles in this and earlier chapters, but care should be taken when calculating a model, which can become very complex. The interested reader is strongly advised to check the detailed literature as it is easy to become very confused when analysing such types of design, although it is important not to be put off; as many problems in chemistry involve both types of variables and since there are often interactions between mixture and process variables (a simple example is that the pH dependence of a reaction depends on solvent composition), such situations can be fairly common.

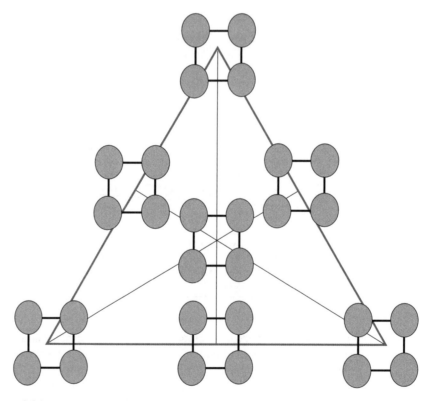

Figure 2.36
Mixture design with process variables

2.6   Simplex Optimisation

Experimental designs can be employed for a large variety of purposes, one of the most successful being optimisation. Traditional statistical approaches normally involve forming a mathematical model of a process, and then, either computationally or algebraically, optimising this model to determine the best conditions. There are many applications, however, in which a mathematical relationship between the response and the factors that influence it is not of primary interest. Is it necessary to model precisely how pH and temperature influence the yield of a reaction? When shimming an NMR machine, is it really important to know the precise relationship between field homogeneity and resolution? In engineering, especially, methods for optimisation have been developed which do not require a mathematical model of the system. The philosophy is to perform a series of experiments, changing the values of the control parameters, until a desired response is obtained. Statisticians may not like this approach as it is not normally possible to calculate confidence in the model and the methods may fail when experiments are highly irreproducible, but in practice sequential optimisation has been very successfully applied throughout chemistry.

One of the most popular approaches is called simplex optimisation. A simplex is the simplest possible object in N-dimensional space, e.g. a line in one dimension and a triangle in two dimensions, as introduced previously (Figure 2.32). Simplex optimisation implies that a series of experiments are performed on the corners of such a figure. Most simple descriptions are of two factor designs, where the simplex is a triangle, but, of course, there is no restriction on the number of factors.

2.6.1   Fixed Sized Simplex

The most common, and easiest to understand, method of simplex optimisation is called the fixed sized simplex. It is best described as a series of rules.

The main steps are as follows, exemplified by a two factor experiment.

1. Define how many factors are of interest, which we will call k.
2. Perform $k + 1(=3)$ experiments on the vertices of a simplex (or triangle for two factors) in factor space. The conditions for these experiments depend on the step size. This defines the final 'resolution' of the optimum. The smaller the step size, the better the optimum can be defined, but the more the experiments are necessary. A typical initial simplex using the step size above might consist of the three experiments, for example
   (a) pH 3, temperature 30 °C;
   (b) pH 3.01, temperature 31 °C;
   (c) pH 3.02, temperature 30 °C.
   Such a triangle is illustrated in Figure 2.37. It is important to establish sensible initial conditions, especially the spacing between the experiments; in this example one is searching very narrow pH and temperature ranges, and if the optimum is far from these conditions, the optimisation will take a long time.
3. Rank the response (e.g. the yield of rate of the reaction) from 1 (worst) to $k + 1$ (best) over each of the initial conditions. Note that the response does not need to be quantitative, it could be qualitative, e.g. which food tastes best. In vector form the conditions for the nth response are given by x_n, where the higher the value of

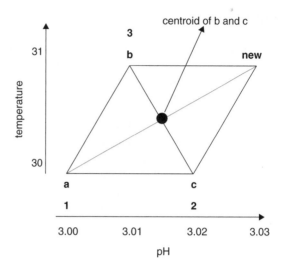

Figure 2.37
Initial experiments (a, b and c) on the edge of a simplex: two factors, and the new conditions
if experiment a results in the worst response

   n the better is the response, e.g. $x_3 = (3.01\ 31)$ implies that the best response was
   at pH 3.01 and 31 °C.
4. Establish new conditions for the next experiment as follows:

$$x_{new} = c + c - x_1$$

   where c is the centroid of the responses 2 to $k + 1$ (excluding the worst response),
   defined by the average of these responses represented in vector form, an alternative
   expression for the new conditions is $x_{new} = x_2 + x_3 - x_1$ when there are two factors.
   In the example above
   • if the worst response is at $x_1 = (3.00\ 30)$,
   • the centroid of the remaining responses is $c = [(3.01 + 3.02)/2\ (30 + 31)/2] =$
     $(3.015\ 30.5)$,
   • so the new response is $x_{new} = (3.015\ 30.5) + (30.015\ 30.5) - (3.00\ 30) =$
     $(30.03\ 31)$.
   This is illustrated in Figure 2.37, with the centroid indicated. The new experimental
   conditions are often represented by reflection of the worst conditions in the centroid
   of the remaining conditions. Keep the points x_{new} and the kth (=2) best responses
   from the previous simplex, resulting in $k + 1$ new responses. The worst response
   from the previous simplex is rejected.
5. Continue as in steps 3 and 4 unless the new conditions result in a response that is
   worst than the remaining $k(=2)$ conditions, i.e. $y_{new} < y_2$, where y is the corresponding response and the aim is maximisation. In this case return to the previous
   conditions, and calculate

$$x_{new} = c + c - x_2$$

   where c is the centroid of the responses 1 and 3 to $k + 1$ (excluding the second
   worst response) and can also be expressed by $x_{new} = x_1 + x_3 - x_2$, for two factors.

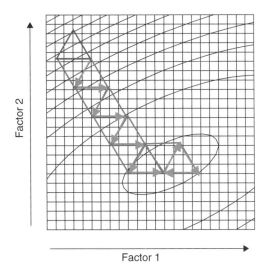

Figure 2.38
Progress of a fixed sized simplex

In the case illustrated in Figure 2.37, this would simply involve reflecting point **2** in the centroid of points **1** and **3**. Keep these new conditions together with the worst and the $k-1$ best responses from the previous simplex. The second worst response from the previous simplex is rejected, so in the case of three factors, we keep old responses 1, 3 and the new one, rather than old responses 2, 3 and the new one.

6. Check for convergence. When the simplex is at an optimum it normally oscillates around in a triangle or hexagon. If the same conditions reappear, stop. There are a variety of stopping rules, but it should generally be obvious when optimisation has been achieved. If you are writing a robust package you will be need to take a lot of rules into consideration, but if you are doing the experiments manually it is normal simply to check what is happening.

The progress of a fixed sized simplex is illustrated in Figure 2.38.

2.6.2 Elaborations

There are many elaborations that have been developed over the years. One of the most important is the $k+1$ rule. If a vertex has remained part of the simplex for $k+1$ steps, perform the experiment again. The reason for this is that response surfaces may be noisy, so an unduly optimistic response could have been obtained because of experimental error. This is especially important when the response surface is flat near the optimum. Another important issue relates to boundary conditions. Sometimes there are physical reasons why a condition cannot cross a boundary, an obvious case being a negative concentration. It is not always easy to deal with such situations, but it is possible to use step 5 rather than step 4 above under such circumstances. If the simplex constantly tries to cross a boundary either the constraints are slightly unrealistic and so should be changed, or the behaviour near the boundary needs further investigation. Starting a new simplex near the boundary with a small step size may solve the problem.

2.6.3   Modified Simplex

A weakness with the standard method for simplex optimisation is a dependence on the initial step size, which is defined by the initial conditions. For example, in Figure 2.37 we set a very small step size for both variables; this may be fine if we are sure we are near the optimum, but otherwise a bigger triangle would reach the optimum quicker, the problem being that the bigger step size may miss the optimum altogether. Another method is called the modified simplex algorithm and allows the step size to be altered, reduced as the optimum is reached, or increased when far from the optimum.

For the modified simplex, step 4 of the fixed sized simplex (Section 2.6.1) is changed as follows. A new response at point x_{test} is determined, where the conditions are obtained as for fixed sized simplex. The four cases below are illustrated in Figure 2.39.

(a) If the response is better than all the other responses in the previous simplex, i.e. $y_{test} > y_{k+1}$ then *expand* the simplex, so that

$$x_{new} = c + \alpha(c - x_1)$$

where α is a number greater than 1, typically equal to 2.

(b) If the response is better than the worst of the other responses in the previous simplex, but worst than the second worst, i.e. $y_1 < y_{test} < y_2$, then *contract* the

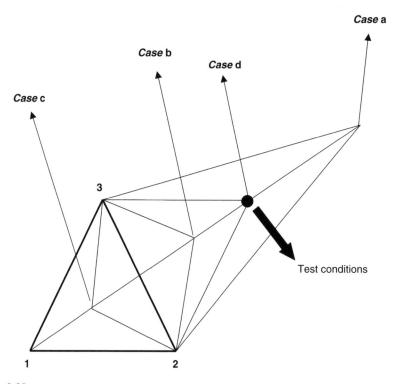

Figure 2.39
Modified simplex. The original simplex is indicated in bold, with the responses ordered from 1 (worst) to 3 (best). The test conditions are indicated

simplex but in the direction of this new response:

$$x_{new} = c + \beta(c - x_1)$$

where β is a number less than 1, typically equal to 0.5.

(c) If the response is worse than the other responses, i.e. $y_{test} < y_1$, then *contract* the simplex but in the opposite direction of this new response:

$$x_{new} = c - \beta(c - x_1)$$

where β is a number less than 1, typically equal to 0.5.

(d) In all other cases simply calculate

$$x_{new} = x_{test} = c + c - x_1$$

as in the normal (fixed-sized) simplex.

Then perform another experiment at x_{new} and keep this new experiment plus the $k(=2)$ best experiments from the previous simplex to give a new simplex.

Step 5 still applies: if the value of the response at the new vertex is less than that of the remaining k responses, return to the original simplex and reject the second best response, repeating the calculation as above.

There are yet further sophistications such as the supermodified simplex, which allows mathematical modelling of the shape of the response surface to provide guidelines as to the choice of the next simplex. Simplex optimisation is only one of several computational approaches to optimisation, including evolutionary optimisation and steepest ascent methods. However, it has been much used in chemistry, largely owing to the work of Deming and colleagues, being one of the first systematic approaches applied to the optimisation of real chemical data.

2.6.4 Limitations

In many well behaved cases, simplex performs well and is an efficient approach for optimisation. There are, however, a number of limitations.

- If there is a large amount of experimental error, then the response is not very reproducible. This can cause problems, for example, when searching a fairly flat response surface.
- Sensible initial conditions and scaling (coding) of the factors are essential. This can only come form empirical chemical knowledge.
- If there are serious discontinuities in the response surface, this cannot always be taken into account.
- There is no modelling information. Simplex does not aim to predict an unknown response, produce a mathematical model or test the significance of the model using ANOVA. There is no indication of the size of interactions or related effects.

There is some controversy as to whether simplex methods should genuinely be considered as experimental designs, rather than algorithms for optimisation. Some statisticians often totally ignore this approach and, indeed, many books and courses of experimental design in chemistry will omit simplex methods altogether, concentrating exclusively on approaches for mathematical modelling of the response surface. However, engineers and programmers have employed simplex and related approaches for optimisation for many years, and these methods have been much used, for example, in spectroscopy and chromatography, and so should be considered by the chemist. As a practical tool where the detailed mathematical relationship between response and underlying variables is not of primary concern, the methods described above are very valuable. They are also easy to implement computationally and to automate and simple to understand.

Problems

Problem 2.1 A Two Factor, Two Level Design

Section 2.2.3   Section 2.3.1

The following represents the yield of a reaction recorded at two catalyst concentrations and two reaction times:

Concentration (mM)	Time (h)	Yield (%)
0.1	2	29.8
0.1	4	22.6
0.2	2	32.6
0.2	4	26.2

1. Obtain the design matrix from the raw data, D, containing four coefficients of the form

$$y = b_0 + b_1 x_1 + b_2 x_2 + b_{12} x_1 x_2$$

2. By using this design matrix, calculate the relationship between the yield (y) and the two factors from the relationship

$$b = D^{-1}.y$$

3. Repeat the calculations in question 2 above, but using the coded values of the design matrix.

Problem 2.2 Use of a Fractional Factorial Design to Study Factors That Influence NO Emissions in a Combustor

Section 2.2.3   Section 2.3.2

It is desired to reduce the level of NO in combustion processes for environmental reasons. Five possible factors are to be studied. The amount of NO is measured as mg MJ^{-1} fuel. A fractional factorial design was performed. The following data were obtained, using coded values for each factor:

Load	Air: fuel ratio	Primary air (%)	NH_3 $(dm^3\,h^{-1})$	Lower secondary air (%)	NO $(mg\,MJ^{-1})$
−1	−1	−1	−1	1	109
1	−1	−1	1	−1	26
−1	1	−1	1	−1	31
1	1	−1	−1	1	176
−1	−1	1	1	1	41
1	−1	1	−1	−1	75
−1	1	1	−1	−1	106
1	1	1	1	1	160

1. Calculate the coded values for the intercept, the linear and all two factor interaction terms. You should obtain a matrix of 16 terms.
2. Demonstrate that there are only eight unique possible combinations in the 16 columns and indicate which terms are confounded.
3. Set up the design matrix inclusive of the intercept and five linear terms.
4. Determine the six terms arising from question 3 using the pseudo-inverse. Interpret the magnitude of the terms and comment on their significance.
5. Predict the eight responses using $\hat{y} = D.b$ and calculate the percentage root mean square error, adjusted for degrees of freedom, relative to the average response.

Problem 2.3 Equivalence of Mixture Models

Section 2.5.2.2

The following data are obtained for a simple mixture design:

Factor 1	Factor 2	Factor 3	Response
1	0	0	41
0	1	0	12
0	0	1	18
0.5	0.5	0	29
0.5	0	0.5	24
0	0.5	0.5	17

1. The data are to be fitted to a model of the form

$$y = b_1 x_1 + b_2 x_2 + b_3 x_3 + b_{12} x_1 x_2 + b_{13} x_1 x_3 + b_{23} x_2 x_3$$

Set up the design matrix, and by calculating $D^{-1}.y$ determine the six coefficients.
2. An alternative model is of the form

$$y = a_0 + a_1 x_1 + a_2 x_2 + a_{11} x_1^2 + a_{22} x_2^2 + a_{12} x_1 x_2$$

Calculate the coefficients for this model.
3. Show, algebraically, the relationship between the two sets of coefficients, by substituting

$$x_3 = 1 - x_1 - x_2$$

into the equation for the model 1 above. Verify that the numerical terms do indeed obey this relationship and comment.

Problem 2.4 Construction of Mixture Designs

Section 2.5.3   Section 2.5.4

1. How many experiments are required for $\{5, 1\}$, $\{5, 2\}$ and $\{5, 3\}$ simplex lattice designs?
2. Construct a $\{5, 3\}$ simplex lattice design.
3. How many combinations are required in a full five factor simplex centroid design? Construct this design.
4. Construct a $\{3, 3\}$ simplex lattice design.
5. Repeat the above design using the following lower bound constraints:

$$x_1 \geq 0.0$$
$$x_2 \geq 0.3$$
$$x_3 \geq 0.4$$

Problem 2.5 Normal Probability Plots

Section 2.2.4.5

The following is a table of responses of eight experiments at coded levels of three variables, A, B and C:

A	B	C	response
−1	−1	−1	10
1	−1	−1	9.5
−1	1	−1	11
1	1	−1	10.7
−1	−1	1	9.3
1	−1	1	8.8
−1	1	1	11.9
1	1	1	11.7

1. It is desired to model the intercept and all single, two and three factor coefficients. Show that there are only eight coefficients and explain why squared terms cannot be taken into account.
2. Set up the design matrix and calculate the coefficients. Do this without using the pseudo-inverse.
3. Excluding the intercept term, there are seven coefficients. A normal probability plot can be obtained as follows. First, rank the seven coefficients in order. Then, for each coefficient of rank p calculate a probability $(p - 0.5)/7$. Convert these probabilities into expected proportions of the normal distribution for a reading of appropriate rank using an appropriate function in Excel. Plot the values of each of the seven effects (horizontal axis) against the expected proportion of normal distribution for a reading of given rank.

4. From the normal probability plot, several terms are significant. Which are they?
5. Explain why normal probability plots work.

Problem 2.6 Use of a Saturated Factorial Design to Study Factors in the Stability of a Drug

Section 2.3.1   Section 2.2.3

The aim of the study is to determine factors that influence the stability of a drug, diethylpropion, as measured by HPLC after 24 h. The higher the percentage, the better is the stability. Three factors are considered:

Factor	Level ($-$)	Level ($+$)
Moisture (%)	57	75
Dosage form	Powder	Capsule
Clorazepate (%)	0	0.7

A full factorial design is performed, with the following results, using coded values for each factor:

Factor 1	Factor 2	Factor 3	Response
-1	-1	-1	90.8
1	-1	-1	88.9
-1	1	-1	87.5
1	1	-1	83.5
-1	-1	1	91.0
1	-1	1	74.5
-1	1	1	91.4
1	1	1	67.9

1. Determine the design matrix corresponding to the model below, using coded values throughout:

$$y = b_0 + b_1 x_1 + b_2 x_2 + b_3 x_3 + b_{12} x_1 x_2 + b_{13} x_1 x_3 + b_{23} x_2 x_3 + b_{123} x_1 x_2 x_3$$

2. Using the inverse of the design matrix, determine the coefficients $b = D^{-1}.y$.
3. Which of the coefficients do you feel are significant? Is there any specific interaction term that is significant?
4. The three main factors are all negative, which, without considering the interaction terms, would suggest that the best response is when all factors are at their lowest level. However, the response for the first experiment is not the highest, and this suggests that for best performance at least one factor must be at a high level. Interpret this in the light of the coefficients.
5. A fractional factorial design could have been performed using four experiments. Explain why, in this case, such a design would have missed key information.
6. Explain why the inverse of the design matrix can be used to calculate the terms in the model, rather than using the pseudo-inverse $b = (D'.D)^{-1}.D'.y$. What changes in the design or model would require using the pseudo-inverse in the calculations?

7. Show that the coefficients in question 2 could have been calculated by multiplying the responses by the coded value of each term, summing all eight values, and dividing by eight. Demonstrate that the same answer is obtained for b_1 using both methods of calculation, and explain why.
8. From this exploratory design it appears that two major factors and their interaction are most significant. Propose a two factor central composite design that could be used to obtain more detailed information. How would you deal with the third original factor?

Problem 2.7 Optimisation of Assay Conditions for tRNAs Using a Central Composite Design

Section 2.4   Section 2.2.3   Section 2.2.2   Section 2.2.4.4   Section 2.2.4.3

The influence of three factors, namely pH, enzyme concentration and amino acid concentration, on the esterification of tRNA arginyl-tRNA synthetase is to be studied by counting the radioactivity of the final product, using ^{14}C-labelled arginine. The higher is the count, the better are the conditions.

The factors are coded at five levels as follows:

	Level				
	−1.7	−1	0	1	1.7
Factor 1: enzyme (μg protein)	3.2	6.0	10.0	14.0	16.8
Factor 2: arginine (pmol)	860	1000	1200	1400	1540
Factor 3: pH	6.6	7.0	7.5	8.0	8.4

The results of the experiments are as follows:

Factor 1	Factor 2	Factor 3	Counts
1	1	1	4930
1	1	−1	4810
1	−1	1	5128
1	−1	−1	4983
−1	1	1	4599
−1	1	−1	4599
−1	−1	1	4573
−1	−1	−1	4422
1.7	0	0	4891
−1.7	0	0	4704
0	1.7	0	4566
0	−1.7	0	4695
0	0	1.7	4872
0	0	−1.7	4773
0	0	0	5063
0	0	0	4968
0	0	0	5035

Factor 1	Factor 2	Factor 3	Counts
0	0	0	5122
0	0	0	4970
0	0	0	4925

1. Using a model of the form

$$\hat{y} = b_0 + b_1 x_1 + b_2 x_2 + b_3 x_3 + b_{11} x_1{}^2 + b_{22} x_2{}^2 + b_{33} x_3{}^2$$
$$+ b_{12} x_1 x_2 + b_{13} x_1 x_3 + b_{23} x_2 x_3$$

   set up the design matrix D.
2. How many degrees-of-freedom are required for the model? How many are available for replication and so how many left to determine the significance of the lack-of-fit?
3. Determine the coefficients of the model using the pseudo-inverse $b = (D'.D)^{-1}.D'.y$ where y is the vector of responses.
4. Determine the 20 predicted responses by $\hat{y} = D.b$, and so the overall sum of square residual error, and the root mean square residual error (divide by the residual degrees of freedom). Express the latter error as a percentage of the standard deviation of the measurements. Why is it more appropriate to use a standard deviation rather than a mean in this case?
5. Determine the sum of square replicate error, and so, from question 4, the sum of square lack-of-fit error. Divide the sum of square residual, lack-of-fit and replicate errors by their appropriate degrees of freedom and so construct a simple ANOVA table with these three errors, and compute the F-ratio.
6. Determine the variance of each of the 10 parameters in the model as follows. Compute the matrix $(D'.D)^{-1}$ and take the diagonal elements for each parameter. Multiply these by the mean square residual error obtained in question 5.
7. Calculate the t-statistic for each of the 10 parameters in the model, and so determine which are most significant.
8. Select the intercept and five other most significant coefficients and determine a new model. Calculate the new sum of squares residual error, and comment.
9. Using partial derivatives, determine the optimum conditions for the enzyme assay using coded values of the three factors. Convert these to the raw experimental conditions.

Problem 2.8 Simplex Optimisation

Section 2.6

Two variables, a and b, influence a response y. These variables may, for example, correspond to pH and temperature, influencing level of impurity. It is the aim of optimisation to find the values of a and b that give the minimum value of y.

The theoretical dependence of the response on the variables is

$$y = 2 + a^2 - 2a + 2b^2 - 3b + (a - 2)(b - 3)$$

Assume that this dependence is unknown in advance, but use it to generate the response for any value of the variables. Assume there is no noise in the system.

1. Using partial derivatives show that the minimum value of y is obtained when $a = 15/7$ and compute the value of b and y at this minimum.
2. Perform simplex optimisation using as a starting point

a	b
0	0
1	0
0.5	0.866

This is done by generating the equation for y and watching how y changes with each new set of conditions a and b. You should reach a point where the response oscillates; although the oscillation is not close to the minimum, the values of a and b giving the best overall response should be reasonable. Record each move of the simplex and the response obtained.

3. What are the estimated values of a, b and y at the minimum and why do they differ from those in question 1?
4. Perform a simplex using a smaller step-size, namely starting at

a	b
0	0
0.5	0
0.25	0.433

What are the values of a, b and y and why are they much closer to the true minimum?

Problem 2.9 Error Analysis for Simple Response Modelling

Section 2.2.2   Section 2.2.3

The follow represents 12 experiments involving two factors x_1 and x_2, together with the response y:

x_1	x_2	y
0	0	5.4384
0	0	4.9845
0	0	4.3228
0	0	5.2538
-1	-1	8.7288
-1	1	0.7971
1	-1	10.8833

x_1	x_2	y
1	1	11.1540
1	0	12.4607
−1	0	6.3716
0	−1	6.1280
0	1	2.1698

1. By constructing the design matrix and then using the pseudo-inverse, calculate the coefficients for the best fit model given by the equation

$$y = b_0 + b_1 x_1 + b_2 x_2 + b_{11} x_1^2 + b_{22} x_2^2 + b_{12} x_1 x_2$$

2. From these coefficients, calculate the 12 predicted responses, and so the residual (modelling) error as the sum of squares of the residuals.
3. Calculate the contribution to this error of the replicates simply by calculating the average response over the four replicates, and then subtracting each replicate response, and summing the squares of these residuals.
4. Calculate the sum of square lack-of-fit error by subtracting the value in question 3 from that in question 2.
5. Divide the lack-of-fit and replicate errors by their respective degrees of freedom and comment.

Problem 2.10 The Application of a Plackett–Burman Design to the Screening of Factors Influencing a Chemical Reaction

Section 2.3.3

The yield of a reaction of the form

$$A + B \longrightarrow C$$

is to be studied as influenced by 10 possible experimental conditions, as follows:

Factor		Units	Low	High
x_1	% NaOH	%	40	50
x_2	Temperature	°C	80	110
x_3	Nature of catalyst		A	B
x_4	Stirring		Without	With
x_5	Reaction time	min	90	210
x_6	Volume of solvent	ml	100	200
x_7	Volume of NaOH	ml	30	60
x_8	Substrate/NaOH ratio	mol/ml	0.5×10^{-3}	1×10^{-3}
x_9	Catalyst/substrate ratio	mol/ml	4×10^{-3}	6×10^{-3}
x_{10}	Reagent/substrate ratio	mol/mol	1	1.25

The design, including an eleventh dummy factor, is as follows, with the observed yields:

Expt No.	x_1	x_2	x_3	x_4	x_5	x_6	x_7	x_8	x_9	x_{10}	x_{11}	Yield (%)
1	−	−	−	−	−	−	−	−	−	−	−	15
2	+	+	−	+	+	+	−	−	−	+	−	42
3	−	+	+	−	+	+	+	−	−	−	+	3
4	+	−	+	+	−	+	+	+	−	−	−	57
5	−	+	−	+	+	−	+	+	+	−	−	38
6	−	−	+	−	+	+	−	+	+	+	−	37
7	−	−	−	+	−	+	+	−	+	+	+	74
8	+	−	−	−	+	−	+	+	−	+	+	54
9	+	+	−	−	−	+	−	+	+	−	+	56
10	+	+	+	−	−	−	+	−	+	+	−	64
11	−	+	+	+	−	−	−	+	−	+	+	65
12	+	−	+	+	+	−	−	−	+	−	+	59

1. Why is a dummy factor employed? Why is a Plackett–Burman design more desirable than a two level fractional factorial in this case?
2. Verify that all the columns are orthogonal to each other.
3. Set up a design matrix, D, and determine the coefficients b_0 to b_{11}.
4. An alternative method for calculating the coefficients for factorial designs such as the Plackett–Burman design is to multiply the yields of each experiment by the levels of the corresponding factor, summing these and dividing by 12. Verify that this provides the same answer for factor 1 as using the inverse matrix.
5. A simple method for reducing the number of experimental conditions for further study is to look at the size of the factors and eliminate those that are less than the dummy factor. How many factors remain and what are they?

Problem 2.11 Use of a Constrained Mixture Design to Investigate the Conductivity of a Molten Salt System

Section 2.5.4   Section 2.5.2.2

A molten salt system consisting of three components is prepared, and the aim is to investigate the conductivity according to the relative proportion of each component. The three components are as follows:

Component		Lower limit	Upper limit
x_1	NdCl$_3$	0.2	0.9
x_2	LiCl	0.1	0.8
x_3	KCl	0.0	0.7

The experiment is coded to give pseudo-components so that a value of 1 corresponds to the upper limit, and a value of 0 to the lower limit of each component. The experimental results are as follows:

z_1	z_2	z_3	Conductivity ($\Omega^{-1}\,cm^{-1}$)
1	0	0	3.98
0	1	0	2.63
0	0	1	2.21
0.5	0.5	0	5.54
0.5	0	0.5	4.00
0	0.5	0.5	2.33
0.3333	0.3333	0.3333	3.23

1. Represent the constrained mixture space, diagrammatically, in the original mixture space. Explain why the constraints are possible and why the new reduced mixture space remains a triangle.
2. Produce a design matrix consisting of seven columns in the true mixture space as follows. The true composition of a component 1 is given by $Z_1(U_1 - L_1) + L_1$, where U and L are the upper and lower bounds for the component. Convert all three columns of the matrix above using this equation and then set up a design matrix, containing three single factor terms, and all possible two and three factor interaction terms (using a Sheffé model).
3. Calculate the model linking the conductivity to the proportions of the three salts.
4. Predict the conductivity when the proportion of the salts is 0.209, 0.146 and 0.645.

Problem 2.12 Use of Experimental Design and Principal Components Analysis for Reduction of Number of Chromatographic Tests

Section 2.4.5   Section 4.3.6.4   Section 4.3   Section 4.4.1

The following table represents the result of a number of tests performed on eight chromatographic columns, involving performing chromatography on eight compounds at pH 3 in methanol mobile phase, and measuring four peakshape parameters. Note that you may have to transpose the matrix in Excel for further work. The aim is to reduce the number of experimental tests necessary using experimental design. Each test is denoted by a mnemonic. The first letter (e.g. P) stands for a compound, the second part of the name, k, N, N(df), or As standing for four peakshape/retention time measurements.

	Inertsil ODS	Inertsil ODS-2	Inertsil ODS-3	Kromasil C18	Kromasil C8	Symmetry C18	Supelco ABZ+	Purospher
Pk	0.25	0.19	0.26	0.3	0.28	0.54	0.03	0.04
PN	10 200	6930	7420	2980	2890	4160	6890	6960
PN(df)	2650	2820	2320	293	229	944	3660	2780
PAs	2.27	2.11	2.53	5.35	6.46	3.13	1.96	2.08
Nk	0.25	0.12	0.24	0.22	0.21	0.45	0	0
NN	12 000	8370	9460	13 900	16 800	4170	13 800	8260
NN(df)	6160	4600	4880	5330	6500	490	6020	3450

	Inertsil ODS	Inertsil ODS-2	Inertsil ODS-3	Kromasil C18	Kromasil C8	Symmetry C18	Supelco ABZ+	Purospher
NAs	1.73	1.82	1.91	2.12	1.78	5.61	2.03	2.05
Ak	2.6	1.69	2.82	2.76	2.57	2.38	0.67	0.29
AN	10 700	14 400	11 200	10 200	13 800	11 300	11 700	7160
AN(df)	7790	9770	7150	4380	5910	6380	7000	2880
AAs	1.21	1.48	1.64	2.03	2.08	1.59	1.65	2.08
Ck	0.89	0.47	0.95	0.82	0.71	0.87	0.19	0.07
CN	10 200	10 100	8500	9540	12 600	9690	10 700	5300
CN(df)	7830	7280	6990	6840	8340	6790	7250	3070
CAs	1.18	1.42	1.28	1.37	1.58	1.38	1.49	1.66
Qk	12.3	5.22	10.57	8.08	8.43	6.6	1.83	2.17
QN	8800	13 300	10 400	10 300	11 900	9000	7610	2540
QN(df)	7820	11 200	7810	7410	8630	5250	5560	941
QAs	1.07	1.27	1.51	1.44	1.48	1.77	1.36	2.27
Bk	0.79	0.46	0.8	0.77	0.74	0.87	0.18	0
BN	15 900	12 000	10 200	11 200	14 300	10 300	11 300	4570
BN(df)	7370	6550	5930	4560	6000	3690	5320	2060
BAs	1.54	1.79	1.74	2.06	2.03	2.13	1.97	1.67
Dk	2.64	1.72	2.73	2.75	2.27	2.54	0.55	0.35
DN	9280	12 100	9810	7070	13 100	10 000	10 500	6630
DN(df)	5030	8960	6660	2270	7800	7060	7130	3990
DAs	1.71	1.39	1.6	2.64	1.79	1.39	1.49	1.57
Rk	8.62	5.02	9.1	9.25	6.67	7.9	1.8	1.45
RN	9660	13 900	11 600	7710	13 500	11 000	9680	5140
RN(df)	8410	10 900	7770	3460	9640	8530	6980	3270
RAs	1.16	1.39	1.65	2.17	1.5	1.28	1.41	1.56

1. Transpose the data so that the 32 tests correspond to columns of a matrix (variables) and the eight chromatographic columns to the rows of a matrix (objects). Standardise each column by subtracting the mean and dividing by the population standard deviation (Chapter 4, Section 4.3.6.4). Why is it important to standardise these data?

2. Perform PCA (principal components analysis) on these data and retain the first three loadings (methods for performing PCA are discussed in Chapter 4, Section 4.3; see also Appendix A.2.1 and relevant sections of Appendices A.4 and A.5 if you are using Excel or Matlab).

3. Take the three loadings vectors and transform to a common scale as follows. For each loadings vector select the most positive and most negative values, and code these to $+1$ and -1, respectively. Scale all the intermediate values in a similar fashion, leading to a new scaled loadings matrix of 32 columns and 3 rows. Produce the new scaled loadings vectors.

4. Select a factorial design as follows, with one extra point in the centre, to obtain a range of tests which is a representative subset of the original tests:

Design point	PC1	PC2	PC3
1	−	−	−
2	+	−	−
3	−	+	−
4	+	+	−
5	−	−	+
6	+	−	+
7	−	+	+
8	+	+	+
9	0	0	0

Calculate the Euclidean distance of each of the 32 scaled loadings from each of the nine design points; for example, the first design point calculates the Euclidean distance of the loadings scaled as in question 3 from the point $(-1,-1,-1)$, by the equation

$$d_1 = \sqrt{(p_{11} + 1)^2 + (p_{12} + 1)^2 + (p_{13} + 1)^2}$$

(Chapter 4, Section 4.4.1).

5. Indicate the chromatographic parameters closest to the nine design points. Hence recommend a reduced number of chromatographic tests and comment on the strategy.

Problem 2.13 A Mixture Design with Constraints

Section 2.5.4

It is desired to perform a three factor mixture design with constraints on each factor as follows:

	x_1	x_2	x_3
Lower	0.0	0.2	0.3
Upper	0.4	0.6	0.7

1. The mixture design is normally represented as an irregular polygon, with, in this case, six vertices. Calculate the percentage of each factor at the six coordinates.
2. It is desired to perform 13 experiments, namely on the six corners, in the middle of the six edges and in the centre. Produce a table of the 13 mixtures.
3. Represent the experiment diagrammatically.

Problem 2.14 Construction of Five Level Calibration Designs

Section 2.3.4

The aim is to construct a five level partial factorial (or calibration) design involving 25 experiments and up to 14 factors, each at levels -2, -1, 0, 1 and 2. Note that this design is only one of many possible such designs.

1. Construct the experimental conditions for the first factor using the following rules.
   - The first experiment is at level −2.
   - This level is repeated for experiments 2, 8, 14 and 20.
   - The levels for experiments 3–7 are given as follows (0, 2, 0, 0, 1).
   - A cyclic permuter of the form $0 \longrightarrow -1 \longrightarrow 1 \longrightarrow 2 \longrightarrow 0$ is then used. Each block of experiments 9–13, 15–19 and 21–25 are related by this permuter, each block being one permutation away from the previous block, so experiments 9 and 10 are at levels −1 and 0, for example.
2. Construct the experimental conditions for the other 13 factors as follows.
   - Experiment 1 is always at level −2 for all factors.
   - The conditions for experiments 2–24 for the other factors are simply the cyclic permutation of the previous factor as explained in Section 2.3.4. So produce the matrix of experimental conditions.
3. What is the difference vector used in this design?
4. Calculate the correlation coefficients between all pairs of factors 1–14. Plot the two graphs of the levels of factor 1 versus factors 2 and 7. Comment.

Problem 2.15 A Four Component Mixture Design Used for Blending of Olive Oils

Section 2.5.2.2

Fourteen blends of olive oils from four cultivars A–D are mixed together in the design below presented together with a taste panel score for each blend. The higher the score the better the taste of the olive oil.

A	B	C	D	Score
1	0	0	0	6.86
0	1	0	0	6.50
0	0	1	0	7.29
0	0	0	1	5.88
0.5	0.5	0	0	7.31
0.5	0	0.5	0	6.94
0.5	0	0	0.5	7.38
0	0.5	0.5	0	7.00
0	0.5	0	0.5	7.13
0	0	0.5	0.5	7.31
0.333 33	0.333 33	0.333 33	0	7.56
0.333 33	0.333 33	0	0.333 33	7.25
0.333 33	0	0.333 33	0.333 33	7.31
0	0.333 33	0.333 33	0.333 33	7.38

1. It is desired to produce a model containing 14 terms, namely four linear, six two component and four three component terms. What is the equation for this model?
2. Set up the design matrix and calculate the coefficients.
3. A good way to visualise the data is via contours in a mixture triangle, allowing three components to vary and constraining the fourth to be constant. Using a step size of 0.05, calculate the estimated responses from the model in question 2 when

D is absent and $A + B + C = 1$. A table of 231 numbers should be produced. Using a contour plot, visualise these data. If you use Excel, the upper right-hand half of the plot may contain meaningless data; to remove these, simply cover up this part of the contour plot with a white triangle. In modern versions of Matlab and some other software packages, triangular contour plots can be obtained straightforwardly Comment on the optimal blend using the contour plot when D is absent.

4. Repeat the contour plot in question 3 for the following: (i) $A + B + D = 1$, (ii) $B + C + D = 1$ and (iii) $A + C + D = 1$, and comment.
5. Why, in this example, is a strategy of visualisation of the mixture contours probably more informative than calculating a single optimum?

Problem 2.16 Central Composite Design Used to Study the Extraction of Olive Seeds in a Soxhlet

Section 2.4   Section 2.2.2

Three factors, namely (1) irradiation power as a percentage, (2) irradiation time in seconds and (3) number of cycles, are used to study the focused microwave assisted Soxhlet extraction of olive oil seeds, the response measuring the percentage recovery, which is to be optimised. A central composite design is set up to perform the experiments.

The results are as follows, using coded values of the variables:

Factor 1	Factor 2	Factor 3	Response
−1	−1	−1	46.64
−1	−1	1	47.23
−1	1	−1	45.51
−1	1	1	48.58
1	−1	−1	42.55
1	−1	1	44.68
1	1	−1	42.01
1	1	1	43.03
−1	0	0	49.18
1	0	0	44.59
0	−1	0	49.22
0	1	0	47.89
0	0	−1	48.93
0	0	1	49.93
0	0	0	50.51
0	0	0	49.33
0	0	0	49.01
0	0	0	49.93
0	0	0	49.63
0	0	0	50.54

1. A 10 parameter model is to be fitted to the data, consisting of the intercept, all single factor linear and quadratic terms and all two factor interaction terms. Set up the design matrix, and by using the pseudo-inverse, calculate the coefficients of the model using coded values.

2. The true values of the factors are as follows:

Variable	-1	$+1$
Power (%)	30	60
Time (s)	20	30
Cycles	5	7

Re-express the model in question 1 in terms of the true values of each variable, rather than the coded values.

3. Using the model in question 1 and the coded design matrix, calculate the 20 predicted responses and the total error sum of squares for the 20 experiments.

4. Determine the sum of squares replicate error as follows: (i) calculate the mean response for the six replicates; (ii) calculate the difference between the true and average response, square these and sum the six numbers.

5. Determine the sum of squares lack-of-fit error as follows: (i) replace the six replicate responses by the average response for the replicates; (ii) using the 20 responses (with the replicates averaged) and the corresponding predicted responses, calculate the differences, square them and sum them.

6. Verify that the sums of squares in questions 4 and 5 add up to the total error obtained in question 3.

7. How many degrees of freedom are available for assessment of the replicate and lack-of-fit errors? Using this information, comment on whether the lack-of-fit is significant, and hence whether the model is adequate.

8. The significance each term can be determined by omitting the term from the overall model. Assess the significance of the linear term due to the first factor and the interaction term between the first and third factors in this way. Calculate a new design matrix with nine rather than ten columns, removing the relevant column, and also remove the corresponding coefficients from the equation. Determine the new predicted responses using nine factors, and calculate the increase in sum of square error over that obtained in question 3. Comment on the significance of these two terms.

9. Using coded values, determine the optimum conditions as follows. Discard the two interaction terms that are least significant, resulting in eight remaining terms in the equation. Obtain the partial derivatives with respect to each of the three variables, and set up three equations equal to zero. Show that the optimum value of the third factor is given by $-b_3/(2b_{33})$, where the coefficients correspond to the linear and quadratic terms in the equations. Hence calculate the optimum coded values for each of the three factors.

10. Determine the optimum true values corresponding to the conditions obtained in question 9. What is the percentage recovery at this optimum? Comment.

Problem 2.17 A Three Component Mixture Design

Section 2.5.2

A three factor mixture simplex centroid mixture design is performed, with the following results:

x_1	x_2	x_3	Response
1	0	0	9
0	1	0	12
0	0	1	17
0.5	0.5	0	3
0.5	0	0.5	18
0	0.5	0.5	14
0.3333	0.3333	0.3333	11

1. A seven term model consisting of three linear terms, three two factor interaction terms and one three factor interaction term is fitted to the data. Give the equation for this model, compute the design matrix and calculate the coefficients.

2. Instead of seven terms, it is decided to fit the model only to the three linear terms. Calculate these coefficients using only three terms in the model employing the pseudo-inverse. Determine the root mean square error for the predicted responses, and comment on the difference in the linear terms in question 1 and the significance of the interaction terms.

3. It is possible to convert the model of question 1 to a seven term model in two independent factors, consisting of two linear terms, two quadratic terms, two linear interaction terms and a quadratic term of the form $x_1 x_2 (x_1 + x_2)$. Show how the models relate algebraically.

4. For the model in question 3, set up the design matrix, calculate the new coefficients and show how these relate to the coefficients calculated in question 1 using the relationship obtained in question 3.

5. The matrices in questions 1, 2 and 4 all have inverses. However, a model that consisted of an intercept term and three linear terms would not, and it is impossible to use regression analysis to fit the data under such circumstances. Explain these observations.

3 Signal Processing

3.1 Sequential Signals in Chemistry

Sequential signals are surprisingly widespread in chemistry, and require a large number of methods for analysis. Most data are obtained via computerised instruments such as those for NIR, HPLC or NMR, and raw information such as peak integrals, peak shifts and positions is often dependent on how the information from the computer is first processed. An appreciation of this step is essential prior to applying further multivariate methods such as pattern recognition or classification. Spectra and chromatograms are examples of series that are sequential in time or frequency. However, time series also occur very widely in other areas of chemistry, for example in the area of industrial process control and natural processes.

3.1.1 Environmental and Geological Processes

An important source of data involves recording samples regularly with time. Classically such time series occur in environmental chemistry and geochemistry. A river might be sampled for the presence of pollutants such as polyaromatic hydrocarbons or heavy metals at different times of the year. Is there a trend, and can this be related to seasonal factors? Different and fascinating processes occur in rocks, where depth in the sediment relates to burial time. For example, isotope ratios are a function of climate, as relative evaporation rates of different isotopes are temperature dependent: certain specific cyclical changes in the Earth's rotation have resulted in the Ice Ages and so climate changes, leave a systematic chemical record. A whole series of methods for time series analysis based primarily on the idea of correlograms (Section 3.4) can be applied to explore such types of cyclicity, which are often hard to elucidate. Many of these approaches were first used by economists and geologists who also encounter related problems.

One of the difficulties is that long-term and interesting trends are often buried within short-term random fluctuations. Statisticians distinguish between various types of noise which interfere with the signal as discussed in Section 3.2.3. Interestingly, the statistician Herman Wold, who is known among many chemometricians for the early development of the partial least squares algorithm, is probably more famous for his work on time series, studying this precise problem.

In addition to obtaining correlograms, a large battery of methods are available to smooth time series, many based on so-called 'windows', whereby data are smoothed over a number of points in time. A simple method is to take the average reading over five points in time, but sometimes this could miss out important information about cyclicity especially for a process that is sampled slowly compared to the rate of oscillation. A number of linear filters have been developed which are applicable to this time of data (Section 3.3), this procedure often being described as convolution.

3.1.2 Industrial Process Control

In industry, a time series may occur in the manufacturing process of a product. It could be crucial that a drug has a certain well defined composition, otherwise an entire batch is unmarketable. Sampling the product regularly in time is essential, for two reasons. The first is monitoring, simply to determine whether the quality is within acceptable limits. The second is for control, to predict the future and check whether the process is getting out of control. It is costly to destroy a batch, and not economically satisfactory to obtain information about acceptability several days after the event. As soon as the process begins to go wrong it is often advisable to stop the plant and investigate. However, too many false alarms can be equally inefficient. A whole series of methods have been developed for the control of manufacturing processes, an area where chemometrics can often play a key and crucial role. In this text we will not be discussing statistical control charts in detail, the whole topic being worthy of a book in its own right, concentrating primarily on areas of interest to the practising chemist, but a number of methods outlined in this chapter are useful for the handling of such sequential processes, especially to determine if there are long-term trends that are gradually influencing the composition or nature of a manufactured product. Several linear filters together with modifications such as running median smoothing and reroughing (Section 3.3) can be employed under such circumstances. Chemometricians are specially interested in the extension to multivariate methods, for example, monitoring a spectrum as recorded regularly in time, which will be outlined in detail in later chapters.

3.1.3 Chromatograms and Spectra

The most common applications of methods for handling sequential series in chemistry arise in chromatography and spectroscopy and will be emphasized in this chapter. An important aim is to smooth a chromatogram. A number of methods have been developed here such as the Savitsky–Golay filter (Section 3.3.1.2). A problem is that if a chromatogram is smoothed too much the peaks become blurred and lose resolution, negating the benefits, so optimal filters have been developed that remove noise without broadening peaks excessively.

Another common need is to increase resolution, and sometimes spectra are routinely displayed in the derivative mode (e.g. electron spin resonance spectroscopy): there are a number of rapid computational methods for such calculations that do not emphasize noise too much (Section 3.3.2). Other approaches based on curve fitting and Fourier filters are also very common.

3.1.4 Fourier Transforms

The Fourier transform (FT) has revolutionised spectroscopy such as NMR and IR over the past two decades. The raw data are not obtained as a comprehensible spectrum but as a time series, where all spectroscopic information is muddled up and a mathematical transformation is required to obtain a comprehensible spectrum. One reason for performing FT spectroscopy is that a spectrum of acceptable signal to noise ratio is recorded much more rapidly then via conventional spectrometers, often 100 times more rapidly. This has allowed the development of, for example, ^{13}C NMR as a routine analytical tool, because the low abundance of ^{13}C is compensated by faster data acquisition. However, special methods are required to convert this 'time domain' information (called a free induction decay in NMR parlance) to a 'frequency domain' spectrum, which can be interpreted directly (see Section 3.5.1).

Parallel with Fourier transform spectroscopy have arisen a large number of approaches for enhancement of the quality of such data, often called Fourier deconvolution, involving manipulating the time series prior to Fourier transformation (see Section 3.5.2). Many of these filters have their origins in engineering and are often described as digital filters. These are quite different to the classical methods for time series analysis used in economics or geology. Sometimes it is even possible to take non-Fourier data, such as a normal spectrum, and Fourier transform it back to a time series, then use deconvolution methods and Fourier transform back again, often called Fourier self-deconvolution.

Fourier filters can be related to linear methods in Section 3.3 by an important principle called the convolution theorem as discussed in Section 3.5.3.

3.1.5   Advanced Methods

In data analysis there will always be new computational approaches that promote great interest among statisticians and computer scientists. To the computer based chemist such methods are exciting and novel. Much frontline research in chemometrics is involved in refining such methods, but it takes several years before the practical worth or otherwise of novel data analytical approaches is demonstrated. The practising chemist, in many cases, may often obtain just as good results using an extremely simple method rather than a very sophisticated algorithm. To be fair, the originators of many of these methods never claimed they will solve every problem, and often presented the first theoretical descriptions within well defined constraints. The pressure for chemometricians to write original research papers often exaggerates the applicability of some methods, and after the initial enthusiasm and novelty has worn off some approaches tend to receive an unfairly bad press as other people find they can obtain just as good results using extremely basic methods. The original advocates would argue that this is because their algorithms are designed only for certain situations, and have simply been misapplied to inappropriate problems, but if applied in appropriate circumstances can be quite powerful.

Methods for so-called non-linear deconvolution have been developed over the past few years, one of the best known being maximum entropy (Section 3.6.3). This latter approach was first used in IR astronomy to deblur weak images of the sky, and has been successfully applied to police photography to determine car number plates from poor photographic images of a moving car in a crime. Enhancing the quality of a spectrum can also be regarded as a form of image enhancement and so use similar computational approaches. A very successful application is in NMR imaging for medical tomography. The methods are called non-linear because they do not insist that the improved image is a linear function of the original data. A number of other approaches are also available in the literature, but maximum entropy has received much publicity largely because of the readily available software.

Wavelet transforms (Section 3.6.2) are a hot topic, and involve fitting a spectrum or chromatogram to a series of functions based upon a basic shape called a wavelet, of which there are several in the literature. These transforms have the advantage that, instead of storing, for example, 1024 spectral datapoints, it may be possible to retain only a few most significant wavelets and still not lose much information. This can result in both data decompression and denoising of data.

Rapid algorithms for real-time filtering have attracted much interest among engineers, and can be used to follow a process by smoothing the data as they occur. The Kalman filter is one such method (Section 3.6.1) that has been reported extensively in the analytical chemistry literature. It was an interesting challenge to programmers, representing a numerical method that is not particularly difficult to implement but of sufficient sophistication to involve a few afternoons' work especially on computer systems with limited memory and no direct matrix manipulation functions. Such approaches captured the imagination of the more numerate chemists and so form the basis of a large number of papers. With faster and more powerful computers, such filters (which are computationally very complex) are not universally useful, but many chemometricians of the 1980s and early 1990s cut their teeth on Kalman filters and in certain situations there still is a need for these techniques.

3.2 Basics

3.2.1 Peakshapes

Chromatograms and spectra are normally considered to consist of a series of peaks, or lines, superimposed upon noise. Each peak arises from either a characteristic absorption or a characteristic compound. In most cases the underlying peaks are distorted for a variety of reasons such as noise, blurring, or overlap with neighbouring peaks. A major aim of chemometric methods is to obtain the underlying, undistorted, information.

Peaks can be characterised in a number of ways, but a common approach, as illustrated in Figure 3.1, is to characterise each peak by

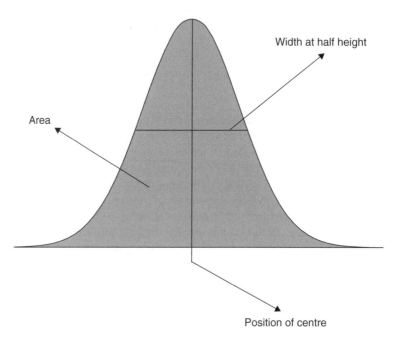

Figure 3.1
Main parameters that characterise a peak

1. a position at the centre (e.g. the elution time or spectral frequency),
2. a width, normally at half-height, and
3. an area.

The relationship between area and peak height is dependent on the peakshape, as discussed below, although heights are often easier to measure experimentally. If all peaks have the same shape, then the ratios of heights are proportional to ratios of areas. However, area is usually a better measure of chemical properties such as concentration and it is important to obtain precise information relating to peakshapes before relying on heights, for example as raw data for pattern recognition programs.

Sometimes the width at a different percentage of the peak height is cited rather than the half-width. A further common measure is when the peak has decayed to a small percentage of the overall height (for example 1 %), which is often taken as the total width of the peak, or alternatively has decayed to a size that relates to the noise.

In many cases of spectroscopy, peakshapes can be very precisely predicted, for example from quantum mechanics, such as in NMR or visible spectroscopy. In other situations, the peakshape is dependent on complex physical processes, for example in chromatography, and can only be modelled empirically. In the latter situation it is not always practicable to obtain an exact model, and a number of closely similar empirical estimates will give equally useful information.

Three common peakshapes cover most situations. If these general peakshapes are not suitable for a particular purpose, it is probably best to consult specialised literature on the particular measurement technique.

3.2.1.1 Gaussians

These peakshapes are common in most types of chromatography and spectroscopy. A simplified equation for a Gaussian is

$$x_i = A \exp[-(x_i - x_0)^2/s^2]$$

where A is the height at the centre, x_0 is the position of the centre and s relates to the peak width.

Gaussians are based on a normal distribution where x_0 corresponds to the mean of a series of measurements and $s/\sqrt{2}$ to the standard deviation.

It can be shown that the width at half-height of a Gaussian peak is given by $\Delta_{1/2} = 2s\sqrt{\ln 2}$ and the area by $\sqrt{\pi}As$ using the equation presented above; note that this depends on both the height and the width.

Note that Gaussians are also the statistical basis of the normal distribution (see Appendix A.3.2), but the equation is normally scaled so that the area under the curve equals one. For signal analysis, we will use this simplified expression.

3.2.1.2 Lorentzians

The Lorentzian peakshape corresponds to a statistical function called the Cauchy distribution. It is less common but often arises in certain types of spectroscopy such as NMR. A simplified equation for a Lorentzian is

$$x_i = A/[1 + (x_i - x_0)^2/s^2]$$

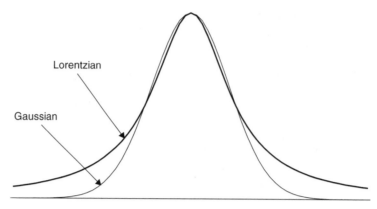

Figure 3.2
Gaussian and Lorentzian peakshapes of equal half-heights

where A is the height at the centre, x_0 is the position of the centre and s relates to the peak width.

It can be shown that the width at half-height of a Lorentzian peak is given by $\Delta_{1/2} = 2s$ and the area by πAs; note this depends on both the height and the width.

The main difference between Gaussian and Lorentzian peakshapes is that the latter has a bigger tail, as illustrated in Figure 3.2 for two peaks with identical half-widths and heights.

3.2.1.3 Asymmetric Peakshapes

In many forms of chromatography it is hard to obtain symmetrical peakshapes. Although there are a number of sophisticated models available, a very simple first approximation is that of a Lorentzian/Gaussian peakshape. Figure 3.3(a) represents a tailing peakshape, in which the left-hand side is modelled by a Gaussian and the right-hand side by a Lorentzian. A fronting peak is illustrated in Figure 3.3(b); such peaks are much rarer.

3.2.1.4 Use of Peakshape Information

Peakshape information can be employed in two principal ways.

1. Curve fitting is fairly common. There are a variety of computational algorithms, most involving some type of least squares minimisation. If there are suspected (or known) to be three peaks in a cluster, of Gaussian shape, then nine parameters need to be found, namely the three peak positions, peak widths and peak heights. In any curve fitting it is important to determine whether there is certain knowledge of the peakshapes, and of certain features, for example the positions of each component. It is also important to appreciate that many chemical data are not of sufficient quality for very detailed models. In chromatography an empirical approach is normally adequate: over-modelling can be dangerous. The result of the curve fitting can be a better description of the system; for example, by knowing peak areas, it may be possible to determine relative concentrations of components in a mixture.
2. Simulations also have an important role in chemometrics. Such simulations are a way of trying to understand a system. If the result of a chemometric method

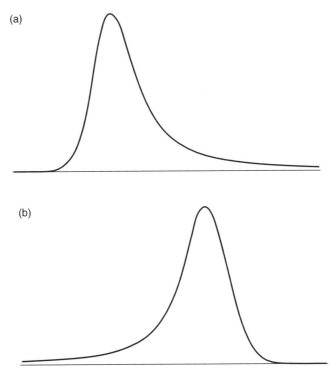

Figure 3.3
Asymmetric peakshapes often described by a Gaussian/Lorentzian model. (a) Tailing: left is Gaussian and right Lorentzian. (b) Fronting: left is Lorentzian and right Gaussian

(such as multivariate curve resolution – see Chapter 6) results in reconstructions of peaks that are close to the real data, then the underlying peakshapes provide a good description. Simulations are also used to explore how well different techniques work, and under what circumstances they break down.

A typical chromatogram or spectrum consists of several peaks, at different positions, of different intensities and sometimes of different shapes. Figure 3.4 represents a cluster of three peaks, together with their total intensity. Whereas the right-hand side peak pair is easy to resolve visually, this is not true for the left-hand side peak pair, and it would be especially hard to identify the position and intensity of the first peak of the cluster without using some form of data analysis.

3.2.2 Digitisation

Almost all modern laboratory based data are now obtained via computers, and are acquired in a digitised rather than analogue form. It is always important to understand how digital resolution influences the ability to resolve peaks.

Many techniques for recording information result in only a small number of datapoints per peak. A typical NMR peak may be only a few hertz at half-width, especially using well resolved instrumentation. Yet a spectrum recorded at 500 MHz, where 8K (=8192) datapoints are used to represent 10 ppm (or 5000 Hz) involves each datapoint

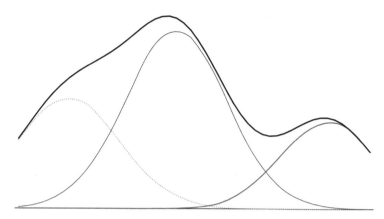

Figure 3.4
Three peaks forming a cluster

representing $1.64 = 8192/5000$ Hz. A 2 Hz peakwidth is represented by only 3.28 datapoints. In chromatography a typical sampling rate is 2 s, yet peak half-widths may be 20 s or less, and interesting compounds separated by 30 s. Poor digital resolution can influence the ability to obtain information. It is useful to be able to determine how serious these errors are.

Consider a Gaussian peak, with a true width at half-height of 30 units, and a height of 1 unit. The theoretical area can be calculated using the equations in Section 3.2.1.1:

- the width at half-height is given by $2s\sqrt{\ln 2}$, so that $s = 30/(2\sqrt{\ln 2}) = 18.017$ units;
- the area is given by $\sqrt{\pi}As$, but $A = 1$, so that the area is $\sqrt{\pi}30/(2\sqrt{\ln 2}) = 31.934$ units.

Typical units for area might be AU.s if the sampling time is in seconds and the intensity in absorption units.

Consider the effect of digitising this peak at different rates, as indicated in Table 3.1, and illustrated in Figure 3.5. An easy way of determining integrated intensities is simply to sum the product of the intensity at each datapoint (x_i) by the sampling interval (δ) over a sufficiently wide range, i.e. to calculate $\delta \Sigma x_i$. The estimates are given in Table 3.1 and it can be seen that for the worst digitised peak (at 24 units, or once per half-height), the estimated integral is 31.721, an error of 0.67 %.

A feature of Table 3.1 is that the acquisition of data starts at exactly two datapoints in each case. In practice, the precise start of acquisition cannot easily be controlled and is often irreproducible, and it is easy to show that when poorly digitised, estimated integrals and apparent peakshapes will depend on this offset. In practice, the instrumental operator will notice a bigger variation in estimated integrals if the digital resolution is low. Although peakwidths must approach digital resolution for there to be significant errors in integration, in some techniques such as GC–MS or NMR this condition is often obtained. In many situations, instrumental software is used to smooth or interpolate the data, and many users are unaware that this step has automatically taken place. These simple algorithms can result in considerable further distortions in quantitative parameters.

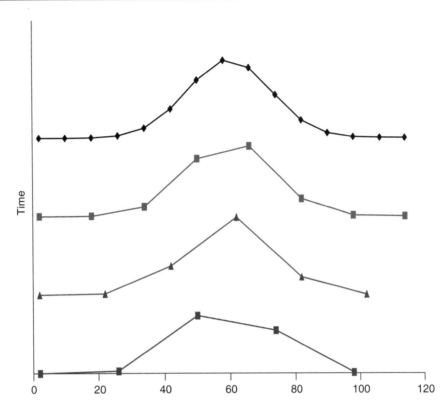

Figure 3.5
Influence on the appearance of a peak as digital resolution is reduced

Table 3.1 Reducing digital resolution.

8 units		16 units		20 units		24 units	
Time	Intensity	Time	Intensity	Time	Intensity	Time	Intensity
2	0.000	2	0.000	2	0.000	2	0.000
10	0.000	18	0.004	22	0.012	26	0.028
18	0.004	34	0.125	42	0.369	50	0.735
26	0.028	50	0.735	62	0.988	74	0.547
34	0.125	66	0.895	82	0.225	98	0.012
42	0.369	82	0.225	102	0.004		
50	0.735	98	0.012				
58	0.988	114	0.000				
66	0.895						
74	0.547						
82	0.225						
90	0.063						
98	0.012						
106	0.001						
114	0.000						
Integral	31.934		31.932		31.951		31.721

A second factor that can influence quantitation is digital resolution in the intensity direction (or vertical scale in the graph). This is due to the analogue to digital converter (ADC) and sometimes can be experimentally corrected by changing the receiver gain. However, for most modern instrumentation this limitation is not so serious and, therefore, will not be discussed in detail below, but is illustrated in Problem 3.7.

3.2.3 Noise

Imposed on signals is noise. In basic statistics, the nature and origin of noise are often unknown, and assumed to obey a normal distribution. Indeed, many statistical tests such as the t-test and F-test (see Appendices A.3.3 and A.3.4) assume this, and are only approximations in the absence of experimental study of such noise distributions. In laboratory based chemistry, there are two fundamental sources of error in instrumental measurements.

1. The first involves sample preparation, for example dilution, weighing and extraction efficiency. We will not discuss these errors in this chapter, but many of the techniques of Chapter 2 have relevance.
2. The second is inherent to the measurement technique. No instrument is perfect, so the signal is imposed upon noise. The observed signal is given by

$$x = \tilde{x} + e$$

where $\tilde{x}$ is the 'perfect' or true signal, and e is a noise function. The aim of most signal processing techniques is to obtain information on the true underlying signal in the absence of noise, i.e. to separate the signal from the noise. The 'tilde' notation is to be distinguished from the 'hat' notation, which refers to the estimated signal, often obtained from regression techniques including methods described in this chapter. Note that in this chapter, x will be used to denote the analytical signal or instrumental response, not y as in Chapter 2. This is so as to introduce a notation that is consistent with most of the open literature. Different investigators working in different areas of science often independently developed incompatible notation, and in a overview such as this text it is preferable to stick reasonably closely to the generally accepted conventions to avoid confusion.

There are two main types of measurement noise.

3.2.3.1 Stationary Noise

The noise at each successive point (normally in time) does not depend on the noise at the previous point. In turn, there are two major types of stationary noise.

1. *Homoscedastic noise.* This is the simplest to envisage. The features of the noise, normally the mean and standard deviation, remain constant over the entire data series. The most common type of noise is given by a normal distribution, with mean zero, and standard deviation dependent on the instrument used. In most real world situations, there are several sources of instrumental noise, but a combination of different symmetric noise distributions often tends towards a normal distribution. Hence this is a good approximation in the absence of more detailed knowledge of a system.

2. *Heteroscedastic noise*. This type of noise is dependent on signal intensity, often proportional to intensity. The noise may still be represented by a normal distribution, but the standard deviation of that distribution is proportional to intensity. A form of heteroscedastic noise often appears to arise if the data are transformed prior to processing, a common method being a logarithmic transform used in many types of spectroscopy such as UV/vis or IR spectroscopy, from transmittance to absorbance. The true noise distribution is imposed upon the raw data, but the transformed information distorts this.

Figure 3.6 illustrates the effect of both types of noise on a typical signal. It is important to recognise that several detailed models of noise are possible, but in practice it is not easy or interesting to perform sufficient experiments to determine such distributions. Indeed, it may be necessary to acquire several hundred or thousand spectra to obtain an adequate noise model, which represents overkill in most real world situations. It is not possible to rely too heavily on published studies of noise distribution because each instrument is different and the experimental noise distribution is a balance between several sources, which differ in relative importance in each instrument. In fact, as the manufacturing process improves, certain types of noise are reduced in size and new effects come into play, hence a thorough study of noise distributions performed say 5 years ago is unlikely to be correct in detail on a more modern instrument.

In the absence of certain experimental knowledge, it is best to stick to a fairly straightforward distribution such as a normal distribution.

3.2.3.2 Correlated Noise

Sometimes, as a series is sampled, the level of noise in each sample depends on that of the previous sample. This is common in process control. For example, there may be problems in one aspect of the manufacturing procedure, an example being the proportion of an ingredient. If the proportion is in error by 0.5 % at 2 pm, does this provide an indication of the error at 2.30 pm?

Many such sources cannot be understood in great detail, but a generalised approach is that of *autoregressive moving average* (ARMA) noise.

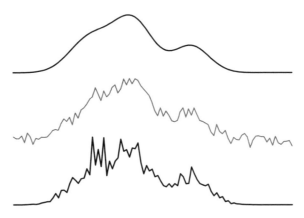

Figure 3.6
Examples of noise. From the top: noise free, homoscedastic, heteroscedastic

1. The *moving average* (MA) component relates the noise at time i to the values of the noise at previous times. A model of order p is given by

$$e_i = \sum_{t=0}^{t=p} c_{i-t} e_{i-t}$$

where e_{i-t} is the noise at time $i - t$ and c_{i-t} is a corresponding coefficient. A simple approach for simulating or modelling this type of noise is to put $p = 1$ and set the coefficient to 1. Under such circumstances

$$e_i = g_i + e_{i-1}$$

where g_i may be generated using a normal distribution. Table 3.2 illustrates a stationary noise distribution and an MA distribution generated simply by adding successive values of the noise, so that, for example, the noise at time $= 4$ is given by $0.00547 = -0.04527 + 0.05075$.

2. The *autoregressive* component relates the noise to the observed value of the response at one or more previous times. A model of order p is given by

$$x_i = \sum_{t=0}^{t=p} c_{i-t} x_{i-t} + e_i$$

Note that in a full ARMA model, e_i itself is dependent on past values of noise.

There is a huge literature on ARMA processes, which are particularly important in the analysis of long-term trends such as in economics: it is likely that an underlying factor

Table 3.2 Stationary and moving average noise.

Time	Stationary	Moving average
1	−0.12775	
2	0.14249	0.01474
3	−0.06001	−0.04527
4	0.05075	0.00548
5	0.06168	0.06716
6	−0.14433	−0.07717
7	−0.10591	−0.18308
8	0.06473	−0.11835
9	0.05499	−0.06336
10	−0.00058	−0.06394
11	0.04383	−0.02011
12	−0.08401	−0.10412
13	0.21477	0.11065
14	−0.01069	0.09996
15	−0.08397	0.01599
16	−0.14516	−0.12917
17	0.11493	−0.01424
18	0.00830	−0.00595
19	0.13089	0.12495
20	0.03747	0.16241

causing errors in estimates changes with time rather than fluctuating completely randomly. A battery of specialised techniques to cope with such situations have been developed. The chemist must be aware of these noise models, especially when studying natural phenomena such as in environmental chemistry, but also to a lesser extent in instrumental analysis. However, there is rarely sufficient experimental evidence to establish highly sophisticated noise models. It is well advised, however, when studying a process, to determine whether a stationary noise distribution is adequate, especially if the results of simulations are to be relied upon, so an appreciation of basic methods for modelling noise is important. Very elaborate models are unlikely to be easy to verify experimentally.

3.2.3.3 Signal to Noise Ratio

The signal to noise ratio is a useful parameter to measure. The higher this number, the more intense the signal is relative to the background. This measurement is essentially empirical, and involves dividing the height of a relevant signal (normally the most intense if there are several in a dataset) by the root mean square of the noise, measured in a region of the data where there is known to be no signal.

3.2.4 Sequential Processes

Not all chemometric data arise from spectroscopy or chromatography, some are from studying processes evolving over time, ranging from a few hours (e.g. a manufacturing process) to thousands of years (e.g. a geological process). Many techniques for studying such processes are common to those developed in analytical instrumentation.

In some cases cyclic events occur, dependent, for example, on time of day, season of the year or temperature fluctuations. These can be modelled using sine functions, and are the basis of time series analysis (Section 3.4). In addition, cyclicity is also observed in Fourier spectroscopy, and Fourier transform techniques (Section 3.5) may on occasions be combined with methods for time series analysis.

3.3 Linear Filters

3.3.1 Smoothing Functions

A key need is to obtain as informative a signal as possible after removing the noise from a dataset. When data are sequentially obtained, such as in time or frequency, the underlying signals often arise from a sum of smooth, monotonic functions, such as are described in Section 3.2.1, whereas the underlying noise is a largely uncorrelated function. An important method for revealing the signals involves smoothing the data; the principle is that the noise will be smoothed away using mild methods, whilst the signal will remain. This approach depends on the peaks having a half-width of several datapoints: if digital resolution is very poor signals will appear as spikes and may be confused with noise.

It is important to determine the optimum filter for any particular application. Too much smoothing and the signal itself is reduced in intensity and resolution. Too little smoothing, and noise remains. The optimum smoothing function depends on peak-widths (in datapoints) as well as noise characteristics.

3.3.1.1 Moving Averages

Conceptually, the simplest methods are linear filters whereby the resultant smoothed data are a linear function of the raw data. Normally this involves using the surrounding datapoints, for example, using a function of the three points in Figure 3.7 to recalculate a value for point i. Algebraically, such functions are expressed by

$$x_{i,new} = \sum_{j=-p}^{p} c_j x_{i+j}$$

One of the simplest is a three point moving average (MA). Each point is replaced by the average of itself and the points before and after, so in the equation above $p = 1$ and $c_j = 1/3$ for all three points.

The filter can be extended to a five ($p = 2$, $c = 1/5$), seven, etc., point MA:

- the more the points in the filter, the greater is the reduction in noise, but the higher is the chance of blurring the signal;
- the number of points in the filter is often called a 'window'.

The filter is moved along the time series or spectrum, each datapoint being replaced successively by the corresponding filtered datapoint. The optimal filter depends on the noise distribution and signal width. It is best to experiment with a number of different filter widths.

3.3.1.2 Savitsky–Golay Filters, Hanning and Hamming Windows

MA filters have the disadvantage in that they use a linear approximation for the data. However, peaks are often best approximated by curves, e.g. a polynomial. This is particularly true at the centre of a peak, where a linear model will always underestimate the intensity. Quadratic, cubic or even quartic models provide better approximations. The principle of moving averages can be extended. A seven point cubic filter, for example, is used to fit

$$\hat{x}_i = b_0 + b_1 i + b_2 i^2 + b_3 i^3$$

using a seven point window, replacing the centre point by its best fit estimate. The window is moved along the data, point by point, the calculation being repeated each time.

However, regression is computationally intense and it would be time consuming to perform this calculation in full simply to improve the appearance of a spectrum or chromatogram, which may consist of hundreds of datapoints. The user wants to be

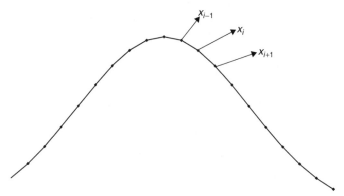

Figure 3.7
Selection of points to use in a three point moving average filter

able to select a menu item or icon on a screen and almost instantaneously visualise an improved picture. Savitsky and Golay in 1964 presented an alternative, and simplified, method of determining the new value of x_i simply by re-expressing the calculation as a sum of coefficients. These Savitsky–Golay filters are normally represented in tabular form (see Table 3.3). Both quadratic and cubic models result in identical coefficients, as do quartic and quintic models. To determine a coefficient c_j,

1. decide on the order of the model (quadratic and cubic models give identical results as do quartic and quintic models);
2. decide on the window size;
3. determine c_j by selecting the appropriate number from Table 3.3 and dividing by the normalisation constant.

Several other MA methods have been proposed in the literature, two of the best known being the Hanning window (named after Julius Von Hann) (which for 3 points has weights 0.25, 0.5 and 0.25), and the Hamming window (named after R. W. Hamming) (which for 5 points has weights 0.0357, 0.2411, 0.4464, 0.2411, 0.0357) – not to be confused in name but very similar in effects. These windows can be calculated for any size, but we recommend these two filter sizes.

Note that although quadratic, cubic or higher approximations of the data are employed, the filters are still called linear because each filtered point is a linear combination of the original data.

3.3.1.3 Calculation of Linear Filters

The calculation of moving average and Savitsky–Golay filters is illustrated in Table 3.4.

• The first point of the three point moving average (see column 2) is simply given by

$$-0.049 = (0.079 - 0.060 - 0.166)/3$$

• The first point of the seven point Savitsky–Golay quadratic/cubic filtered data can be calculated as follows:
   — From Table 3.3, obtain the seven coefficients, namely $c_{-3} = c_3 = -2/21 = -0.095$, $c_{-2} = c_2 = 3/21 = 0.143$, $c_{-1} = c_1 = 6/21 = 0.286$ and $c_0 = 7/21 = 0.333$.

Table 3.3 Savitsky–Golay coefficients c_{i+j} for smoothing.

Window size j	5	7	9	7	9
		Quadratic/cubic		Quartic/quintic	
−4			−21		15
−3		−2	14	5	−55
−2	−3	3	39	−30	30
−1	12	6	54	75	135
0	17	7	59	131	179
1	12	6	54	75	135
2	−3	3	39	−30	30
3		−2	14	5	−55
4			−21		15
Normalisation constant	35	21	231	231	429

Table 3.4   Results of various filters on a dataset.

Raw data	Moving average			Quadratic/cubic Savitsky–Golay		
	3 point	5 point	7 point	5 point	7 point	9 point
0.079						
−0.060	−0.049					
−0.166	−0.113	−0.030		−0.156		
−0.113	−0.056	−0.017	0.030	−0.081	−0.069	
0.111	0.048	0.038	0.067	0.061	0.026	−0.005
0.145	0.156	0.140	0.168	0.161	0.128	0.093
0.212	0.233	0.291	0.338	0.206	0.231	0.288
0.343	0.400	0.474	0.477	0.360	0.433	0.504
0.644	0.670	0.617	0.541	0.689	0.692	0.649
1.024	0.844	0.686	0.597	0.937	0.829	0.754
0.863	0.814	0.724	0.635	0.859	0.829	0.765
0.555	0.651	0.692	0.672	0.620	0.682	0.722
0.536	0.524	0.607	0.650	0.491	0.539	0.628
0.482	0.538	0.533	0.553	0.533	0.520	0.540
0.597	0.525	0.490	0.438	0.550	0.545	0.474
0.495	0.478	0.395	0.381	0.516	0.445	0.421
0.342	0.299	0.330	0.318	0.292	0.326	0.335
0.061	0.186	0.229	0.242	0.150	0.194	0.219
0.156	0.102	0.120	0.157	0.103	0.089	0.081
0.090	0.065	0.053	0.118	0.074	0.016	0.041
−0.050	0.016	0.085	0.081	−0.023	0.051	0.046
0.007	0.059	0.070	0.080	0.047	0.055	0.070
0.220	0.103	0.063	0.071	0.136	0.083	0.072
0.081	0.120	0.091	0.063	0.126	0.122	0.102
0.058	0.076	0.096	0.054	0.065	0.114	0.097
0.089	0.060	0.031	0.051	0.077	0.033	0.054
0.033	0.005	0.011	0.015	0.006	0.007	
−0.107	−0.030	−0.007		−0.051		
−0.016	−0.052					
−0.032						

— Multiply these coefficients by the raw data and sum to obtain the smoothed value of the data:

$$x_{i,new} = -0.095 \times 0.079 + 0.143 \times -0.060 + 0.286 \times -0.166 + 0.333$$

$$\times -0.113 + 0.286 \times 0.111 + 0.143 \times 0.145 - 0.095 \times 0.212 = -0.069$$

Figure 3.8(a) is a representation of the raw data. The result of using MA filters is shown in Figure 3.8(b). A three point MA preserves the resolution (just), but a five point MA loses this and the cluster appears to be composed of only one peak. In contrast, the five and seven point quadratic/cubic Savitsky–Golay filters [Figure 3.8(c)] preserve resolution whilst reducing noise and only starts to lose resolution when using a nine point function.

3.3.1.4 Running Median Smoothing

Most conventional filters involve computing local multilinear models, but in certain areas, such as process analysis, there can be spikes (or outliers) in the data which are unlikely to be part of a continuous process. An alternative method involves using running median smoothing (RMS) functions which calculate the median rather than the mean over a window. An example of a process is given in Table 3.5. A five point MA and five point RMS smoothing function are compared. A check on the calculation of the two different filters is as follows.

- The five point MA filter at time 4 is -0.010, calculated by taking the mean values for times 2–6, i.e.

$$-0.010 = (0.010 - 0.087 - 0.028 + 0.021 + 0.035)/5$$

- The five point RMS filter at time 4 is 0.010. This is calculating by arranging the readings for times 2–6 in the order -0.087, -0.028, 0.010, 0.021, 0.035, and selecting the middle value.

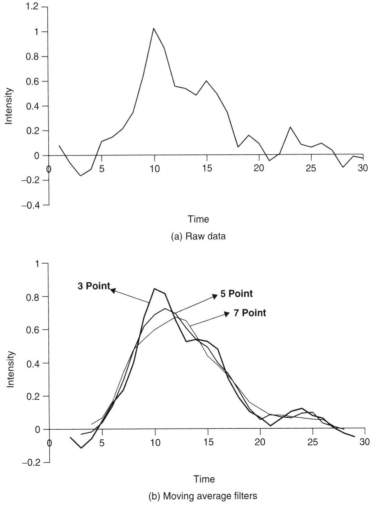

(a) Raw data

(b) Moving average filters

Figure 3.8
Filtering of data

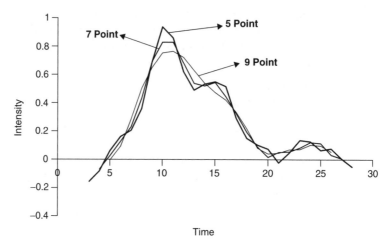

(c) Quadratic/cubic Savitsky-Golay filters

Figure 3.8
(*continued*)

Table 3.5 A sequential process: illustration of moving average and median smoothing.

Time	Data	5 point MA	5 point RMS
1	0.133		
2	0.010		
3	−0.087	0.010	0.010
4	−0.028	−0.010	0.010
5	0.021	0.048	0.021
6	0.035	0.047	0.021
7	0.298	0.073	0.035
8	−0.092	0.067	0.035
9	0.104	0.109	0.104
10	−0.008	0.094	0.104
11	0.245	0.207	0.223
12	0.223	0.225	0.223
13	0.473	0.251	0.223
14	0.193	0.246	0.223
15	0.120	0.351	0.223
16	0.223	0.275	0.193
17	0.745	0.274	0.190
18	0.092	0.330	0.223
19	0.190	0.266	0.190
20	0.398	0.167	0.190
21	−0.095	0.190	0.207
22	0.250	0.200	0.239
23	0.207	0.152	0.207
24	0.239		
25	0.160		

The results are presented in Figure 3.9. Underlying trends are not obvious from inspection of the raw data. Of course, further mathematical analysis might reveal a systematic trend, but in most situations the first inspection is graphical. In real time situations, such as process control, on-line graphical inspection is essential. The five point MA

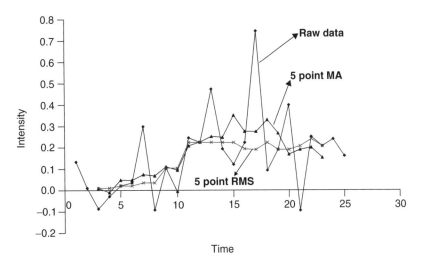

Figure 3.9
Comparison of moving average (MA) and running median smoothing (RMS)

does suggest a systematic process, but it is not at all clear whether the underlying process increases monotonically with time, or increases and then decreases. The five point RMS suggests an increasing process, and is much smoother than the result of a MA filter.

Each type of smoothing function removes different features in the data and often a combination of several approaches is recommended especially for real world problems. Dealing with outliers is an important issue: sometimes these points are due to measurement errors. Many processes take time to deviate from the expected value, and a sudden glitch in the system unlikely to be a real effect. Often a combination of filters is recommend, for example a five point median smoothing followed by a three point Hanning window. These methods are very easy to implement computationally and it is possible to view the results of different filters simultaneously.

3.3.1.5 Reroughing

Finally, brief mention will be made of the technique of reroughing. The 'rough' is given by

$$Rough = Original - Smooth$$

where the smoothed data value is obtained by one of the methods described above. The rough represents residuals but can in itself be smoothed. A new original data value is calculated by

$$Reroughed = Smooth + Smoothed\ rough$$

This is useful if there is suspected to be a number of sources of noise. One type of noise may genuinely reflect difficulties in the underlying data, the other may be due to outliers that do not reflect a long term trend. Smoothing the rough can remove one of these sources.

3.3.2 Derivatives

The methods in Section 3.3.1 are concerned primarily with removing noise. Most methods leave peakwidths either unchanged or increased, equivalent to blurring. In signal analysis an important separate need is to increase resolution. In Section 3.5.2 we will discuss the use of filters combined with Fourier transformation. In Chapter 6 we will discuss how to improve resolution when there is an extra dimension to the data (multivariate curve resolution). However, a simple and frequently used approach is to calculate derivatives. The principle is that inflection points in close peaks become turning points in the derivatives. The first and second derivatives of a pure Gaussian are presented in Figure 3.10.

- The first derivative equals zero at the centre of the peak, and is a good way of accurately pinpointing the position of a broad peak. It exhibits two turning points.
- The second derivative is a minimum at the centre of the peak, crosses zero at the positions of the turning points for the first derivative and exhibits two further turning points further apart than in the first derivative.
- The apparent peak width is reduced using derivatives.

The properties are most useful when there are several closely overlapping peaks, and higher order derivatives are commonly employed, for example in electron spin resonance and electronic absorption spectroscopy, to improve resolution. Figure 3.11 illustrates the first and second derivatives of two closely overlapping peaks. The second derivative clearly indicates two peaks and fairly accurately pinpoints their positions. The appearance of the first derivative would suggest that the peak is not pure but, in this case, probably does not provide definitive evidence. It is, of course, possible to continue and calculate the third, fourth, etc., derivatives.

There are, however, two disadvantages of derivatives. First, they are computationally intense, as a fresh calculation is required for each datapoint in a spectrum or chromatogram. Second, and most importantly, they amplify noise substantially, and, therefore, require low signal to noise ratios. These limitations can be overcome by using Savitsky–Golay coefficients similar to those described in Section 3.3.1.2, which involve rapid calculation of smoothed higher derivatives. The coefficients for a number of window sizes and approximations are presented in Table 3.6. This is a common method for the determination of derivatives and is implemented in many software packages.

3.3.3 Convolution

Common principles occur in different areas of science, often under different names, and are introduced in conceptually radically different guises. In many cases the driving force is the expectations of the audience, who may be potential users of techniques, customers on courses or even funding bodies. Sometimes even the marketplace forces different approaches: students attend courses with varying levels of background knowledge and will not necessarily opt (or pay) for courses that are based on certain requirements. This is especially important in the interface between mathematical and experimental science.

Smoothing functions can be introduced in a variety of ways, for example, as sums of coefficients or as a method for fitting local polynomials. In the signal analysis literature, primarily dominated by engineers, linear filters are often reported as a form

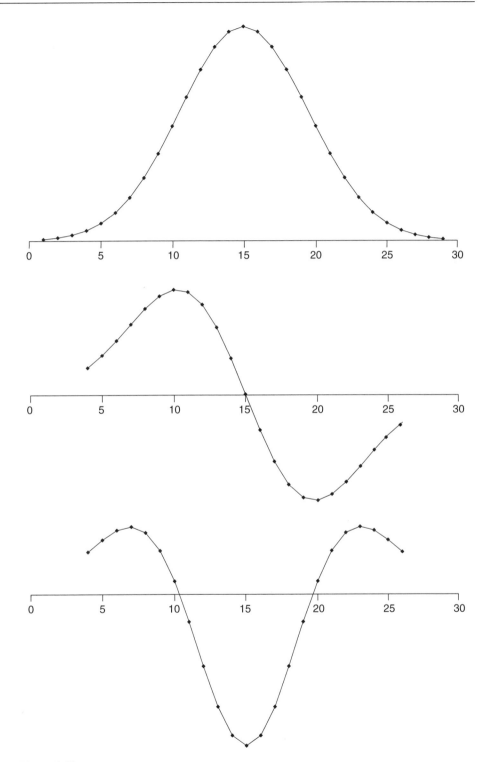

Figure 3.10
A Gaussian together with its first and second derivatives

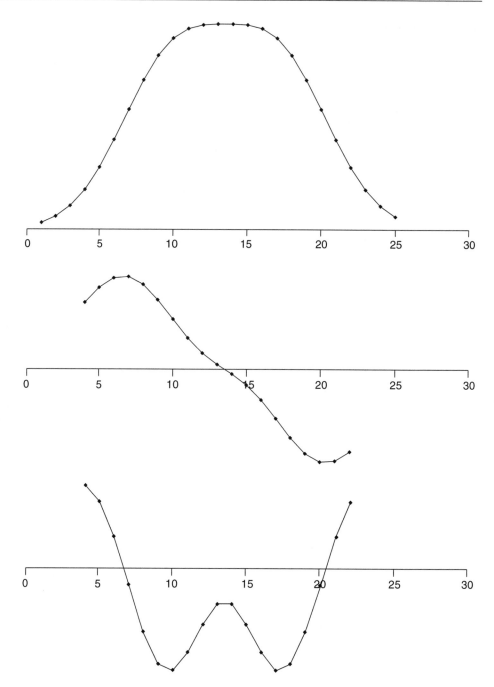

Figure 3.11
Two closely overlapping peaks together with their first and second derivatives

Table 3.6 Savitsky–Golay coefficients c_{i+j} for derivatives.

Window size j	5	7	9	5	7	9
First derivatives	Quadratic			Cubic/quartic		
−4			−4			86
−3		−3	−3		22	−142
−2	−2	−2	−2	1	−67	−193
−1	−1	−1	−1	−8	−58	−126
0	0	0	0	0	0	0
1	1	1	1	8	58	126
2	2	2	2	−1	67	193
3		3	3		−22	142
4			4			−86
Normalisation	10	28	60	12	252	1188
Second derivatives	Quadratic/cubic			Quartic/quintic		
−4			28			−4158
−3		5	7		−117	12243
−2	2	0	−8	−3	603	4983
−1	−1	−3	−17	48	−171	−6963
0	−2	−4	−20	−90	−630	−12210
1	−1	−3	−17	48	−171	−6963
2	2	0	−8	−3	603	4983
3		5	7		−117	12243
4			28			−4158
Normalisation		7	42	462	36	1188	56628

of convolution. The principles of convolution are straightforward. Two functions, f and g, are convoluted to give h if

$$h_i = \sum_{j=-p}^{j=p} f_j g_{i+j}$$

Sometimes this operation is written using a convolution operator denoted by an asterisk, so that

$$h(i) = f(i) * g(i)$$

This process of convolution is exactly equivalent to digital filtering, in the example above:

$$x_{new}(i) = x(i) * g(i)$$

where $g(i)$ is a filter function. It is, of course, possible to convolute any two functions with each other, provided that each is of the same size. It is possible to visualise these filters graphically. Figure 3.12 illustrates the convolution function for a three point MA, a Hanning window and a five point Savitsky–Golay second derivative quadratic/cubic filter. The resulted spectrum is the convolution of such functions with the raw data.

Convolution is a convenient general mathematical way of dealing with a number of methods for signal enhancement.

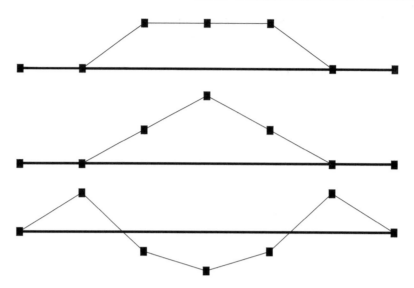

Figure 3.12
From top to bottom: a three point moving average, a Hanning window and a five point Savitsky–Golay quadratic derivative window

3.4 Correlograms and Time Series Analysis

Time series analysis has a long statistical vintage, with major early applications in economics and engineering. The aim is to study cyclical trends. In the methods in Section 3.3, we were mainly concerned with peaks arising from chromatography or spectroscopy or else processes such as occur in manufacturing. There were no underlying cyclical features. However, in certain circumstances features can reoccur at regular intervals. These could arise from a geological process, a manufacturing plant or environmental monitoring, the cyclic changes being due to season of the year, time of day or even hourly events.

The aim of time series analysis is to reveal mainly the cyclical trends in a dataset. These will be buried within noncyclical phenomena and also various sources of noise. In spectroscopy, where the noise distributions are well understood and primarily stationary, Fourier transforms are the method of choice. However, when studying natural processes, there are likely to be a much larger number of factors influencing the response, including often correlated (or ARMA) noise. Under such circumstances, time series analysis is preferable and can reveal even weak cyclicities. The disadvantage is that original intensities are lost, the resultant information being primarily about how strong the evidence is that a particular process reoccurs regularly. Most methods for time series analysis involve the calculation of a correlogram at some stage.

3.4.1 Auto-correlograms

The most basic calculation is that of an auto-correlogram. Consider the information depicted in Figure 3.13, which represents a process changing with time. It appears that there is some cyclicity but this is buried within the noise. The data are presented in Table 3.7.

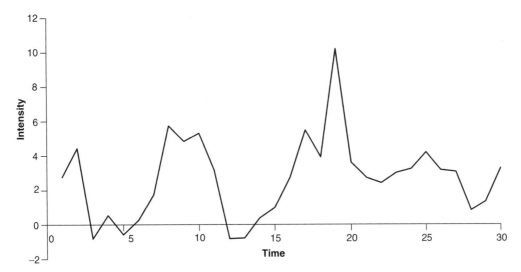

Figure 3.13
A time series

Table 3.7 Data of Figure 3.13 together with the data lagged by five points in time.

i	Data, $l = 0$	Data, $l = 5$
1	2.768	0.262
2	4.431	1.744
3	−0.811	5.740
4	0.538	4.832
5	−0.577	5.308
6	0.262	3.166
7	1.744	−0.812
8	5.740	−0.776
9	4.832	0.379
10	5.308	0.987
11	3.166	2.747
12	−0.812	5.480
13	−0.776	3.911
14	0.379	10.200
15	0.987	3.601
16	2.747	2.718
17	5.480	2.413
18	3.911	3.008
19	10.200	3.231
20	3.601	4.190
21	2.718	3.167
22	2.413	3.066
23	3.008	0.825
24	3.231	1.338
25	4.190	3.276
26	3.167	
27	3.066	
28	0.825	
29	1.338	
30	3.276	

A correlogram involves calculating the correlation coefficient between a time series and itself, shifted by a given number of datapoints called a 'lag'. If there are I datapoints in the original time series, then a correlation coefficient for a lag of l points will consist of $I - l$ datapoints. Hence, in Table 3.7, there are 30 points in the original dataset, but only 25 points in the dataset for which $l = 5$. Point number 1 in the shifted dataset corresponds to point number 6 in the original dataset. The correlation coefficient for lag l is given by

$$r_l = \frac{\sum_{i=1}^{I-l} x_i x_{i+l} - \frac{1}{I-l} \sum_{i=1}^{I-l} x_i \sum_{i=l}^{I} x_i}{\sqrt{\sum_{i=1}^{I-l} x_i^2 - \frac{1}{I-l} \sum_{i=1}^{I-l} x_i} \sqrt{\sum_{i=l}^{I} x_i^2 - \frac{1}{I-l} \sum_{i=l}^{I} x_i}}$$

Sometimes a simplified equation is employed:

$$r_l = \frac{\left(\sum_{i=1}^{I-l} x_i x_{i+p} - \frac{1}{I-l} \sum_{i=1}^{I-l} x_i \sum_{i=l}^{I} x_i \right) \Big/ (I - l)}{\left(\sum_{i=1}^{I} x_i^2 - \frac{1}{I} \sum_{i=1}^{I} x_i \right) \Big/ I}$$

The latter equation is easier for repetitive computations because the term at the bottom needs to be calculated only once, and such shortcuts were helpful prior to the computer age. However, using modern packages, it is not difficult to use the first equation, which will be employed in this text. It is important, though, always to understand and check different methods. In most cases there is little difference between the two calculations.

There are a number of properties of the correlogram:

1. for a lag of 0, the correlation coefficient is 1;
2. it is possible to have both negative and positive lags, but for an auto-correlogram, $r_l = r_{-l}$, and sometimes only one half of the correlogram is displayed;
3. the closer the correlation coefficient is to 1, the more similar are the two series; if a high correlation is observed for a large lag, this indicates cyclicity;
4. as the lag increases, the number of datapoints used to calculate the correlation coefficient decreases, and so r_l becomes less informative and more dependent on noise. Large values of l are not advisable; a good compromise is to calculate the correlogram for values of l up to $I/2$, or half the points in the original series.

The resultant correlogram is presented in Figure 3.14. The cyclic pattern is now much clearer than in the original data. Note that the graph is symmetric about the origin, as expected, and the maximum lag used in this example equals 14 points.

An auto-correlogram emphasizes only cyclical features. Sometimes there are non-cyclical trends superimposed over the time series. Such situations regularly occur in economics. Consider trying to determine the factors relating to expenditure in a seaside

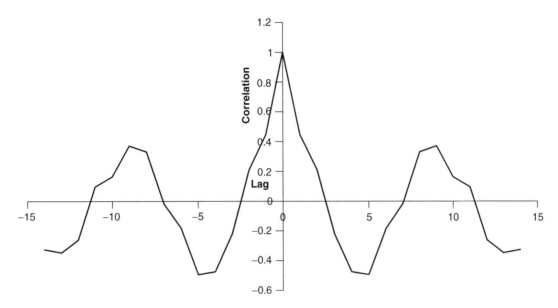

Figure 3.14
Auto-correlogram of the data in Figure 3.13

resort. A cyclical factor will undoubtedly be seasonal, there being more business in the summer. However, other factors such as interest rates and exchange rates will also come into play and the information will be mixed up in the resultant statistics. Expenditure can also be divided into food, accommodation, clothes and so on. Each will be influenced to a different extent by seasonality. Correlograms specifically emphasise the cyclical causes of expenditure. In chemistry, they are most valuable when time dependent noise interferes with stationary noise, for example in a river where there may be specific types of pollutants or changes in chemicals that occur spasmodically but, once discharged, take time to dissipate.

The correlogram can be processed further either by Fourier transformation or smoothing functions, or a combination of both; these techniques are discussed in Sections 3.3 and 3.5. Sometimes the results can be represented in the form of probabilities, for example the chance that there really is a genuine underlying cyclical trend of a given frequency. Such calculations, though, make certain definitive assumptions about the underlying noise distributions and experimental error and cannot always be generalised.

3.4.2 Cross-correlograms

It is possible to extend these principles to the comparison of two independent time series. Consider measuring the levels of Ag and Ni in a river with time. Although each may show a cyclical trend, are there trends common to both metals? The cross-correlation function between x and y can be calculated for a lag of l:

$$r_l = \frac{c_{xy,l}}{s_x s_y}$$

where $c_{xy,l}$ is the covariance between the functions at lag l, given by

$$c_{xy,l} = \sum_{i=1}^{I-l}(x_i - \bar{x})(y_{i+l} - \bar{y})/(I - l) \quad \text{for } l \geqslant 0$$

$$c_{xy,l} = \sum_{i=1}^{I-l}(x_{i+l} - \bar{x})(y_i - \bar{y})/(I - l) \quad \text{for } l < 0$$

and s corresponds to the appropriate standard deviations (see Appendix A.3.1.3 for more details about the covariance). Note that the average of x and y should strictly be recalculated according to the number of datapoints in the window but, in practice, provided that the window is not too small the overall average is acceptable.

The cross-correlogram is no longer symmetric about zero, so a negative lag does not give the same result as a positive lag. Table 3.8 is for two time series, 1 and 2. The raw time series and the corresponding cross-correlogram are presented in Figure 3.15. The raw time series appear to exhibit a long-term trend to increase, but it is not entirely obvious that there are common cyclical features. The correlogram suggests that both contain a cyclical trend of around eight datapoints, since the correlogram exhibits a strong minimum at $l = \pm 8$.

3.4.3 Multivariate Correlograms

In the real world there may be a large number of variables that change with time, for example the composition of a manufactured product. In a chemical plant the resultant material could depend on a huge number of factors such as the quality of the raw material, the performance of the apparatus and even the time of day, which could relate to who is on shift or small changes in power supplies. Instead of monitoring each factor individually, it is common to obtain an overall statistical indicator, often related to a principal component (see Chapter 4). The correlogram is computed of this mathematical summary of the raw data rather than the concentration of an individual constituent.

Table 3.8 Two time series, for which the cross-correlogram is presented in Figure 3.15.

Time	Series 1	Series 2	Time	Series 1	Series 2
1	2.768	1.061	16	3.739	2.032
2	2.583	1.876	17	4.192	2.485
3	0.116	0.824	18	1.256	0.549
4	−0.110	1.598	19	2.656	3.363
5	0.278	1.985	20	1.564	3.271
6	2.089	2.796	21	3.698	5.405
7	1.306	0.599	22	2.922	3.629
8	2.743	1.036	23	4.136	3.429
9	4.197	2.490	24	4.488	2.780
10	5.154	4.447	25	5.731	4.024
11	3.015	3.722	26	4.559	3.852
12	1.747	3.454	27	4.103	4.810
13	0.254	1.961	28	2.488	4.195
14	1.196	1.903	29	2.588	4.295
15	3.298	2.591	30	3.625	4.332

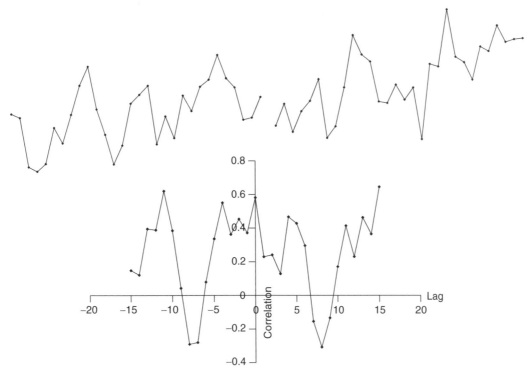

Figure 3.15
Two time series (top) and their corresponding cross-correlogram (bottom)

3.5 Fourier Transform Techniques

The mathematics of Fourier transformation (FT) has been well established for two centuries, but early computational algorithms were first applied in the 1960s, a prime method being the Cooley–Tukey algorithm. Originally employed in physics and engineering, FT techniques are now essential tools of the chemist. Modern NMR, IR and X-ray spectroscopy, among others, depend on FT methods. FTs have been extended to two-dimensional time series, plus a wide variety of modifications, for example phasing, resolution enhancement and applications to image analysis have been developed over the past two decades.

3.5.1 Fourier Transforms

3.5.1.1 General Principles

The original literature on Fourier series and transforms involved applications to continuous datasets. However, in chemical instrumentation, data are not sampled continuously but at regular intervals in time, so all data are digitised. The discrete Fourier transform (DFT) is used to process such data and will be described below. It is important to recognise that DFTs have specific properties that distinguish them from continuous FTs.

DFTs involve transformation between two types of data. In FT-NMR the raw data are acquired at regular intervals in time, often called the *time domain*, or more specifically a free induction decay (FID). FT-NMR has been developed over the years because it is much quicker to obtain data than using conventional (continuous wave) methods. An entire spectrum can be sampled in a few seconds, rather than minutes, speeding up the procedure of data acquisition by one to two orders of magnitude. This has meant that it is possible to record spectra of small quantities of compounds or of natural abundance of isotopes such as ^{13}C, now routine in modern chemical laboratories.

The trouble with this is that the time domain is not easy to interpret, and here arises the need for DFTs. Each peak in a spectrum can be described by three parameters, namely a height, width and position, as in Section 3.2.1. In addition, each peak has a shape; in NMR this is Lorentzian. A spectrum consists of a sum of peaks and is often referred to as the *frequency domain*. However, raw data, e.g. in NMR are recorded in the *time domain* and each frequency domain peak corresponds to a time series characterised by

- an initial intensity;
- an oscillation rate; and
- a decay rate.

The time domain consists of a sum of time series, each corresponding to a peak in the spectrum. Superimposed on this time series is noise. Fourier transforms convert the time series to a recognisable spectrum as indicated in Figure 3.16. Each parameter in the time domain corresponds to a parameter in the frequency domain as indicated in Table 3.9.

- The faster the rate of oscillation in the time series, the further away the peak is from the origin in the spectrum.
- The faster the rate of decay in the time series, the broader is the peak in the spectrum.
- The higher the initial intensity in the time series, the greater is the area of the transformed peak.

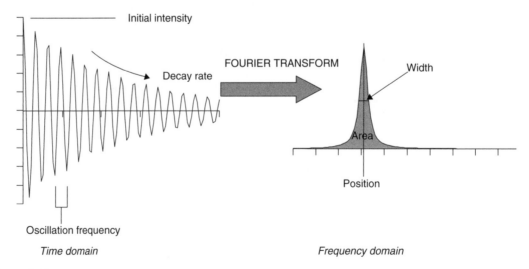

Figure 3.16
Fourier transformation from a time domain to a frequency domain

Table 3.9 Equivalence between parameters in the time domain and frequency domain.

Time domain	Frequency domain
Initial intensity	Peak area
Oscillation frequency	Peak position
Decay rate	Peak width

The peakshape in the frequency domain relates to the decay curve (or mechanism) in the time domain. The time domain equivalent of a Lorentzian peak is

$$f(t) = A \cos(\omega t) e^{-t/s}$$

where A is the initial height (corresponding to the area in the transform), ω is the oscillation frequency (corresponding to the position in the transform) and s is the decay rate (corresponding to the peak width in the transform). The key to the lineshape is the exponential decay mechanism, and it can be shown that a decaying exponential transforms into a Lorentzian. Each type of time series has an equivalent in peakshape in the frequency domain, and together these are called a *Fourier pair*. It can be shown that a Gaussian in the frequency domain corresponds to a Gaussian in the time domain, and an infinitely sharp spike in the frequency domain to a nondecaying signal in the time domain.

In real spectra, there will be several peaks, and the time series appear much more complex than in Figure 3.16, consisting of several superimposed curves, as exemplified in Figure 3.17. The beauty of Fourier transform spectroscopy is that all the peaks can

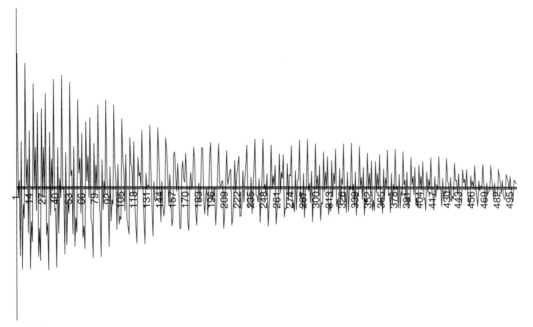

Figure 3.17
Typical time series consisting of several components

be observed simultaneously, so allowing rapid acquisition of data, but a mathematical transform is required to make the data comprehensible.

3.5.1.2 Fourier Transform Methods

The process of Fourier transformation converts the raw data (e.g. a time series) to two frequency domain spectra, one which is called a real spectrum and the other imaginary (this terminology comes from complex numbers). The true spectrum is represented only by half the transformed data as indicated in Figure 3.18. Hence if there are 1000 datapoints in the original time series, 500 will correspond to the real transform and 500 to the imaginary transform.

The mathematics of Fourier transformation is not too difficult to understand, but it is important to realise that different authors use slightly different terminology and definitions, especially with regard to constants in the transform. When reading a paper or text, consider these factors very carefully and always check that the result is realistic. We will adopt a number of definitions as follows.

The *forward* transform converts a purely *real* series into both a real and an imaginary transform, which spectrum may be defined by

$$F(\omega) = \mathrm{RL}(\omega) - i\ \mathrm{IM}(\omega)$$

where F is the Fourier transform, ω the frequency in the spectrum, i the square root of -1 and RL and IM the real and imaginary halves of the transform, respectively.

The *real* part is obtained by performing a *cosine* transform on the original data, given by (in its simplest form)

$$\mathrm{RL}(n) = \sum_{m=0}^{M-1} f(m)\cos(nm/M)$$

and the *imaginary* part by performing a *sine* transform:

$$\mathrm{IM}(n) = \sum_{m=0}^{M-1} f(m)\sin(nm/M)$$

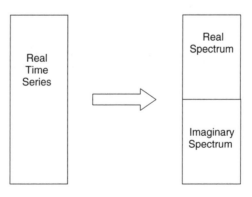

Figure 3.18
Transformation of a real time series to real and imaginary pairs

These terms need some explanation:

- there are M datapoints in the original data;
- m refers to each point in the original data;
- n is a particular point in the transform;
- the angles are in cycles per second.

If one uses radians, one must multiply the angles by 2π and if degrees divide by $360°$, but the equations above are presented in a simple way. There are a number of methods for determining the units of the transformed data, but provided that we are transforming a purely real time series to a real spectrum of half the size ($M/2$), then if the sampling interval in the time domain is δt s, the interval of each datapoint in the frequency domain is $\delta\omega = 1/(M\delta t)$ Hz (= cycles per second). To give an example, if we record 8000 datapoints in total in the time domain at intervals of 0.001 s (so the total acquisition time is 8 s), then the real spectrum will consist of 4000 datapoints at intervals of $1/(8000 \times 0.001) = 0.125$ Hz. The rationale behind these numbers will be described in Section 3.5.1.4. Some books contain equations that appear more complicated than those presented here because they transform from time to frequency units rather than from datapoints.

An *inverse* transform converts the real and imaginary pairs into a real series and is of the form

$$f(t) = RL(t) + i\ IM(t)$$

Note the + sign. Otherwise the transform is similar to the forward transform, the real part involving the multiplication of a cosine wave with the spectrum. Sometimes a factor of $1/N$, where there are N datapoints in the transformed data, is applied to the inverse transform, so that a combination of forward and inverse transforms gives the starting answer.

FTs are best understood by a simple numerical example. For simplicity we will give an example where there is a purely real spectrum and both real and imaginary time series – the opposite to normal but perfectly reasonable: in the case of Fourier self-convolution (Section 3.5.2.3) this indeed is the procedure. We will show only the real half of the transformed time series. Consider a spike as pictured in Figure 3.19. The spectrum is of zero intensity except at one point, $m = 2$. We assume there are $M(=20)$ points numbered from 0 to 19 in the spectrum.

What happens to the first 10 points of the transform? The values are given by

$$RL(n) = \sum_{m=0}^{19} f(m) \cos(nm/M)$$

Since $f(m) = 0$ except where $m = 2$, when it $f(m) = 10$, the equation simplifies still further so that

$$RL(n) = 10 \cos(2n/20)$$

The angular units of the cosine are cycles per unit time, so this angle must be multiplied by 2π to convert to radians (when employing computer packages for trigonometry, *always* check whether units are in degrees, radians or cycles; this is simple to do: the cosine of $360°$ equals the cosine of 2π radians which equals the cosine of 1 cycle

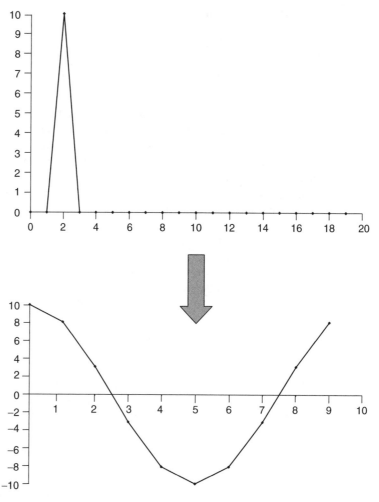

Figure 3.19
Fourier transform of a spike

and equals 1). As shown in Figure 3.19, there is one cycle every 10 datapoints, since $2 \times 10/20 = 1$, and the initial intensity equals 10 because this is the area of the spike (obtaining by summing the intensity in the spectrum over all datapoints). It should be evident that the further the spike is from the origin, the greater the number of cycles in the transform. Similar calculations can be employed to demonstrate other properties of Fourier transforms as discussed above.

3.5.1.3 Real and Imaginary Pairs

In the Fourier transform of a real time series, the peakshapes in the real and imaginary halves of the spectrum differ. Ideally, the real spectrum corresponds to an *absorption* lineshape, and the imaginary spectrum to a *dispersion* lineshape, as illustrated in Figure 3.20. The absorption lineshape is equivalent to a pure peakshape such as a Lorentzian or Gaussian, whereas the dispersion lineshape is a little like a derivative.

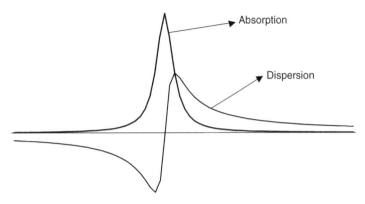

Figure 3.20
Absorption and dispersion lineshapes

However, often these two peakshapes are mixed together in the real spectrum owing to small imperfections in acquiring the data, called phase errors. The reason for this is that data acquisition does not always start exactly at the top of the cosine wave, and in practice, the term $\cos(\omega t)$ is substituted by $\cos(\omega t + \phi)$, where the angle ϕ is the phase angle. Since a phase angle in a time series of $-90°$ converts a cosine wave into a sine wave, the consequence of phase errors is to mix the sine and cosine components of the real and imaginary transforms for a perfect peakshape. As this angle changes, the shape of the real spectrum gradually distorts, as illustrated in Figure 3.21. There are various different types of phase errors. A zero-order phase error is one which is constant through a spectrum, whereas a first-order phase error varies linearly from one end of a spectrum to the other, so that $\phi = \phi_0 + \phi_1\omega$ and is dependent on ω. Higher order phase errors are possible for example when looking at images of the body or food.

There are a variety of solutions to this problem, a common one being to correct this by adding together proportions of the real and imaginary data until an absorption peakshape is achieved using an angle ψ so that

$$ABS = \cos(\psi)RL + \sin(\psi)IM$$

Ideally this angle should equal the phase angle, which is experimentally unknown. Sometimes phasing is fairly tedious experimentally, and can change across a spectrum. For complex problems such as two-dimensional Fourier transforms, phasing can be difficult.

An alternative is to take the absolute value, or magnitude, spectrum, which is defined by

$$MAG = \sqrt{RL^2 + IM^2}$$

Although easy to calculate and always positive, it is important to realise that it is not quantitative: the peak area of a two-component mixture is not equal to the sum of peak areas of each individual component, the reason being that the sum of squares of two numbers is not equal to the square of their sum. Because sometimes spectroscopic peak

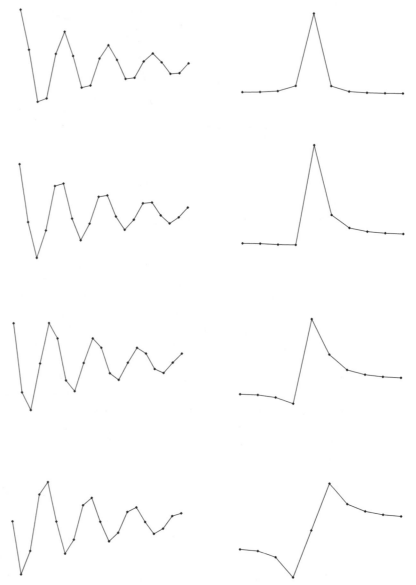

Figure 3.21
Illustration of phase errors (time series on left and real transform on the right)

areas (or heights) are used for chemometric pattern recognition studies, it is important to appreciate this limitation.

3.5.1.4 Sampling Rates and Nyquist Frequency

An important property of DFTs relates to the rate which data are sampled. Consider the time series in Figure 3.22, each cross indicating a sampling point. If it is sampled at half the rate, it will appear that there is no oscillation, as every alternative datapoint

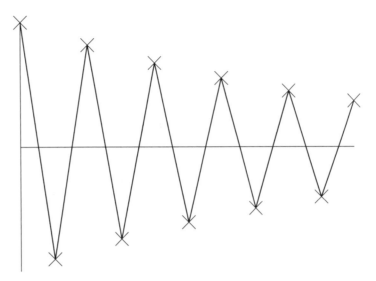

Figure 3.22
A sparsely sampled time series

will be eliminated. Therefore, there is no way of distinguishing such a series from a zero frequency series. The oscillation frequency in Figure 3.22 is called the Nyquist frequency. Anything that oscillates faster than this frequency will appear to be at a lower frequency. The rate of sampling establishes the range of observable frequencies. The higher the rate, the greater is the range of observable frequencies. In order to increase the spectral width, a higher sampling rate is required, and so more datapoints must be collected per unit time. The equation

$$M = 2ST$$

links the number of datapoints acquired (e.g. $M = 4000$), the range of observable frequencies (e.g. $S = 500$ Hz) and the acquisition time (e.g. $T = 4$ s). Higher frequencies are 'folded over' or 'aliased', and appear to be at lower frequencies, as they are indistinguishable. If $S = 500$ Hz, a peak oscillating at 600 Hz will appear at 400 Hz in the transform. Note that this relationship determines how a sampling rate in the time domain results in a digital resolution in the frequency or spectral domain (see Section 3.5.1.3). In the time domain, if samples are taken every $\delta t = T/M$ s, in the frequency domain we obtain a datapoint every $\delta\omega = 2S/M = 1/T = 1/(M\delta t)$ Hz. Note that in certain types of spectroscopy (such as quadrature detection FT-NMR) it is possible to record two time domain signals (treated mathematically as real and imaginary time series) and transform these into real and imaginary spectra. In such cases, only $M/2$ points are recorded in time, so the sampling frequency in the time domain is halved.

The Nyquist frequency is not only important in instrumental analysis. Consider sampling a geological core where depth relates to time, to determine whether the change in concentrations of a compound, or isotopic ratios, display cyclicity. A finite amount of core is needed to obtain adequate quality samples, meaning that there is a limitation in samples per unit length of core. This, in turn, limits the maximum frequency that can be observed. More intense sampling may require a more sensitive analytical technique, so for a given method there is a limitation to the range of frequencies that can be observed.

3.5.1.5 Fourier Algorithms

A final consideration relates to algorithms used for Fourier transforms. DFT methods became widespread in the 1960s partly because Cooley and Tukey developed a rapid computational method, the fast Fourier transform (FFT). This method required the number of sampling points to be a power of two, e.g. 1024, 2048, etc., and many chemists still associate powers of two with Fourier transformation. However, there is no special restriction on the number of data points in a time series, the only consideration relating to the speed of computation. The method for Fourier transformation introduced above is slow for large datasets, and early computers were much more limited in capabilities, but it is not always necessary to use rapid algorithms in modern day applications unless the amount of data is really large. There is a huge technical literature on Fourier transform algorithms, but it is important to recognise that an algorithm is simply a means to an end, and not an end in itself.

3.5.2 Fourier Filters

In Section 3.3 we discussed a number of linear filter functions that can be used to enhance the quality of spectra and chromatograms. When performing Fourier transforms, it is possible to apply filters to the raw (time domain) data prior to Fourier transformation, and this is a common method in spectroscopy to enhance resolution or signal to noise ratio, as an alternative to applying filters directly to the spectral data.

3.5.2.1 Exponential Filters

The width of a peak in a spectrum depends primarily on the decay rate in the time domain. The faster the decay, the broader is the peak. Figure 3.23 illustrates a broad peak together with its corresponding time domain. If it is desired to increase resolution, a simple approach is to change the shape of the time domain function so that the decay is slower. In some forms of spectroscopy (such as NMR), the time series contains a term due to exponential decay and can be characterised by

$$f(t) = A\cos(\omega t)e^{-t/s} = A\cos(\omega t)e^{-\lambda t}$$

as described in Section 3.5.1.1. The larger the magnitude of λ, the more rapid the decay, and hence the broader the peak. Multiplying the time series by a positive exponential of the form

$$g(t) = e^{+\kappa t}$$

changes the decay rate to give a new time series:

$$h(t) = f(t) \cdot g(t) = A\cos(\omega t)e^{-\lambda t}e^{+\kappa t}$$

The exponential decay constant is now equal to $-\lambda + \kappa$. Provided that $\kappa < \lambda$, the rate of decay is reduced and, as indicated in Figure 3.24, results in a narrower linewidth in the transform, and so improved resolution.

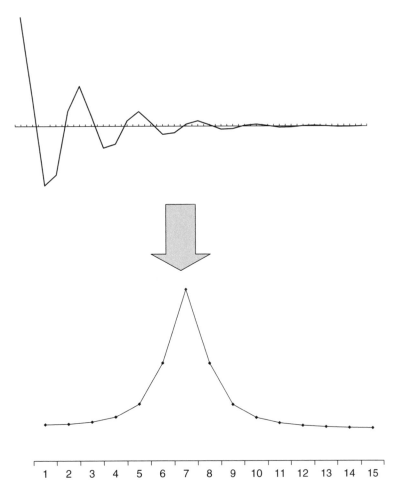

Figure 3.23
Fourier transformation of a rapidly decaying time series

3.5.2.2 *Influence of Noise*

Theoretically, it is possible to conceive of multiplying the original time series by
increasingly positive exponentials until peaks are one datapoint wide. Clearly there is
a flaw in our argument, as otherwise it would be possible to obtain indefinitely narrow
peaks and so achieve any desired resolution.

The difficulty is that real spectra always contain noise. Figure 3.25 represents a
noisy time series, together with the exponentially filtered data. The filtered time series
amplifies noise substantially, which can interfere with signals. Although the peak width
of the new transform has indeed decreased, the noise has increased. In addition to
making peaks hard to identify, noise also reduces the ability to determine integrals and
so concentrations and sometimes to accurately pinpoint peak positions.

How can this be solved? Clearly there are limits to the amount of peak sharpening
that is practicable, but the filter function can be improved so that noise reduction and
resolution enhancement are applied simultaneously. One common method is to multiply

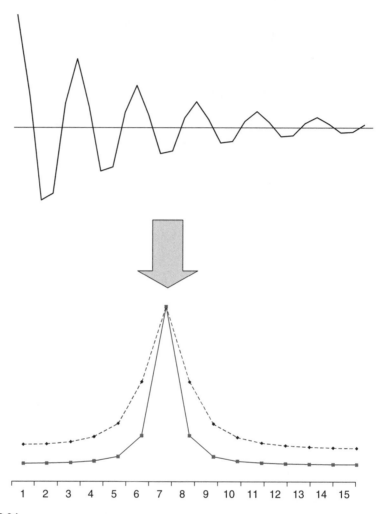

Figure 3.24
Result of multiplying the time series in Figure 3.23 by a positive exponential (original signal is dotted line)

the time series by a double exponential filter of the form

$$g(t) = e^{+\kappa t - \nu t^2}$$

where the first (linear) term of the exponential increases with time and enhances resolution, and the second (quadratic) term decreases noise. Provided that the values of κ and ν are chosen correctly, the result will be increased resolution without increased noise. The main aim is to emphasize the middle of the time series whilst reducing the end. These two terms can be optimised theoretically if peak widths and noise levels are known in advance but, in most practical cases, they are chosen empirically. The effect on the noisy data in Figure 3.25 is illustrated in Figure 3.26, for a typical double exponential filter, the dotted line representing the result of the single exponential filter.

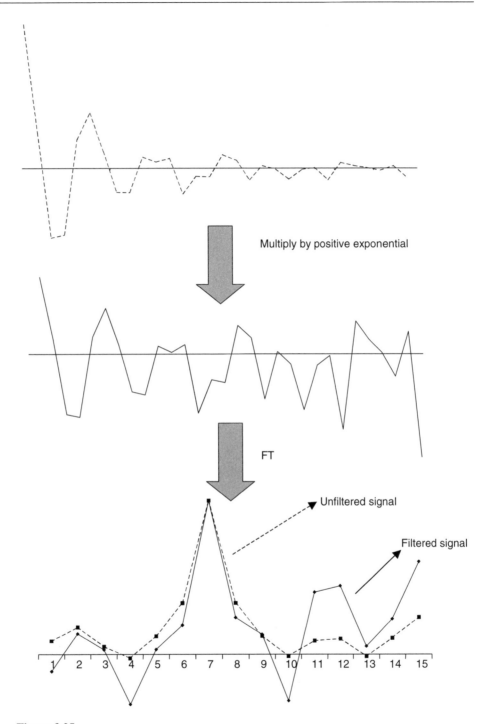

Figure 3.25
Result of multiplying a noisy time series by a positive exponential and transforming the new signal

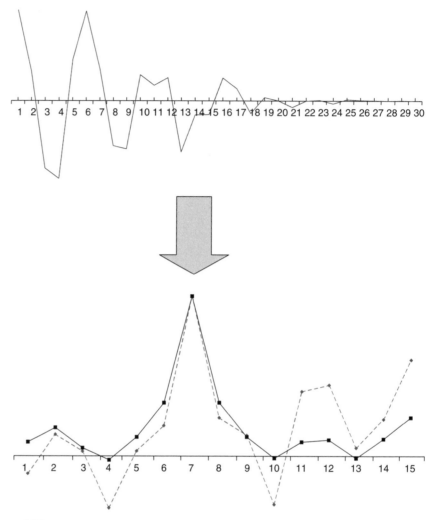

Figure 3.26
Multiplying the data in Figure 3.25 by a double exponential

The time series decays more slowly than the original, but there is not much increase in noise. The peakshape in the transform is almost as narrow as that obtained using a single exponential, but noise is dramatically reduced.

A large number of so-called matched or optimal filters have been proposed in the literature, many specific to a particular kind of data, but the general principles are to obtain increased resolution without introducing too much noise. In some cases pure noise reduction filters (e.g. negative exponentials) can be applied where noise is not a serious problem. It is important to recognise that these filters can distort peakshapes. Although there is a substantial literature on this subject, the best approach is to tackle the problem experimentally rather than rely on elaborate rules. Figure 3.27 shows the result of applying a simple double exponential function to a typical time series. Note the bell-shaped function, which is usual. The original spectrum suggests a cluster of

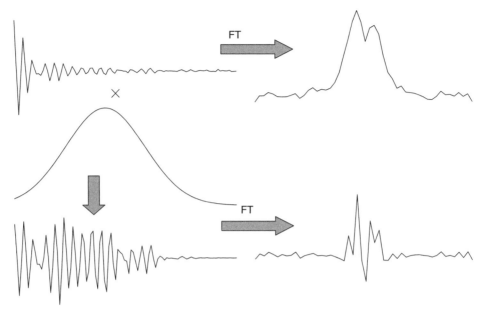

Figure 3.27
Use of a double exponential filter

peaks, but only two clear peaks are visible. Applying a filter function suggests that there are at least four underlying peaks in the spectrum, although there is some distortion of the data in the middle, probably a result of a function that is slightly too severe.

3.5.2.3 Fourier Self-deconvolution

In many forms of spectroscopy such as NMR and IR, data are acquired directly as a time series, and must be Fourier transformed to obtain an interpretable spectrum. However, any spectrum or chromatogram can be processed using methods described in this section, even if not acquired as a time series. The secret is to inverse transform (see Section 3.5.1) back to a time series.

Normally, three steps are employed, as illustrated in Figure 3.28:

1. transform the spectrum into a time series: this time series does not physically exist but can be handled by a computer;
2. then apply a Fourier filter to the time series;
3. finally, transform the spectrum back, resulting in improved quality.

This procedure is called Fourier self-deconvolution, and is an alternative to the digital filters in Section 3.3.

3.5.3 Convolution Theorem

Some people are confused by the difference between Fourier filters and linear smoothing and resolution functions. In fact, both methods are equivalent and are related

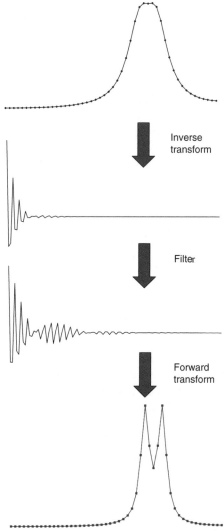

Figure 3.28
Fourier self-deconvolution of a peak cluster

by the convolution theorem, and both have similar aims, to improve the quality of spectroscopic or chromatographic or time series data.

The principles of convolution have been discussion in Section 3.3.3. Two functions, f and g, are said to be convoluted to give h if

$$h_i = \sum_{j=-p}^{j=p} f_j g_{i+j}$$

Convolution involves moving a window or digital filter function (such as a Savit-sky–Golay or moving average) along a series of data such as a spectrum, multiplying the data by that function at each successive datapoint. A three point moving average involves multiplying each set of three points in a spectrum by a function containing the values (1/3, 1/3, 1/3), and the spectrum is said to be convoluted by the moving average filter function.

Filtering a time series, using Fourier time domain filters, however, involves multiplying the *entire* time series by a single function, so that

$$H_i = F_i.G_i$$

The convolution theorem states that f, g and h are Fourier transforms of F, G and H. Hence linear filters as applied directly to spectroscopic data have their equivalence as Fourier filters in the time domain; in other words, convolution in one domain is equivalent to multiplication in the other domain. Which approach is best depends largely on computational complexity and convenience. For example, both moving averages and exponential Fourier filters are easy to apply, and so are simple approaches, one applied direct to the frequency spectrum and the other to the raw time series. Convoluting a spectrum with the Fourier transform of an exponential decay is a difficult procedure and so the choice of domain is made according to how easy the calculations are.

3.6 Topical Methods

There are a number of more sophisticated methods that have been developed over the past two decades. In certain instances a more specialised approach is appropriate and also generates much interest in the literature. There are particular situations, for example, where data are very noisy or incomplete, or where rapid calculations are required, which require particular solutions. The three methods listed below are topical and implemented within a number of common software packages. They do not represent a comprehensive review but are added for completion, as they are regularly reported in the chemometrics literature and are often available in common software packages.

3.6.1 Kalman Filters

The Kalman filter has its origin in the need for rapid on-line curve fitting. In some situations, such as chemical kinetics, it is desirable to calculate a model whilst the reaction is taking place rather than wait until the end. In on-line applications such as process control, it may be useful to see a smoothed curve as the process is taking place, in real time, rather than later. The general philosophy is that, as something evolves with time, more information becomes available so the model can be refined. As each successive sample is recorded, the model improves. It is possible to predict the response from information provided at previous sample times and see how this differs from the observed response, so changing the model.

Kalman filters are fairly complex to implement computationally, but the principles are as follows, and will be illustrated by the case where a single response (y) depends on a single factor (x). There are three main steps:

1. Model the current datapoint (i), for example, calculate $\hat{y}_{i|i-1} = x_i \cdot b_{i-1}$ using a polynomial in x, and methods introduced in Chapter 2. The parameters b_{i-1} are initially guesses, which are refined with time. The | symbol means that the model

of y_i is based on the first $i - 1$ datapoints, x_i is a row vector consisting of the terms in a model (usually, but not exclusively, polynomial) and b is a column vector. For example, if $x_i = 2$, then a three parameter quadratic model of the form $y_i = b_0 + b_1 x + b_2 x_2^2$ gives $x_i = (1, 2, 4)$.

2. This next step is to see how well this model predicts the current datapoint and calculate $d_i = y_i - \hat{y}_{i|i-1}$, which is called the *innovation*. The closer these values are, the better is the model.

3. Finally, refine the model by recalculating the coefficients

$$b_i = b_{i-1} + k_i d_i$$

If the estimated and observed values of y are identical, the value of b will be unchanged. If the observed value is more than the estimated value, it makes sense to increase the size of the coefficients to compensate. The column vector k_i is called the *gain* vector. There are a number of ways of calculating this, but the larger it is the greater is the uncertainty in the data.

A common (but complicated way) of calculating the gain vector is as follows.

1. Start with a matrix V_{i-1} which represents the variance of the coefficients. This is a square matrix, with the number or rows and columns equal to the number of coefficients in the model. Hence if there are five coefficients, there will be 25 elements in the matrix. The higher these numbers are, the less certain is the prediction of the coefficients. Start with a diagonal matrix containing some high numbers.

2. Guess a number r that represents the approximate error at each point. This could be the root mean square replicate error. This number is not too crucial, and it can be set as a constant throughout the calculation.

3. The vector k_i is given by

$$k_i = \frac{V_{i-1}.x'_i}{x_i.V_{i-1}.x'_i - r} = \frac{V_{i-1}.x'_i}{q - r}$$

4. The new matrix V_i is given by

$$V_i = V_{i-1} - k_i.x_i.V_{i-1}$$

The magnitude of the elements of this matrix should reduce with time, as the measurements become more certain, meaning a consequential reduction in k and so the coefficient b converging (see step 3 of the main algorithm).

Whereas it is not always necessary to understand the computational details, it is important to appreciate the application of the method. Table 3.10 represents the progress of such a calculation.

- A model of the form $y_i = b_0 + b_1 x + b_2 x^2$ is to be set up, there being three coefficients.
- The initial guess of the three coefficients is 0.000. Therefore, the guess of the response when $x = 0$ is 0, and the innovation is $0.840 - 0.000$ (or the observed minus the predicted using the initial model).

Table 3.10 Kalman filter calculation.

x_i	y_i	b_0	b_1	b_2	$\hat{x}_i$		k	
0.000	0.840	0.841	0.000	0.000	0.000	1.001	0.000	0.000
1.000	0.737	0.841	−0.052	−0.052	0.841	−0.001	0.501	0.501
2.000	0.498	0.841	−0.036	−0.068	0.530	0.001	−0.505	0.502
3.000	0.296	0.849	−0.114	−0.025	0.124	0.051	−0.451	0.250
4.000	0.393	0.883	−0.259	0.031	0.003	0.086	−0.372	0.143
5.000	0.620	0.910	−0.334	0.053	0.371	0.107	−0.304	0.089
6.000	0.260	0.842	−0.192	0.020	0.829	0.119	−0.250	0.060
7.000	0.910	0.898	−0.286	0.038	0.458	0.125	−0.208	0.042
8.000	0.124	0.778	−0.120	0.010	1.068	0.127	−0.176	0.030
9.000	0.795	0.817	−0.166	0.017	0.490	0.127	−0.150	0.023
10.000	0.436	0.767	−0.115	0.010	0.831	0.126	−0.129	0.017
11.000	0.246	0.712	−0.064	0.004	0.693	0.124	−0.113	0.014
12.000	0.058	0.662	−0.024	−0.001	0.469	0.121	−0.099	0.011
13.000	−0.412	0.589	0.031	−0.006	0.211	0.118	−0.088	0.009
14.000	0.067	0.623	0.007	−0.004	−0.236	0.115	−0.078	0.007
15.000	−0.580	0.582	0.033	−0.006	−0.210	0.112	−0.070	0.006
16.000	−0.324	0.605	0.020	−0.005	−0.541	0.108	−0.063	0.005
17.000	−0.896	0.575	0.036	−0.007	−0.606	0.105	−0.057	0.004
18.000	−1.549	0.510	0.069	−0.009	−0.919	0.102	−0.052	0.004
19.000	−1.353	0.518	0.065	−0.009	−1.426	0.099	−0.047	0.003
20.000	−1.642	0.521	0.064	−0.009	−1.675	0.097	−0.043	0.003
21.000	−2.190	0.499	0.073	−0.009	−1.954	0.094	−0.040	0.002
22.000	−2.206	0.513	0.068	−0.009	−2.359	0.091	−0.037	0.002

- Start with a matrix

$$V_i = \begin{bmatrix} 100 & 0 & 0 \\ 0 & 100 & 0 \\ 0 & 0 & 100 \end{bmatrix}$$

the diagonal numbers representing high uncertainty in measurements of the parameters, given the experimental numbers.
- Use a value of r of 0.1. Again this is a guess, but given the scatter of the experimental points, it looks as if this is a reasonable number. In fact, values 10-fold greater or smaller do not have a major impact on the resultant model, although they do influence the first few estimates.

As more samples are obtained it can be seen that

- the size of k decreases;
- the values of the coefficients converge;
- there is a better fit to the experimental data.

Figure 3.29 shows the progress of the filter. The earlier points are very noisy and deviate considerably from the experimental data, whereas the later points represent a fairly smooth curve. In Figure 3.30, the progress of the three coefficients is presented, the graphs being normalised to a common scale for clarity. Convergence takes about 20 iterations. A final answer of $y_i = 0.513 + 0.068x − 0.009x^2$ is obtained in this case.

It is important to recognise that Kalman filters are computationally elaborate and are not really suitable unless there is a special reason for performing on-line calculations.

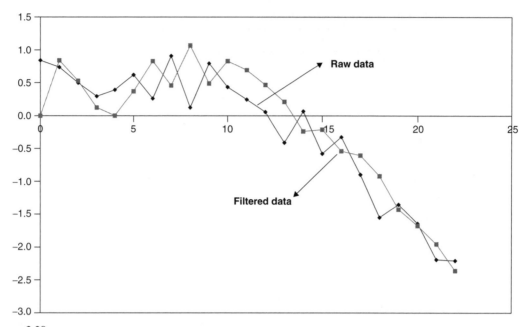

Figure 3.29
Progress of the Kalman filter, showing the filtered and raw data

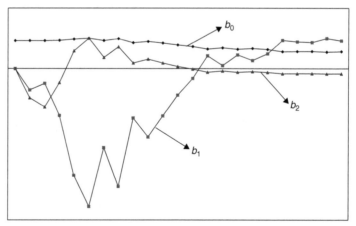

Figure 3.30
Change in the three coefficients predicted by the Kalman filter with time

It is possible to take the entire X and y data of Table 3.10 and perform multiple linear regression as discussed in Chapter 2, so that

$$y = X.b$$

or

$$b = (X'.X)^{-1}.X'.y$$

using the standard equation for the pseudo-inverse, giving an equation, $y_i = 0.512 + 0.068x - 0.009x^2$, only very slightly different to the prediction by Kalman filters. If all the data are available there is little point in using Kalman filters in this case; the method is mainly useful for on-line predictions. On the whole, with modern computers, speed is no longer a very important consideration for curve fitting if all the data are available in one block, so some of the major interest in this area is historical; nevertheless, for real time graphical applications, Kalman filters are still useful, especially if one wants to look at a process evolving.

Kalman filters can be extended to more complex situations with many variables and many responses. The model does not need to be multilinear but, for example, may be exponential (e.g. in kinetics). Although the equations increase considerably in complexity, the basic ideas are the same.

3.6.2 Wavelet Transforms

Another topical method in chemical signal processing is the wavelet transform. The general principles are discussed below, without providing detailed information about algorithms, and provide a general understanding of the approach; wavelets are implemented in several chemometric packages, so many people have come across them.

Wavelet transforms are normally applied to datasets whose size is a power of two, for example consisting of 512 or 1024 datapoints. If a spectrum or chromatogram is longer, it is conventional simply to clip the data to a conveniently sized window.

A wavelet is a general function, usually, but by no means exclusively, of time, $g(t)$, which can be modified by *translation* (b) or dilation (expansion/contraction) (a). The function should add up to 0, and can be symmetric around its mid-point. A very simple example the first half of which has the value $+1$ and the second half -1. Consider a small spectrum eight datapoints in width. A very simple basic wavelet function consists of four $+1$s followed by four -1s. This covers the entire spectrum and is said to be a wavelet of level 0. It is completely expanded and there is no room to translate this function as it covers the entire spectrum. The function can be halved in size ($a = 2$), to give a wavelet of level 1. This can now be translated (changing b), so there are two possible wavelets of level 1. The wavelets may be denoted by $\{n, m\}$ where n is the level and m the translation.

Seven wavelets for an eight point series are presented in Table 3.11. The smallest is a two point wavelet. It can be seen that for a series consisting of 2^N points,

- there will be N levels numbered from 0 to $N - 1$;
- there will be 2^n wavelets at level n;
- there will be $2^N - 1$ wavelets in total if all levels are employed.

Table 3.11 Wavelets.

Level 0	1	1	1	1	-1	-1	-1	-1	$\{0, 1\}$
Level 1	1	1	-1	-1					$\{1, 1\}$
					1	1	-1	-1	$\{1, 2\}$
Level 2	1	-1							$\{2, 1\}$
			1	-1					$\{2, 2\}$
					1	-1			$\{2, 3\}$
							1	-1	$\{2, 4\}$

The key to the usefulness of wavelet transforms is that it is possible to express the data in terms of a sum of wavelets. For a spectrum 512 datapoints long, there will be 511, plus associated scaling factors. This transform is sometimes expressed by

$$h = W . f$$

where

- f represents the raw data (e.g. a spectrum of 512 points in time);
- W is a square matrix (with dimensions 512×512 in our example);
- h are the coefficients of the wavelets, the calculation determining the best fit coefficients.

It is beyond the scope of this text to provide details as to how to obtain W, but many excellent papers exist on this topic.

Of course, a function such as that in Table 3.11 is not always ideal or particularly realistic in many cases, so much interest attaches to determining optimum wavelet functions, there being many proposed and often exotically named wavelets.

There are two principal uses for wavelets.

1. The first involves smoothing. If the original data consist of 512 datapoints, and are exactly fitted by 511 wavelets, choose the most significant wavelets (those with the highest coefficients), e.g. the top 50. In fact, if the nature of the wavelet function is selected with care only a small number of such wavelets may be necessary to model a spectrum which, in itself, consists of only a small number of peaks. Replace the spectrum simply with that obtained using the most significant wavelets.
2. The second involves data compression. Instead of storing all the raw data, store simply the coefficients of the most significant wavelets. This is equivalent to saying that if a spectrum is recorded over 1024 datapoints but consists of only five overlapping Gaussians, it is more economical (and, in fact, useful to the chemist) to store the parameters for the Gaussians rather than the raw data. In certain areas such as LC–MS there is a huge redundancy of data, most mass numbers having no significance and many data matrices being extremely sparse. Hence it is useful to reduce the amount of information.

Wavelets are a computationally sophisticated method for achieving these two facilities and are an area of active research within the data analytical community.

3.6.3 Maximum Entropy (Maxent) and Bayesian Methods

Over the past two decades there has been substantial scientific interest in the application of maximum entropy techniques with notable successes, for the chemist, in areas such as NMR spectroscopy and crystallography. Maxent has had a long statistical vintage, one of the modern pioneers being Jaynes, but the first significant scientific applications were in the area of deblurring of infrared images of the sky, involving the development of the first modern computational algorithm, in the early 1980s. Since then, there has been an explosion of interest and several implementations are available within commercial instrumentation. The most spectacular successes have been in the area of image analysis, for example NMR tomography, as well as forensic applications such as obtaining clear car number plates from hazy police photographs. In addition, there has been a very solid and large literature in the area of analytical chemistry.

3.6.3.1 Bayes' Theorem

The fundamental application of maximum entropy techniques requires an understanding of Bayes' theorem. Various definitions are necessary:

- *Data* are experimental observations, for example the measurement of a time series of free induction decay prior to Fourier transformation. *Data space* contains a dataset for each experiment.
- A *map* is the desired result, for example a clean and noise free spectrum, or the concentration of several compounds in a mixture. *Map space* exists in a similar fashion to data space.
- An operation or *transformation* links these two spaces, such as Fourier transformation or factor analysis.

The aim of the experimenter is to obtain as good an estimate of map space as possible, consistent with his or her knowledge of the system. Normally there are two types of knowledge:

1. *Prior* knowledge is available before the experiment. There is almost always some information available about chemical data. An example is that a true spectrum will always be positive: we can reject statistical solutions that result in negative intensities. Sometimes much more detailed information such as lineshapes or compound concentrations is known.
2. Experimental information, which refines the prior knowledge to give a *posterior* model of the system.

The theorem is often presented in the following form:

probability (answer given new information)

$\propto$ probability (answer given prior information)

$\times$ probability (new information given answer)

or

$$p(\text{map}|\text{experiment}) \propto p(\text{map}|\text{prior information}) \times p(\text{experiment}|\text{map})$$

where the | symbol stands for 'given by' and p is probability.

Many scientists ignore the prior information, and for cases where data are fairly good, this can be perfectly acceptable. However, chemical data analysis is most useful where the answer is not so obvious, and the data are difficult to analyse. The Bayesian method allows prior information or measurements to be taken into account. It also allows continuing experimentation, improving a model all the time.

3.6.3.2 Maximum Entropy

Maxent is one method for determining the probability of a model. A simple example is that of the toss of a six sided unbiassed die. What is the most likely underlying frequency distribution, and how can each possible distribution be measured? Figure 3.31 illustrates a flat distribution and Figure 3.32 a skew distribution (expressed

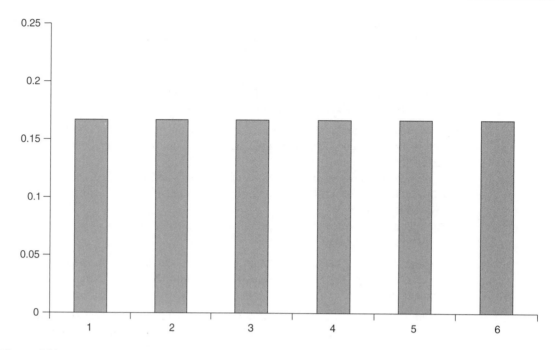

Figure 3.31
Frequency distribution for the toss of a die

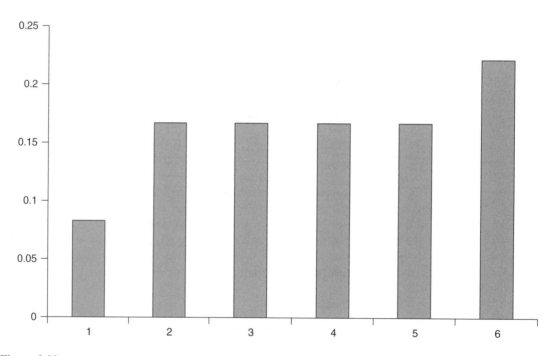

Figure 3.32
Another, but less likely, frequency distribution for toss of a die

as proportions). The concept of entropy can be introduced in a simple form is defined by

$$S = -\sum_{i=1}^{I} p_i \log(p_i)$$

where p_i is the probability of outcome i. In the case of our die, there are six outcomes, and in Figure 3.31 each outcome has a probability of 1/6. The distribution with maximum entropy is the most likely underlying distribution. Table 3.12 presents the entropy calculation for the two distributions and demonstrates that the even distribution results in the highest entropy and so is best, given the evidence available. In the absence of experimental information, a flat distribution is indeed the most likely. There is no reason why any one number on the die should be favoured above other numbers. These distributions can be likened to spectra sampled at six datapoints – if there is no other information, the spectrum with maximum entropy is a flat distribution.

However, constraints can be added. For example, it might be known that the die is actually a biassed die with a mean of 4.5 instead of 3.5, and lots of experiments suggest this. What distribution is expected now? Consider distributions A and B in Table 3.13. Which is more likely? Maximum entropy will select distribution B. It is rather unlikely (unless we know something) that the numbers 1 and 2 will never appear. Note that the value of $0\log(0)$ is 0, and that in this example, logarithms are calculated to the base 10, although using natural logarithms is equally acceptable. Of course, in this simple example we do not include any knowledge about the distribution of the faces of the

Table 3.12 Maximum entropy calculation for unbiased die.

	p		$p \log (p)$	
	Figure 3.31	Figure 3.32	Figure 3.31	Figure 3.32
1	0.167	0.083	0.130	0.090
2	0.167	0.167	0.130	0.130
3	0.167	0.167	0.130	0.130
4	0.167	0.167	0.130	0.130
5	0.167	0.167	0.130	0.130
6	0.167	0.222	0.130	0.145
Entropy			0.778	0.754

Table 3.13 Maximum entropy calculation for biased die.

	p		$p\log(p)$	
	A	B	A	B
1	0.00	0.0238	0.000	0.039
2	0.00	0.0809	0.000	0.088
3	0.25	0.1380	0.151	0.119
4	0.25	0.1951	0.151	0.138
5	0.25	0.2522	0.151	0.151
6	0.25	0.3093	0.151	0.158
Entropy			0.602	0.693

die, and it might be that we suspect that uneven weight causes this deviation. We could then include more information, perhaps that the further a face is from the weight the less likely it is to land upwards, this could help refine the distributions further.

A spectrum or chromatogram can be considered as a probability distribution. If the data are sampled at 1000 different points, then the intensity at each datapoint is a probability. For a flat spectrum, the intensity at each point in the spectrum equals 0.001, so the entropy is given by

$$S = - \sum_{i=1}^{1000} 0.001 \log(0.001) = -1000 \times 0.001 \times (-3) = 3$$

This, in fact, is the maximum entropy solution but does not yet take account of experimental data, but is the most likely distribution in the absence of more information.

It is important to realise that there are a number of other definitions of entropy in the literature, only the most common being described in this chapter.

3.6.3.3 Modelling

In practice, there are an infinite, or at least very large, number of statistically identical models that can be obtained from a system. If I know that a chromatographic peak consists of two components, I can come up with any number of ways of fitting the chromatogram all with identical least squares fits to the data. In the absence of further information, a smoother solution is preferable and most definitions of entropy will pick such an answer.

Although, in the absence of any information at all, a flat spectrum or chromatogram is the best answer, experimentation will change the solutions considerably, and should pick two underlying peaks that fit the data well, consistent with maximum entropy. Into the entropic model information can be built relating to knowledge of the system. Normally a parameter calculated as

entropy function − statistical fit function

High entropy is good, but not at the cost of a numerically poor fit to the data; however, a model that fits the original (and possibly very noisy) experiment well is not a good model if the entropy is too low. The statistical fit can involve a least squares function such as χ^2 which it is hoped to minimise. In practice, what we are saying is that for a number of models with identical fit to the data, the one with maximum entropy is the most likely. Maximum entropy is used to calculate a *prior* probability (see discussion on Bayes' theorem) and experimentation refines this to give a *posterior* probability. Of course, it is possible to refine the model still further by performing yet more experiments, using the posterior probabilities of the first set of experiments as prior probabilities for the next experiments. In reality, this is what many scientists do, continuing experimentation until they reach a desired level of confidence, the Bayesian method simply refining the solutions.

For relatively sophisticated applications it is necessary to implement the method as a computational algorithm, there being a number of packages available in instrumental software. One of the biggest successes has been the application to FT-NMR, the implementation being as follows.

1. Guess the solution, e.g. a spectrum using whatever knowledge is available. In NMR it is possible to start with a flat spectrum.
2. Take this guess and transform it into data space, for example Fourier transforming the spectrum to a time series.
3. Using a statistic such as the χ^2 statistic, see how well this guess compares with the experimental data.
4. Refine this guess of data space and try to reduce the statistic by a set amount. There will, of course, be a large number of possible solutions; select the solution with maximum entropy.
5. Then repeat the cycle but using the new solution until a good fit to the data is available.

It is important to realise that least squares and maximum entropy solutions often provide different best answers and move the solution in opposite directions, hence a balance is required. Maximum entropy algorithms are often regarded as a form of *nonlinear deconvolution*. For linear methods the new (improved) data set can be expressed as linear functions of the original data as discussed in Section 3.3, whereas nonlinear solutions cannot. Chemical knowledge often favours nonlinear answers: for example, we know that most underlying spectra are all positive, yet solutions involving sums of coefficients may often produce negative answers.

Problems

Problem 3.1 Savitsky–Golay and Moving Average Smoothing Functions

Section 3.3.1.2   Section 3.3.1.1

A dataset is recorded over 26 sequential points to give the following data:

0.0168	0.7801
0.0591	0.5595
−0.0009	0.6675
0.0106	0.7158
0.0425	0.5168
0.0236	0.1234
0.0807	0.1256
0.1164	0.0720
0.7459	−0.1366
0.7938	−0.1765
1.0467	0.0333
0.9737	0.0286
0.7517	−0.0582

1. Produce a graph of the raw data. Verify that there appear to be two peaks, but substantial noise. An aim is to smooth away the noise but preserving resolution.
2. Smooth the data in the following five ways: (a) five point moving average; (b) seven point moving average; (c) five point quadratic Savitsky–Golay filter; (d) seven point

quadratic Savitsky–Golay filter; and (e) nine point quadratic Savitsky–Golay filter. Present the results numerically and in the form of two graphs, the first involving superimposing (a) and (b) and the second involving superimposing (c), (d) and (e).

3. Comment on the differences between the five smoothed datasets in question 2. Which filter would you choose as the optimum?

Problem 3.2 Fourier Functions

Section 3.5

The following represent four real functions, sampled over 32 datapoints, numbered from 0 to 31:

Sample	A	B	C	D
0	0	0	0	0
1	0	0	0	0
2	0	0	0.25	0
3	0	0	0.5	0
4	0	0	0.25	0.111
5	0	0.25	0	0.222
6	1	0.5	0	0.333
7	0	0.25	0	0.222
8	0	0	0	0.111
9	0	0	0	0
10	0	0	0	0
11	0	0	0	0
12	0	0	0	0
13	0	0	0	0
14	0	0	0	0
15	0	0	0	0
16	0	0	0	0
17	0	0	0	0
18	0	0	0	0
19	0	0	0	0
20	0	0	0	0
21	0	0	0	0
22	0	0	0	0
23	0	0	0	0
24	0	0	0	0
25	0	0	0	0
26	0	0	0	0
27	0	0	0	0
28	0	0	0	0
29	0	0	0	0
30	0	0	0	0
31	0	0	0	0

1. Plot graphs of these functions and comment on the main differences.
2. Calculate the real transform over the points 0 to 15 of each of the four functions by using the following equation:

$$RL(n) = \sum_{m=0}^{M-1} f(m) \cos(nm/M)$$

where $M = 32$, n runs from 0 to 15 and m runs from 0 to 31 (if you use angles in radians you should include the factor of 2π in the equation).
3. Plot the graphs of the four real transforms.
4. How many oscillations are in the transform for A? Why is this? Comment on the reason why the graph does not decay.
5. What is the main difference between the transforms of A, B and D, and why is this so?
6. What is the difference between the transforms of B and C, and why?
7. Calculate the imaginary transform of A, replacing cosine by sine in the equation above and plot a graph of the result. Comment on the difference in appearance between the real and imaginary transforms.

Problem 3.3 Cross-correlograms

Section 3.4.2

Two time series, A and B are recorded as follows: (over 29 points in time, the first two columns represent the first 15 points in time and the last two columns the remaining 14 points)

A	B	A	B
6.851	3.721	2.149	1.563
2.382	0.024	−5.227	−4.321
2.629	5.189	−4.980	0.517
3.047	−1.022	−1.655	−3.914
−2.598	−0.975	−2.598	−0.782
−0.449	−0.194	4.253	2.939
−0.031	−4.755	7.578	−0.169
−7.578	1.733	0.031	5.730
−4.253	−1.964	0.449	−0.154
2.598	0.434	2.598	−0.434
1.655	2.505	−3.047	−0.387
4.980	−1.926	−2.629	−5.537
5.227	3.973	−2.382	0.951
−2.149	−0.588	−6.851	−2.157
0.000	0.782		

1. Plot superimposed graphs of each time series.
2. Calculate the cross-correlogram of these time series, by lagging the second time series by between −20 and 20 points relative to the first time series. To perform

this calculation for a lag of +5 points, shift the second time series so that the first point in time (3.721) is aligned with the sixth point (−0.449) of the first series, shifting all other points as appropriate, and calculate the correlation coefficient of the 25 lagged points of series B with the first 25 points of series A.

3. Plot a graph of the cross-correlogram. Are there any frequencies common to both time series?

Problem 3.4 An Introduction to Maximum Entropy

Section 3.6.3.2

The value of entropy can be defined, for a discrete distribution, by $-\sum_{i=1}^{I} p_i \log p_i$, where there are i states and p_i is the probability of each state. In this problem, use probabilities to the base 10 for comparison.

The following are three possible models of a spectrum, recorded at 20 wavelengths:

A	B	C
0.105	0.000	0.118
0.210	0.000	0.207
0.368	0.000	0.332
0.570	0.000	0.487
0.779	0.002	0.659
0.939	0.011	0.831
1.000	0.044	0.987
0.939	0.135	1.115
0.779	0.325	1.211
0.570	0.607	1.265
0.368	0.882	1.266
0.210	1.000	1.201
0.105	0.882	1.067
0.047	0.607	0.879
0.018	0.325	0.666
0.006	0.135	0.462
0.002	0.044	0.291
0.001	0.011	0.167
0.000	0.002	0.087
0.000	0.000	0.041

1. The spectral models may be regarded as a series of 20 probabilities of absorbance at each wavelength. Hence if the total absorbance over 20 wavelengths is summed to x, then the probability at each wavelength is simply the absorbance divided by x. Convert the three models into three probability vectors.
2. Plot a graph of the three models.
3. Explain why only positive values of absorbance are expected for ideal models.
4. Calculate the entropy for each of the three models.
5. The most likely model, in the absence of other information, is one with the most positive entropy. Discuss the relative entropies of the three models.

6. What other information is normally used when maximum entropy is applied to chromatography or spectroscopy?

Problem 3.5 Some Simple Smoothing Methods for Time Series

Section 3.3.1.4   Section 3.3.1.2   Section 3.3.1.5

The following represents a time series recorded at 40 points in time. (The first column represents the first 10 points in time, and so on.) The aim of this problem is to look at a few smoothing functions.

16.148	16.628	16.454	17.176
17.770	16.922	16.253	17.229
16.507	17.655	17.140	17.243
16.760	16.479	16.691	17.176
16.668	16.578	16.307	16.682
16.433	16.240	17.487	16.557
16.721	17.478	17.429	17.463
16.865	17.281	16.704	17.341
15.456	16.625	16.124	17.334
17.589	17.111	17.312	16.095

1. Plot a graph of the raw data.
2. Calculate three and five point median smoothing functions (denoted by '3' and '5') on the data (to do this, replace each point by the median of a span of N points), and plot the resultant graphs.
3. Re-smooth the three point median smoothed data by a further three point median smoothing function (denoted by '33') and then further by a Hanning window of the form $\hat{x}_i = 0.25x_{i-j} + 0.5x_i + 0.25x_{i+j}$ (denoted by '33H'), plotting both graphs as appropriate.
4. For the four smoothed datasets in points 2 and 3, calculate the 'rough' by subtracting the smooth from the original data, and plot appropriate graphs.
5. Smooth the rough obtained from the '33' dataset in point 2 by a Hanning window, and plot a graph.
6. If necessary, superimpose selected graphs computed above on top of the graph of the original data, comment on the results, and state where you think there may be problems with the process, and whether these are single discontinuities or deviations over a period of time.

Problem 3.6 Multivariate Correlograms

Section 3.4.1   Section 3.4.3   Section 4.3

The following data represent six measurements (columns) on a sample taken at 30 points in time (rows).

0.151	0.070	1.111	−3.179	−3.209	−0.830
−8.764	−0.662	−10.746	−0.920	−8.387	−10.730
8.478	−1.145	−3.412	1.517	−3.730	11.387
−10.455	−8.662	−15.665	−8.423	−8.677	−5.209
−14.561	−12.673	−24.221	−7.229	−15.360	−13.078

0.144	−8.365	−7.637	1.851	−6.773	−5.180
11.169	−7.514	4.265	1.166	−2.531	5.570
−0.169	−10.222	−6.537	0.643	−7.554	−7.441
−5.384	10.033	−3.394	−2.838	−1.695	1.787
−16.154	−1.021	−19.710	−7.255	−10.494	−8.910
−12.160	−11.327	−20.675	−13.737	−14.167	−4.166
−13.621	−7.623	−14.346	−4.428	−8.877	−15.555
4.985	14.100	3.218	14.014	1.588	−0.403
−2.004	0.032	−7.789	1.958	−8.476	−5.601
−0.631	14.561	5.529	1.573	4.462	4.209
0.120	4.931	−2.821	2.159	0.503	4.237
−6.289	−10.162	−14.459	−9.184	−9.207	−0.314
12.109	9.828	4.683	5.089	2.167	17.125
13.028	0.420	14.478	9.405	8.417	4.700
−0.927	−9.735	−0.106	−3.990	−3.830	−6.613
3.493	−3.541	−0.747	6.717	−1.275	−3.854
−4.282	−3.337	2.726	−4.215	4.459	−2.810
−16.353	0.135	−14.026	−7.458	−5.406	−9.251
−12.018	−0.437	−7.208	−5.956	−2.120	−8.024
10.809	3.737	8.370	6.779	3.963	7.699
−8.223	6.303	2.492	−5.042	−0.044	−7.220
−10.299	8.805	−8.334	−3.614	−7.137	−6.348
−17.484	6.710	0.535	−9.090	4.366	−11.242
5.400	−4.558	10.991	−7.394	9.058	11.433
−10.446	−0.690	1.412	−11.214	4.081	−2.988

1. Perform PCA (uncentred) on the data, and plot a graph of the scores of the first four PCs against time (this technique is described in Chapter 4 in more detail).
2. Do you think there is any cyclicity in the data? Why?
3. Calculate the correlogram of the first PC, using lags of 0–20 points. In order to determine the correlation coefficient of a lag of two points, calculate the correlation between points 3–30 and 1–28. The correlogram is simply a graph of the correlation coefficient against lag number.
4. From the correlogram, if there is cyclicity, determine the approximate frequency of this cyclicity, and explain.

Problem 3.7 Simple Integration Errors When Digitisation is Poor

Section 3.2.1.1   Section 3.2.2

A Gaussian peak, whose shape is given by

$$A = 2e^{-(7-x)^2}$$

where A is the intensity as a function of position (x) is recorded.

1. What is the expected integral of this peak?
2. What is the exact width of the peak at half-height?

3. The peak is recorded at every x value between 1 and 20. The integral is computed simply by adding the intensity at each of these values. What is the estimated integral?
4. The detector is slightly misaligned, so that the data are not recorded at integral values of x. What are the integrals if the detector records the data at (a) $1.2, 2.2, \ldots, 20.2$, (b) $1.5, 2.5, \ldots, 20.5$ and (c) $1.7, 2.7, \ldots, 20.7$.
5. There is a poor ADC resolution in the intensity direction: five bits represent a true reading of 2, so that a true value of 2 is represented by 11111 in the digital recorder. The true reading is always rounded down to the nearest integer. This means that possible levels are 0/31 (=binary 00000), 2/31 (=00001), 3/31, etc. Hence a true reading of 1.2 would be rounded down to 18/31 or 1.1613. Explain the principle of ADC resolution and show why this is so.
6. Calculate the estimated integrals for the case in question 3 and the three cases in question 4 using the ADC of question 5 (*hint*: if using Excel you can use the INT function).

Problem 3.8 First and Second Derivatives of UV/VIS Spectra Using the Savitsky–Golay method

Section 3.3.1.2

Three spectra have been obtained, A consisting of pure compound 1, B of a mixture of compounds 1 and 2 and C of pure compound 2. The data, together with wavelengths, scaled to a maximum intensity of 1, are as follows:

Wavelength (nm)	A	B	C
220	0.891	1.000	1.000
224	1.000	0.973	0.865
228	0.893	0.838	0.727
232	0.592	0.575	0.534
236	0.225	0.288	0.347
240	0.108	0.217	0.322
244	0.100	0.244	0.370
248	0.113	0.267	0.398
252	0.132	0.262	0.376
256	0.158	0.244	0.324
260	0.204	0.251	0.306
264	0.258	0.311	0.357
268	0.334	0.414	0.466
272	0.422	0.536	0.595
276	0.520	0.659	0.721
280	0.621	0.762	0.814
284	0.711	0.831	0.854
288	0.786	0.852	0.834
292	0.830	0.829	0.763
296	0.838	0.777	0.674
300	0.808	0.710	0.589
304	0.725	0.636	0.529

Wavelength (nm)	A	B	C
308	0.606	0.551	0.480
312	0.477	0.461	0.433
316	0.342	0.359	0.372
320	0.207	0.248	0.295
324	0.113	0.161	0.226
328	0.072	0.107	0.170
332	0.058	0.070	0.122
336	0.053	0.044	0.082
340	0.051	0.026	0.056
344	0.051	0.016	0.041
348	0.051	0.010	0.033

1. Produce and superimpose the graphs of the raw spectra. Comment.
2. Calculate the five point Savitsky–Golay quadratic first and second derivatives of A. Plot the graphs, and interpret them; compare both first and second derivatives and discuss the appearance in terms of the number and positions of the peaks.
3. Repeat this for spectrum C. Why is the pattern more complex? Interpret the graphs.
4. Calculate the five point Savitsky–Golay quadratic second derivatives of all three spectra and superimpose the resultant graphs. Repeat for the seven point derivatives. Which graph is clearer, five or seven point derivatives? Interpret the results for spectrum B. Do the derivatives show it is clearly a mixture? Comment on the appearance of the region between 270 and 310 nm, and compare with the original spectra.

Problem 3.9 Fourier Analysis of NMR Signals

Section 3.5.1.4   Section 3.5.1.2   Section 3.5.1.3

The data below consists of 72 sequential readings in time (organised in columns for clarity), which represent a raw time series (or FID) acquired over a region of an NMR spectrum. The first column represents the first 20 points in time, the second points 21 to 40, and so on.

−2732.61	−35.90	−1546.37	267.40
−14083.58	845.21	−213.23	121.18
−7571.03	−1171.34	1203.41	11.60
5041.98	−148.79	267.88	230.14
5042.45	2326.34	−521.55	−171.80
2189.62	611.59	45.08	−648.30
1318.62	−2884.74	−249.54	−258.94
−96.36	−2828.83	−1027.97	264.47
−2120.29	−598.94	−39.75	92.67
−409.82	1010.06	1068.85	199.36
3007.13	2165.89	160.62	−330.19
5042.53	1827.65	−872.29	991.12
3438.08	−786.26	−382.11	

−2854.03	−2026.73	−150.49
−9292.98	−132.10	−460.37
−6550.05	932.92	256.68
3218.65	−305.54	989.48
7492.84	−394.40	−159.55
1839.61	616.13	−1373.90
−2210.89	−306.17	−725.96

1. The data were acquired at intervals of 0.008124 s. What is the spectral width of the Fourier transform, taking into account that only half the points are represented in the transform? What is the digital resolution of the transform?

2. Plot a graph of the original data, converting the horizontal axis to seconds.

3. In a simple form, the real transform can be expressed by

$$RL(n) = \sum_{m=0}^{M-1} f(m)\cos(nm/M)$$

Define the parameters in the equation in terms of the dataset discussed in this problem. What is the equivalent equation for the imaginary transform?

4. Perform the real and imaginary transforms on this data (note you may have to write a small program to do this, but it can be laid out in a spreadsheet without a program). Notice that n and m should start at 0 rather than 1, and if angles are calculated in radians it is necessary to include a factor of 2π. Plot the real and imaginary transforms using a scale of hertz for the horizontal axis.

5. Comment on the phasing of the transform and produce a graph of the absolute value spectrum.

6. Phasing involves finding an angle ψ such that

$$ABS = \cos(\psi)RL + \sin(\psi)IM$$

A first approximation is that this angle is constant throughout a spectrum. By looking at the phase of the imaginary transform, obtained in question 4, can you produce a first guess of this angle? Produce the result of phasing using this angle and comment.

7. How might you overcome the remaining problem of phasing?

4 Pattern Recognition

4.1 Introduction

One of the first and most publicised success stories in chemometrics is pattern recognition. Much chemistry involves using data to determine patterns. For example, can infrared spectra be used to classify compounds into ketones and esters? Is there a pattern in the spectra allowing physical information to be related to chemical knowledge? There have been many spectacular successes of chemical pattern recognition. Can a spectrum be used in forensic science, for example to determine the cause of a fire? Can a chromatogram be used to decide on the origin of a wine and, if so, what main features in the chromatogram distinguish different wines? And is it possible to determine the time of year the vine was grown? Is it possible to use measurements of heavy metals to discover the source of pollution in a river?

There are several groups of methods for chemical pattern recognition.

4.1.1 Exploratory Data Analysis

Exploratory data analysis (EDA) consists mainly of the techniques of principal components analysis (PCA) and factor analysis (FA). The statistical origins are in biology and psychology. Psychometricians have for many years had the need to translate numbers such as answers to questions in tests into relationships between individuals. How can verbal ability, numeracy and the ability to think in three dimensions be predicted from a test? Can different people be grouped by these abilities? And does this grouping reflect the backgrounds of the people taking the test? Are there differences according to educational background, age, sex or even linguistic group?

In chemistry, we too need to ask similar questions, but the raw data are often chromatographic or spectroscopic. An example is animal pheromones: animals recognise each other more by smell than by sight, and different animals often lay scent trails, sometimes in their urine. The chromatogram of a urine sample may containing several hundred compounds, and it is often not obvious to the untrained observer which are the most significant. Sometimes the most potent compounds are present in only minute quantities. Yet animals can often detect through scent marking whether there is one of the opposite sex in-heat looking for a mate, or whether there is a dangerous intruder entering his or her territory. Exploratory data analysis of chromatograms of urine samples can highlight differences in chromatograms of different social groups or different sexes, and give a simple visual idea as to the main relationships between these samples. Sections 4.2 and 4.3 cover these approaches.

4.1.2 Unsupervised Pattern Recognition

A more formal method of treating samples is unsupervised pattern recognition, mainly consisting of cluster analysis. Many methods have their origins in numerical taxonomy.

Biologists measure features in different organisms, for example various body length parameters. Using a couple of dozen features, it is possible to see which species are most similar and draw a picture of these similarities, called a dendrogram, in which more closely related species are closer to each other. The main branches of the dendrogram can represent bigger divisions, such as subspecies, species, genera and families.

These principles can be directly applied to chemistry. It is possible to determine similarities in amino acid sequences in myoglobin in a variety of species. The more similar the species, the closer is the relationship: chemical similarity mirrors biological similarity. Sometimes the amount of information is so huge, for example in large genomic or crystallographic databases, that cluster analysis is the only practicable way of searching for similarities.

Unsupervised pattern recognition differs from exploratory data analysis in that the aim of the methods is to detect similarities, whereas using EDA there is no particular prejudice as to whether or how many groups will be found. Cluster analysis is described in more detail in Section 4.4.

4.1.3 Supervised Pattern Recognition

There are a large number of methods for supervised pattern recognition, mostly aimed at classification. Multivariate statisticians have developed many discriminant functions, some of direct relevance to chemists. A classical application is the detection of forgery of banknotes. Can physical measurements such as width and height of a series of banknotes be used to identify forgeries? Often one measurement is not enough, so several parameters are required before an adequate mathematical model is available.

So in chemistry, similar problems occur. Consider using a chemical method such as IR spectroscopy to determine whether a sample of brain tissue is cancerous or not. A method can be set up in which the spectra of two groups, cancerous and noncancerous tissues, are recorded. Then some form of mathematical model is set up. Finally, the diagnosis of an unknown sample can be predicted.

Supervised pattern recognition requires a training set of known groupings to be available in advance, and tries to answer a precise question as to the class of an unknown sample. It is, of course, always necessary first to establish whether chemical measurements are actually good enough to fit into the predetermined groups. However, spectroscopic or chromatographic methods for diagnosis are often much cheaper than expensive medical tests, and provide a valuable first diagnosis. In many cases chemical pattern recognition can be performed as a type of screening, with doubtful samples being subjected to more sophisticated tests. In areas such as industrial process control, where batches of compounds might be produced at hourly intervals, a simple on-line spectroscopic test together with chemical data analysis is often an essential first step to determine the possible acceptability of a batch.

Section 4.5 describes a variety of such techniques and their applications.

4.2 The Concept and Need for Principal Components Analysis

PCA is probably the most widespread multivariate chemometric technique, and because of the importance of multivariate measurements in chemistry, it is regarded by many as the technique that most significantly changed the chemist's view of data analysis.

4.2.1 History

There are numerous claims to the first use of PCA in the literature. Probably the most famous early paper was by Pearson in 1901. However, the fundamental ideas are based on approaches well known to physicists and mathematicians for much longer, namely those of eigenanalysis. In fact, some school mathematics syllabuses teach ideas about matrices which are relevant to modern chemistry. An early description of the method in physics was by Cauchy in 1829. It has been claimed that the earliest nonspecific reference to PCA in the chemical literature was in 1878, although the author of the paper almost certainly did not realise the potential, and was dealing mainly with a simple problem of linear calibration.

It is generally accepted that the revolution in the use of multivariate methods took place in psychometrics in the 1930s and 1940s, of which Hotelling's work is regarded as a classic. Psychometrics is well understood by most students of psychology and one important area involves relating answers in tests to underlying factors, for example, verbal and numerical ability as illustrated in Figure 4.1. PCA relates a data matrix consisting of these answers to a number of psychological 'factors'. In certain areas of statistics, ideas of factor analysis and PCA are intertwined, but in chemistry the two approaches have different implications: PCA involves using abstract functions of the data to look at patterns whereas FA involves obtaining information such as spectra that can be directly related to the chemistry.

Natural scientists of all disciplines, including biologists, geologists and chemists, have caught on to these approaches over the past few decades. Within the chemical community, the first major applications of PCA were reported in the 1970s, and form the foundation of many modern chemometric methods described in this chapter.

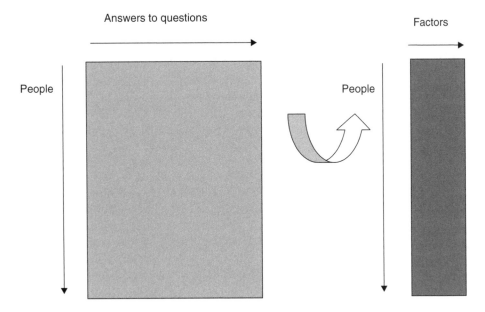

Figure 4.1
Factor analysis in psychology

4.2.2   Case Studies

In order to illustrate the main concepts of PCA, we will introduce two case studies, both from chromatography (although there are many other applications in the problems at the end of the chapter). It is not necessary to understand the detailed chemical motivations behind the chromatographic technique. The first case studies represents information sequentially related in time, and the second information where there is no such relationship but variables are on very different scales.

4.2.2.1   Case Study 1: Resolution of Overlapping Peaks

This case study involves a chromatogram obtained by high-performance liquid chromatography with diode array detection (HPLC–DAD) sampled at 30 points in time (each at 1 s intervals) and 28 wavelengths of approximately 4.8 nm intervals as presented in Table 4.1 (note that the wavelengths are rounded to the nearest nanometre for simplicity, but the original data were not collected at exact nanometre intervals). Absorbances are presented in AU (absorbance units). For readers not familiar with this application, the dataset can be considered to consist of a series of 30 spectra recorded sequentially in time, arising from a mixture of compounds each of which has its own characteristic underlying unimodal time profile (often called an 'elution profile').

The data can be represented by a 30×28 matrix, the rows corresponding to elution times and the columns wavelengths. Calling this matrix X, and each element x_{ij}, the profile chromatogram

$$X_i = \sum_{j=1}^{28} x_{ij}$$

is given in Figure 4.2, and consists of at least two co-eluting peaks.

4.2.2.2   Case Study 2: Chromatographic Column Performance

This case study is introduced in Table 4.2. The performances of eight commercial chromatographic columns are measured. In order to do this, eight compounds are tested, and the results are denoted by a letter (P, N, A, C, Q, B, D, R). Four peak characteristics are measured, namely, k' (which relates to elution time), N (relating to peak width), N(df) (another peak width parameter) and A_s (asymmetry). Each measurement is denoted by a mnemonic of two halves, the first referring to the compound and the second to the nature of the test, k being used for k' and As for asymmetry. Hence the measurement CN refers to a peak width measurement on compound C. The matrix is transposed in Table 4.2, for ease of presentation, but is traditionally represented by an 8×32 matrix, each of whose rows represents a chromatographic column and whose columns represent a measurement. Again for readers not familiar with this type of case study, the aim is to ascertain the similarities between eight objects (chromatographic columns – not be confused with columns of a matrix) as measured by 32 parameters (related to the quality of the chromatography).

One aim is to determine which columns behave in a similar fashion, and another which tests measure similar properties, so to reduce the number of tests from the original 32.

Table 4.1 Case study 1: a chromatogram recorded at 30 points in time and 28 wavelengths (nm).

	220	225	230	234	239	244	249	253	258	263	268	272	277	282	287	291	296	301	306	310	315	320	325	329	334	339	344	349
1	0.006	0.004	0.003	0.003	0.003	0.004	0.004	0.003	0.002	0.003	0.004	0.005	0.006	0.006	0.005	0.004	0.003	0.002	0.002	0.003	0.003	0.002	0.002	0.001	0.001	0.000	0.000	0.000
2	0.040	0.029	0.023	0.021	0.026	0.030	0.029	0.023	0.018	0.021	0.029	0.038	0.045	0.046	0.040	0.030	0.021	0.017	0.017	0.019	0.019	0.017	0.013	0.009	0.005	0.002	0.001	0.000
3	0.159	0.115	0.091	0.085	0.101	0.120	0.117	0.090	0.071	0.084	0.116	0.153	0.178	0.182	0.158	0.120	0.083	0.066	0.069	0.075	0.075	0.067	0.053	0.035	0.019	0.008	0.003	0.001
4	0.367	0.267	0.212	0.198	0.236	0.280	0.271	0.209	0.165	0.194	0.270	0.354	0.413	0.422	0.368	0.279	0.194	0.155	0.160	0.173	0.174	0.157	0.123	0.081	0.043	0.019	0.008	0.003
5	0.552	0.405	0.321	0.296	0.352	0.419	0.405	0.312	0.247	0.291	0.405	0.532	0.621	0.635	0.555	0.422	0.296	0.237	0.244	0.262	0.262	0.234	0.183	0.120	0.065	0.029	0.011	0.004
6	0.634	0.477	0.372	0.323	0.377	0.449	0.435	0.338	0.271	0.321	0.447	0.588	0.689	0.710	0.629	0.490	0.356	0.289	0.301	0.325	0.292	0.255	0.196	0.129	0.069	0.031	0.012	0.005
7	0.687	0.553	0.412	0.299	0.329	0.391	0.382	0.305	0.255	0.310	0.433	0.571	0.676	0.713	0.657	0.546	0.430	0.362	0.340	0.378	0.290	0.234	0.172	0.111	0.060	0.026	0.010	0.004
8	0.795	0.699	0.494	0.270	0.263	0.311	0.311	0.262	0.240	0.304	0.426	0.565	0.682	0.744	0.731	0.662	0.571	0.497	0.438	0.441	0.317	0.209	0.139	0.087	0.046	0.021	0.008	0.003
9	0.914	0.854	0.586	0.251	0.207	0.243	0.252	0.230	0.233	0.308	0.432	0.576	0.708	0.798	0.824	0.793	0.724	0.643	0.547	0.469	0.320	0.191	0.111	0.066	0.035	0.016	0.006	0.003
10	0.960	0.928	0.628	0.232	0.165	0.193	0.206	0.202	0.222	0.301	0.424	0.568	0.705	0.809	0.860	0.855	0.802	0.719	0.602	0.450	0.299	0.174	0.091	0.051	0.026	0.012	0.005	0.002
11	0.924	0.902	0.606	0.206	0.133	0.155	0.170	0.174	0.201	0.277	0.391	0.524	0.655	0.760	0.819	0.826	0.786	0.707	0.587	0.402	0.264	0.154	0.074	0.040	0.021	0.009	0.004	0.002
12	0.834	0.815	0.544	0.178	0.108	0.126	0.140	0.147	0.174	0.243	0.342	0.460	0.576	0.672	0.729	0.741	0.709	0.638	0.528	0.373	0.242	0.132	0.061	0.032	0.016	0.007	0.003	0.002
13	0.725	0.704	0.468	0.150	0.089	0.103	0.115	0.123	0.148	0.206	0.291	0.391	0.490	0.573	0.624	0.636	0.610	0.550	0.455	0.345	0.225	0.111	0.050	0.026	0.013	0.006	0.003	0.001
14	0.615	0.596	0.395	0.125	0.073	0.085	0.095	0.102	0.123	0.173	0.244	0.327	0.411	0.481	0.525	0.535	0.514	0.463	0.383	0.290	0.189	0.093	0.042	0.021	0.011	0.005	0.002	0.001
15	0.519	0.500	0.331	0.105	0.061	0.071	0.080	0.086	0.103	0.144	0.203	0.273	0.343	0.402	0.438	0.447	0.429	0.387	0.320	0.242	0.158	0.077	0.035	0.018	0.009	0.004	0.002	0.001
16	0.437	0.419	0.277	0.088	0.052	0.060	0.067	0.072	0.086	0.121	0.171	0.229	0.288	0.337	0.367	0.374	0.359	0.324	0.268	0.203	0.132	0.065	0.029	0.015	0.008	0.004	0.002	0.001
17	0.369	0.354	0.234	0.075	0.044	0.051	0.057	0.061	0.073	0.102	0.144	0.193	0.243	0.288	0.315	0.309	0.302	0.272	0.225	0.171	0.112	0.055	0.025	0.013	0.007	0.003	0.002	0.002
18	0.314	0.300	0.198	0.064	0.038	0.044	0.049	0.052	0.062	0.087	0.122	0.165	0.206	0.241	0.263	0.267	0.256	0.231	0.191	0.145	0.095	0.047	0.021	0.011	0.006	0.003	0.001	0.001
19	0.269	0.257	0.170	0.055	0.033	0.038	0.042	0.045	0.053	0.074	0.105	0.141	0.177	0.207	0.225	0.229	0.219	0.197	0.163	0.124	0.081	0.041	0.019	0.010	0.005	0.003	0.001	0.001
20	0.233	0.222	0.147	0.048	0.029	0.033	0.037	0.039	0.046	0.064	0.091	0.122	0.153	0.179	0.194	0.197	0.189	0.170	0.141	0.107	0.070	0.035	0.016	0.009	0.004	0.002	0.001	0.001
21	0.203	0.193	0.127	0.042	0.025	0.029	0.033	0.034	0.040	0.056	0.079	0.107	0.133	0.156	0.169	0.171	0.164	0.147	0.122	0.093	0.061	0.031	0.014	0.008	0.004	0.002	0.001	0.000
22	0.178	0.169	0.112	0.037	0.022	0.026	0.029	0.030	0.036	0.049	0.070	0.094	0.117	0.137	0.148	0.150	0.143	0.129	0.107	0.082	0.054	0.027	0.013	0.007	0.004	0.002	0.001	0.000
23	0.157	0.149	0.099	0.032	0.020	0.023	0.026	0.027	0.031	0.044	0.062	0.083	0.104	0.121	0.131	0.132	0.126	0.114	0.094	0.072	0.048	0.024	0.011	0.006	0.003	0.002	0.001	0.000
24	0.140	0.132	0.088	0.029	0.018	0.021	0.023	0.024	0.028	0.039	0.055	0.074	0.092	0.107	0.116	0.118	0.112	0.101	0.084	0.064	0.042	0.021	0.010	0.005	0.003	0.002	0.001	0.000
25	0.125	0.118	0.078	0.026	0.016	0.019	0.021	0.021	0.025	0.035	0.049	0.066	0.082	0.096	0.104	0.105	0.100	0.090	0.075	0.057	0.038	0.019	0.009	0.005	0.003	0.002	0.001	0.000
26	0.112	0.106	0.070	0.023	0.015	0.017	0.019	0.019	0.023	0.031	0.044	0.059	0.073	0.086	0.093	0.094	0.090	0.081	0.067	0.051	0.034	0.017	0.008	0.004	0.002	0.002	0.000	0.000
27	0.101	0.096	0.063	0.021	0.013	0.015	0.017	0.018	0.020	0.028	0.040	0.054	0.067	0.078	0.084	0.085	0.081	0.073	0.061	0.046	0.031	0.016	0.008	0.004	0.002	0.001	0.001	0.000
28	0.092	0.087	0.057	0.019	0.012	0.014	0.015	0.016	0.019	0.026	0.036	0.049	0.061	0.071	0.077	0.076	0.073	0.066	0.055	0.042	0.028	0.014	0.007	0.004	0.002	0.002	0.000	0.000
29	0.084	0.079	0.052	0.017	0.011	0.013	0.014	0.015	0.017	0.023	0.033	0.044	0.055	0.064	0.070	0.070	0.067	0.060	0.050	0.038	0.025	0.013	0.006	0.003	0.002	0.001	0.000	0.000
30	0.076	0.072	0.048	0.016	0.010	0.012	0.013	0.013	0.015	0.021	0.030	0.041	0.051	0.059	0.064	0.064	0.061	0.055	0.046	0.035	0.023	0.012	0.006	0.003	0.002	0.001	0.000	0.000

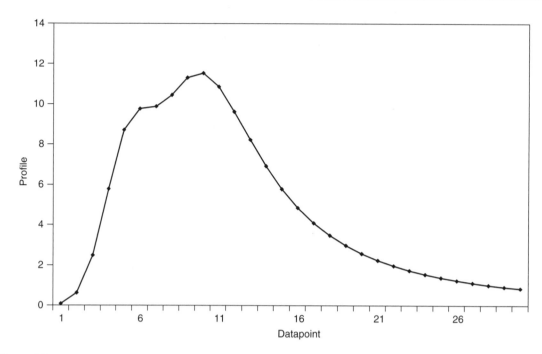

Figure 4.2
Case study 1: chromatographic peak profiles

4.2.3   Multivariate Data Matrices

A key idea is that most chemical measurements are inherently *multivariate*. This means that more than one measurement can be made on a single sample. An obvious example is spectroscopy: we can record a spectrum at hundreds of wavelength on a single sample. Many traditional chemical approaches are *univariate*, in which only one wavelength (or measurement) is used per sample, but this misses much information. Another important application is quantitative structure–property–activity relationships, in which many physical measurements are available on a number of candidate compounds (bond lengths, dipole moments, bond angles, etc.). Can we predict, *statistically*, the biological activity of a compound? Can this assist in pharmaceutical drug development? There are several pieces of information available. PCA is one of several multivariate methods that allows us to explore patterns in these data, similar to exploring patterns in psychometric data. Which compounds behave similarly? Which people belong to a similar group? How can this behaviour be predicted from available information?

As an example, consider a chromatogram in which a number of compounds are detected with different elution times, at the same time as a their spectra (such as UV or mass spectra) are recorded. Coupled chromatography, such as high-performance chromatography–diode array detection (HPLC–DAD) or liquid chromatography–mass spectrometry (LC–MS), is increasingly common in modern laboratories, and represents a rich source of multivariate data. These data can be represented as a matrix as in Figure 4.3.

What might we want to ask about the data? How many compounds are in the chromatogram would be useful information. Partially overlapping peaks and minor

Table 4.2 Case study 2: 32 performance parameters and eight chromatographic columns.

Parameter	Inertsil ODS	Inertsil ODS-2	Inertsil ODS-3	Kromasil C18	Kromasil C8	Symmetry C18	Supelco ABZ+	Purospher
Pk	0.25	0.19	0.26	0.3	0.28	0.54	0.03	0.04
PN	10 200	6 930	7 420	2 980	2 890	4 160	6 890	6 960
PN(df)	2 650	2 820	2 320	293	229	944	3 660	2 780
PAs	2.27	2.11	2.53	5.35	6.46	3.13	1.96	2.08
Nk	0.25	0.12	0.24	0.22	0.21	0.45	0	0
NN	12 000	8 370	9 460	13 900	16 800	4 170	13 800	8 260
NN(df)	6 160	4 600	4 880	5 330	6 500	490	6 020	3 450
NAs	1.73	1.82	1.91	2.12	1.78	5.61	2.03	2.05
Ak	2.6	1.69	2.82	2.76	2.57	2.38	0.67	0.29
AN	10 700	14 400	11 200	10 200	13 800	11 300	11 700	7 160
AN(df)	7 790	9 770	7 150	4 380	5 910	6 380	7 000	2 880
AAs	1.21	1.48	1.64	2.03	2.08	1.59	1.65	2.08
Ck	0.89	0.47	0.95	0.82	0.71	0.87	0.19	0.07
CN	10 200	10 100	8 500	9 540	12 600	9 690	10 700	5 300
CN(df)	7 830	7 280	6 990	6 840	8 340	6 790	7 250	3 070
CAs	1.18	1.42	1.28	1.37	1.58	1.38	1.49	1.66
Qk	12.3	5.22	10.57	8.08	8.43	6.6	1.83	2.17
QN	8 800	13 300	10 400	10 300	11 900	9 000	7 610	2 540
QN(df)	7 820	11 200	7 810	7 410	8 630	5 250	5 560	941
QAs	1.07	1.27	1.51	1.44	1.48	1.77	1.36	2.27
Bk	0.79	0.46	0.8	0.77	0.74	0.87	0.18	0
BN	15 900	12 000	10 200	11 200	14 300	10 300	11 300	4 570
BN(df)	7 370	6 550	5 930	4 560	6 000	3 690	5 320	2 060
BAs	1.54	1.79	1.74	2.06	2.03	2.13	1.97	1.67
Dk	2.64	1.72	2.73	2.75	2.27	2.54	0.55	0.35
DN	9 280	12 100	9 810	7 070	13 100	10 000	10 500	6 630
DN(df)	5 030	8 960	6 660	2 270	7 800	7 060	7 130	3 990
DAs	1.71	1.39	1.6	2.64	1.79	1.39	1.49	1.57
Rk	8.62	5.02	9.1	9.25	6.67	7.9	1.8	1.45
RN	9 660	13 900	11 600	7 710	13 500	11 000	9 680	5 140
RN(df)	8 410	10 900	7 770	3 460	9 640	8 530	6 980	3 270
RAs	1.16	1.39	1.65	2.17	1.5	1.28	1.41	1.56

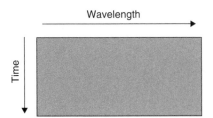

Figure 4.3
Matrix representation of coupled chromatographic data

impurities are the bug-bears of modern chromatography. What are the spectra of these compounds? Can we reliably determine these spectra which may be useful for library searching? Finally, what are the quantities of each component? Some of this information could undoubtedly be obtained by better chromatography, but there is a limit, especially with modern trends towards recording more and more data, more and more rapidly. And in many cases the identities and amounts of unknowns may not be available in advance. PCA is one tool from multivariate statistics that can help sort out these data. We will discuss the main principles in this chapter but deal with this type of application in greater depth in Chapter 6.

4.2.4 Aims of PCA

There are two principal needs in chemistry. In the case of the example of case study 1, we would like to extract information from the two way chromatogram.

- The number of significant PCs is ideally equal to the number of significant components. If there are three components in the mixture, then we expect that there are only three PCs.
- Each PC is characterised by two pieces of information, the *scores*, which, in the case of chromatography, relate to the elution profiles, and the *loadings*, which relate to the spectra.

Below we will look in more detail how this information is obtained. However, the ultimate information has a physical meaning to chemists.

Figure 4.4 represents the result of performing PCA (standardised as discussed in Section 4.3.6.4) on the data of case study 2. Whereas in case study 1 we can often relate PCs to chemical factors such as spectra of individual compounds, for the second example there is no obvious physical relationship and PCA is mainly employed to see the main trends in the data more clearly. One aim is to show which columns behave

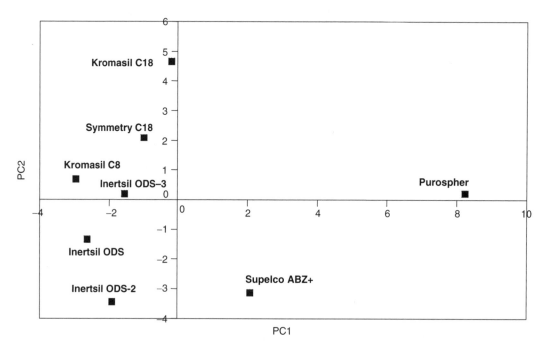

Figure 4.4
Plot of scores of PC2 versus PC1 after standardisation for case study 2

in a similar fashion. The picture suggests that the three Inertsil columns behave very similarly whereas Kromasil C18 and Supelco ABZ+ behave in a diametrically opposite manner. This could be important, for example, in the determination of which columns are best for different types of separations; if columns appear in opposite ends of the graph they are likely to fulfil different functions. The resultant picture is a principal component plot. We will discuss below how to use these plots, which can be regarded as a form of map or graphical representation of the data.

4.3 Principal Components Analysis: the Method

4.3.1 Chemical Factors

As an illustration, we will use the case of coupled chromatography, such as HPLC–DAD, as in case study 1. For a simple chromatogram, the underlying dataset can be described as a sum of responses for each significant compound in the data, which are characterised by (a) an elution profile and (b) a spectrum, plus noise or instrumental error. In matrix terms, this can be written as

$$X = C.S + E$$

where

- X is the original data matrix or coupled chromatogram;
- C is a matrix consisting of the elution profiles of each compound;
- S is a matrix consisting of the spectra of each compound;
- E is an error matrix.

This is illustrated in Figure 4.5.

Consider the matrix of case study 1, a portion of a chromatogram recorded over 30 and 28 wavelengths, consisting of two partially overlapping compounds:

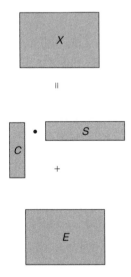

Figure 4.5
Chemical factors

- X is a matrix of 30 rows and 28 columns;
- C is a matrix of 30 rows and 2 columns, each column corresponding to the elution profile of a single compound;
- S is a matrix of 2 rows and 28 columns, each row corresponding to the spectrum of a single compound;
- E is a matrix of the same size as X.

If we observe X, can we then predict C and S? In previous chapters we have used a 'hat' notation to indicate a *prediction*, so it is also possible to write the equation above as

$$X \approx \hat{C}.\hat{S}$$

Ideally, the predicted spectra and chromatographic elution profiles are close to the true ones, but it is important to realise that we can *never directly or perfectly* observe the underlying data. There will always be measurement error, even in practical spectroscopy. Chromatographic peaks may be partially overlapping or even embedded, meaning that chemometric methods will help resolve the chromatogram into individual components.

One aim of chemometrics is to obtain these predictions after first treating the chromatogram as a multivariate data matrix, and then performing PCA. Each compound in the mixture is a 'chemical' factor with its associated spectra and elution profile, which can be related to principal components, or 'abstract' factors, by a mathematical transformation.

4.3.2 Scores and Loadings

PCA, however, results in an abstract mathematical transformation of the original data matrix, which takes the form

$$X = T.P + E$$

where

- T are called the scores, and have as many rows as the original data matrix;
- P are the loadings, and have as many columns as the original data matrix;
- the number of columns in the matrix T equals the number of rows in the matrix P.

It is possible to calculate scores and loadings matrices as large as desired, provided that the 'common' dimension is no larger than the smallest dimension of the original data matrix, and corresponds to the number of PCs that are calculated.

Hence if the original data matrix is dimensions 30×28 (or $I \times J$), no more than 28 (nonzero) PCs can be calculated. If the number of PCs is denoted by A, then this number can be no larger than 28.

- The dimensions of T will be $30 \times A$;
- the dimensions of P will be $A \times 28$.

Each scores matrix consists of a series of column vectors, and each loadings matrix a series of row vectors. Many authors denote these vectors by t_a and p_a, where a is the number of the principal component (1, 2, 3 up to A). The scores matrices T and P are composed of several such vectors, one for each principal component. The first

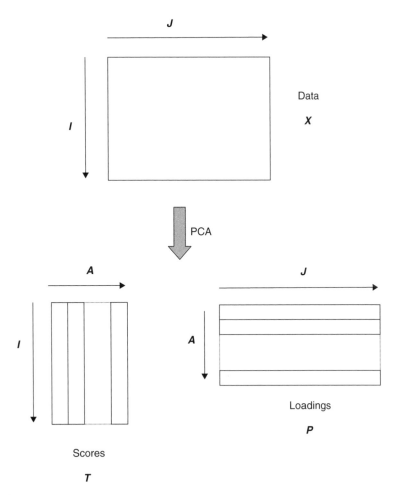

Figure 4.6
Principal components analysis

scores vector and the first loadings vector are often called the eigenvectors of the first principal component. This is illustrated in Figure 4.6. Each successive component is characterised by a pair of eigenvectors.

The first three scores and loadings vectors for the data of Table 4.1 (case study 1) are presented in Table 4.3 for the first three PCs ($A = 3$).

There are a number of important features of scores and loadings. It is important to recognise that the aim of PCA involves finding mathematical functions which contain certain properties which can then be related to chemical factors, and in themselves PCs are simply abstract mathematical entities.

- All scores and loadings vectors have the following property:
  — the sums $\Sigma_{i=1}^{I} t_{ia}.t_{ib} = 0$ and $\Sigma_{j=1}^{J} p_{aj}.p_{bj} = 0$
  where $a \neq b$, and t and p correspond to the elements of the corresponding eigenvectors. Some authors state that the scores and loadings vectors are mutually *orthogonal*, since some of the terminology of chemometrics arises from multivariate statistics,

Table 4.3 Scores and loadings for case study 1.

Scores			Loadings		
0.017	0.006	−0.001	0.348	−0.103	−0.847
0.128	0.046	0.000	0.318	−0.254	−0.214
0.507	0.182	−0.002	0.220	−0.110	−0.011
1.177	0.422	−0.001	0.101	0.186	−0.022
1.773	0.626	0.001	0.088	0.312	−0.028
2.011	0.639	0.000	0.104	0.374	−0.031
2.102	0.459	−0.004	0.106	0.345	−0.018
2.334	0.180	−0.003	0.094	0.232	0.008
2.624	−0.080	0.007	0.093	0.132	0.041
2.733	−0.244	0.018	0.121	0.123	0.048
2.602	−0.309	0.016	0.170	0.166	0.060
2.320	−0.310	0.006	0.226	0.210	0.080
1.991	−0.280	−0.004	0.276	0.210	0.114
1.676	−0.241	−0.009	0.308	0.142	0.117
1.402	−0.202	−0.012	0.314	−0.002	0.156
1.176	−0.169	−0.012	0.297	−0.166	0.212
0.991	−0.141	−0.012	0.267	−0.284	0.213
0.842	−0.118	−0.011	0.236	−0.290	0.207
0.721	−0.100	−0.009	0.203	−0.185	0.149
0.623	−0.086	−0.009	0.166	−0.052	0.107
0.542	−0.073	−0.008	0.123	0.070	0.042
0.476	−0.063	−0.007	0.078	0.155	0.000
0.420	−0.055	−0.006	0.047	0.158	−0.018
0.373	−0.049	−0.006	0.029	0.111	−0.018
0.333	−0.043	−0.005	0.015	0.061	−0.021
0.299	−0.039	−0.005	0.007	0.027	−0.013
0.271	−0.034	−0.004	0.003	0.010	−0.017
0.245	−0.031	−0.004	0.001	0.003	−0.003
0.223	−0.028	−0.004			
0.204	−0.026	−0.003			

where people like to think of PCs as vectors in multidimensional space, each variable representing an axis, so some of the geometric analogies have been incorporated into the mainstream literature. If the columns are mean-centred, then also:

— the correlation coefficient between any two scores vectors is equal to 0.

• Each loadings vector is also *normalised*. There are various different definitions of a normalised vector, but we use $\Sigma_{j=1}^{J} p_{aj}^2 = 1$. Note that there are several algorithms for PCA; using the SVD (singular value decomposition) method, the scores are also normalised, but in this text we will restrict the calculations to the NIPALS methods. It is sometimes stated that the loadings vectors are *orthonormal*.

• Some people use the square matrix $T'.T$, which has the properties that all elements are zero except along the diagonals, the value of the diagonal elements relating to the size (or importance) of each successive PC. The square matrix $P.P'$ has the special property that it is an identity matrix, with the dimensions equal to the number of PCs.

After PCA, the original variables (e.g. absorbances recorded at 28 wavelengths) are reduced to a number of significant principal components (e.g. three). PCA can be used as a form of variable reduction, reducing the large original dataset (recorded at

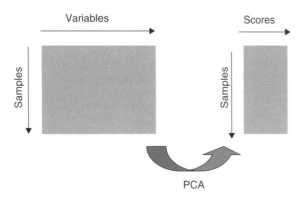

Figure 4.7
PCA as a form of variable reduction

28 wavelengths) to a much smaller more manageable dataset (e.g. consisting of three principal components) which can be interpreted more easily, as illustrated in Figure 4.7. The loadings represent the means to this end.

The original data are said to be mathematically modelled by the PCs. Using A PCs, it is possible to establish a model for each element of X of the form

$$x_{ij} = \sum_{a=1}^{A} t_{ia} p_{aj} + e_{ij} = \hat{x}_{ij} + e_{ij}$$

which is the nonmatrix version of the fundamental PCA equation at the start of this section. Hence the estimated value of x for the data of Table 4.1 at the tenth wavelength (263 nm) and eighth point in time (true value of 0.304) is given by

- $2.334 \times 0.121 = 0.282$ for a one component model and
- $2.334 \times 0.121 + 0.180 \times 0.123 = 0.304$ for a two component model,

suggesting that two PCs provide a good estimate of this datapoint.

4.3.3 Rank and Eigenvalues

A fundamental next step is to determine the number of significant factors or PCs in a matrix. In a series of mixture spectra or portion of a chromatogram, this should, ideally, correspond to the number of compounds under observation.

The *rank* of a matrix is a mathematical concept that relates to the number of significant compounds in a dataset, in chemical terms to the number of compounds in a mixture. For example, if there are six compounds in a chromatogram, the rank of the data matrix from the chromatogram should ideally equal 6. However, life is never so simple. What happens is that noise distorts this ideal picture, so even though there may be only six compounds, either it may appear that the rank is 10 or more, or else the apparent rank might even be reduced if the distinction between the profiles for certain compounds are indistinguishable from the noise. If a 15×300 X matrix (which may correspond to 15 UV/vis spectra recorded at 1 nm intervals between 201 and 500 nm) has a rank of 6, the scores matrix T has six columns and the loadings matrix P has six rows.

Note that there is sometimes a difference in definition in the literature between the exact mathematical rank (which involves finding the number of PCs that precisely model a dataset with no residual error) and the approximate (or chemical) rank which involves defining the number of significant components in a dataset; we will be concerned with this later in the chapter.

4.3.3.1 Eigenvalues

Normally after PCA, the *size* of each component can be measured. This is often called an *eigenvalue*: the earlier (and more significant) the components, the larger their size. There are a number of definitions in the literature, but a simple one defines the eigenvalue of a PC as the sum of squares of the scores, so that

$$g_a = \sum_{i=1}^{I} t_{ia}^2$$

where g_a is the ath eigenvalue.

The *sum* of all nonzero eigenvalues for a datamatrix equals the sum of squares of the entire data-matrix, so that

$$\sum_{a=1}^{K} g_a = \sum_{i=1}^{I} \sum_{j=1}^{J} x_{ij}^2$$

where K is the smaller of I or J. Note that if the data are preprocessed prior to PCA, x must likewise be preprocessed for this property to hold; if mean centring has been performed, K cannot be larger than $I - 1$, where I equals the number of samples.

Frequently eigenvalues are presented as percentages, for example of the sum of squares of the entire (preprocessed) dataset, or

$$V_a = 100 \frac{g_a}{\displaystyle\sum_{i=1}^{I} \sum_{j=1}^{J} x_{ij}^2}$$

Successive eigenvalues correspond to smaller percentages.

The *cumulative* percentage eigenvalue is often used to determine (approximately) what proportion of the data has been modelled using PCA and is given by $\Sigma_{a=1}^{A} g_a$. The closer to 100 %, the more faithful is the model. The percentage can be plotted against the number of eigenvalues in the PC model.

It is an interesting feature that the *residual sum of squares*

$$RSS_A = \sum_{i=1}^{I} \sum_{j=1}^{J} x_{ij}^2 - \sum_{a=1}^{A} g_a$$

after A eigenvalues also equals the sum of squares for the error matrix, between the PC model and the raw data, whose elements are defined by

$$e_{ij} = x_{ij} - \hat{x}_{ij} = x_{ij} - \sum_{a=1}^{A} t_{ia} p_{aj}$$

or

$$\sum_{i=1}^{I}\sum_{j=1}^{J}x_{ij}^{2} - \sum_{a=1}^{A}g_{a} = \sum_{i=1}^{I}\sum_{j=1}^{J}e_{ij}^{2}$$

This is because the product of any two different eigenvectors is 0 as discussed in Section 4.3.2.

The first three eigenvalues for the data of Table 4.1 are presented in Table 4.4. The total sum of squares of the entire dataset is 61.00, allowing the various percentages to be calculated. It can be seen that two eigenvalues represent more than 99.99 % of the data in this case. In fact the interpretation of the size of an eigenvalue depends, in part, on the nature of the preprocessing. However, since the chromatogram probably consists of only two compounds in the cluster, this conclusion is physically reasonable.

The data in Table 4.2 provide a very different story: first these need to be standardised (see Section 4.3.6.4 for a discussion about this). There are only eight objects but one degree of freedom is lost on standardising, so seven nonzero eigenvalues can be calculated as presented in Table 4.5. Note now that there could be several (perhaps four) significant PCs, and the cut-off is not so obvious as for case study 1. Another interesting feature is that the sum of all nonzero eigenvalues comes to exactly 256 or the number of elements in the X matrix. It can be shown that this is a consequence of standardising.

Using the size of eigenvalues we can try to estimate the number of significant components in the dataset.

A simple rule might be to reject PCs whose cumulative eigenvalues account for less that a certain percentage (e.g. 5 %) of the data; in the case of Table 4.5 this would suggest that only the first four components are significant. For Table 4.4, this would suggest that only one PC should be retained. However, in the latter case we would be incorrect, as the original information was not centred prior to PCA and the first component is mainly influenced by the overall size of the dataset. Centring the columns reduces the total sum of squares of the dataset from 61.00 to 24.39. The

Table 4.4 Eigenvalues for case study 1 (raw data).

g_a	V_a	Cumulative %
59.21	97.058	97.058
1.79	2.939	99.997
0.0018	0.003	100.000

Table 4.5 Eigenvalues for case study 2 (standardised).

g_a	V_a	Cumulative %
108.59	42.42	42.42
57.35	22.40	64.82
38.38	14.99	79.82
33.98	13.27	93.09
8.31	3.25	96.33
7.51	2.94	99.27
1.87	0.73	100.00

Table 4.6 Size of eigenvalues for case study 1 after column centring.

g_a	V_a	Cumulative %
22.6	92.657	92.657
1.79	7.338	99.995
0.0103	0.004	100.000

eigenvalues from the mean-centred principal components are presented in Table 4.6 and now the first eigenvalue contributes less, so suggesting that two components are required to model 95 % of the data. There is no general guidance as to whether to use centred or raw data when determining the number of significant components, the most appropriate method being dependent on the nature of the data, and one's experience.

More elaborate information can be obtained by looking at the size of the error matrix as defined above. The sum of squares of the matrix E can be expressed as the difference between the sum of squares of the matrices X and $\hat{X}$. Consider Table 4.5: after three components are calculated

- the sum of squares of $\hat{X}$ equals 204.32 (or the sum of the first three eigenvalues = 108.59 + 57.35 + 38.38). However,
- the sum of the square of the original data X equals 256 since the data have been standardised and there are 32×8 measurements. Therefore,
- the sum of squares of the error matrix E equals $256 - 204.32$ or 51.68; this number is also equal to the sum of eigenvalues 4–7.

Sometime the eigenvalues can be interpreted in physical terms. For example,

- the dataset of Table 4.1 consists of 30 spectra recorded in time at 28 wavelengths;
- the error matrix is of size 30×28, consisting of 840 elements;
- but the error sum of squares after $a = 1$ PC as been calculated equals $60.99 - 59.21 = 1.78$ AU^2;
- so the root mean square error is equal to $\sqrt{1.78/840} = 0.046$ (in fact some chemometricians adjust this for the loss of degrees of freedom due to the calculation of one PC, but because 840 is a large number this adjustment is small and we will stick to the convention in this book of dividing x errors simply by the number of elements in the data matrix).

Is this a physically sensible number? This depends on the original units of measurement and what the instrumental noise characteristics are. If it is known that the root mean square noise is about 0.05 units, then it seems sensible. If the noise level, however, is substantially lower, then not enough PCs have been calculated. In fact, most modern chromatographic instruments can determine peak intensities much more accurately than 0.05 AU, so this would suggest a second PC is required. Many statisticians do not like these approaches, but in most areas of instrumentally based measurements it is possible to measure noise levels. In psychology or economics we cannot easily consider performing experiments in the absence of signals.

The principle of examining the size of successive eigenvalues can be extended, and in spectroscopy a large number of so-called 'indicator' functions have been proposed, many by Malinowski, whose text on factor analysis is a classic. Most functions involve producing graphs of functions of eigenvalues and predicting the number of significant components using various criteria. Over the past decade, several new functions have been proposed, some based on distributions such as the F-test. For more statistical applications such as QSAR these indicator functions are not so applicable, but in spectroscopy and certain forms of chromatography where there are normally a physically defined number of factors and well understood error (or noise) distributions, such approaches are valuable. From a simple rule of thumb, knowing (or estimating) the noise distribution, e.g. from a portion of a chromatogram or spectrum where there is known to be no compounds (or a blank), and then determining how many eigenvalues are required to reduce the level of error, allowing an estimate of rank.

4.3.3.2 Cross-Validation

A complementary series of methods for determining the number of significant factors are based on cross-validation. It is assumed that significant components model 'data', whilst later (and redundant) components model 'noise'. Autopredictive models involve fitting PCs to the entire dataset, and always provide a closer fit to the data the more the components are calculated. Hence the residual error will be smaller if 10 rather than nine PCs are calculated. This does not necessarily indicate that it is correct to retain all 10 PCs; the later PCs may model noise which we do not want.

The significance of the each PC can be tested out by see how well an 'unknown' sample is predicted. In many forms of cross-validation, each sample is removed once from the dataset and then the remaining samples are predicted. For example, if there are 10 samples, perform PC on nine samples, and see how well the remaining sample is predicted. This can be slightly tricky, the following steps are normally employed.

1. Initially leave out sample 1 ($=i$).
2. Perform PCA on the remaining $I - 1$ samples, e.g. samples 2–9. For efficiency it is possible to calculate several PCs ($=A$) simultaneously. Obtain the scores T and loadings P. Note that there will be different scores and loadings matrices according to which sample is removed.
3. Next determine the what the scores would be for sample i simply by

$$\hat{t}_i = x_i.P'$$

Note that this equation is simple, and is obtained from standard multiple linear regression $\hat{t}_i = x_i.P'.(P.P')^{-1}$, but the loadings are orthonormal, so $(P.P')^{-1}$ is a unit matrix.

4. Then calculate the model for sample i for a PCs by

$$^{a,cv}\hat{x}_i = {}^a\hat{t}_i.{}^a P$$

where the superscript a refers to the model using the first a PCs, so $^a\hat{x}_i$ has the dimension $1 \times J$, $^a\hat{t}_i$ has dimensions $1 \times a$ (i.e. is a scalar if only one PC is retained) and consists of the first a scores obtained in step 3, and $^a P$ has dimensions $a \times J$ and consists of the first a rows of the loadings matrix.

5. Next, repeat this by leaving another sample out and going to step 2 until all samples have been removed once.

6. The errors, often called the *predicted residual error sum of squares* or PRESS, is then calculated

$$\text{PRESS}_a = \sum_{i=1}^{I} \sum_{j=1}^{J} (^{a,cv}\hat{x}_{ij} - x_{ij})^2$$

This is simply the sum of square difference between the observed and true values for each object using an *a* PC model.

The PRESS errors can then be compared with the RSS (residual sum of square) errors for each object for straight PCA (sometimes called the autoprediction error), given by

$$\text{RSS}_a = \sum_{i=1}^{I} \sum_{j}^{J} x_{ij}^2 - \sum_{k=1}^{a} g_k$$

or

$$\text{RSS}_a = \sum_{i=1}^{I} \sum_{j=1}^{J} (^{a,auto}\hat{x}_{ij} - x_{ij})^2$$

All equations presented above assume no mean centring or data preprocessing, and further steps are required involving subtracting the mean of $I - 1$ samples each time a sample is left out. If the data are preprocessed prior to cross-validation, it is essential that both PRESS and RSS are presented on the same scale. A problem is that if one takes a subset of the original data, the mean and standard deviation will differ for each group, so it is safest to convert all the data to the original units for calculation of errors. The computational method can be complex and there are no generally accepted conventions, but we recommend the following.

1. Preprocess the entire dataset.
2. Perform PCA on the entire dataset, to give predicted $\hat{X}$ in preprocessed units (e.g. mean centred or standardised).
3. Convert this matrix back to the original units.
4. Determine the RSS in the original units.
5. Next take one sample out and determine statistics such as means or standard deviations for the remaining $I - 1$ samples.
6. Then preprocess these remaining samples and perform PCA on these data.
7. Scale the remaining Ith sample using the mean and standard deviation (as appropriate) obtained in step 5.
8. Obtain the predicted scores $\hat{t}_i$ for this sample, using the loadings in step 6 and the scaled vector x_i obtained in step 7.
9. Predict the vector $\hat{x}_i$ by multiplying $\hat{t}_i.p$ where the loadings have been determined from the $I - 1$ preprocessed samples in step 6.
10. Now rescale the predicted vector to the original units.
11. Next remove another sample, and repeat steps 6–11 until each sample is removed once.
12. Finally, calculate PRESS values in the original units.

There are a number of variations in cross-validation, especially methods for calculating errors, and each group or programmer has their own favourite. For brevity we recommend a single approach. Note that although some steps are common, it is normal to use different criteria when using cross-validation in multivariate calibration as described in Chapter 5, Section 5.6.2. Do not be surprised if different packages provide what appear to be different numerical answers for the estimation of similar parameters – always try to understand what the developer of the software has intended; normally extensive documentation is available.

There are various ways of interpreting these two errors numerically, but a common approach is to compare the PRESS error using $a + 1$ PCs with the RSS using a PCs. If the latter error is significantly larger, then the extra PC is modelling only noise, and so is not significant. Sometimes this is mathematically defined by computing the ratio $\text{PRESS}_a/\text{PRESS}_{a-1}$ and if this exceeds 1, use $a - 1$ PCs in the model. If the errors are close in size, it is safest to continue checking further components, and normally there will be a sharp difference when sufficient components have been computed. Often PRESS will start to increase after the optimum number of components have been calculated.

It is easiest to understand the principle using a small numerical example. Because the datasets of case studies 1 and 2 are rather large, a simulation will be introduced.

Table 4.7 is for a dataset consisting of 10 objects and eight measurements. In Table 4.8(a) the scores and loadings for eight PCs (the number is limited by the measurements) using only samples 2–10 are presented. Table 4.8(b) Illustrates the calculation of the sum of square cross-validated error for sample 1 as increasing number of PCs are calculated. In Table 4.9(a) these errors are summarised for all samples. In Table 4.9(b) eigenvalues are calculated, together with the residual sum of squares as increasing number of PCs are computed for both autopredictive and cross-validated models. The latter can be obtained by summing the rows in Table 4.9(a). RSS decreases continuously, whereas PRESS levels off. This information can be illustrated graphically (see Figure 4.8): the vertical sale is usually presented logarithmically, which takes into account the very high first eigenvalues, usual in cases where the data are uncentred, and so the first eigenvalue is mainly one of size and can appear (falsely) to dominate the data. Using the criteria above, the PRESS value of the fourth PC is greater than the RSS of the third PC, so an optimal model would appear to consist of three PCs. A simple graphical approach, taking the optimum number of PCs to be where the graph of PRESS levels off or increases, would likewise suggest that there are three PCs in the model. Sometimes PRESS values increase after the optimum number of components has been calculated, but this is not so in this example.

There are, of course, many other modifications of cross-validation, and two of the most common are listed below.

1. Instead of removing one object at a time, remove a block of objects, for example four objects, and then cycle round so that each object is part of a group. This can speed up the cross-validation algorithm. However, with modern fast computers, this enhancement is less needed.
2. Remove portions of the data rather than individual samples. Statisticians have developed a number of approaches, and some traditional chemometrics software uses this method. This involves removing a certain number of measurements and replacing them by guesses, e.g. the standard deviation of the column, performing PCA and then determining how well these measurements are predicted. If too many PCs have been employed the measurements are not predicted well.

Table 4.7   Cross-validation example.

	A	B	C	D	E	F	G	H
1	89.821	59.760	68.502	48.099	56.296	95.478	71.116	95.701
2	97.599	88.842	95.203	71.796	97.880	113.122	72.172	92.310
3	91.043	79.551	104.336	55.900	107.807	91.229	60.906	97.735
4	30.015	22.517	60.330	21.886	53.049	23.127	12.067	37.204
5	37.438	38.294	50.967	29.938	60.807	31.974	17.472	35.718
6	83.442	48.037	59.176	47.027	43.554	84.609	67.567	81.807
7	71.200	47.990	86.850	35.600	86.857	57.643	38.631	67.779
8	37.969	15.468	33.195	12.294	32.042	25.887	27.050	37.399
9	34.604	68.132	63.888	48.687	86.538	63.560	35.904	40.778
10	74.856	36.043	61.235	37.381	53.980	64.714	48.673	73.166

Table 4.8   Calculation of cross-validated error for sample 1.

(a) Scores and loadings for first 8 PCs on 9 samples, excluding sample 1.

Scores

259.25	9.63	20.36	2.29	−3.80	0.04	−2.13	0.03
248.37	−8.48	−5.08	−3.38	1.92	−5.81	0.53	−0.46
96.43	−24.99	−20.08	8.34	2.97	0.12	0.33	0.29
109.79	−23.52	−3.19	−0.38	−5.57	0.38	3.54	1.41
181.87	46.76	4.34	2.51	2.44	0.30	0.65	1.63
180.04	−16.41	−20.74	−2.09	−1.57	1.91	−3.55	0.16
80.31	8.27	−13.88	−5.92	2.75	2.54	0.60	1.17
157.45	−34.71	27.41	−1.10	4.03	2.69	0.80	−0.46
161.67	23.85	−12.29	0.32	−1.12	2.19	2.14	−2.63

Loadings

0.379	0.384	−0.338	−0.198	−0.703	0.123	−0.136	0.167
0.309	−0.213	0.523	−0.201	−0.147	−0.604	−0.050	0.396
0.407	−0.322	−0.406	0.516	0.233	−0.037	−0.404	0.286
0.247	−0.021	0.339	0.569	−0.228	0.323	0.574	0.118
0.412	−0.633	−0.068	−0.457	−0.007	0.326	0.166	−0.289
0.388	0.274	0.431	0.157	0.064	0.054	−0.450	−0.595
0.263	0.378	0.152	−0.313	0.541	0.405	0.012	0.459
0.381	0.291	−0.346	−0.011	0.286	−0.491	0.506	−0.267

(b) Predictions for sample 1

Predicted scores			PC1				PC7
207.655	43.985	4.453	−1.055	4.665	−6.632	0.329	

Predictions	A	B	C	D	E	F	G	H	Sum of square error
PC1	78.702	64.124	84.419	51.361	85.480	80.624	54.634	79.109	2025.934
	95.607	54.750	70.255	50.454	57.648	92.684	71.238	91.909	91.235
	94.102	57.078	68.449	51.964	57.346	94.602	71.916	90.369	71.405
	94.310	57.290	67.905	51.364	57.828	94.436	72.245	90.380	70.292
	91.032	56.605	68.992	50.301	57.796	94.734	74.767	91.716	48.528
	90.216	60.610	69.237	48.160	55.634	94.372	72.078	94.972	4.540
PC7	90.171	60.593	69.104	48.349	55.688	94.224	72.082	95.138	4.432

Table 4.9 Calculation of RSS and PRESS.

(a) Summary of cross-validated sum of square errors

Object	1	2	3	4	5	6	7	8	9	10
PC1	2025.9	681.1	494.5	1344.6	842	2185.2	1184.2	297.1	2704	653.5
	91.2	673	118.1	651.5	66.5	67.4	675.4	269.8	1655.4	283.1
	71.4	91.6	72.7	160.1	56.7	49.5	52.5	64.6	171.6	40.3
	70.3	89.1	69.7	159	56.5	36.2	51.4	62.1	168.5	39.3
	48.5	59.4	55.5	157.4	46.7	36.1	39.4	49.9	160.8	29.9
	4.5	40.8	8.8	154.5	39.5	19.5	38.2	18.9	148.4	26.5
PC7	4.4	0.1	2.1	115.2	30.5	18.5	27.6	10	105.1	22.6

(b) RSS and PRESS calculations

Eigenvalues	RSS	PRESS	$PRESS_a/RSS_{a-1}$
316522.1	10110.9	12412.1	
7324.6	2786.3	4551.5	0.450
2408.7	377.7	830.9	0.298
136.0	241.7	802.2	2.124
117.7	123.9	683.7	2.829
72.9	51.1	499.7	4.031
36.1	15.0	336.2	6.586
15.0	0.0	n/a	

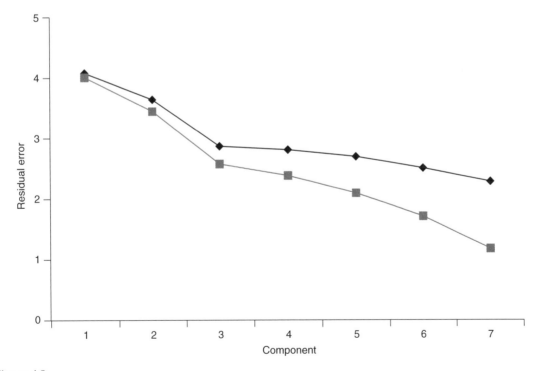

Figure 4.8
Graph of log of PRESS (top) and RSS (bottom) for dataset in Table 4.7

As in the case of most chemometric methods, there are innumerable variations on the theme, and it is important to be careful to check every author in detail. However, the 'leave one sample out at a time' method described above is popular, and relatively easy to implement and understand.

4.3.4 Factor Analysis

Statisticians do not always distinguish between factor analysis and principal components analysis, but for chemists factors often have a physical significance, whereas PCs are simply abstract entities. However, it is possible to relate PCs to chemical information, such as elution profiles and spectra in HPLC–DAD by

$$\hat{X} = T.P = \hat{C}.\hat{S}$$

The conversion from 'abstract' to 'chemical' factors is sometimes called a rotation or transformation and will be discussed in more detail in Chapter 6, and is illustrated in Figure 4.9. Note that factor analysis is by no means restricted to chromatography. An example is the pH titration profile of a number of species containing different numbers of protons together with their spectra. Each equilibrium species has a pH titration profile and a characteristic spectrum.

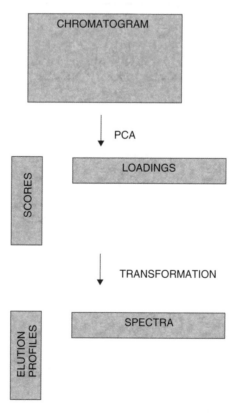

Figure 4.9
Relationship between PCA and factor analysis in coupled chromatography

Factor analysis is often called by a number of alternative names such as 'rotation' or 'transformation', but is a procedure used to relate the abstract PCs to meaningful chemical factors, and the influence of Malinowski in the 1980s introduced this terminology into chemometrics.

4.3.5 Graphical Representation of Scores and Loadings

Many revolutions in chemistry relate to the graphical presentation of information. For example, fundamental to the modern chemist's way of thinking is the ability to draw structures on paper in a convenient and meaningful manner. Years of debate preceded the general acceptance of the Kekulé structure for benzene: today's organic chemist can write down and understand complex structures of natural products without the need to plough through pages of numbers of orbital densities and bond lengths. Yet, underlying these representations are quantum mechanical probabilities, so the ability to convert from numbers to a simple diagram has allowed a large community to think clearly about chemical reactions.

So with statistical data, and modern computers, it is easy to convert from numbers to graphs. Many modern multivariate statisticians think geometrically as much as numerically, and concepts such as principal components are often treated as objects in an imaginary space rather than mathematical entities. The algebra of multidimensional space is the same as that of multivariate statistics. Older texts, of course, were written before the days of modern computing, so the ability to produce graphs was more limited. However, now it is possible to obtain a large number of graphs rapidly using simple software, and much is even possible using Excel. There are many ways of visualising PCs. Below we will look primarily at graphs of first two PCs, for simplicity.

4.3.5.1 Scores Plots

One of the simplest plots is that of the score of one PC against the other. Figure 4.10 illustrates the PC plot of the first two PCs obtained from case study 1, corresponding to plotting a graph of the first two columns of Table 4.3. The horizontal axis represents the scores for the first PC and the vertical axis those for the second PC. This 'picture' can be interpreted as follows:

- the linear regions of the graph represent regions of the chromatogram where there are pure compounds;
- the curved portion represents a region of co-elution;
- the closer to the origin, the lower the intensity.

Hence the PC plot suggests that the region between 6 and 10 s (approximately) is one of co-elution. The reason why this method works is that the spectrum over the chromatogram changes with elution time. During co-elution the spectral appearance changes most, and PCA uses this information.

How can these graphs help?

- The pure regions can inform us about the spectra of the pure compounds.
- The shape of the PC plot informs us of the amount of overlap and quality of chromatography.

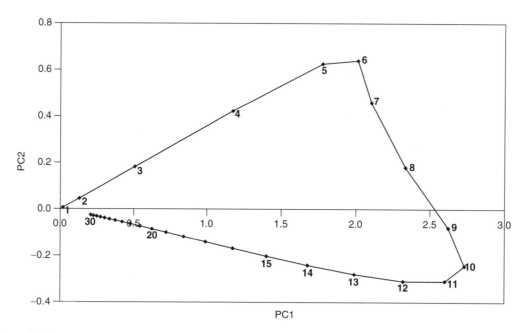

Figure 4.10
Scores of PC2 (vertical axis) versus PC1 (horizontal axis) for case study 1

- The number of bends in a PC plot can provide information about the number of different compounds in a complex multipeak cluster.

In cases where there is a meaningful sequential order to a dataset, as in spectroscopy or chromatography, but also, for example, where objects are related in time or pH, it is possible to plot the scores against sample number (see Figure 4.11). From this it appears that the first PC primarily relates to the magnitude of the measurements, whereas the second discriminates between the two components in the mixture, being positive for the fastest eluting component and negative for the slowest. Note that the appearance and interpretation of such plots depend crucially on data scaling, as will be discussed in Section 4.3.6. This will be described in more detail in Chapter 6, Section 6.2 in the context of evolutionary signals.

The scores plot for the first two PCs of case study 2 has already been presented in Figure 4.4. Unlike case study 1, there is no sequential significance in the order of the chromatographic columns. However, several deductions are possible.

- Closely clustering objects, such as the three Inertsil columns, behave very similarly.
- Objects that are diametrically opposed are 'negatively correlated', for example Kromasil C8 and Purospher. This means that a parameter that has a high value for Purospher is likely to have a low value for Kromasil C8 and vice versa. This would suggest that each column has a different purpose.

Scores plots can be used to answer many different questions about the relationship between objects and more examples are given in the problems at the end of this chapter.

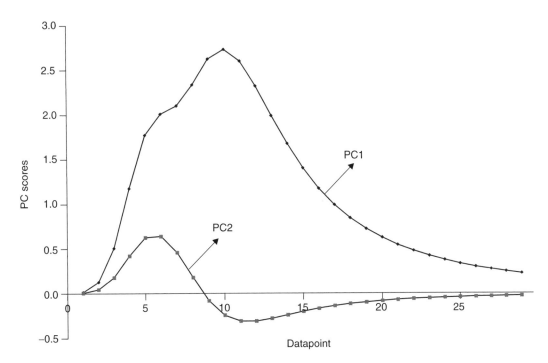

Figure 4.11
Scores of the first two PCs of case study 1 versus sample number

4.3.5.2 Loadings Plots

It is not only the scores, however, that are of interest, but also sometimes the loadings. Exactly the same principles apply in that the value of the loadings at one PC can be plotted against that at the other PC. The result for the first two PCs for case study 1 is shown in Figure 4.12. This figure looks complicated, which is because both spectra overlap and absorb at similar wavelengths. The pure spectra are presented in Figure 4.13. Now we can understand a little more about these graphs.

We can see that the top of the scores plot corresponds to the direction for the fastest eluting compound (=A), whereas the bottom corresponds to that for the slowest eluting compound (=B) (see Figure 4.10). Similar interpretation can be obtained from the loadings plots. Wavelengths in the bottom half of the graph correspond mainly to B, for example 301 and 225 nm. In Figure 4.13, these wavelengths are indicated and represent the maximum ratio of the spectral intensities of B to A. In contrast, high wavelengths, above 325 nm, belong to A, and are displayed in the top half of the graph. The characteristic peak for A at 244 nm is also obvious in the loadings plot.

Further interpretation is possible, but it can easily be seen that the loadings plots provide detailed information about which wavelengths are most associated with which compound. For complex multicomponent clusters or spectral of mixtures, this information can be very valuable, especially if the pure components are not available.

The loadings plot for case study 2 is especially interesting, and is presented in Figure 4.14. What can we say about the tests?

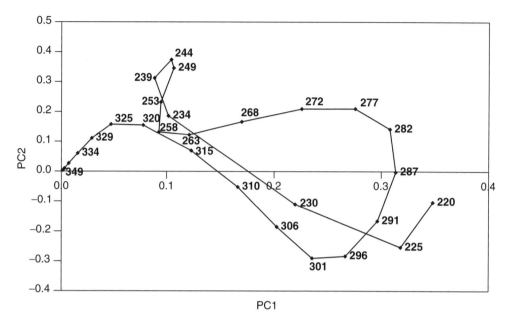

Figure 4.12
Loadings plot of PC2 (vertical axis) against PC1 (horizontal axis) for case study 1

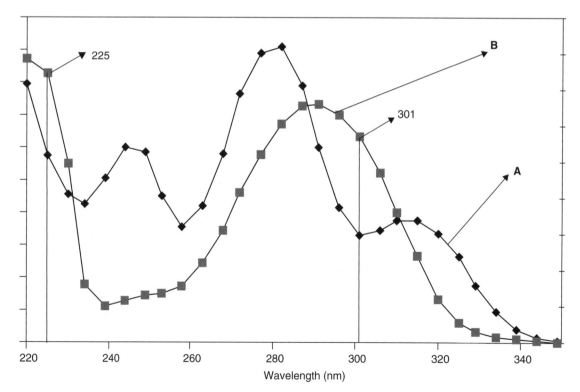

Figure 4.13
Pure spectra of compounds in case study 1

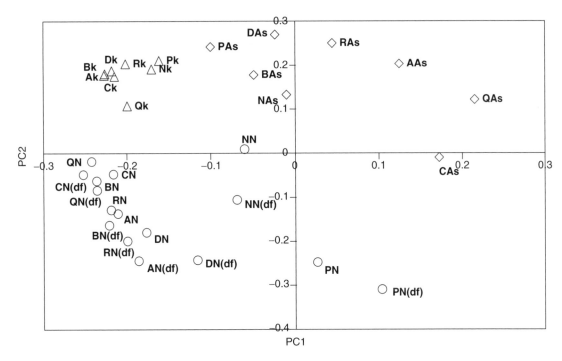

Figure 4.14
Loadings plots for the first two (standardised) PCs of case study 2

- The k loadings are very closely clustered, suggesting that this parameter does not vary much according to compound or column. As, N and N(df) show more discrimination. N and N(df) are very closely correlated.
- As and N are almost diametrically opposed, suggesting that they measure opposite properties, i.e. a high As corresponds to a low N [or N(df)] value.
- Some parameters are in the middle of the loadings plots, such as NN. These behave atypically and are probably not useful indicators of column performance.
- Most loadings are on an approximate circle. This is because standardisation is used, and suggests that we are probably correct in keeping only two principal components.
- The order of the compounds for both As and N reading clockwise around the circle are very similar, with P, D and N at one extreme and Q and C at the other extreme. This suggests that behaviour is grouped according to chemical structure, and also that it is possible to reduce the number of test compounds by selecting one compound in each group.

These conclusions can provide very valuable experimental clues as to which tests are most useful. For example, it might be impracticable to perform large numbers of tests, so can we omit some compounds? Should we measure all these parameters, or are some of them useful and some not? Are some measurements misleading, and not typical of the overall pattern?

Loadings plots can be used to answer a lot of questions about the data, and are a very flexible facility available in almost all chemometrics software.

4.3.5.3 Extensions

In many cases, more than two significant principal components characterise the data, but the concepts above can be employed, except that many more possible graphs can be computed. For example, if four significant components are calculated, we can produce *six* possible graphs, of each possible combination of PCs, for example, PC 4 versus 2, or PC 3 versus 1, and so on. If there are A significant PCs, there will be $\sum_{a=1}^{A-1} a$ possible PC plots of one PC against another. Each graph could reveal interesting trends. It is also possible to produce three-dimensional PC plots, the axes of which consist of three PCs (normally the first three) and so visualise relationships between and clusters of variables in three-dimensional space.

4.3.6 Preprocessing

All chemometric methods are influenced by the method for data preprocessing, or preparing information prior to application of mathematical algorithms. An understanding is essential for correct interpretation from multivariate data packages, but will be illustrated with reference to PCA, and is one of the first steps in data preparation. It is often called scaling and the most appropriate choice can relate to the chemical or physical aim of the analysis. Scaling is normally performed prior to PCA, but in this chapter it is introduced afterwards as it is hard to understand how preprocessing influences the resultant models without first appreciating the main concepts of PCA.

4.3.6.1 Example

As an example, consider a data matrix consisting of 10 rows (labelled from 1 to 10) and eight columns (labelled from A to H), as in Table 4.10. This could represent a portion of a two-way HPLC–DAD data matrix, the elution profile of which in given in Figure 4.15, but similar principles apply to all multivariate data matrices. We choose a small example rather than case study 1 for this purpose, in order to be able to demonstrate all the steps numerically. The calculations are illustrated with reference to the first two PCs, but similar ideas are applicable when more components are computed.

4.3.6.2 Raw Data

The resultant principal components scores and loadings plots are given in Figure 4.16. Several conclusions are possible.

- There are probably two main compounds, one which has a region of purity between points 1 and 3, and the other between points 8 and 10.
- Measurements (e.g. spectral wavelengths) A, B, G and H correspond mainly to the first chemical component, whereas measurements D and E to the second chemical component.

PCA has been performed directly on the raw data, something statisticians in other disciplines very rarely do. It is important to be very careful when using for chemical

Table 4.10   Raw data for Section 4.3.6.

	A	B	C	D	E	F	G	H
1	0.318	0.413	0.335	0.196	0.161	0.237	0.290	0.226
2	0.527	0.689	0.569	0.346	0.283	0.400	0.485	0.379
3	0.718	0.951	0.811	0.521	0.426	0.566	0.671	0.526
4	0.805	1.091	0.982	0.687	0.559	0.676	0.775	0.611
5	0.747	1.054	1.030	0.804	0.652	0.695	0.756	0.601
6	0.579	0.871	0.954	0.841	0.680	0.627	0.633	0.511
7	0.380	0.628	0.789	0.782	0.631	0.505	0.465	0.383
8	0.214	0.402	0.583	0.635	0.510	0.363	0.305	0.256
9	0.106	0.230	0.378	0.440	0.354	0.231	0.178	0.153
10	0.047	0.117	0.212	0.257	0.206	0.128	0.092	0.080

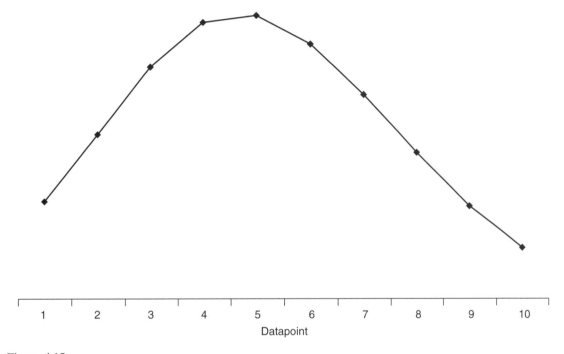

Datapoint

Figure 4.15
Profile for data in Table 4.10

data packages that have been designed primarily by statisticians. What is mainly interesting in traditional studies is the deviation around a mean, for example, how do the
mean characteristics of a forged banknote vary? What is an 'average' banknote? In
chemistry, however, we are often (but by no means exclusively) interested in the deviation above a baseline, such as in spectroscopy. It is also crucial to recognise that some
of the traditional properties of principal components, such as the correlation coefficient
between two score vectors being equal to zero, are no longer valid for raw data. Despite
this, there is often good chemical reason for using applying PCA to the raw data.

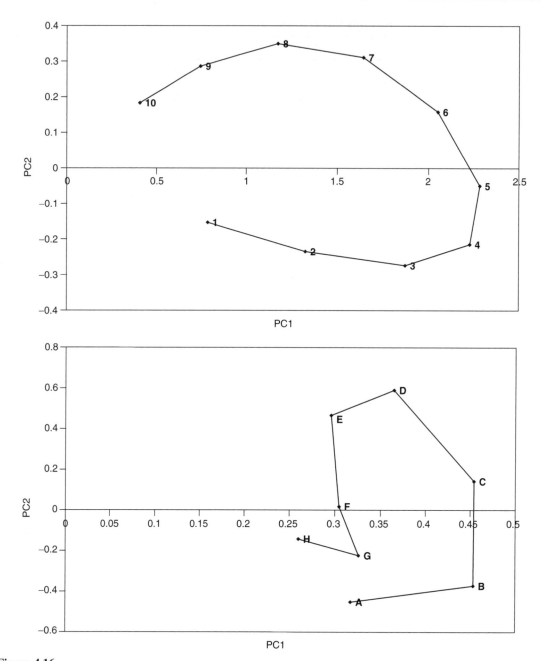

Figure 4.16
Scores and loadings plots of first two PCs of data in Table 4.10

4.3.6.3 Mean Centring

It is, however, possible to mean centre the columns by subtracting the mean of each column (or variable) so that

$$^{cen}x_{ij} = x_{ij} - \overline{x}_j$$

Table 4.11 Mean-centred data corresponding to Table 4.10.

	A	B	C	D	E	F	G	H
1	−0.126	−0.231	−0.330	−0.355	−0.285	−0.206	−0.175	−0.146
2	0.083	0.045	−0.095	−0.205	−0.163	−0.042	0.020	0.006
3	0.273	0.306	0.146	−0.030	−0.020	0.123	0.206	0.153
4	0.360	0.446	0.318	0.136	0.113	0.233	0.310	0.238
5	0.303	0.409	0.366	0.253	0.206	0.252	0.291	0.229
6	0.135	0.226	0.290	0.291	0.234	0.185	0.168	0.139
7	−0.064	−0.017	0.125	0.231	0.184	0.062	0.000	0.010
8	−0.230	−0.243	−0.081	0.084	0.064	−0.079	−0.161	−0.117
9	−0.338	−0.414	−0.286	−0.111	−0.093	−0.212	−0.287	−0.220
10	−0.397	−0.528	−0.452	−0.294	−0.240	−0.315	−0.373	−0.292

The result is presented in Table 4.11. Note that the sum of each column is now zero. Almost all traditional statistical packages perform this operation prior to PCA, whether desired or not. The PC plots are presented in Figure 4.17.

The most obvious difference is that the scores plot is now centred around the origin. However, the relative positions of the points in both graphs change slightly, the largest effect being on the loadings in this case. In practice, mean centring can have a large influence, for example if there are baseline problems or only a small region of the data is recorded. The reasons why the distortion is not dramatic in this case is that the averages of the eight variables are comparable in size, varying from 0.38 to 0.64. If there is a much wider variation this can change the patterns of both scores and loadings plots significantly, so the scores of the mean-centred data are not necessarily the 'shifted' scores of the original dataset.

Mean centring often has a significant influence on the relative size of the first eigenvalue, which is reduced dramatically in size, and can influence the apparent number of significant components in a dataset. However, it is important to recognise that in signal analysis the main feature of interest is variation above a baseline, so mean centring is not always appropriate in a physical sense in certain areas of chemistry.

4.3.6.4 Standardisation

Standardisation is another common method for data scaling and occurs after mean centring; each variable is also divided by its standard deviation:

$$^{stn}x_{ij} = \frac{x_{ij} - \bar{x}_j}{\sqrt{\sum_{i=1}^{I}(x_{ij} - \bar{x}_j)^2/I}}$$

see Table 4.12 for our example. Note an interesting feature that the sum of squares of each column equals 10 (which is the number of objects in the dataset), and note also that the 'population' rather than 'sample' standard deviation (see Appendix A.3.1.2) is employed. Figure 4.18 represents the new graphs. Whereas the scores plot hardly changes in appearance, there is a dramatic difference in the appearance of the loadings. The reason is that standardisation puts all the variables on approximately the same scale. Hence variables (such as wavelengths) of low intensity assume equal significance

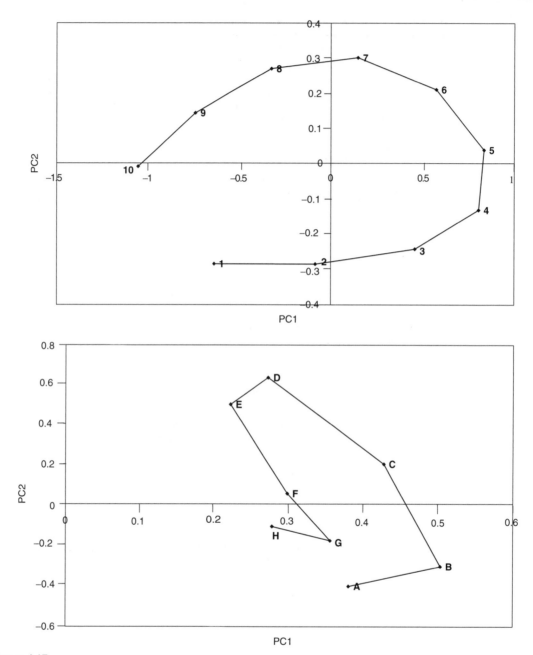

Figure 4.17
Scores and loadings plots of first two PCs of data in Table 4.11

to those of high intensity. Note that now the variables are roughly the same distance away from the origin, on an approximate circle (this looks distorted simply because the horizontal axis is longer than the vertical axis), because there are only two significant components. If there were three significant components, the loadings would fall roughly on a sphere, and so on as the number of components is increased. This simple

Table 4.12 Standardised data corresponding to Table 4.10.

	A	B	C	D	E	F	G	H
1	−0.487	−0.705	−1.191	−1.595	−1.589	−1.078	−0.760	−0.818
2	0.322	0.136	−0.344	−0.923	−0.909	−0.222	0.087	0.035
3	1.059	0.933	0.529	−0.133	−0.113	0.642	0.896	0.856
4	1.396	1.361	1.147	0.611	0.629	1.218	1.347	1.330
5	1.174	1.248	1.321	1.136	1.146	1.318	1.263	1.277
6	0.524	0.690	1.046	1.306	1.303	0.966	0.731	0.774
7	−0.249	−0.051	0.452	1.040	1.026	0.326	0.001	0.057
8	−0.890	−0.740	−0.294	0.376	0.357	−0.415	−0.698	−0.652
9	−1.309	−1.263	−1.033	−0.497	−0.516	−1.107	−1.247	−1.228
10	−1.539	−1.608	−1.635	−1.321	−1.335	−1.649	−1.620	−1.631

visual technique is also a good method for confirming that there are two significant components in a dataset.

Standardisation can be important in real situations. Consider, for example, a case where the concentrations of 30 metabolites are monitored in a series of organisms. Some metabolites might be abundant in all samples, but their variation is not very significant. The change in concentration of the minor compounds might have a significant relationship to the underlying biology. If standardisation is not performed, PCA will be dominated by the most intense compounds.

In some cases standardisation (or closely related scaling) is an essential first step in data analysis. In case study 2, each type of chromatographic measurement is on a different scale. For example, the N values may exceed 10 000, whereas k' rarely exceeds 2. If these two types of information were not standardised, PCA will be dominated primarily by changes in N, hence all analysis of case study 2 in this chapter involves preprocessing via standardisation. Standardisation is also useful in areas such as quantitative structure–property relationships, where many different pieces of information are measured on very different scales, such as bond lengths and dipoles.

4.3.6.5 Row Scaling

Scaling the rows to a constant total, usually 1 or 100, is a common procedure, for example

$$^{cs}x_{ij} = \frac{x_{ij}}{\sum\limits_{j=1}^{J} x_{ij}}$$

Note that some people use term 'normalisation' in the chemometrics literature to define this operation, but others define normalised vectors (see Section 4.3.2 and Chapter 6, Section 6.2.2) as those whose sum of squares rather than sum of elements equals one. In order to reduce confusion, in this book we will restrict the term normalisation to the transformation that sets the sum of squares to one.

Scaling the rows to a constant total is useful if the absolute concentrations of samples cannot easily be controlled. An example might be biological extracts: the precise amount of material might vary unpredictably, but the *relative* proportions of each chemical can be measured. This method of scaling introduces a constraint which is often called *closure*. The numbers in the multivariate data matrix are proportions and

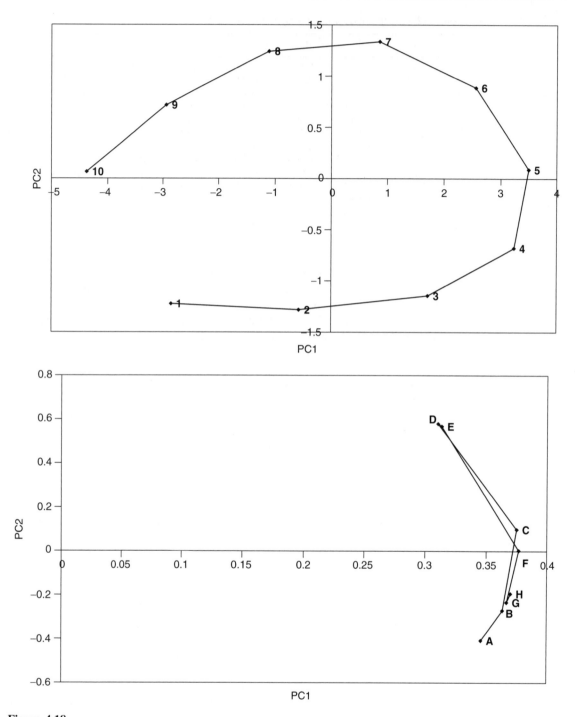

Figure 4.18
Scores and loadings plots of first two PCs of data in Table 4.12

Table 4.13 Scaling rows to constant total of 1 for the data in Table 4.10.

	A	B	C	D	E	F	G	H
1	0.146	0.190	0.154	0.090	0.074	0.109	0.133	0.104
2	0.143	0.187	0.155	0.094	0.077	0.109	0.132	0.103
3	0.138	0.183	0.156	0.100	0.082	0.109	0.129	0.101
4	0.130	0.176	0.159	0.111	0.090	0.109	0.125	0.099
5	0.118	0.166	0.162	0.127	0.103	0.110	0.119	0.095
6	0.102	0.153	0.167	0.148	0.119	0.110	0.111	0.090
7	0.083	0.138	0.173	0.171	0.138	0.111	0.102	0.084
8	0.066	0.123	0.178	0.194	0.156	0.111	0.093	0.078
9	0.051	0.111	0.183	0.213	0.171	0.112	0.086	0.074
10	0.041	0.103	0.186	0.226	0.181	0.112	0.081	0.071

some of the properties are closely analogous to properties of compositional mixtures (Chapter 2, Section 2.5).

The result is presented in Table 4.13 and Figure 4.19. The scores plot appears very different from those in previous figures. The datapoints now lie on a straight line (this is a consequence of there being exactly two components in this particular dataset). The 'mixed' points are in the centre of the straight line, with the pure regions at the extreme ends. Note that sometimes, if extreme points are primarily influenced by noise, the PC plot can be distorted, and it is important to select carefully an appropriate region of the data.

4.3.6.6 Further Methods

There is a very large battery of methods for data preprocessing, although those described above are the most common.

- It is possible to combine approaches, for example, first to scale the rows to a constant total and then standardise a dataset.
- Weighting of each variable according to any external criterion of importance is sometimes employed.
- Logarithmic scaling of measurements is often useful if there are large variations in intensities. If there are a small number of negative or zero numbers in the raw dataset, these can be handled by setting them to small a positive number, for example, equal to half the lowest positive number in the existing dataset (or for each variable as appropriate). Clearly, if a dataset is dominated by negative numbers or has many zero or missing values, logarithmic scaling is inappropriate.
- Selectively scaling some of the variables to a constant total is also useful, and it is even possible to divide the variables into blocks and perform scaling separately on each block. This could be useful if there were several types of measurement, for example two spectra and one chromatogram, each constituting a single block. If one type of spectrum is recorded at 100 wavelengths and a second type at 20 wavelengths, it may make sense to scale each type of spectrum to ensure equal significance of each block.

Undoubtedly, however, the appearance and interpretation not only of PC plots but also of almost all chemometric techniques, depend on data preprocessing. The influence of preprocessing can be dramatic, so it is essential for the user of chemometric software to understand and question how and why the data have been scaled prior to interpreting the result from a package. More consequences are described in Chapter 6.

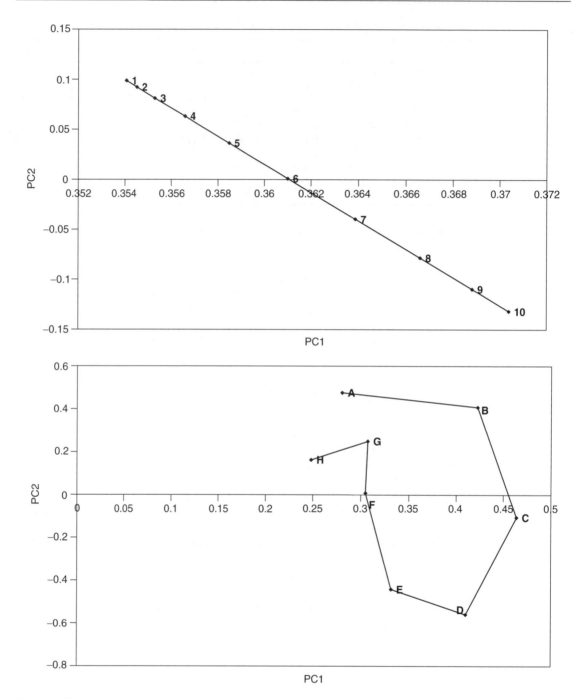

Figure 4.19
Scores and loadings plots of first two PCs of data in Table 4.13

4.3.7   Comparing Multivariate Patterns

PC plots are often introduced only by reference to the independent loadings or scores plot of a single dataset. However, there are common patterns within different graphs. Consider taking measurements of the concentration of a mineral in a geochemical deposit. This information could be presented as a table of sampling sites and observed concentrations, but a much more informative approach would be to produce a picture in which physical location and mineral concentration are superimposed, such as a coloured map, each different colour corresponding to a concentration range of the mineral. Two pieces of information are connected, namely geography and concentration. Hence in many applications of multivariate analysis, one aim may be to connect the samples (e.g. geographical location/sampling site) represented by scores, to the variables (e.g. chemical measurements) represented by loadings. Graphically this requires the superimposition of two types of information.

Another common need is to compare two independent types of measurements. Consider recording the results of a taste panel for a type of food. Their scores relate to the underlying chemical or manufacturing process. A separate measurement could be chemical, such as a chromatographic or spectroscopic profile. Ideally the chemical measurements will relate to the taste: can each type of measurement give similar information and, so, can we predict the taste by using analytical chemical techniques?

4.3.7.1   Biplots

A biplot involves the superimposition of a scores and a loadings plot. In order to superimpose each plot on a sensible scale, one approach is to divide the scores as follows:

$$^{new}t_{ia} = \frac{t_{ia}}{\sum\limits_{i=1}^{I} t_{ia}^2 / I}$$

Note that if the scores are mean centred, the denominator equals the variance. Some authors us the expression in the denominator of this equation to denote an eigenvalue, so in certain articles it is stated that the scores of each PC are divided by their eigenvalue. As is usual in chemometrics, it is important to recognise that there are many different schools of thought and incompatible definitions.

Consider superimposing these plots for case study 2 to give Figure 4.20. What can we deduce from this?

- We see that Purospher lies at the extreme position along the horizontal axis, as does CAs. Hence we would expect CAs to have a high value for Purospher, which can be verified by examining Table 4.2 (1.66). A similar comment can be made concerning DAs and Kromasil C18. These tests are good specific markers for particular columns.
- Likewise, parameters at opposite corners to chromatographic columns will exhibit characteristically low values, for example, QN has a value of 2540 for Purospher.
- The chromatographic columns Supelco ABZ+ and Symmetry C18 are almost diametrically opposed, and good discriminating parameters are the measurements on the peaks corresponding to compound P (pyridine), PAs and PN(df). Hence to distinguish the behaviour between columns lying on this line, one of the eight compounds can be employed for the tests.

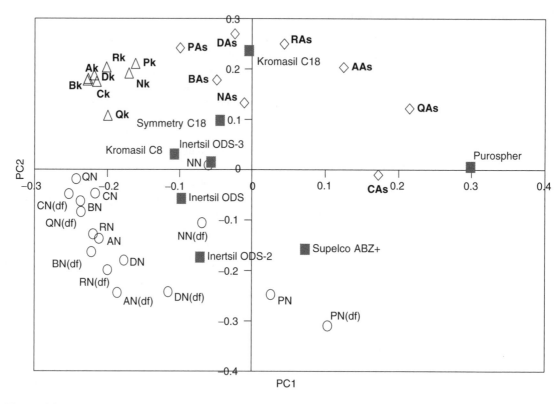

Figure 4.20
Biplot

Many other deductions can be made from Figure 4.20, but biplots provide a valuable pictorial representation of the relationship between measurements (in this case chromatographic tests) and objects (in this case chromatographic columns).

It is not necessary to restrict biplots to two PCs but, of course, when more than three are used graphical representation becomes difficult, and numerical measures are often employed, using statistical software.

4.3.7.2 Procrustes Analysis

Another important facility is to be able to compare different types of measurements. For example, the mobile phase in the example in Figure 4.4 is methanol. How about using a different mobile phase? A statistical method called procrustes analysis will help us here.

Procrustes was a Greek god who kept a house by the side of the road where he offered hospitality to passing strangers, who were invited in for a meal and a night's rest in his very special bed which Procrustes described as having the unique property that its length exactly matched whomsoever lay down upon it. What he did not say was the method by which this 'one-size-fits-all' was achieved: as soon as the guest lay down Procrustes went to work upon them, stretching them if they were too short for the bed or chopping off their legs if they were too long.

Similarly, procrustes analysis in chemistry involves comparing two diagrams, such as two PC scores plots. One such plot is the reference and a second plot is manipulated to resemble the reference plot as closely as possible. This manipulation is done mathematically involving up to three main transformations.

1. *Reflection*. This transformation is a consequence of the inability to control the sign of a principal component.
2. *Rotation*.
3. *Scaling* (or stretching). This transformation is used because the scales of the two types of measurements may be very different.
4. *Translation*.

If two datasets are already standardised, transformation 3 may not be necessary, and the fourth transformation is not often used.

The aim is to reduce the root mean square difference between the scores of the reference dataset and the transformed dataset:

$$r = \sqrt{\sum_{i=1}^{I}\sum_{a=1}^{A} ({}^{ref}t_{ia} - {}^{trans}t_{ia})^2 / I}$$

For case study 2, it might be of interest to compare performances using different mobile phases (solvents). The original data were obtained using methanol: are similar separations achievable using acetonitrile or THF? The experimental measurements are presented in Tables 4.14 and 4.15. PCA is performed on the standardised data (transposing the matrices as appropriate). Figure 4.21 illustrates the two scores plots using

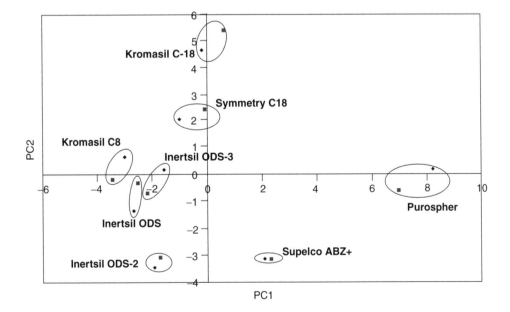

Figure 4.21
Comparison of scores plots for methanol and acetonitrile (case study 2)

Table 4.14 Chromatographic parameters corresponding to case study 2, obtained using acetonitrile as mobile phase.

Parameter	Inertsil ODS	Inertsil ODS-2	Inertsil ODS-3	Kromasil C18	Kromasil C8	Symmetry C18	Supelco ABZ+	Purospher
Pk	0.13	0.07	0.11	0.15	0.13	0.37	0	0
PN	7 340	10 900	13 500	7 450	9 190	9 370	18 100	8 990
PN(df)	5 060	6 650	6 700	928	1 190	3 400	7 530	2 440
PAs	1.55	1.31	1.7	4.39	4.36	1.92	2.16	2.77
Nk	0.19	0.08	0.16	0.16	0.15	0.39	0	0
NN	15 300	11 800	10 400	13 300	16 800	5 880	16 100	10 700
NN(df)	7 230	6 020	5 470	3 980	7 860	648	6 780	3 930
NAs	1.81	1.91	1.81	2.33	1.83	5.5	2.03	2.2
Ak	2.54	1.56	2.5	2.44	2.48	2.32	0.62	0.2
AN	15 500	16 300	14 900	11 600	16 300	13 500	13 800	9 700
AN(df)	9 100	10 400	9 480	3 680	8 650	7 240	7 060	4 600
AAs	1.51	1.62	1.67	2.6	1.85	1.72	1.85	1.8
Ck	1.56	0.85	1.61	1.39	1.32	1.43	0.34	0.11
CN	14 600	14 900	13 500	13 200	18 100	13 100	18 000	9 100
CN(df)	13 100	12 500	12 200	10 900	15 500	10 500	11 700	5 810
CAs	1.01	1.27	1.17	1.2	1.17	1.23	1.67	1.49
Qk	7.34	3.62	7.04	5.6	5.48	5.17	1.4	0.92
QN	14 200	16 700	13 800	14 200	16 300	11 100	10 500	4 200
QN(df)	12 800	13 800	11 400	10 300	12 600	5 130	7 780	2 220
QAs	1.03	1.34	1.37	1.44	1.41	2.26	1.35	2.01
Bk	0.67	0.41	0.65	0.64	0.65	0.77	0.12	0
BN	15 900	12 000	12 800	14 100	19 100	12 900	13 600	5 370
BN(df)	8 100	8 680	6 210	5 370	8 820	5 290	6 700	2 470
BAs	1.63	1.5	1.92	2.11	1.9	1.97	1.82	1.42
Dk	5.73	4.18	6.08	6.23	6.26	5.5	1.27	0.75
DN	14 400	20 200	17 700	11 800	18 500	15 600	14 600	11 800
DN(df)	10 500	15 100	13 200	3 870	12 600	10 900	10 400	8 950
DAs	1.39	1.51	1.54	2.98	1.65	1.53	1.49	1.3
Rk	14.62	10.8	15.5	15.81	14.57	13.81	3.41	2.22
RN	12 100	19 400	17 500	10 800	16 600	15 700	14 000	10 200
RN(df)	9 890	13 600	12 900	3 430	12 400	11 600	10 400	7 830
RAs	1.3	1.66	1.62	3.09	1.52	1.54	1.49	1.32

methanol and acetonitrile, as procrustes rotation has been performed to ensure that they agree as closely as possible, whereas Figure 4.22 is for methanol and THF. It appears that acetonitrile has similar properties to methanol as a mobile phase, but THF is very different.

It is not necessary to restrict each measurement technique to two PCs; indeed, in many practical cases four or five PCs are employed. Computer software is available to compare scores plots and provide a numerical indicator of the closeness of the fit, but it is not easy to visualise. Because PCs do not often have a physical meaning, it is important to recognise that in some cases it is necessary to include several PCs for a meaningful result. For example, if two datasets are characterised by four PCs, and each one is of approximately equal size, the first PC for the reference dataset may correlate most closely with the third for the comparison dataset, so including only the

Table 4.15 Chromatographic parameters corresponding to case study 2, obtained using THF as mobile phase.

Parameter	Inertsil ODS	Inertsil ODS-2	Inertsil ODS-3	Kromasil C18	Kromasil C8	Symmetry C18	Supelco ABZ+	Purospher
Pk	0.05	0.02	0.02	0.04	0.04	0.31	0	0
PN	17 300	12 200	10 200	14 900	18 400	12 400	16 600	13 700
PN(df)	11 300	7 080	6 680	7 560	11 400	5 470	10 100	7 600
PAs	1.38	1.74	1.53	1.64	1.48	2.01	1.65	1.58
Nk	0.05	0.02	0.01	0.03	0.03	0.33	0	0
NN	13 200	9 350	7 230	11 900	15 800	3 930	14 300	11 000
NN(df)	7 810	4 310	4 620	7 870	11 500	543	7 870	6 140
NAs	1.57	2	1.76	1.43	1.31	5.58	1.7	1.7
Ak	2.27	1.63	1.79	1.96	2.07	2.37	0.66	0.19
AN	14 800	15 300	13 000	13 500	18 300	12 900	14 200	11 000
AN(df)	10 200	11 300	10 700	9 830	14 200	9 430	9 600	7 400
AAs	1.36	1.46	1.31	1.36	1.32	1.37	1.51	1.44
Ck	0.68	0.45	0.52	0.53	0.54	0.84	0.15	0
CN	14 600	11 800	9 420	11 600	16 500	11 000	12 700	8 230
CN(df)	12 100	9 170	7 850	9 820	13 600	8 380	9 040	5 100
CAs	1.12	1.34	1.22	1.14	1.15	1.28	1.42	1.48
Qk	4.67	2.2	3.03	2.78	2.67	3.28	0.9	0.48
QN	10 800	12 100	9 150	10 500	13 600	8 000	10 100	5 590
QN(df)	8 620	9 670	7 450	8 760	11 300	3 290	7 520	4 140
QAs	1.17	1.36	1.3	1.19	1.18	2.49	1.3	1.38
Bk	0.53	0.39	0.42	0.46	0.51	0.77	0.11	0
BN	14 800	12 100	11 900	14 300	19 100	13 700	15 000	4 290
BN(df)	8 260	6 700	8 570	9 150	14 600	9 500	9 400	3 280
BAs	1.57	1.79	1.44	1.44	1.3	1.37	1.57	1.29
Dk	3.11	3.08	2.9	2.89	3.96	4.9	1.36	0.66
DN	10 600	15 000	10 600	9 710	14 400	10 900	11 900	8 440
DN(df)	8 860	12 800	8 910	7 800	11 300	7 430	8 900	6 320
DAs	1.15	1.32	1.28	1.28	1.28	1.6	1.37	1.31
Rk	12.39	12.02	12.01	11.61	15.15	19.72	5.6	3.08
RN	9 220	17 700	13 000	10 800	13 800	12 100	12 000	9 160
RN(df)	8 490	13 900	11 000	8 820	12 000	8 420	9 630	7 350
RAs	1.07	1.51	1.32	1.31	1.27	1.64	1.32	1.25

first two components in the model could result in very misleading conclusions. It is usually a mistake to compare PCs of equivalent significance to each other, especially when their sizes are fairly similar.

Procrustes analysis can be used to answer fairly sophisticated questions. For example, in sensory research, are the results of a taste panel comparable to chemical measurements? If so, can the rather expensive and time-consuming taste panel be replaced by chromatography? A second use of procrustes analysis is to reduce the number of tests, an example being clinical trials. Sometimes 50 or more bacteriological tests are performed, but can these be reduced to 10 or fewer? A way to check this is by performing PCA on the results of all 50 tests and compare the scores plot when using a subset of 10 tests. If the two scores plots provide comparable information, the 10 selected tests are just as good as the full set of tests. This can be of significant economic benefit.

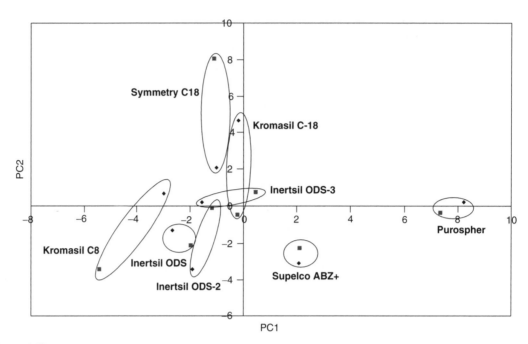

Figure 4.22
Comparison of scores plots for methanol and THF (case study 2)

4.4 Unsupervised Pattern Recognition: Cluster Analysis

Exploratory data analysis such as PCA is used primarily to determine general relationships between data. Sometimes more complex questions need to be answered, such as, do the samples fall into groups? Cluster analysis is a well established approach that was developed primarily by biologists to determine similarities between organisms. Numerical taxonomy emerged from a desire to determine relationships between different species, for example genera, families and phyla. Many textbooks in biology show how organisms are related using family trees.

The chemist also wishes to relate samples in a similar manner. Can protein sequences from different animals be related and does this tell us about the molecular basis of evolution? Can the chemical fingerprint of wines be related and does this tell us about the origins and taste of a particular wine? Unsupervised pattern recognition employs a number of methods, primarily cluster analysis, to group different samples (or objects) using chemical measurements.

4.4.1 Similarity

The first step is to determine the similarity between objects. Table 4.16 consists of six objects, 1–6, and five measurements, A–E. What are the similarities between the objects? Each object has a relationship to the remaining five objects. How can a numerical value of similarity be defined? A similarity matrix can be obtained, in which the similarity between each pair of objects is calculated using a numerical indicator. Note that it is possible to preprocess data prior to calculation of a number of these measures (see Section 4.3.6).

Table 4.16 Example of cluster analysis.

	A	B	C	D	E
1	0.9	0.5	0.2	1.6	1.5
2	0.3	0.2	0.6	0.7	0.1
3	0.7	0.2	0.1	0.9	0.1
4	0.1	0.4	1.1	1.3	0.2
5	1.0	0.7	2.0	2.2	0.4
6	0.3	0.1	0.3	0.5	0.1

Table 4.17 Correlation matrix.

	1	2	3	4	5	6
1	1.000					
2	−0.041	1.000				
3	0.503	0.490	1.000			
4	−0.018	0.925	0.257	1.000		
5	−0.078	0.999	0.452	0.927	1.000	
6	0.264	0.900	0.799	0.724	0.883	1.000

Four of the most popular ways of determining how similar objects are to each other are as follows.

1. *Correlation coefficient* between samples. A correlation coefficient of 1 implies that samples have identical characteristics, which all objects have with themselves. Some workers use the square or absolute value of a correlation coefficient, and it depends on the precise physical interpretation as to whether negative correlation coefficients imply similarity or dissimilarity. In this text we assume that the more negative is the correlation coefficient, the less similar are the objects. The correlation matrix is presented in Table 4.17. Note that the top right-hand side is not presented as it is the same as the bottom left-hand side. The higher is the correlation coefficient, the more similar are the objects.
2. *Euclidean distance*. The distance between two samples samples k and l is defined by

$$d_{kl} = \sqrt{\sum_{j=1}^{J} (x_{kj} - x_{lj})^2}$$

where there are j measurements and x_{ij} is the jth measurement on sample i, for example, x_{23} is the third measurement on the second sample, equalling 0.6 in Table 4.16. The smaller is this value, the more similar are the samples, so this distance measure works in an opposite manner to the correlation coefficient and strictly is a dissimilarity measure. The results are presented in Table 4.18. Although correlation coefficients vary between -1 and $+1$, this is not true for the Euclidean distance, which has no limit, although it is always positive. Sometimes the equation is presented in matrix format:

$$d_{kl} = \sqrt{(x_k - x_l) . (x_k - x_l)'}$$

Table 4.18 Euclidean distance matrix.

	1	2	3	4	5	6
1	0.000					
2	1.838	0.000				
3	1.609	0.671	0.000			
4	1.800	0.837	1.253	0.000		
5	2.205	2.245	2.394	1.600	0.000	
6	1.924	0.374	0.608	1.192	2.592	0.000

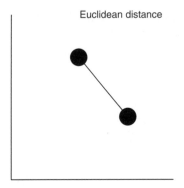

Euclidean distance

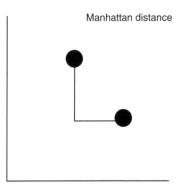

Manhattan distance

Figure 4.23
Euclidean and Manhattan distances

where the objects are row vectors as in Table 4.16; this method is easy to implement in Excel or Matlab.

3. *Manhattan distance.* This is defined slightly differently to the Euclidean distance and is given by

$$d_{kl} = \sum_{j=1}^{J} |x_{kj} - x_{lj}|$$

The difference between the Euclidean and Manhattan distances is illustrated in Figure 4.23. The values are given in Table 4.19; note the Manhattan distance will always be greater than (or in exceptional cases equal to) the Euclidean distance.

Table 4.19 Manhattan distance matrix.

	1	2	3	4	5	6
1	0					
2	3.6	0				
3	2.7	1.1	0			
4	3.4	1.6	2.3	0		
5	3.8	4.4	4.3	3.2	0	
6	3.6	0.6	1.1	2.2	5	0

4. *Mahalanobis distance.* This method is popular with many chemometricians and, whilst superficially similar to the Euclidean distance, it takes into account that some variables may be correlated and so measure more or less the same properties. The distance between objects k and l is best defined in matrix terms by

$$d_{kl} = \sqrt{(x_k - x_l).C^{-1}.(x_k - x_l)'}$$

where C is the variance–covariance matrix of the variables, a matrix symmetric about the diagonal, whose elements represent the covariance between any two variables, of dimensions $J \times J$. See Appendix A.3.1 for definitions of these parameters; note that one should use the population rather than sample statistics. This measure is very similar to the Euclidean distance except that the inverse of the variance–covariance matrix is inserted as a scaling factor. However, this method cannot easily be applied where the number of measurements (or variables) exceeds the number of objects, because the variance–covariance matrix would not have an inverse. There are some ways around this (e.g. when calculating spectral similarities where the number of wavelengths far exceeds the number of spectra), such as first performing PCA and then retaining the first few PCs for subsequent analysis. For a meaningful measure, the number of objects must be significantly greater than the number of variables, otherwise there are insufficient degrees of freedom for measurement of this parameter. In the case of Table 4.16, the Mahalanobis distance would be an inappropriate measure unless either the number of samples is increased or the number of variables decreased. This distance metric does have its uses in chemometrics, but more commonly in the areas of supervised pattern recognition (Section 4.5) where its properties will be described in more detail. Note in contrast that if the number of variables is small, although the Mahalanobis distance is an appropriate measure, correlation coefficients are less useful.

There are several other related distance measures in the literature, but normally good reasons are required if a very specialist distance measure is to be employed.

4.4.2 Linkage

The next step is to link the objects. The most common approach is called *agglomerative* clustering whereby single objects are gradually connected to each other in groups. Any similarity measure can be used in the first step, but for simplicity we will illustrate this using only the correlation coefficients of Table 4.17. Similar considerations apply to all the similarity measures introduced in Section 4.4.1, except that in the other cases the lower the distance the more similar the objects.

1. From the raw data, find the two objects that are most similar (closest together). According to Table 4.17, these are objects 2 and 5, as they have the highest correlation coefficient (=0.999) (remember that because only five measurements have been recorded there are only four degrees of freedom for calculation of correlation coefficients, meaning that high values can be obtained fairly easily).

2. Next, form a 'group' consisting of the two most similar objects. Four of the original objects (1, 3 and 6) and a group consisting of objects 2 and 5 together remain, leaving a total of five new groups, four consisting of a single original object and one consisting of two 'clustered' objects.

3. The tricky bit is to decide how to represent this new grouping. As in the case of distance measures, there are several approaches. The main task is to recalculate the numerical similarity values between the new group and the remaining objects. There are three principal ways of doing this.

   (a) *Nearest neighbour.* The similarity of the new group from all other groups is given by the *highest* similarity of either of the original objects to each other object. For example, object 6 has a correlation coefficient of 0.900 with object 2, and 0.883 with object 5. Hence the correlation coefficient with the new combined group consisting of objects 2 and 5 is 0.900.

   (b) *Furthest neighbour.* This is the opposite to nearest neighbour, and the *lowest* similarity is used, 0.883 in our case. Note that the furthest neighbour method of linkage refers only to the calculation of similarity measures after new groups are formed, and the two groups (or objects) with highest similarity are always joined first.

   (c) *Average linkage.* The average similarity is used, 0.892 in our case. There are, in fact, two different ways of doing this, according to the size of each group being joined together. Where they are of equal size (e.g. each consists of one object), both methods are equivalent. The two different ways are as follows.

   - *Unweighted.* If group A consists of N_A objects and group B of N_B objects, the new similarity measure is given by

   $$s_{AB} = (N_A s_A + N_B s_B)/(N_A + N_B)$$

   - *Weighted.* The new similarity measure is given by

   $$s_{AB} = (s_A + s_B)/2$$

The terminology indicates that for the unweighted method, the new similarity measure takes into consideration the number of objects in a group, the conventional terminology possibly being the opposite to what is expected. For the first link, each method provides identical results.

There are numerous further linkage methods, but it would be rare that a chemist needs to use too many combination of similarity and linkage methods, however, a good rule of thumb is to check the result of using a combination of approaches.

The new data matrix using nearest neighbour clustering is presented in Table 4.20, with the new values shaded. Remember that there are many similarity measures and methods for linking, so this table represents only one possible way of handling the information.

4.4.3 Next Steps

The next steps consist of continuing to group the data just as above, until all objects have joined one large group. Since there are six original objects, there will be five steps before this is achieved. At each step, the most similar pair of objects or clusters are identified, then they are combined into one new cluster, until all objects have been joined. The calculation is illustrated in Table 4.20, using nearest neighbour linkage, with the most similar objects at each step indicated in bold type, and the new similarity measures shaded. In this particular example, all objects ultimately belong to the same cluster, although arguably object 1 (and possibly 3) does not have a very high similarity to the main group. In some cases, several clusters can be formed, although ultimately one large group is usually formed.

It is normal then to determine at what similarity measure each object joined a larger group, and so which objects resemble each other most.

4.4.4 Dendrograms

Often the result of hierarchical clustering is presented in the form of a dendrogram (sometimes called a 'tree diagram'). The objects are organised in a row, according

Table 4.20 Nearest neighbour cluster analysis, using correlation coefficients for similarity measures, and data in Table 4.16.

	1	2	3	4	5	6
1	1.000					
2	−0.041	1.000				
3	0.503	0.490	1.000			
4	−0.018	0.925	0.257	1.000		
5	−0.078	**0.999**	0.452	0.927	1.000	
6	0.264	0.900	0.799	0.724	0.883	1.000

	1	2&5	3	4	6
1	1.000				
2&5	−0.041	1.000			
3	0.503	0.490	1.000		
4	−0.018	**0.927**	0.257	1.000	
6	0.264	0.900	0.799	0.724	1.000

	1	2&5&4	3	6
1	1.000			
2&5&4	−0.018	1.000		
3	0.503	0.490	1.000	
6	0.264	**0.900**	0.799	1.000

	1	2&5&4&6	3
1	1.000		
2&5&4&6	0.264	1.000	
3	0.503	**0.799**	1.000

	1	2&5&4&6&3
1	1.000	
2&5&4&6&3	**0.503**	1.000

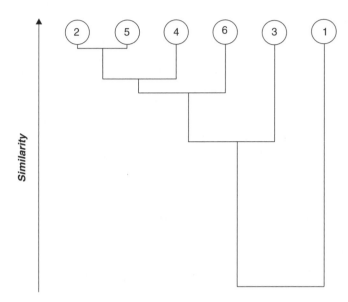

Figure 4.24
Dendrogram for cluster analysis example

to their similarities: the vertical axis represents the similarity measure at which each successive object joins a group. Using nearest neighbour linkage and correlation coefficients for similarities, the dendrogram for Table 4.20 is presented in Figure 4.24. It can be seen that object 1 is very different from the others. In this case all the other objects appear to form a single group, but other clustering methods may give slightly different results. A good approach is to perform several different methods of cluster analysis and compare the results. If similar clusters are obtained, no matter which method is employed, we can rely on the results.

4.5 Supervised Pattern Recognition

Classification (often called supervised pattern recognition) is at the heart of chemistry. Mendeleev's periodic table, grouping of organic compounds by functionality and listing different reaction types all involve classification. Much of traditional chemistry involves grouping chemical behaviour. Most early texts in organic, inorganic and analytical chemistry are systematically divided into chapters according to the behaviour or structure of the underlying compounds or techniques.

So the modern chemist also has a significant need for classification. Can a spectrum be used to determine whether a compound is a ketone or an ester? Can the chromatogram of a tissue sample be used to determine whether a patient is cancerous or not? Can we record the spectrum of an orange juice and decide its origin? Is it possible to monitor a manufacturing process by near-infrared spectroscopy and decide whether the product is acceptable or not? Supervised pattern recognition is used to assign samples to a number of groups (or classes). It differs from cluster analysis where, although the relationship between samples is important, there are no predefined groups.

4.5.1   General Principles

Although there are numerous algorithms in the literature, chemists and statisticians often use a common strategy for classification no matter what algorithm is employed.

4.5.1.1   Modelling the Training Set

The first step is normally to produce a mathematical model between some measurements (e.g. spectra) on a series of objects and their known groups. These objects are called a *training set*. For example, a training set might consist of the near-infrared spectra of 30 orange juices, 10 known to be from Spain, 10 known to be from Brazil and 10 known to be adulterated. Can we produce a mathematical equation that predicts the class to which an orange juice belongs from its spectrum?

Once this has been done, it is usual to determine how well the model predicts the groups. Table 4.21 illustrates a possible scenario. Of the 30 spectra, 24 are correctly classified, as indicated along the diagonals. Some classes are modelled better than others, for example, nine out of 10 of the Spanish orange juices are correctly classified, but only seven of the Brazilians. A parameter representing the percentage correctly classified (%CC) can be calculated. After application of the algorithm, the origin (or class) of each spectrum is predicted. In this case, the overall value of %CC is 80 %. Note that some groups appear to be better modelled than others, but also that the training set is fairly small, so it may not be particularly significant that seven out of 10 are correctly classified in one group compared with nine in another. A difficulty in many real situations is that it can be expensive to perform experiments that result in large training sets. There appears to be some risk of making a mistake, but many spectroscopic techniques are used for screening, and there is a high chance that suspect orange juices (e.g. those adulterated) would be detected, which could then be subject to further detailed analysis. Chemometrics combined with spectroscopy acts like a 'sniffer dog' in a customs checkpoint trying to detect drugs. The dog may miss some cases, and may even get excited when there are no drugs, but there will be a good chance the dog is correct. Proof, however, only comes when the suitcase is opened. Spectroscopy is a common method for screening. Further investigations might involve subjecting a small number of samples to intensive, expensive, and in some cases commercially or politically sensitive tests, and a laboratory can only afford to look in detail at a portion of samples, just as customs officers do not open every suitcase.

4.5.1.2   Test Sets and Cross-Validation

However, normally training sets give fairly good predictions, because the model itself has been formed using these datasets, but this does not mean that the method is yet

Table 4.21 Predictions from a training set.

Known	Predicted			Correct	%CC
	Spain	Brazil	Adulterated		
Spain	9	0	1	9	90
Brazil	1	7	2	7	70
Adulterated	0	2	8	8	80
Overall				24	80

Table 4.22 Predictions from a test set.

	Predicted			Correct	%CC
	Spain	Brazil	Adulterated		
Spain	5	3	2	5	50
Brazil	1	6	3	6	60
Adulterated	4	2	4	4	40
Overall				15	50

safe to use in practical situations. A recommended next step is to test the quality of predictions using an independent *test* set. This is a series of samples that has been left out of the original calculations, and is like a 'blind test'. These samples are assumed to be of unknown class membership at first, then the model from the training set is applied to these extra samples. Table 4.22 presents the predictions from a test set (which does not necessarily need to be the same size as the training set), and we see that now only 50 % are correctly classified so the model is not particularly good. The %CC will almost always be lower for the test set.

Using a test set to determine the quality of predictions is a form of *validation*. The test set could be obtained, experimentally, in a variety of ways, for example 60 orange juices might be analysed in the first place, and then randomly divided into 30 for the training set and 30 for the test set. Alternatively, the test set could have been produced in an independent laboratory.

A second approach is *cross-validation*. This technique was introduced in the context of PCA in Section 4.3.3.2, and other applications will be described in Chapter 5 Section 5.6.2, and so will be introduced only briefly below. Only a single training set is required, but what happens is that one (or a group) of objects is removed at a time, and a model determined on the remaining samples. Then the predicted class membership of the object (or set of objects) left out is tested. This procedure is repeated until all objects have been left out. For example, it would be possible to produce a class model using 29 orange juices. Is the 30th orange juice correctly classified? If so, this counts towards the percentage correctly classified. Then, instead of removing the 30th orange juice, we decide to remove the 29th and see what happens. This is repeated 30 times, which leads to a value of %CC for cross-validation. Normally the cross-validated %CC is lower (worse) than that for the training set. In this context cross-validation is not used to obtain a numerical error, unlike in PCA, but the proportion assigned to correct groups. However, if the %CC of the training and test set and cross-validation are all very similar, the model is considered to be a good one. Where alarm bells ring is if the %CC is high for the training set but significantly lower when using one of the two methods for validation. It is recommended that all classification methods be validated in some way, but sometimes there can be limitations on the number of samples available. Note that there are many very different types of cross-validation available according to methods, so the use in this section differs strongly from other applications discussed in this text.

4.5.1.3 Improving the Data

If the model is not very satisfactory, there are a number of ways to improve it. The first is to use a different computational algorithm. The second is to modify the existing method – a common approach might involve wavelength selection in spectroscopy; for example, instead of using an entire spectrum, many wavelengths which are not very meaningful, can we select the most diagnostic parts of the spectrum? Finally, if all else fails, the analytical technique might not be up to scratch. Sometimes a low %CC may be acceptable in the case of screening; however, if the results are to be used to make a judgement (for example in the validation of the quality of pharmaceutical products into 'acceptable' and 'unacceptable' groups), a higher %CC of the validation sets is mandatory. The limits of acceptability are not primarily determined statistically, but according to physical needs.

4.5.1.4   Applying the Model

Once a satisfactory model is available, it can then be applied to unknown samples, using analytical data such as spectra or chromatograms, to make predictions. Usually by this stage, special software is required that is tailor-made for a specific application and measurement technique. The software will also have to determine whether a new sample really fits into the training set or not. One major difficulty is the detection of outliers that belong to none of the previously studied groups, for example if a Cypriot orange juice sample was measured when the training set consists just of Spanish and Brazilian orange juices. In areas such as clinical and forensic science, outlier detection can be important, indeed an incorrect conviction or inaccurate medical diagnosis could be obtained otherwise.

Another important consideration is the stability of the method over time; for example, instruments tend to perform slightly differently every day. Sometimes this can have a serious influence on the classification ability of chemometrics algorithms. One way round this is to perform a small test on the instrument on a regular basis.

However, there have been some significant successes, a major area being in industrial process control using near-infrared spectroscopy. A manufacturing plant may produce samples on a continuous basis, but there are a large number of factors that could result in an unacceptable product. The implications of producing substandard batches may be economic, legal and environmental, so continuous testing using a quick and easy method such as on-line spectroscopy is valuable for the rapid detection of whether a process is going wrong. Chemometrics can be used to classify the spectra into acceptable or otherwise, and so allow the operator to close down a manufacturing plant in real time if it looks as if a batch can no longer be assigned to the group of acceptable samples.

4.5.2   Discriminant Analysis

Most traditional approaches to classification in science are called *discriminant analysis* and are often also called forms of 'hard modelling'. The majority of statistically based software packages such as SAS, BMDP and SPSS contain substantial numbers of procedures, referred to by various names such as linear (or Fisher) discriminant analysis and canonical variates analysis. There is a substantial statistical literature in this area.

4.5.2.1   Univariate Classification

The simplest form of classification is *univariate*, where one measurement or variable is used to divide objects into groups. An example may be a blood alcohol reading.

If a reading on a meter in a police station is above a certain level, then the suspect will be prosecuted for drink driving, otherwise not. Even in such a simple situation, there can be ambiguities, for example measurement errors and metabolic differences between people.

4.5.2.2 Multivariate Models

More often, several measurements are required to determine the group to which a sample belongs. Consider performing two measurements and producing a graph of the values of these measurements for two groups, as in Figure 4.25. The objects represented

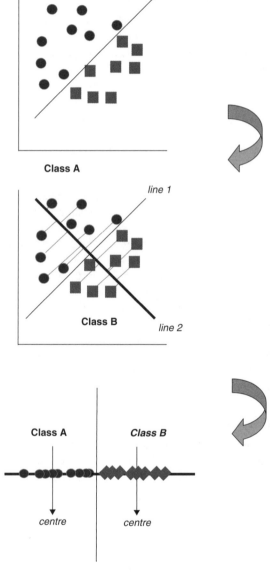

Figure 4.25
Discrimination between two classes, and projections

by circles are clearly distinct from those represented by squares, but neither of the two measurements alone can discriminate between these groups, and therefore both are essential for classification. It is possible, however, to draw a line between the two groups. If above the line, an object belongs to class A, otherwise to class B.

Graphically this can be represented by *projecting* the objects on to a line at right angles to the discriminating line, as demonstrated in the figure. The projection can now be converted to a position along line 2 of the figure. The distance can be converted to a number, analogous to a 'score'. Objects with lower values belong to class A, whereas those with higher values belong to class B. It is possible to determine class membership simply according to whether the value is above or below a divisor. Alternatively, it is possible to determine the centre of each class along the projection and if the distance to the centre of class A is greater than that to class B, the object is placed in class A, and vice versa, but this depends on each class being roughly equally diffuse.

It is not always possible to divide the classes exactly into two groups by this method (see Figure 4.26), but the misclassified samples are far from the centre of both classes, with two class distances that are approximately equal. It would be possible to define a boundary towards the centre of the overall dataset, where classification is deemed to be ambiguous.

The data can also be presented in the form of a distance plot, where the two axes are the distances to the centres of the projections of each class as presented in Figure 4.27. This figure probably does not tell us much that cannot be derived from Figure 4.26. However, the raw data actually consist of more than one measurement, and it is possible to calculate the Euclidean class distance using the raw two-dimensional information, by computing the centroids of each class in the raw data rather than one-dimensional projection. Now the points can fall anywhere on a plane, as illustrated in Figure 4.28. This graph is often called a class distance plot and can still be divided into four regions:

1. top left: almost certainly class A;
2. bottom left: possibly a member of both classes, but it might be that we do not have enough information;
3. bottom right: almost certainly class B;
4. top right: unlikely to be a member of either class, sometimes called an outlier.

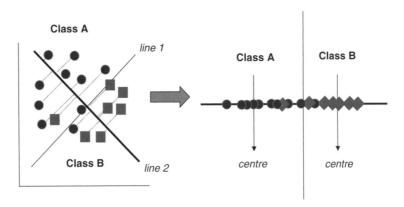

Figure 4.26
Discrimination where exact cut-off is not possible

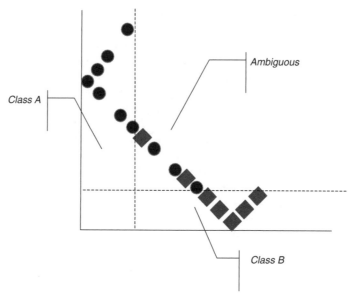

Figure 4.27
Distance plot to class centroids of the projection in Figure 4.26

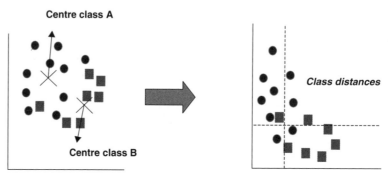

Figure 4.28
Class distance plot

In chemistry, these four divisions are perfectly reasonable. For example, if we try to use spectra to classify compounds into ketones and esters, there may be some compounds that are both or neither. If, on the other hand, there are only two possible classifications, for example whether a manufacturing sample is acceptable or not, a conclusion about objects in the bottom left or top right is that the analytical data is insufficiently good to allow us to assign conclusively a sample to a graph. This is a valuable conclusion, for example it is helpful to tell a laboratory that their clinical diagnosis or forensic test is inconclusive and that if they want better evidence they should perform more experiments or analyses.

4.5.2.3 Mahalanobis Distance and Linear Discriminant Functions

Previously we discussed the use of different similarity measures in cluster analysis (Section 4.4.1), including various approaches for determining the distance between objects. Many chemometricians use the Mahalanobis distance, sometimes called the 'statistical' distance, between objects, and we will expand on the concept below.

In areas such as spectroscopy it is normal that some wavelengths or regions of the spectra are more useful than others for discriminant analysis. This is especially true in near-infrared (NIR) spectroscopy. Also, different parts of a spectrum might be of very different intensities. Finally, some classes are more diffuse than others. A good example is in forensic science, where forgeries often have a wider dispersion to legitimate objects. A forger might work in his or her back room or garage, and there can be a considerable spread in quality, whereas the genuine article is probably manufactured under much stricter specifications. Hence a large deviation from the mean may not be significant in the case of a class of forgeries. The Mahalanobis distance takes this information into account. Using a Euclidean distance each measurement assumes equal significance, so correlated variables, which may represent an irrelevant feature, can have a disproportionate influence on the analysis.

In supervised pattern recognition, a major aim is to define the distance of an object from the centre of a class. There are two principle uses of statistical distances. The first is to obtain a measurement analogous to a score, often called the linear discriminant function, first proposed by the statistician R A Fisher. This differs from the distance above in that it is a single number if there are only two classes. It is analogous to the distance along line 2 in Figure 4.26, but defined by

$$f_i = (\overline{x}_A - \overline{x}_B).C_{AB}^{-1}.x_{i'}$$

where

$$C_{AB} = \frac{(N_A - 1)C_A + (N_B - 1)C_B}{(N_A + N_B - 2)}$$

which is often called the *pooled* variance–covariance matrix, and can be extended to any number of groups; N_A represents the number of objects in group A, and C_A the variance–covariance matrix for this group (whose diagonal elements correspond to the variance of each variable and the off-diagonal elements the covariance – use the population rather than sample formula), with $\overline{x}_A$ the corresponding centroid. Note that the mathematics becomes more complex if there are more than two groups. This function can take on negative values.

The second is to determine the Mahalanobis distance to the centroid of any given group, a form of *class distance*. There will be a separate distance to the centre of each group defined, for class A, by

$$d_{iA} = \sqrt{(x_i - \overline{x}_A).C_A^{-1}.(x_i - \overline{x}_A)'}$$

where x_i is a row vector for sample i and $\overline{x}_A$ is the *mean* measurement (or centroid) for class A. This measures the scaled distance to the centroid of a class analogous to Figure 4.28, but scaling the variables using the Mahalanobis rather than Euclidean criterion.

An important difficulty with using this distance is that the number of objects much be significantly larger than the number of measurements. Consider the case of Mahalanobis distance being used to determine within group distances. If there are J measurements than there must be at least $J + 2$ objects for there to be any discrimination. If there are less than $J + 1$ measurements, the variance–covariance matrix will not have an inverse. If there are $J + 1$ objects, the estimated squared distance to the centre of the cluster will equal J for each object no matter what its position in the group, and discrimination will only be possible if the class consists of at least $J + 2$ objects, unless some measurements are discarded or combined. This is illustrated in Table 4.23 (as can be verified computationally), for a simple dataset.

Note how the average squared distance from the mean of the dataset over all objects always equals 5 ($=J$) no matter how large the group. It is important always to understand the fundamental properties of this distance measure, especially in spectroscopy or chromatography where there are usually a large number of potential variables which must first be reduced, sometimes by PCA.

We will illustrate the methods using a simple numerical example (Table 4.24), consisting of 19 samples, the first nine of which are members of group A, and the second 10 of group B. The data are presented in Figure 4.29. Although the top left-hand

Table 4.23 Squared Mahalanobis distance from the centre of a dataset as increasing number of objects are included.

	A	B	C	D	E	6 objects	7 objects	8 objects	9 objects
1	0.9	0.5	0.2	1.6	1.5	5	5.832	6.489	7.388
2	0.3	0.3	0.6	0.7	0.1	5	1.163	1.368	1.659
3	0.7	0.7	0.1	0.9	0.5	5	5.597	6.508	6.368
4	0.1	0.4	1.1	1.3	0.2	5	5.091	4.457	3.938
5	1	0.7	2.6	2.1	0.4	5	5.989	6.821	7.531
6	0.3	0.1	0.5	0.5	0.1	5	5.512	2.346	2.759
7	0.9	0.1	0.5	0.6	0.7		5.817	5.015	4.611
8	0.3	1.2	0.7	0.1	1.4			6.996	7.509
9	1	0.7	0.6	0.5	0.9				3.236

Table 4.24 Example of discriminant analysis.

Class	Sample	x_1	x_2
A	1	79	157
A	2	77	123
A	3	97	123
A	4	113	139
A	5	76	72
A	6	96	88
A	7	76	148
A	8	65	151
A	9	32	88
B	10	128	104
B	11	65	35
B	12	77	86
B	13	193	109
B	14	93	84
B	15	112	76
B	16	149	122
B	17	98	74
B	18	94	97
B	19	111	93

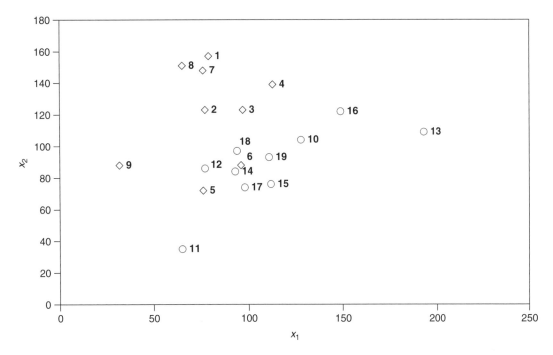

Figure 4.29
Graph of data in Table 4.24: class A is indicated by diamonds and class B by circles

corner corresponds mainly to group A, and the bottom right-hand corner to group B, no single measurement is able to discriminate and there is a region in the centre where both classes are represented, and it is not possible to draw a line that unambiguously distinguishes the two classes.

The calculation of the linear discriminant function is presented in Table 4.25 and the values are plotted in Figure 4.30. It can be seen that objects 5, 6, 12 and 18 are not easy to classify. The centroids of each class in this new dataset using the linear discriminant function can be calculated, and the distance from these values could be calculated; however, this would result in a diagram comparable to Figure 4.27, missing information obtained by taking two measurements.

The class distances using both variables and the Mahalanobis method are presented in Table 4.26. The predicted class for each object is the one whose centroid it is closest to. Objects 5, 6, 12 and 18 are still misclassified, making a %CC of 79 %. However, it is not always a good idea to make hard and fast deductions, as discussed above, as in certain situations an object could belong to two groups simultaneously (e.g. a compound having two functionalities), or the quality of the analytical data may be insufficient for classification. The class distance plot is presented in Figure 4.31. The data are better spread out compared with Figure 4.29 and there are no objects in the top right-hand corner; the misclassified objects are in the bottom left-hand corner. The boundaries for each class can be calculated using statistical considerations, and are normally available from most packages. Depending on the aim of the analysis, it is possible to select samples that are approximately equally far from the centroids of both classes and either reject them or subject them to more measurements.

Table 4.25 Calculation of discriminant function for data in Table 4.24.

	Class A		Class B	
Covariance matrix				
	466.22	142.22	1250.2	588.1
	142.22	870.67	588.1	512.8
Centroid				
	79	121	112	88

$$(\overline{x}_A - \overline{x}_B)$$

−33	33	

$$C_{AB}$$

881.27	378.28
378.28	681.21

$$C_{AB}^{-1}$$

0.00149	−0.00083
−0.0008	0.00198

$$(\overline{x}_A - \overline{x}_B).C_{AB}^{-1}$$

−0.0765	0.0909

Linear discriminant function

1	8.23	11	−1.79
2	5.29	12	1.93
3	3.77	13	−4.85
4	4.00	14	0.525
5	0.73	15	−1.661
6	0.66	16	−0.306
7	7.646	17	−0.77
8	8.76	18	1.63
9	5.556	19	−0.03
10	−0.336		

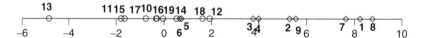

Figure 4.30
Linear discriminant function

Whereas the results in this section could probably be obtained fairly easily by inspecting the original data, numerical values of class membership have been obtained which can be converted into probabilities, assuming that the measurement error is normally distributed. In most real situations, there will be a much larger number of measurements, and discrimination (e.g. by spectroscopy) is not easy to visualise without further data analysis. Statistics such as %CC can readily be obtained from the data, and it is also possible to classify unknowns or validation samples as discussed in Section 4.5.1 by this means. Many chemometricians use the Mahalanobis distance as defined above, but the normal Euclidean distance or a wide range of other measures can also be employed, if justified by the data, just as in cluster analysis.

Table 4.26 Mahalanobis distances from class centroids.

Object	Distance to Class A	Distance to Class B	True classification	Predicted classification
1	1.25	5.58	A	A
2	0.13	3.49	A	A
3	0.84	2.77	A	A
4	1.60	3.29	A	A
5	1.68	1.02	A	B
6	1.54	0.67	A	B
7	0.98	5.11	A	A
8	1.36	5.70	A	A
9	2.27	3.33	A	A
10	2.53	0.71	B	B
11	2.91	2.41	B	B
12	1.20	1.37	B	A
13	5.52	2.55	B	B
14	1.57	0.63	B	B
15	2.45	0.78	B	B
16	3.32	1.50	B	B
17	2.04	0.62	B	B
18	1.21	1.25	B	A
19	1.98	0.36	B	B

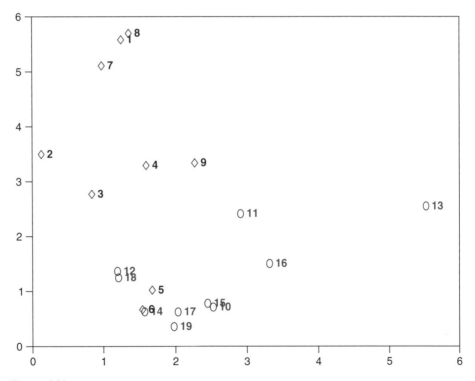

Figure 4.31
Class distance plot: horizontal axis = class A, vertical axis = class B

4.5.2.4 Extending the Methods

There are numerous extensions to the basic method in the section above, some of which are listed below.

- Probabilities of class membership and class boundaries or confidence intervals can be constructed, assuming a multivariate normal distribution.
- Discriminatory power of variables. This is described in a different context in Section 4.5.3, in the context of SIMCA, but similar parameters can be obtained for any method of discrimination. This can have practical consequences; for example, if variables are expensive to measure it is possible to reduce the expense by making only the most useful measurements.
- The method can be extended to any number of classes. It is easy to calculate the Mahalanobis distance to several classes, and determine which is the most appropriate classification of an unknown sample, simply by finding the smallest class distance. More discriminant functions are required, however, and the computation can be rather complicated.
- Instead of using raw data, it is possible to use the PCs of the data. This acts as a form of variable reduction, but also simplifies the distance measures, because the variance–covariance matrix will only contain nonzero elements on the diagonals. The expressions for Mahalanobis distance and linear discriminant functions simplify dramatically.
- One interesting modification involves the use of Bayesian classification functions. Such concepts have been introduced previously (see Chapter 3) in the context of signal analysis. The principle is that membership of each class has a predefined (prior) probability, and the measurements are primarily used to refine this. If we are performing a routine screen of blood samples for drugs, the chances might be very low (less than 1 %). In other situations an expert might already have done preliminary tests and suspects that a sample belongs to a certain class, so the prior probability of class membership is much higher than in the population as a whole; this can be taken into account. Normal statistical methods assume that there is an equal chance of membership of each class, but the distance measure can be modified as follows:

$$d_{iA} = \sqrt{(x - \overline{x}_A).C_A^{-1}.(x_i - \overline{x}_A)' + 2\ln(q_A)}$$

where q_A is the prior probability of membership of class A. The prior probabilities of membership of all relevant classes must add up to 1. It is even possible to use the Bayesian approach to refine class membership as more data are acquired. Screening will provide a probability of class membership which can then be used as prior probability for a more rigorous test, and so on.
- Several other functions such as the *quadratic* and *regularised* discriminant functions have been proposed and are suitable under certain circumstances.

It is always important to examine each specific paper and software package for details on the parameters that have been calculated. In many cases, however, fairly straightforward approaches suffice.

4.5.3 SIMCA

The SIMCA method, first advocated by the S. Wold in the early 1970s, is regarded by many as a form of soft modelling used in chemical pattern recognition. Although there are some differences with linear discriminant analysis as employed in traditional statistics, the distinction is not as radical as many would believe. However, SIMCA has an important role in the history of chemometrics so it is important to understand the main steps of the method.

4.5.3.1 Principles

The acronym SIMCA stands for *soft independent modelling of class analogy*. The idea of soft modelling is illustrated in Figure 4.32. Two classes can overlap (and hence are 'soft'), and there is no problem with an object belonging to both (or neither) class simultaneously. In most other areas of statistics, we insist that an object belongs to a discrete class, hence the concept of hard modelling. For example, a biologist trying to determine the sex of an animal from circumstantial evidence (e.g. urine samples) knows that the animal cannot simultaneously belong to two sexes at the same time, and a forensic scientist trying to determine whether a banknote is forged or not knows

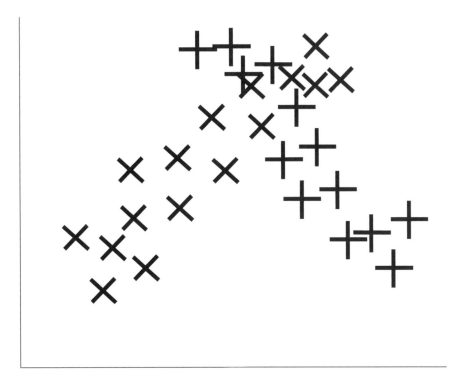

Figure 4.32
Two overlapping classes

that there can be only one true answer: if this appears not to be so, the problem lies with the quality of the evidence. The philosophy of soft modelling is that, in many situations in chemistry, it is entirely legitimate for an object to fit into more than one class simultaneously, for example a compound may have an ester and an alkene group, and so will exhibit spectroscopic characteristics of both functionalities, so a method that assumes that the answer must be either an ester or an alkene is unrealistic. In practice, though, it is possible to calculate class distances from discriminant analysis (Section 4.5.2.3) that are close to two or more groups.

Independent modelling of classes, however, is a more useful feature. After making a number of measurements on ketones and alkenes, we may decide to include amides in the model. Figure 4.33 represents the effect of this additional class which can be added independently to the existing model without any changes. In contrast, using classical discriminant analysis the entire modelling procedure must be repeated if extra numbers of groups are added, since the pooled variance–covariance matrix must be recalculated.

4.5.3.2   Methodology

The main steps of SIMCA are as follows.

Principal Components Analysis

Each group is independently modelled using PCA. Note that each group could be described by a different number of PCs. Figure 4.34 represents two groups, each

Figure 4.33
Three overlapping classes

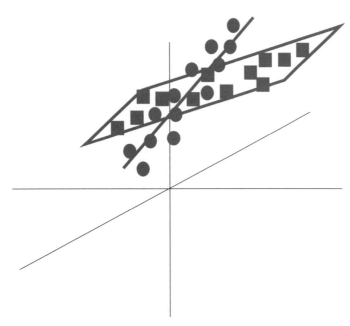

Figure 4.34
Two groups obtained from three measurements

characterised by three raw measurements, e.g. chromatographic peak heights or physical properties. However, one group falls mainly on a straight line, defined as the first
principal component of the group, whereas the other falls roughly on a plane whose
axes are defined by the first two principal components of this group. When we perform
discriminant analysis we can also use PCs prior to classification, but the difference is
that the PCs are of the entire dataset (which may consist of several groups) rather than
of each group separately.

It is important to note that a number of methods have been proposed for determining
how many PCs are most suited to describe a class, which have been described in
Section 4.3.3. The original advocates of SIMCA used cross-validation, but there is no
reason why one cannot 'pick and mix' various steps in different methods.

Class Distance

The class distance can be calculated as the geometric distance from the PC models; see
the illustration in Figure 4.35. The unknown is much closer to the plane formed than
the line, and so is tentatively assigned to this class. A more elaborate approach is often
employed in which each group is bounded by a region of space, which represents 95 %
confidence that a particular object belongs to a class. Hence geometric class distances
can be converted to statistical probabilities.

Modelling Power

The *modelling power* of each variable for each separate class is defined by

$$M_j = 1 - s_{jresid}/s_{jraw}$$

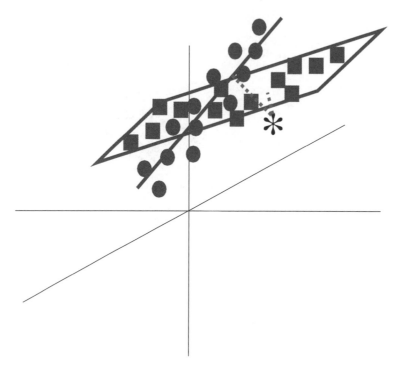

Figure 4.35
Class distance of unknown object (represented by an asterisk) to two classes in Figure 4.34

where s_{jraw} is the standard deviation of the variable in the raw data and s_{jresid} the standard deviation of the variable in the residuals given by

$$E = X - T.P$$

which is the difference between the observed data and the PC model for the class. This is illustrated in Table 4.27 for a single class, with the following steps.

1. The raw data consist of seven objects and six measurements, forming a 7×6 matrix. The standard deviation of all six variables in calculated.
2. PCA is performed on this data and two PCs are retained. The scores (T) and loadings (P) matrices are computed.
3. The original data are estimated by PCA by multiplying T and P together.
4. The residuals, which are the difference between the datasets obtained in steps 1 and 3, are computed (note that there are several alternative ways of doing this). The standard deviation of the residuals for all six variables is computed.
5. Finally, using the standard deviations obtained in steps 1 and 4, the modelling power of the variables for this class is obtained.

The modelling power varies between 1 (excellent) and 0 (no discrimination). Variables with M below 0.5 are of little use. In this example, it can be seen that variables 4 and 6 are not very helpful, whereas variables 2 and 5 are extremely useful. This information can be used to reduce the number of measurements.

Table 4.27 Calculation of modelling power in SIMCA.

1. Raw data

24.990	122.326	68.195	28.452	105.156	48.585
14.823	51.048	57.559	21.632	62.764	26.556
28.612	171.254	114.479	43.287	159.077	65.354
41.462	189.286	113.007	25.289	173.303	50.215
5.168	50.860	37.123	16.890	57.282	35.782
20.291	88.592	63.509	14.383	85.250	35.784
31.760	151.074	78.699	34.493	141.554	56.030
s_{jraw} 10.954	51.941	26.527	9.363	43.142	12.447

2. PCA

Scores		Loadings	
185.319	0.168	0.128	−0.200
103.375	19.730	0.636	−0.509
273.044	9.967	0.396	0.453
288.140	−19.083	0.134	0.453
92.149	15.451	0.594	−0.023
144.788	3.371	0.229	0.539
232.738	−5.179		

3. Data estimated by PCA

23.746	117.700	73.458	24.861	110.006	42.551
9.325	55.652	49.869	22.754	60.906	34.313
33.046	168.464	112.633	41.028	161.853	67.929
40.785	192.858	105.454	29.902	171.492	55.739
8.739	50.697	43.486	19.316	54.342	29.436
17.906	90.307	58.859	20.890	85.871	34.990
30.899	150.562	89.814	28.784	138.280	50.536

4. Residuals

1.244	4.626	−5.263	3.591	−4.850	6.034
5.498	−4.604	7.689	−1.122	1.858	−7.757
−4.434	2.790	1.846	2.259	−2.776	−2.575
0.677	−3.572	7.553	−4.612	1.811	−5.524
−3.570	0.163	−6.363	−2.426	2.940	6.346
2.385	−1.715	4.649	−6.508	−0.621	0.794
0.861	0.512	−11.115	5.709	3.274	5.494
s_{jresid} 3.164	3.068	6.895	4.140	2.862	5.394

5. Modelling power

0.711	0.941	0.740	0.558	0.934	0.567

Discriminatory Power

Another measure is how well a variable discriminates between two classes. This is distinct from modelling power – being able to model one class well does not necessarily imply being able to discriminate two groups effectively. In order to determine this, it is necessary to fit each sample to both class models. For example, fit sample 1 to the PC model of both class A and class B. The residual matrices are then calculated, just as for discriminatory power, but there are now four such matrices:

1. samples in class A fitted to the model of class A;
2. samples in class A fitted to the model of class B;

3. samples in class B fitted to the model of class B;
4. samples in class B fitted to the model of class A.

We would expect matrices 2 and 4 to be a worse fit than matrices 1 and 3. The standard deviations are then calculated for these matrices to give

$$D_j = \sqrt{\frac{{}^{classAmodelB}s_{jresid}{}^2 + {}^{classBmodelA}s_{jresid}{}^2}{{}^{classAmodelA}s_{jresid}{}^2 + {}^{classBmodelB}s_{jresid}{}^2}}$$

The bigger the value, the higher is the discriminatory power. This could be useful information, for example if clinical or forensic measurements are expensive, so allowing the experimenter to choose only the most effective measurements. Discriminatory power can be calculated between any two classes.

4.5.3.3 Validation

Like all methods for supervised pattern recognition, testing and cross-validation are possible in SIMCA. There is a mystique in the chemometrics literature whereby some general procedures are often mistakenly associated with a particular package or algorithm; this is largely because the advocates promote specific strategies that form part of papers or software that is widely used. It also should be recognised that there are two different needs for validation: the first is to determine the optimum number of PCs for a given class model, and the second to determine whether unknowns are classified adequately.

4.5.4 Discriminant PLS

An important variant that is gaining popularity in chemometrics circles is called discriminant PLS (DPLS). We describe the PLS method in more detail in Chapter 5, Section 5.4 and also Appendix A.2, but it is essentially another approach to regression, complementary to MLR with certain advantages in particular situations. The principle is fairly simple. For each class, set up a model

$$\hat{c} = T.q$$

where T are the PLS scores obtained from the original data, q is a vector, the length equalling the number of significant PLS components, and $\hat{c}$ is a class membership function; this is obtained by PLS regression from an original c vector whose elements have values of 1 if an object is a member of a class and 0 otherwise and an X matrix consisting of the original preprocessed data. If there are three classes, then a matrix $\hat{C}$, of dimensions $I \times 3$, can be obtained from a training set consisting of I objects; the closer each element is to 1, the more likely an object is to be a member of the particular class. All the normal procedures of training and test sets, and cross-validation, can be used with DPLS, and various extensions have been proposed in the literature.

MLR regression on to a class membership function would not work well because the aim of conventional regression is to model the data exactly and it is unlikely that a good relationship could be obtained, but PLS does not require an exact fit to the data, so DPLS is more effective. Because we describe the PLS method in detail in

Chapter 5, we will not give a numerical example in this chapter, but if the number of variables is fairly large, approaches such as Mahalanobis distance are not effective unless there is variable reduction either by first using PCA or simply selecting some of the measurements, so the approach discussed in this section is worth trying.

4.5.5 *K* Nearest Neighbours

The methods of SIMCA, discriminant analysis and DPLS involve producing statistical models, such as principal components and canonical variates. Nearest neighbour methods are conceptually much simpler, and do not require elaborate statistical computations.

The KNN (or *K* nearest neighbour) method has been with chemists for over 30 years. The algorithm starts with a number of objects assigned to each class. Figure 4.36 represents five objects belonging to two classes A and class B recorded using two measurements, which may, for example, be chromatographic peak areas or absorption intensities at two wavelengths; the raw data are presented in Table 4.28.

4.5.5.1 Methodology

The method is implemented as follows.

1. Assign a training set to known classes.
2. Calculate the distance of an unknown to all members of the training set (see Table 4.28). Usually a simple 'Euclidean' distance is computed, see Section 4.4.1.
3. Rank these in order (1 = smallest distance, and so on).

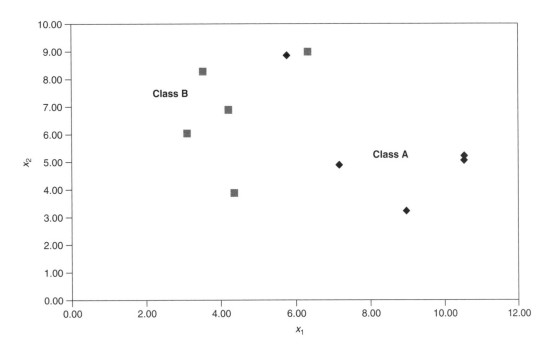

Figure 4.36
Two groups

Table 4.28 Example for KNN classification: the three closest distances are indicated in bold.

Class	x_1	x_2	Distance to unknown	Rank
A	5.77	8.86	3.86	6
A	10.54	5.21	5.76	10
A	7.16	4.89	2.39	4
A	10.53	5.05	5.75	9
A	8.96	3.23	4.60	8
B	3.11	6.04	**1.91**	3
B	4.22	6.89	**1.84**	2
B	6.33	8.99	4.16	7
B	4.36	3.88	**1.32**	1
B	3.54	8.28	3.39	5
Unknown	4.78	5.13		

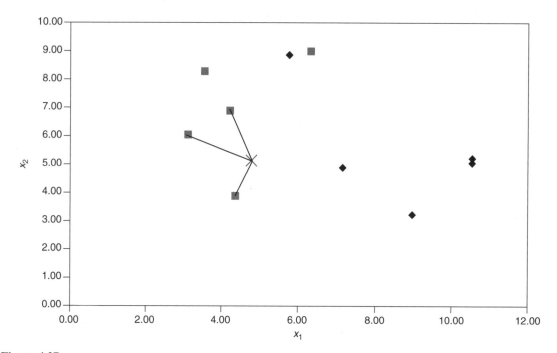

Figure 4.37
Three nearest neighbours to unknown

4. Pick the K smallest distances and see what classes the unknown in closest to; this number is usually a small odd number. The case where $K = 3$ is illustrated in Figure 4.37 for our example. All objects belong to class B.
5. Take the 'majority vote' and use this for classification. Note that if $K = 5$, one of the five closest objects belongs to class A in this case.
6. Sometimes it is useful to perform KNN analysis for a number of values of K, e.g. 3, 5 and 7, and see if the classification changes. This can be used to spot anomalies or artefacts.

If, as is usual in chemistry, there are many more than two measurements, it is simply necessary to extend the concept of distance to one in multidimensional space. Although we cannot visualise more than three dimensions, computers can handle geometry in an indefinite number of dimensions, and the idea of distance is easy to generalise. In the case of Figure 4.36 it is not really necessary to perform an elaborate computation to classify the unknown, but when a large number of measurements have been made, e.g. in spectroscopy, it is often hard to determine the class of an unknown by simple graphical approaches.

4.5.5.2 *Limitations*

This conceptually simple approach works well in many situations, but it is important to understand the limitations.

The first is that the numbers in each class of the training set should be approximately equal, otherwise the 'votes' will be biased towards the class with most representatives. The second is that for the simplest implementations, each variable assumes equal significance. In spectroscopy, we may record hundreds of wavelengths, and some will either not be diagnostic or else be correlated. A way of getting round this is either to select the variables or else to use another distance measure, just as in cluster analysis. Mahalanobis distance is a common alternative measure. The third problem is that ambiguous or outlying samples in the training set can result in major problems in the resultant classification. Fourth, the methods take no account of the spread or variance in a class. For example, if we were trying to determine whether a forensic sample is a forgery, it is likely that the class of forgeries has a much higher variance to the class of nonforged samples. The methods in the Sections 4.5.2 to 4.5.4 would normally take this into account.

However, KNN is a very simple approach that can be easily understood and programmed. Many chemists like these approaches, whereas statisticians often prefer the more elaborate methods involving modelling the data. KNN makes very few assumptions, whereas methods based on modelling often inherently make assumptions such as normality of noise distributions that are not always experimentally justified, especially when statistical tests are then employed to provide probabilities of class membership. In practice, a good strategy is to use several different methods for classification and see if similar results are obtained. Often the differences in performance of various approaches are not entirely due to the algorithm itself but in data scaling, distance measures, variable selection, validation method and so on. Some advocates of certain approaches do not always make this entirely clear.

4.6 Multiway Pattern Recognition

Most traditional chemometrics is concerned with two-way data, often represented by matrices. However, over the past decade there has been increasing interest in three-way chemical data. Instead of organising the information as a two-dimensional array [Figure 4.38(a)], it falls into a three-dimensional 'tensor' or box [Figure 4.38(b)]. Such datasets are surprisingly common. In Chapter 5 we discussed multiway PLS (Section 5.5.3), the discussion in this section being restricted to pattern recognition.

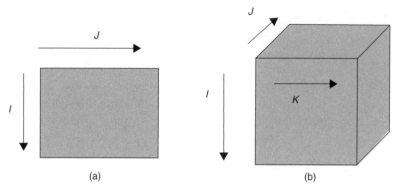

Figure 4.38
(a) Two-way and (b) three-way data

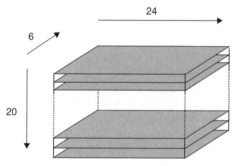

Figure 4.39
Possible method of arranging environmental sampling data

Consider, for example, an environmental chemical experiment in which the concentrations of six elements are measured at 20 sampling sites on 24 days in a year. There will be $20 \times 24 \times 6$ or 2880 measurements; however, these can be organised as a 'box' with 20 planes each corresponding to a sampling site, and of dimensions 24×6 (Figure 4.39). Such datasets have been available for many years to psychologists and in sensory research. A typical example might involve a taste panel assessing 20 food products. Each food could involve the use of 10 judges who score eight attributes, resulting in a $20 \times 10 \times 8$ box. In psychology, we might be following the reactions of 15 individuals to five different tests on 10 different days, possibly each day under slightly different conditions, and so have a $15 \times 5 \times 10$ box. These problems involve finding the main factors that influence the taste of a food or the source of pollutant or the reactions of an individual, and are a form of pattern recognition.

Three-dimensional analogies to principal components are required. There are no direct analogies to scores and loadings as in PCA, so the components in each of the three dimensions are often called 'weights'. There are a number of methods available to tackle this problem.

4.6.1 Tucker3 Models

These models involve calculating weight matrices corresponding to each of the three dimensions (e.g. sampling site, date and metal), together with a 'core' box or array,

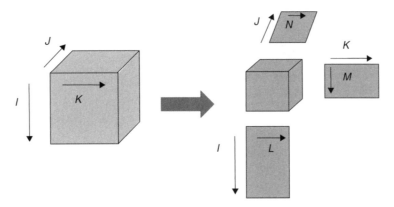

Figure 4.40
Tucker3 decomposition

which provides a measure of magnitude. The three weight matrices do not necessarily have the same dimensions, so the number of significant components for the sampling sites may be different to those for the dates, unlike normal PCA where one of the dimensions of both the scores and loadings matrices must be identical. This model (or decomposition) is represented in Figure 4.40. The easiest mathematical approach is by expressing the model as a summation:

$$x_{ijk} \approx \sum_{l=1}^{L} \sum_{m=1}^{M} \sum_{n=1}^{N} a_{il} b_{jm} c_{kn} z_{lmn}$$

where z represents what is often called a core array and a, b and c are functions relating to each of the three types of variable. Some authors use the concept of 'tensor multiplication', being a 3D analogy to 'matrix multiplication' in two dimensions; however, the details are confusing and conceptually it is probably best to stick to summations, which is what computer programs do.

4.6.2 PARAFAC

PARAFAC (parallel factor analysis) differs from the Tucker3 models in that each of the three dimensions contains the same number of components. Hence the model can be represented as the sum of contributions due to g components, just as in normal PCA, as illustrated in Figure 4.41 and represented algebraically by

$$x_{ijk} \approx \sum_{g=1}^{G} a_{ig} b_{jg} c_{kg}$$

Each component can be characterised by one vector that is analogous to a scores vector and two vectors that are analogous to loadings, but some keep to the notation of 'weights' in three dimensions. Components can, in favourable circumstances, be assigned a physical meaning. A simple example might involve following a reaction by recording a chromatogram from HPLC–DAD at different reaction times. A box

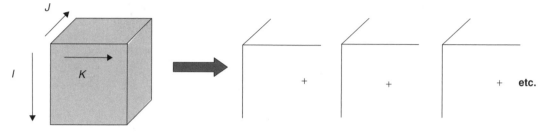

Figure 4.41
PARAFAC

whose dimensions are reaction time × elution time × wavelength is obtained. If there are three factors in the data, this would imply three significant compounds in a cluster in the chromatogram (or three significant reactants), and the weights should correspond to the reaction profile, the chromatogram and the spectrum of each compound.

PARAFAC is difficult to use, however, and, although the results are easy to interpret physically, it is conceptually more complex than PCA. Nevertheless, it can lead to results that are directly interpretable physically, whereas the factors in PCA have a purely abstract meaning.

4.6.3 Unfolding

Another approach is simply to 'unfold' the 'box' to give a long matrix. In the environmental chemistry example, instead of each sample being represented by a 24 × 6 matrix, it could be represented by a vector of length 144, each measurement consisting of the measurement of one element on one date, e.g. the measurement of Cd concentration on July 15. Then a matrix of dimensions 20 (sampling sites) × 144 (variables) is produced (Figure 4.42) and subjected to normal PCA. Note that a box can be subdivided into planes in three different ways (compare Figure 4.39 with Figure 4.42), according to which dimension is regarded as the 'major' dimension. When unfolding it is also important to consider details of scaling and centring which become far more complex in three dimensions as opposed to two. After unfolding, normal PCA can be performed. Components can be averaged over related variables, for example we could take an average loading for Cd over all dates to give an overall picture of its influence on the observed data.

This comparatively simple approach is sometimes sufficient but the PCA calculation neglects to take into account the relationships between the variables. For example, the

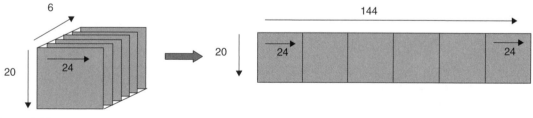

Figure 4.42
Unfolding

relationship between concentration of Cd on July 15 and that on August 1 is considered to be no stronger than the relationship between Cd concentration on July 15 and Hg on November 1 during the calculation of the components. However, after the calculations have been performed it is still possible to regroup the loadings and sometimes an easily understood method such as unfolded PCA can be of value.

Problems

Problem 4.1 Grouping of Elements from Fundamental Properties Using PCA

Section 4.3.2   Section 4.3.3.1   Section 4.3.5   Section 4.3.6.4

The table below lists 27 elements, divided into six groups according to their position in the periodic table together with five physical properties.

Element	Group	Melting point (K)	Boiling point (K)	Density	Oxidation number	Electronegativity
Li	1	453.69	1615	534	1	0.98
Na	1	371	1156	970	1	0.93
K	1	336.5	1032	860	1	0.82
Rb	1	312.5	961	1530	1	0.82
Cs	1	301.6	944	1870	1	0.79
Be	2	1550	3243	1800	2	1.57
Mg	2	924	1380	1741	2	1.31
Ca	2	1120	1760	1540	2	1
Sr	2	1042	1657	2600	2	0.95
F	3	53.5	85	1.7	−1	3.98
Cl	3	172.1	238.5	3.2	−1	3.16
Br	3	265.9	331.9	3100	−1	2.96
I	3	386.6	457.4	4940	−1	2.66
He	4	0.9	4.2	0.2	0	0
Ne	4	24.5	27.2	0.8	0	0
Ar	4	83.7	87.4	1.7	0	0
Kr	4	116.5	120.8	3.5	0	0
Xe	4	161.2	166	5.5	0	0
Zn	5	692.6	1180	7140	2	1.6
Co	5	1765	3170	8900	3	1.8
Cu	5	1356	2868	8930	2	1.9
Fe	5	1808	3300	7870	2	1.8
Mn	5	1517	2370	7440	2	1.5
Ni	5	1726	3005	8900	2	1.8
Bi	6	544.4	1837	9780	3	2.02
Pb	6	600.61	2022	11340	2	1.8
Tl	6	577	1746	11850	3	1.62

1. Standardise the five variables, using the population (rather than sample) standard deviation. Why is this preprocessing necessary to obtain sensible results in this case?

2. Calculate the scores, loadings and eigenvalues of the first two PCs of the standard-ised data. What is the sum of the first two eigenvalues, and what proportion of the overall variability do they represent?

3. Plot a graph of the scores of PC2 versus PC1, labelling the points. Comment on the grouping in the scores plot.

4. Plot a graph of the loadings of PC2 versus PC1, labelling the points. Which variables cluster together and which appears to behave differently? Hence which physical property mainly accounts for PC2?

5. Calculate the correlation matrix between each of the five fundamental parameters. How does this relate to clustering in the loadings plot?

6. Remove the parameter that exhibits a high loading in PC2 and recalculate the scores using only four parameters. Plot the scores. What do you observe, and why?

Problem 4.2 Introductory PCA

Section 4.3.2   Section 4.3.3.1

The following is a data matrix, consisting of seven samples and six variables:

2.7	4.3	5.7	2.3	4.6	1.4
2.6	3.7	7.6	9.1	7.4	1.8
4.3	8.1	4.2	5.7	8.4	2.4
2.5	3.5	6.5	5.4	5.6	1.5
4.0	6.2	5.4	3.7	7.4	3.2
3.1	5.3	6.3	8.4	8.9	2.4
3.2	5.0	6.3	5.3	7.8	1.7

The scores of the first two principal components on the centred data matrix are given as follows:

−4.0863	−1.6700
3.5206	−2.0486
−0.0119	3.7487
−0.7174	−2.3799
−1.8423	1.7281
3.1757	0.6012
−0.0384	0.0206

1. Since $X \approx T.P$, calculate the loadings for the first two PCs using the pseudoinverse, remembering to centre the original data matrix first.

2. Demonstrate that the two scores vectors are orthogonal and the two loadings vectors are orthonormal. Remember that the answer will only to be within a certain degree of numerical accuracy.

3. Determine the eigenvalues and percentage variance of the first two principal com-ponents.

Problem 4.3 Introduction to Cluster Analysis

Section 4.4

The following dataset consists of seven measurements (rows) on six objects A–F (columns)

A	B	C	D	E	F
0.9	0.3	0.7	0.5	1.0	0.3
0.5	0.2	0.2	0.4	0.7	0.1
0.2	0.6	0.1	1.1	2	0.3
1.6	0.7	0.9	1.3	2.2	0.5
1.5	0.1	0.1	0.2	0.4	0.1
0.4	0.9	0.7	1.8	3.7	0.4
1.5	0.3	0.3	0.6	1.1	0.2

1. Calculate the correlation matrix between the six objects.
2. Using the correlation matrix, perform cluster analysis using the furthest neighbour method. Illustrate each stage of linkage.
3. From the results in 2, draw a dendrogram, and deduce which objects cluster closely into groups.

Problem 4.4 Classification Using Euclidean Distance and KNN

Section 4.5.5   Section 4.4.1

The following data represent three measurements, x, y and z, made on two classes of compound:

Object	Class	x	y	z
1	A	0.3	0.4	0.1
2	A	0.5	0.6	0.2
3	A	0.7	0.5	0.3
4	A	0.5	0.6	0.5
5	A	0.2	0.5	0.1
6	B	0.2	0.1	0.6
7	B	0.3	0.4	0.5
8	B	0.1	0.3	0.7
9	B	0.4	0.5	0.7

1. Calculate the centroids of each class (this is done simply by averaging the values of the three measurements over each class).
2. Calculate the Euclidean distance of all nine objects from the centroids of both classes A and B (you should obtain a table of 18 numbers). Verify that all objects do, indeed, belong to their respective classes.
3. An object of unknown origins has measurements (0.5, 0.3, 0.3). What is the distance from the centroids of each class and so to which class is it more likely to belong?
4. The K nearest neighbour criterion can also be used for classification. Find the distance of the object in question 3 from the nine objects in the table above. Which are the three closest objects, and does this confirm the conclusions in question 3?
5. Is there one object in the original dataset that you might be slightly suspicious about?

Problem 4.5 Certification of NIR Filters Using PC Scores Plots

Section 4.3.2   Section 4.3.5.1   Section 4.3.3.1   Section 4.3.6.4

These data were obtained by the National Institute of Standards and Technology (USA) while developing a transfer standard for verification and calibration of the x-axis of NIR spectrometers. Optical filters were prepared from two separate melts, 2035 and 2035a, of a rare earth glass. Filters from both melts provide seven well-suited adsorption bands of very similar but not exactly identical location. One filter, Y, from one of the two melts was discovered to be unlabelled. Four 2035 filters and one 2035a filter were available at the time of this discovery. Six replicate spectra were taken from each filter. Band location data from these spectra are provided below, in cm^{-1}. The expected location uncertainties range from 0.03 to 0.3 cm^{-1}.

Type	No.	P1	P2	P3	P4	P5	P6	P7
2035	18	5138.58	6804.70	7313.49	8178.65	8681.82	9293.94	10245.45
2035	18	5138.50	6804.81	7313.49	8178.71	8681.73	9293.93	10245.49
2035	18	5138.47	6804.87	7313.43	8178.82	8681.62	9293.82	10245.52
2035	18	5138.46	6804.88	7313.67	8178.80	8681.52	9293.89	10245.54
2035	18	5138.46	6804.96	7313.54	8178.82	8681.63	9293.79	10245.51
2035	18	5138.45	6804.95	7313.59	8178.82	8681.70	9293.89	10245.53
2035	101	5138.57	6804.77	7313.54	8178.69	8681.70	9293.90	10245.48
2035	101	5138.51	6804.82	7313.57	8178.75	8681.73	9293.88	10245.53
2035	101	5138.49	6804.91	7313.57	8178.82	8681.63	9293.80	10245.55
2035	101	5138.47	6804.88	7313.50	8178.84	8681.63	9293.78	10245.55
2035	101	5138.48	6804.97	7313.57	8178.80	8681.70	9293.79	10245.50
2035	101	5138.47	6804.99	7313.59	8178.84	8681.67	9293.82	10245.52
2035	102	5138.54	6804.77	7313.49	8178.69	8681.62	9293.88	10245.49
2035	102	5138.50	6804.89	7313.45	8178.78	8681.66	9293.82	10245.54
2035	102	5138.45	6804.95	7313.49	8178.77	8681.65	9293.69	10245.53
2035	102	5138.48	6804.96	7313.55	8178.81	8681.65	9293.80	10245.52
2035	102	5138.47	6805.00	7313.53	8178.83	8681.62	9293.80	10245.52
2035	102	5138.46	6804.97	7313.54	8178.83	8681.70	9293.81	10245.52
2035	103	5138.52	6804.73	7313.42	8178.75	8681.73	9293.93	10245.48
2035	103	5138.48	6804.90	7313.53	8178.78	8681.63	9293.84	10245.48
2035	103	5138.45	6804.93	7313.52	8178.73	8681.72	9293.83	10245.56
2035	103	5138.47	6804.96	7313.53	8178.78	8681.59	9293.79	10245.51
2035	103	5138.46	6804.94	7313.51	8178.81	8681.65	9293.77	10245.52
2035	103	5138.48	6804.98	7313.57	8178.82	8681.51	9293.80	10245.51
2035a	200	5139.26	6806.45	7314.93	8180.19	8682.57	9294.46	10245.62
2035a	200	5139.22	6806.47	7315.03	8180.26	8682.52	9294.35	10245.66
2035a	200	5139.21	6806.56	7314.92	8180.26	8682.61	9294.34	10245.68
2035a	200	5139.20	6806.56	7314.90	8180.23	8682.49	9294.31	10245.69
2035a	200	5139.19	6806.58	7314.95	8180.24	8682.64	9294.32	10245.67
2035a	200	5139.20	6806.50	7314.97	8180.21	8682.58	9294.27	10245.64
Y	201	5138.53	6804.82	7313.62	8178.78	8681.78	9293.77	10245.52

Type	No.	P1	P2	P3	P4	P5	P6	P7
Y	201	5138.49	6804.87	7313.47	8178.75	8681.66	9293.74	10245.52
Y	201	5138.48	6805.00	7313.54	8178.85	8681.67	9293.75	10245.54
Y	201	5138.48	6804.97	7313.54	8178.82	8681.70	9293.79	10245.53
Y	201	5138.47	6804.96	7313.51	8178.77	8681.52	9293.85	10245.54
Y	201	5138.48	6804.97	7313.49	8178.84	8681.66	9293.87	10245.50

1. Standardise the peak positions for the 30 known samples (exclude samples Y).
2. Perform PCA on these data, retaining the first two PCs. Calculate the scores and eigenvalues. What will the sum of squares of the standardised data equal, and so what proportion of the variance is accounted for by the first two PCs?
3. Produce a scores plot of the first two PCs of this data, indicating the two groups using different symbols. Verify that there is a good discrimination using PCA.
4. Determine the origin of Y as follows. (a) For each variable subtract the mean and divide by the standard deviation of the 30 known samples to give a 6×7 matrix $^{stand}X$. (b) Then multiply this standardised data by the overall loadings, for the first PC to give $T = {}^{stand}X.P'$ and predict the scores for these samples. (c) Superimpose the scores of Y on to the scores plot obtained in 3, and so determine the origin of Y.
5. Why is it correct to calculate $T = {}^{stand}X.P'$ rather than using the pseudo-inverse and calculate $T = {}^{stand}X.P'(P.P')^{-1}$?

Problem 4.6 Simple KNN Classification

Section 4.5.5   Section 4.4.1

The following represents five measurements on 16 samples in two classes, a and b:

Sample						Class
1	37	3	56	32	66	a
2	91	84	64	37	50	a
3	27	34	68	28	63	a
4	44	25	71	25	60	a
5	46	60	45	23	53	a
6	25	32	45	21	43	a
7	36	53	99	42	92	a
8	56	53	92	37	82	a
9	95	58	59	35	33	b
10	29	25	30	13	21	b
11	96	91	55	31	32	b
12	60	34	29	19	15	b
13	43	74	44	21	34	b
14	62	105	36	16	21	b
15	88	70	48	29	26	b
16	95	76	74	38	46	b

1. Calculate the 16×16 sample distance matrix, by computing the Euclidean distance between each sample.
2. For each sample, list the classes of the three and five nearest neighbours, using the distance matrix as a guide.
3. Verify that most samples belong to their proposed class. Is there a sample that is most probably misclassified?

Problem 4.7 Classification of Swedes into Fresh and Stored using SIMCA

Section 4.5.3   Section 4.3.6   Section 4.3.5   Section 4.3.2   Section 4.3.3.1

The following consist of a training set of 14 swedes (vegetable) divided into two groups, fresh and stored (indicated by F and S in the names), with the areas of eight GC peaks (A–H) from the extracts indicated. The aim is to set up a model to classify a swede into one of these two groups.

	A	B	C	D	E	F	G	H
FH	0.37	0.99	1.17	6.23	2.31	3.78	0.22	0.24
FA	0.84	0.78	2.02	5.47	5.41	2.8	0.45	0.46
FB	0.41	0.74	1.64	5.15	2.82	1.83	0.37	0.37
FI	0.26	0.45	1.5	4.35	3.08	2.01	0.52	0.49
FK	0.99	0.19	2.76	3.55	3.02	0.65	0.48	0.48
FN	0.7	0.46	2.51	2.79	2.83	1.68	0.24	0.25
FM	1.27	0.54	0.90	1.24	0.02	0.02	1.18	1.22
SI	1.53	0.83	3.49	2.76	10.3	1.92	0.89	0.86
SH	1.5	0.53	3.72	3.2	9.02	1.85	1.01	0.96
SA	1.55	0.82	3.25	3.23	7.69	1.99	0.85	0.87
SK	1.87	0.25	4.59	1.4	6.01	0.67	1.12	1.06
SB	0.8	0.46	3.58	3.95	4.7	2.05	0.75	0.75
SM	1.63	1.09	2.93	6.04	4.01	2.93	1.05	1.05
SN	3.45	1.09	5.56	3.3	3.47	1.52	1.74	1.71

In addition, two test set samples, X and Y, each belonging to one of the groups F and S have also been analysed by GC:

	A	B	C	D	E	F	G	H
FX	0.62	0.72	1.48	4.14	2.69	2.08	0.45	0.45
SY	1.55	0.78	3.32	3.2	5.75	1.77	1.04	1.02

1. Transform the data first by taking logarithms and then standardising over the 14 training set samples (use the population standard deviation). Why are these transformations used?
2. Perform PCA on the transformed PCs of the 14 objects in the training set, and retain the first two PCs. What are the eigenvalues of these PCs and to what percentage

variability do they correspond? Obtain a scores plot, indicating the objects from each class in different symbols. Is there an outlier in the PC plot?

3. Remove this outlier, restandardise the data over 13 objects and perform PCA again. Produce the scores plot of the first two PCs, indicating each class with different symbols. Comment on the improvement.

4. Rescale the data to provide two new datasets, one based on standardising over the first class and the other over the second class (minus the outlier) in all cases using logarithmically transformed data. Hence dataset (a) involves subtracting the mean and dividing by the standard deviation for the fresh swedes and dataset (b) for the stored swedes. Call these datasets $^F X$ and $^S X$, each will be of dimensions 13×8, the superscript relating to the method of preprocessing.

5. For each dataset (a) and (b) perform PCA over the objects belonging only to its own class (six or seven objects as appropriate) and keep the loadings of the first PC in each case. Call these loadings vectors $^F p$ and $^S p$. Two row vectors, consisting of eight numbers, should be obtained.

6. For each dataset calculate the predicted scores for the first PC given by $^F t = {}^F X . {}^F p'$ and $^S t = {}^S X . {}^S p'$. Then recalculate the predicted datasets using models (a) and (b) by multiplying the predicted scores by the appropriate loadings, and call these $^F \hat{X}$ and $^S \hat{X}$.

7. For each of the 13 objects in the training set i, calculate the distance from the PC model of each class c by determine $d_{ic} = \sqrt{\Sigma_{j=1}^{J} ({}^c x_{ij} - {}^c \hat{x}_{ij})^2}$, where $J = 8$ and corresponds to the measurements, and the superscript c indicates a model of for class c. For these objects produce a class distance plot.

8. Extend the class distance plot to include the two samples in the test set using the method of steps 6 and 7 to determine the distance from the PC models. Are they predicted correctly?

Problem 4.8 Classification of Pottery from Pre-classical Sites in Italy, Using Euclidean and Mahalanobis Distance Measures

Section 4.3.6.4   Section 4.3.2   Section 4.3.3.1   Section 4.3.5   Section 4.4.1   Section 4.5.2.3

Measurements of elemental composition was performed on 58 samples of pottery from southern Italy, divided into two groups A (black carbon containing bulks) and B (clayey ones). The data are as follows:

	Ti (%)	Sr (ppm)	Ba (ppm)	Mn (ppm)	Cr (ppm)	Ca (%)	Al (%)	Fe (%)	Mg (%)	Na (%)	K (%)	Class
A1	0.304	181	1007	642	60	1.640	8.342	3.542	0.458	0.548	1.799	A
A2	0.316	194	1246	792	64	2.017	8.592	3.696	0.509	0.537	1.816	A
A3	0.272	172	842	588	48	1.587	7.886	3.221	0.540	0.608	1.970	A
A4	0.301	147	843	526	62	1.032	8.547	3.455	0.546	0.664	1.908	A
A5	0.908	129	913	775	184	1.334	11.229	4.637	0.395	0.429	1.521	A
E1	0.394	105	1470	1377	90	1.370	10.344	4.543	0.408	0.411	2.025	A
E2	0.359	96	1188	839	86	1.396	9.537	4.099	0.427	0.482	1.929	A
E3	0.406	137	1485	1924	90	1.731	10.139	4.490	0.502	0.415	1.930	A
E4	0.418	133	1174	1325	91	1.432	10.501	4.641	0.548	0.500	2.081	A
L1	0.360	111	410	652	70	1.129	9.802	4.280	0.738	0.476	2.019	A
L2	0.280	112	1008	838	59	1.458	8.960	3.828	0.535	0.392	1.883	A

	Ti (%)	Sr (ppm)	Ba (ppm)	Mn (ppm)	Cr (ppm)	Ca (%)	Al (%)	Fe (%)	Mg (%)	Na (%)	K (%)	Class
L3	0.271	117	1171	681	61	1.456	8.163	3.265	0.521	0.509	1.970	A
L4	0.288	103	915	558	60	1.268	8.465	3.437	0.572	0.479	1.893	A
L5	0.253	102	833	415	193	1.226	7.207	3.102	0.539	0.577	1.972	A
C1	0.303	131	601	1308	65	0.907	8.401	3.743	0.784	0.704	2.473	A
C2	0.264	121	878	921	69	1.164	7.926	3.431	0.636	0.523	2.032	A
C3	0.264	112	1622	1674	63	0.922	7.980	3.748	0.549	0.497	2.291	A
C4	0.252	111	793	750	53	1.171	8.070	3.536	0.599	0.551	2.282	A
C5	0.261	127	851	849	61	1.311	7.819	3.770	0.668	0.508	2.121	A
G8	0.397	177	582	939	61	1.260	8.694	4.146	0.656	0.579	1.941	A
G9	0.246	106	1121	795	53	1.332	8.744	3.669	0.571	0.477	1.803	A
G10	1.178	97	886	530	441	6.290	8.975	6.519	0.323	0.275	0.762	A
G11	0.428	457	1488	1138	85	1.525	9.822	4.367	0.504	0.422	2.055	A
P1	0.259	389	399	443	175	11.609	5.901	3.283	1.378	0.491	2.148	B
P2	0.185	233	456	601	144	11.043	4.674	2.743	0.711	0.464	0.909	B
P3	0.312	277	383	682	138	8.430	6.550	3.660	1.156	0.532	1.757	B
P6	0.183	220	435	594	659	9.978	4.920	2.692	0.672	0.476	0.902	B
P7	0.271	392	427	410	125	12.009	5.997	3.245	1.378	0.527	2.173	B
P8	0.203	247	504	634	117	11.112	5.034	3.714	0.726	0.500	0.984	B
P9	0.182	217	474	520	92	12.922	4.573	2.330	0.590	0.547	0.746	B
P14	0.271	257	485	398	955	11.056	5.611	3.238	0.737	0.458	1.013	B
P15	0.236	228	203	592	83	9.061	6.795	3.514	0.750	0.506	1.574	B
P16	0.288	333	436	509	177	10.038	6.579	4.099	1.544	0.442	2.400	B
P17	0.331	309	460	530	97	9.952	6.267	3.344	1.123	0.519	1.746	B
P18	0.256	340	486	486	132	9.797	6.294	3.254	1.242	0.641	1.918	B
P19	0.292	289	426	531	143	8.372	6.874	3.360	1.055	0.592	1.598	B
P20	0.212	260	486	605	123	9.334	5.343	2.808	1.142	0.595	1.647	B
F1	0.301	320	475	556	142	8.819	6.914	3.597	1.067	0.584	1.635	B
F2	0.305	302	473	573	102	8.913	6.860	3.677	1.365	0.616	2.077	B
F3	0.300	204	192	575	79	7.422	7.663	3.476	1.060	0.521	2.324	B
F4	0.225	181	160	513	94	5.320	7.746	3.342	0.841	0.657	2.268	B
F5	0.306	209	109	536	285	7.866	7.210	3.528	0.971	0.534	1.851	B
F6	0.295	396	172	827	502	9.019	7.775	3.808	1.649	0.766	2.123	B
F7	0.279	230	99	760	129	5.344	7.781	3.535	1.200	0.827	2.305	B
D1	0.292	104	993	723	92	7.978	7.341	3.393	0.630	0.326	1.716	B
D2	0.338	232	687	683	108	4.988	8.617	3.985	1.035	0.697	2.215	B
D3	0.327	155	666	590	70	4.782	7.504	3.569	0.536	0.411	1.490	B
D4	0.233	98	560	678	73	8.936	5.831	2.748	0.542	0.282	1.248	B
M1	0.242	186	182	647	92	5.303	8.164	4.141	0.804	0.734	1.905	B
M2	0.271	473	198	459	89	10.205	6.547	3.035	1.157	0.951	0.828	B
M3	0.207	187	205	587	87	6.473	7.634	3.497	0.763	0.729	1.744	B
G1	0.271	195	472	587	104	5.119	7.657	3.949	0.836	0.671	1.845	B
G2	0.303	233	522	870	130	4.610	8.937	4.195	1.083	0.704	1.840	B
G3	0.166	193	322	498	80	7.633	6.443	3.196	0.743	0.460	1.390	B
G4	0.227	170	718	1384	87	3.491	7.833	3.971	0.783	0.707	1.949	B
G5	0.323	217	267	835	122	4.417	9.017	4.349	1.408	0.730	2.212	B
G6	0.291	272	197	613	86	6.055	7.384	3.343	1.214	0.762	2.056	B
G7	0.461	318	42	653	123	6.986	8.938	4.266	1.579	0.946	1.687	B

1. Standardise this matrix, and explain why this transformation is important. Why is it normal to use the population rather than the sample standard deviation? All calculations below should be performed on this standardised data matrix.
2. Perform PCA, initially calculating 11 PCs, on the data of question 1. What is the total sum of the eigenvalues for all 11 components, and to what does this number relate?
3. Plot the scores of PC2 versus PC1, using different symbols for classes A and B. Is there a good separation between classes? One object appears to be an outlier: which one?
4. Plot the loadings of PC2 versus PC1. Label these with the names of the elements.
5. Compare the loadings plot to the scores plot. Pick two elements that appear diagnostic of the two classes: these elements will appear in the loadings plot in the same direction of the classes (there may be more than one answer to this question). Plot the value of the standardised readings these elements against each other, using different symbols and show that reasonable (but not perfect) discrimination is possible.
6. From the loadings plots, choose a pair of elements that are very poor at discriminating (at right angles to the discriminating direction) and show that the resultant graph of the standardised readings of each element against the other is very poor and does not provide good discrimination.
7. Calculate the centroids of class A (excluding the outlier) and class B. Calculate the Euclidean distance of the 58 samples to both these centroids. Produce a class distance plot of distance to centroid of class A against class B, indicating the classes using different symbols, and comment.
8. Determine the variance–covariance matrix for the 11 elements and each of the classes (so there should be two matrices of dimensions 11×11); remove the outlier first. Hence calculate the Mahalanobis distance to each of the class centroids. What is the reason for using Mahalanobis distance rather than Euclidean distance? Produce a class distance plot for this new measure, and comment.
9. Calculate the %CC using the class distances in question 8, using the lowest distance to indicate correct classification.

Problem 4.9 Effect of Centring on PCA

Section 4.3.2   Section 4.3.3.1   Section 4.3.6.3

The following data consist of simulations of a chromatogram sampled at 10 points in time (rows) and at eight wavelengths (columns):

0.131	0.069	0.001	0.364	0.436	0.428	0.419	0.089
0.311	0.293	0.221	0.512	1.005	0.981	0.503	0.427
0.439	0.421	0.713	1.085	1.590	1.595	1.120	0.386
0.602	0.521	0.937	1.462	2.056	2.214	1.610	0.587
1.039	0.689	0.913	1.843	2.339	2.169	1.584	0.815
1.083	1.138	1.539	2.006	2.336	2.011	1.349	0.769
1.510	1.458	1.958	1.812	2.041	1.565	1.075	0.545
1.304	1.236	1.687	1.925	1.821	1.217	0.910	0.341
0.981	1.034	1.336	1.411	1.233	0.721	0.637	0.334
0.531	0.628	0.688	0.812	0.598	0.634	0.385	0.138

1. Perform PCA on the data, both raw and centred. Calculate the first five PCs including scores, and loadings.
2. Verify that the scores and loadings are all orthogonal and that the sum of squares of the loadings equals 1.
3. Calculate the eigenvalues (defined by sum of squares of the scores of the PCs) of the first five PCs for both raw and centred data.
4. For the raw data, verify that the sum of the eigenvalues approximately equals the sum of squares of the data.
5. The sum of the eigenvalues of the column centred data can be roughly related to the sum of the eigenvalues for the uncentred data as follows. Take the mean of each column, square it, multiply by the number of objects in each column (=10) and then add these values for all the eight columns together. This plus the sum of the eigenvalues of the column centred data matrix should be nearly equal to the sum of the eigenvalues of the raw data. Show this numerically, and explain why.
6. How many components do you think are in the data? Explain why the mean centred data, in this case, give answers that are easier to interpret.

Problem 4.10 Linear Discriminant Analysis in QSAR to Study the Toxicity of Polycyclic Aromatic Hydrocarbons (PAHs)

Section 4.3.2   Section 4.3.5.1   Section 4.5.2.3

Five molecular descriptors, A–E, have been calculated using molecular orbital computations for 32 PAHs, 10 of which are have carcinogenic activity (A) and 22 not (I), as given below, the two groups being indicated:

		A	B	C	D	E
(1) Dibenzo[3,4;9,10]pyrene	A	−0.682	0.34	0.457	0.131	0.327
(2) Benzo[3,4]pyrene	A	−0.802	0.431	0.441	0.231	0.209
(3) Dibenzo[3,4;8,9]pyrene	A	−0.793	0.49	0.379	0.283	0.096
(4) Dibenzo[3,4;6,7]pyrene	A	−0.742	0.32	0.443	0.288	0.155
(5) Dibenzo[1,2;3,4]pyrene	A	−0.669	0.271	0.46	0.272	0.188
(6) Naphtho[2,3;3,4]pyrene	A	−0.648	0.345	0.356	0.186	0.17
(7) Dibenz[1,2;5,6]anthracene	A	−0.684	0.21	0.548	0.403	0.146
(8) Tribenzo[3,4;6,7;8,9]pyrene	A	−0.671	0.333	0.426	0.135	0.292
(9) Dibenzo[1,2;3,4]phenanthrene	A	−0.711	0.179	0.784	0.351	0.434
(10) Tribenzo[3,4;6,7;8,9]pyrene	A	−0.68	0.284	0.34	0.648	−0.308
(11) Dibenzo[1,2;5,6]phenanthrene	I	−0.603	0.053	0.308	0.79	−0.482
(12) Benz[1,2]anthracene	I	−0.715	0.263	0.542	0.593	−0.051
(13) Chrysene	I	−0.792	0.272	0.71	0.695	0.016
(14) Benzo[3,4]phenanthrene	I	−0.662	0.094	0.649	0.716	−0.067
(15) Dibenz[1,2;7,8]anthracene	I	−0.618	0.126	0.519	0.5	0.019
(16) Dibenz[1,2;3,4]anthracene	I	−0.714	0.215	0.672	0.342	0.33
(17) Benzo[1,2]pyrene	I	−0.718	0.221	0.541	0.308	0.233
(18) Phenanthrene	I	−0.769	0.164	0.917	0.551	0.366
(19) Triphenylene	I	−0.684	0	0.57	0.763	−0.193
(20) Benzo[1,2]naphthacene	I	−0.687	0.36	0.336	0.706	−0.37

		A	B	C	D	E
(21) Dibenzo[3,4;5,6]phenanthrene	I	−0.657	0.121	0.598	0.452	0.147
(22) Picene	I	−0.68	0.178	0.564	0.393	0.171
(23) Tribenz[1,2;3,4;5,6]anthracene	I	−0.637	0.115	0.37	0.456	−0.087
(24) Dibenzo[1,2;5,6]pyrene	I	−0.673	0.118	0.393	0.395	−0.001
(25) Phenanthr[2,3;1,2]anthracene	I	−0.555	0.126	0.554	0.25	0.304
(26) Benzo[1,2]pentacene	I	−0.618	0.374	0.226	0.581	−0.356
(27) Anthanthrene	I	−0.75	0.459	0.299	0.802	−0.503
(28) Benzene	I	−1	0	2	2	0
(29) Naphthalene	I	−1	0.382	1	1.333	−0.333
(30) Pyrene	I	−0.879	0.434	0.457	0.654	−0.197
(31) Benzo[ghi]perylene	I	−0.684	0.245	0.42	0.492	−0.072
(32) Coronene	I	−0.539	0	0.431	0.45	−0.019

1. Perform PCA on the raw data, and produce a scores plot of PC2 versus PC1. Two compounds appear to be outliers, as evidenced by high scores on PC1. Distinguish the two groups using different symbols.
2. Remove these outliers and repeat the PCA calculation, and produce a new scores plot for the first two PCs, distinguishing the groups. Perform all subsequent steps on the reduced dataset of 30 compounds minus outliers using the raw data.
3. Calculate the variance–covariance matrix for each group (minus the outliers) separately and hence the pooled variance–covariance matrix C_{AB}.
4. Calculate the centroids for each class, and hence the linear discriminant function given by $(\bar{x}_A - \bar{x}_B).C_{AB}^{-1}.x'_i$ for each object i. Represent this graphically. Suggest a cut-off value of this function which will discriminate most of the compounds. What is the percentage correctly classified?
5. One compound is poorly discriminated in question 4: could this have been predicted at an earlier stage in the analysis?

Problem 4.11 Class Modelling Using PCA

Section 4.3.2   Section 4.3.3.1   Section 4.5.3

Two classes of compounds are studied. In each class, there are 10 samples, and eight variables have been measured. The data are as follows, with each column representing a variable and each row a sample:

Class A

−20.1	−13.8	−32.4	−12.1	8.0	−38.3	2.4	−21.0
38.2	3.6	−43.6	2.2	30.8	7.1	−6.2	−5.4
−19.2	1.4	39.3	−7.5	−24.1	−2.9	−0.4	−7.7
9.0	0.2	−15.1	3.0	10.3	2.0	−1.2	2.0
51.3	12.6	−13.3	7.5	20.6	36.7	−32.2	−14.5
−13.9	7.4	61.5	−11.6	−35	7.1	−3	−11.7
−18.9	−2.4	17.6	−8.5	−14.8	−13.5	9.9	−2.7
35.1	10.3	−0.4	6.5	9.9	31.2	−25.4	−9.4

16.6	6.0	5.8	−1.8	−6.4	19.6	−7.1	−1.2
7.1	−2.7	−24.8	7.1	14.9	1.1	−3.0	4.8

Class B

−2.9	−5.4	−12.0	−9.1	3.3	−13.3	−18.9	−30.5
30.7	8.3	−8.0	−39.1	3.8	−25.5	9.0	−47.2
15.1	7.1	10.9	−10.7	16.5	−17.2	−9.0	−34.6
−18.2	−13	−17	6.6	9.1	−9.6	−45.2	−34.6
12.2	2.8	−3.8	−5.2	4.0	1.2	−4.8	−11.2
19.8	19.8	55.0	−30	−26.3	0.3	33.2	−7.1
19.9	5.8	−3.1	−25.3	1.2	−15.6	9.5	−27.0
22.4	4.8	−9.1	−30.6	−3.2	−16.4	12.1	−28.9
5.5	0.6	−7.1	−11.7	−16.0	5.8	18.5	11.4
−36.2	−17.3	−14.1	32.3	2.75	11.2	−39.7	9.1

In all cases perform uncentred PCA on the data. The exercises could be repeated with centred PCA, but only one set of answers is required.

1. Perform PCA on the overall dataset involving all 20 samples.
2. Verify that the overall dataset is fully described by five PCs. Plot a graph of the scores of PC2 versus PC1 and show that there is no obvious distinction between the two classes.
3. Independent class modelling is common in chemometrics, and is the basis of SIMCA. Perform uncentred PCA on classes A and B separately, and verify that class A is described reasonably well using two PCs, but class B by three PCs. Keep only these significant PCs in the data.
4. The predicted fit to a class can be computed as follows. To test the fit to class A, take the loadings of the PC model for class A, including two components (see question 3). Then multiply the observed row vector for each sample in class A by the loadings model, to obtain two scores. Perform the same operation for each sample in class A. Calculate the sum of squares of the scores for each sample, and compare this with the sum of squares original data for this sample. The closer these numbers are, the better. Repeat this for samples of class B, using the model of class A. Perform this operation (a) fitting all samples to the class A model as above and (b) fitting all samples to the class B model using three PCs this time.
5. A table consisting of 40 sums of squares (20 for the model of class A and 16 for the model of class B) should be obtained. Calculate the ratio of the sum of squares of the PC scores for a particular class model to the sum of squares of the original measurements for a given sample. The closer this is to 1, the better the model. A good result will involve a high value (>0.9) for the ratio using its own class model and a low value (<0.1) for the ratio using a different class model.
6. One class seems to be fit much better than the other. Which is it? Comment on the results, and suggest whether there are any samples in the less good class that could be removed from the analysis.

Problem 4.12 Effect of Preprocessing on PCA in LC–MS

Section 4.3.2   Section 4.3.5   Section 4.3.6   Section 4.3.3.1

The intensity of the ion current at 20 masses (96–171) and 27 points in time of an LC–MS chromatogram of two partially overlapping peaks is recorded as follows on page 268:

1. Produce a graph of the total ion current (the sum of intensity over the 20 masses) against time.
2. Perform PCA on the raw data, uncentred, calculating two PCs. Plot the scores of PC2 versus PC1. Are there any trends? Plot the scores of the first two PCs against elution time. Interpret the probable physical meaning of these two principal components. Obtain a loadings plot of PC2 versus PC1, labelling some of the points furthest from the origin. Interpret this graph with reference to the scores plot.
3. Scale the data along the rows, by making each row add up to 1. Perform PCA. Why is the resultant scores plot of little physical meaning?
4. Repeat the PCA in step 4, but remove the first three points in time. Compute the scores and loadings plots of PC2 versus PC1. Why has the scores plot dramatically changed in appearance compared with that obtained in question 2? Interpret this new plot.
5. Return to the raw data, retaining all 27 original points in time. Standardise the columns. Perform PCA on this data, and produce graphs of PC2 versus PC1 for both the loadings and scores. Comment on the patterns in the plots.
6. What are the eigenvalues of the first two PCs of the standardised data? Comment on the size of the eigenvalues and how this relates to the appearance of the loadings plot in question 5.

Problem 4.13 Determining the Number of Significant Components in a Dataset by Cross-validation

Section 4.3.3.2

The following dataset represents six samples (rows) and seven measurements (columns):

62.68	52.17	49.50	62.53	56.68	64.08	59.78
113.71	63.27	94.06	99.50	62.90	98.08	79.61
159.72	115.51	128.46	124.03	76.09	168.02	120.16
109.92	81.11	72.57	72.55	42.82	106.65	87.80
89.42	47.73	68.24	73.68	49.10	78.73	59.86
145.95	96.16	105.36	107.76	48.91	139.58	96.75

The aim is to determine the number of significant factors in the dataset.

1. Perform PCA on the raw data, and calculate the eigenvalues for the six nonzero components. Verify that these eigenvalues add up to the sum of squares of the entire dataset.
2. Plot a graph of eigenvalue against component number. Why is it not clear from this graph how many significant components are in the data? Change the vertical scale to a logarithmic one, and produce a new graph. Comment on the difference in appearance.

96	95	78	155	97	41	154	68	191	113	79	51	172	190	67	156	173	164	112	171
−5.22	21.25	7.06	−5.60	0.76	−1.31	−18.86	5.54	−1.11	9.07	4.78	5.24	−4.10	1.37	3.93	−0.09	−2.27	−0.01	0.47	−0.64
12.32	15.45	24.89	0.68	1.35	3.40	16.44	7.08	1.31	4.79	3.32	3.33	−1.78	−0.96	2.71	2.05	2.73	2.43	0.20	−0.84
115.15	88.47	51.13	28.10	20.01	19.24	15.58	45.09	1.12	1.57	−5.42	−5.08	−1.91	−0.83	4.04	6.03	−4.46	−1.31	−6.54	2.90
544.69	240.76	3.69	107.08	83.66	109.34	44.96	71.50	11.56	2.20	2.18	4.54	9.53	−0.90	28.94	18.46	10.98	5.09	2.72	2.43
1112.09	410.71	13.37	201.24	148.15	231.92	127.30	198.73	30.54	9.15	9.14	3.47	12.18	14.54	61.09	34.53	23.47	6.24	−2.42	6.37
1226.38	557.41	81.54	235.87	204.20	244.35	140.59	192.32	68.94	19.35	8.71	5.81	19.01	36.95	86.36	40.95	22.77	6.39	3.20	11.73
1490.17	622.00	156.30	240.23	164.59	263.74	138.90	113.47	113.68	11.16	19.63	14.14	39.38	58.75	60.70	37.19	21.58	6.83	7.45	11.69
1557.13	756.45	249.16	205.44	212.87	255.27	170.75	87.22	149.62	8.04	39.82	28.36	37.98	51.93	58.07	34.81	23.50	4.11	10.98	17.04
1349.73	640.86	359.48	244.02	186.86	204.96	115.28	64.56	147.56	27.79	43.44	24.52	51.96	83.34	51.66	31.79	22.40	8.07	7.82	33.33
1374.64	454.83	454.54	175.32	154.20	146.15	118.83	73.99	136.44	34.12	39.23	34.82	34.82	73.43	35.39	23.33	21.71	4.44	9.96	14.68
1207.78	528.04	508.69	207.57	172.53	140.39	112.46	21.15	127.40	28.48	46.53	41.21	35.42	75.82	33.93	26.14	25.39	15.19	14.47	15.84
1222.46	270.99	472.10	227.69	183.13	128.84	102.10	66.31	169.38	28.21	61.12	37.59	52.28	82.46	28.28	30.15	17.30	12.04	22.59	14.22
1240.02	531.42	468.20	229.35	166.19	155.55	130.11	78.76	153.54	24.97	41.98	30.96	45.26	74.40	25.23	28.80	23.96	11.88	11.33	15.65
1343.14	505.11	399.50	198.66	138.98	125.46	118.22	68.61	106.16	24.95	46.21	36.62	40.20	49.49	32.09	28.71	22.18	18.22	15.68	11.95
1239.64	620.82	310.26	207.63	136.69	168.99	118.21	61.13	116.28	7.25	54.27	40.56	37.36	49.02	29.62	23.40	21.02	13.04	15.36	9.99
1279.44	573.64	347.99	154.41	169.51	152.62	149.96	50.76	83.85	30.04	47.06	32.87	38.92	46.14	31.16	24.12	27.80	17.04	17.68	11.51
1146.30	380.67	374.81	169.75	138.29	135.18	158.56	32.72	80.89	39.74	38.38	31.24	24.59	35.64	28.30	23.28	18.39	21.93	23.61	11.28
1056.80	474.85	367.77	165.68	142.32	144.29	119.43	73.98	70.94	27.52	38.34	39.10	27.52	46.31	36.38	21.40	25.51	28.87	20.25	11.63
1076.39	433.11	309.50	189.47	141.80	123.69	118.59	74.54	62.70	37.77	36.19	29.90	33.93	39.34	35.33	25.92	13.83	24.03	17.15	10.10
1007.28	383.06	233.38	168.77	125.95	125.74	131.29	52.88	58.22	40.22	39.03	40.17	26.38	36.04	30.01	26.56	23.53	23.54	25.74	7.39
919.18	397.39	218.33	162.08	122.76	95.15	99.53	70.66	57.47	47.40	29.87	32.85	32.88	31.43	28.04	24.83	20.28	22.66	15.74	5.23
656.45	371.74	201.15	119.10	79.72	90.19	84.65	27.95	41.04	31.65	29.91	28.20	25.90	12.08	29.35	19.19	22.23	22.30	7.81	2.63
727.64	294.33	281.14	134.72	90.90	97.64	92.00	56.17	37.36	42.17	25.29	30.48	30.21	23.70	23.59	14.22	20.27	19.46	15.73	9.23
656.70	282.45	308.55	138.62	93.78	87.44	61.85	58.65	46.31	22.80	35.31	22.44	34.28	23.12	15.27	20.85	15.74	16.58	22.28	8.96
745.82	345.59	193.12	144.01	81.15	97.76	80.30	39.74	44.34	35.20	18.83	30.68	30.69	24.05	21.57	22.25	17.86	17.92	14.63	9.35
526.49	268.84	236.18	149.61	88.20	68.65	93.67	81.95	45.02	33.49	27.16	26.23	27.16	20.49	20.77	16.98	14.38	20.79	11.40	12.32
431.35	164.86	190.95	112.79	78.27	43.15	26.48	52.97	23.36	24.21	21.47	19.51	12.29	15.84	15.88	13.76	9.59	13.91	4.51	8.99

3. Remove sample 1 from the dataset, and calculate the five nonzero PCs arising from samples 2–6. What are the loadings? Use these loadings to determine the predicted scores $\hat{t} = x.p'$ for sample 1 using models based on one, two, three, four and five PCs successively, and hence the predictions $\hat{x}$ for each model.

4. Repeat this procedure, leaving each of the samples out once. Hence calculate the residual sum of squares over the entire dataset (all six samples) for models based on 1–5 PCs, and so obtain PRESS values.

5. Using the eigenvalues obtained in question 1, calculate the residual sum of squares error for 1–5 PCs and autoprediction.

6. List the RSS and PRESS values, and calculate the ratio $PRESS_a/RSS_{a-1}$. How many PCs do you think will characterise the data?

5 Calibration

5.1 Introduction

5.1.1 History and Usage

Calibration involves connecting one (or more) sets of variables together. Usually one set (often called a 'block') is a series of physical measurements, such as some spectra or molecular descriptors and the other contains one or more parameter such as the concentrations of a number of compounds or biological activity. Can we predict the concentration of a compound in a mixture spectrum or the properties of a material from its known structural parameters? Calibration provides the answer. In its simplest form, calibration is simply a form of regression as discussed in Chapter 3, in the context of experimental design.

Multivariate calibration has historically been a major cornerstone of chemometrics. However, there are a large number of diverse schools of thought, mainly dependent on people's background and the software with which they are familiar. Many mainstream statistical packages do not contain the PLS algorithm whereas some specialist chemometric software is based around this method. PLS is one of the most publicised algorithms for multivariate calibration that has been widely advocated by many in chemometrics, following the influence of S. Wold, whose father first proposed this in the context of economics. There has developed a mystique surrounding PLS, a technique with its own terminology, conferences and establishment. However, most of its prominent proponents are chemists. There are a number of commercial packages in the market-place that perform PLS calibration and result in a variety of diagnostic statistics. It is important, though, to understand that a major historical (and economic) driving force was near-infrared (NIR) spectroscopy, primarily in the food industry and in process analytical chemistry. Each type of spectroscopy and chromatography has its own features and problems, so much software was developed to tackle specific situations which may not necessarily be very applicable to other techniques such as chromatography, NMR or MS. In many statistical circles, NIR and chemometrics are almost inseparably intertwined. However, other more modern techniques are emerging even in process analysis, so it is not at all certain that the heavy investment on the use of PLS in NIR will be so beneficial in the future. Indeed, as time moves on instruments improve in quality so many of the computational approaches developed one or two decades ago to deal with problems such as background correction, and noise distributions are not so relevant nowadays, but for historical reasons it is often difficult to distinguish between these specialist methods required to prepare data in order to obtain meaningful information from the chemometrics and the actual calibration steps themselves. Despite this, chemometric approaches to calibration have very wide potential applicability throughout all areas of quantitative chemistry and NIR spectroscopists will definitely form an important readership base of this text.

There are very many circumstances in which multivariate calibration methods are appropriate. The difficulty is that to develop a comprehensive set of data analytical techniques for a particular situation takes a huge investment in resources and time, so the applications of multivariate calibration in some areas of science are much less well established than in others. It is important to separate the methodology that has built up around a small number of spectroscopic methods such as NIR from the general principles applicable throughout chemistry. There are probably several hundred favourite diagnostics available to the professional user of PLS, e.g. in NIR spectroscopy, yet each one has been developed with a specific technique or problem in mind, and are not necessarily generally applicable to all calibration problems.

There are a whole series of problems in chemistry for which multivariate calibration is appropriate, but each is very different in nature. Many of the most successful applications have been in the spectroscopy or chromatography of mixtures and we will illustrate this chapter with this example, although several diverse applications are presented in the problems at the end.

1. The simplest is calibration of the concentration of a single compound using a spectroscopic or chromatographic method, an example being the determination of the concentration of chlorophyll by electronic absorption spectroscopy (EAS) – sometimes called UV/vis spectroscopy. Instead of using one wavelength (as is conventional for the determination of molar absorptivity or extinction coefficients), multivariate calibration involves using all or several of the wavelengths. Each variable measures the same information, but better information is obtained by considering all the wavelengths.
2. A more complex situation is a multi-component mixture where all pure standards are available. It is possible to control the concentration of the reference compounds, so that a number of carefully designed mixtures can be produced in the laboratory. Sometimes the aim is to see whether a spectrum of a mixture can be employed to determine individual concentrations and, if so, how reliably. The aim may be to replace a slow and expensive chromatographic method by a rapid spectroscopic approach. Another rather different aim might be impurity monitoring: how well the concentration of a small impurity can be determined, for example, buried within a large chromatographic peak.
3. A different approach is required if only the concentration of a portion of the components is known in a mixture, for example, polyaromatic hydrocarbons within coal tar pitch volatiles. In natural samples there may be tens or hundreds of unknowns, but only a few can be quantified and calibrated. The unknown interferents cannot necessarily be determined and it is not possible to design a set of samples in the laboratory containing all the potential components in real samples. Multivariate calibration is effective providing the range of samples used to develop the model is sufficiently representative of all future samples in the field. If it is not, the predictions from multivariate calibration could be dangerously inaccurate. In order to protect against samples not belonging to the original dataset, a number of approaches for determination of outliers and experimental design have been developed.
4. A final case is where the aim of calibration is not so much to determine the concentration of a particular compound but to determine a statistical parameter. There will no longer be pure standards available, and the training set must consist of a sufficiently representative group. An example is to determine the concentration of a class of compounds in food, such as protein in wheat. It is not possible (or desirable) to isolate each single type of protein, and we rely on the original samples

being sufficiently representative. This situation also occurs, for example, in quantitative structure–property relationships (QSPR) or quantitative structure–activity relationships (QSAR).

There are many pitfalls in the use of calibration models, perhaps the most serious being variability in instrument performance over time. Each measurement technique has different characteristics and on each day and even hour the response can vary. How serious this is for the stability of the calibration model should be assessed before investing a large effort. Sometimes it is necessary to reform the calibration model on a regular basis, by running a standard set of samples, possibly on a daily or weekly basis. In other cases multivariate calibration gives only a rough prediction, but if the quality of a product or the concentration of a pollutant appears to exceed a certain limit, then other more detailed approaches can be used to investigate the sample. For example, on-line calibration in NIR can be used for screening a manufactured sample, and any dubious batches investigated in more detail using chromatography.

This chapter will describe the main algorithms and principles of calibration. We will concentrate on situations in which there is a direct linear relationship between blocks of variables. It is possible to extend the methods to include multilinear (such as squared) terms simply by extended the X matrix, for example, in the case of spectroscopy at high concentrations or nonlinear detection systems.

5.1.2 Case Study

It is easiest to illustrate the methods in this chapter using a small case study, involving recording

- 25 EAS spectra at
- 27 wavelengths (from 220 to 350 nm at 5 nm intervals) and
- consisting of a mixture of 10 compounds [polyaromatic hydrocarbons (PAHs)].

In reality the spectra might be obtained at higher digital resolution, but for illustrative purposes we reduce the sampling rate. The aim is to predict the concentrations of individual PAHs from the mixture spectra. The spectroscopic data are presented in Table 5.1 and the concentrations of the compounds in Table 5.2.

The methods in this chapter will be illustrated as applied to the spectroscopy of mixtures as this is a common and highly successful application of calibration in chemistry. However, similar principles apply to a wide variety of calibration problems.

5.1.3 Terminology

We will refer to physical measurements of the form in Table 5.1 as the 'x' block and those in Table 5.2 as the 'c' block. One area of confusion is that users of different techniques in chemometrics tend to employ incompatible notation. In the area of experimental design it is usual to call the measured response 'y', e.g. the absorbance in a spectrum, and the concentration or any related parameter 'x'. In traditional multivariate calibration this notation is swapped around. For the purpose of a coherent text it would be confusing to use two opposite notations; however, some compatibility with the established literature is desirable. Figure 5.1 illustrates the notation used in this text.

Table 5.1 Case study consisting of 25 spectra recorded at 27 wavelengths (nm) (absorbances in AU).

No.	220	225	230	235	240	245	250	255	260	265	270	275	280	285	290	295	300	305	310	315	320	325	330	335	340	345	350
1	0.771	0.714	0.658	0.537	0.587	0.671	0.768	0.837	0.673	0.678	0.741	0.755	0.682	0.633	0.706	0.290	0.208	0.161	0.135	0.137	0.162	0.130	0.127	0.165	0.110	0.075	0.053
2	0.951	0.826	0.737	0.638	0.738	0.911	1.121	1.162	0.869	0.870	0.965	1.050	0.993	0.934	1.008	0.405	0.254	0.185	0.157	0.159	0.180	0.155	0.150	0.178	0.140	0.105	0.077
3	0.912	0.847	0.689	0.514	0.504	0.622	0.805	0.892	0.697	0.728	0.790	0.728	0.692	0.639	0.717	0.292	0.224	0.168	0.134	0.123	0.136	0.115	0.095	0.102	0.089	0.068	0.048
4	0.688	0.679	0.662	0.558	0.655	0.738	0.838	0.883	0.670	0.656	0.704	0.668	0.562	0.516	0.573	0.295	0.212	0.172	0.138	0.144	0.179	0.130	0.134	0.191	0.107	0.060	0.046
5	0.873	0.801	0.732	0.640	0.750	0.820	0.907	0.955	0.692	0.681	0.775	0.866	0.798	0.749	0.827	0.311	0.213	0.170	0.151	0.162	0.201	0.158	0.170	0.239	0.146	0.094	0.067
6	0.953	0.850	0.732	0.612	0.724	0.882	1.013	1.087	0.860	0.859	0.925	0.938	0.869	0.810	0.868	0.390	0.241	0.175	0.138	0.141	0.167	0.129	0.135	0.178	0.115	0.078	0.056
7	0.613	0.577	0.547	0.448	0.508	0.463	0.473	0.549	0.510	0.560	0.604	0.493	0.369	0.343	0.363	0.215	0.178	0.165	0.136	0.142	0.183	0.122	0.129	0.193	0.089	0.041	0.030
8	0.927	0.866	0.799	0.605	0.618	0.671	0.771	0.842	0.646	0.658	0.716	0.829	0.783	0.731	0.794	0.292	0.220	0.155	0.120	0.123	0.144	0.122	0.127	0.164	0.113	0.078	0.056
9	0.585	0.577	0.543	0.450	0.515	0.614	0.734	0.815	0.624	0.642	0.724	0.716	0.680	0.649	0.713	0.298	0.213	0.167	0.136	0.126	0.137	0.109	0.104	0.129	0.098	0.074	0.057
10	0.835	0.866	0.803	0.550	0.563	0.577	0.642	0.732	0.675	0.713	0.836	0.938	0.911	0.847	0.910	0.358	0.226	0.173	0.153	0.160	0.186	0.156	0.157	0.193	0.134	0.093	0.066
11	0.477	0.454	0.450	0.433	0.538	0.593	0.661	0.724	0.554	0.561	0.616	0.469	0.355	0.322	0.345	0.211	0.154	0.139	0.114	0.118	0.154	0.100	0.100	0.154	0.071	0.030	0.016
12	0.496	0.450	0.402	0.331	0.389	0.504	0.613	0.599	0.383	0.344	0.353	0.366	0.338	0.309	0.334	0.194	0.123	0.094	0.076	0.070	0.077	0.065	0.056	0.065	0.053	0.036	0.025
13	0.594	0.512	0.441	0.392	0.489	0.562	0.637	0.647	0.415	0.387	0.421	0.461	0.396	0.361	0.424	0.161	0.103	0.076	0.062	0.076	0.110	0.082	0.094	0.144	0.078	0.043	0.028
14	0.512	0.478	0.409	0.319	0.375	0.395	0.436	0.487	0.439	0.449	0.474	0.438	0.362	0.331	0.360	0.204	0.158	0.122	0.089	0.087	0.107	0.075	0.079	0.114	0.064	0.040	0.028
15	0.662	0.583	0.586	0.564	0.687	0.708	0.748	0.757	0.611	0.581	0.641	0.727	0.638	0.596	0.658	0.277	0.171	0.133	0.113	0.128	0.168	0.125	0.143	0.211	0.114	0.067	0.044
16	0.768	0.588	0.501	0.386	0.429	0.539	0.671	0.740	0.607	0.627	0.693	0.772	0.751	0.710	0.781	0.287	0.174	0.116	0.093	0.089	0.095	0.087	0.081	0.087	0.081	0.069	0.047
17	0.635	0.557	0.487	0.377	0.404	0.488	0.597	0.669	0.603	0.625	0.647	0.576	0.515	0.477	0.524	0.260	0.188	0.139	0.105	0.096	0.104	0.084	0.071	0.077	0.061	0.045	0.031
18	0.575	0.489	0.432	0.375	0.408	0.449	0.525	0.569	0.442	0.451	0.501	0.519	0.474	0.458	0.500	0.219	0.151	0.119	0.095	0.092	0.107	0.085	0.081	0.106	0.072	0.047	0.032
19	0.768	0.713	0.644	0.528	0.599	0.807	1.009	1.035	0.750	0.722	0.790	0.907	0.921	0.866	0.924	0.375	0.219	0.144	0.118	0.116	0.123	0.120	0.114	0.119	0.115	0.096	0.070
20	0.811	0.655	0.593	0.496	0.601	0.694	0.810	0.835	0.669	0.671	0.718	0.641	0.566	0.524	0.573	0.290	0.182	0.148	0.123	0.119	0.141	0.103	0.098	0.130	0.080	0.051	0.036
21	0.827	0.714	0.660	0.535	0.601	0.660	0.729	0.775	0.619	0.601	0.640	0.667	0.571	0.525	0.602	0.242	0.160	0.120	0.103	0.122	0.164	0.127	0.133	0.182	0.105	0.059	0.037
22	0.673	0.492	0.427	0.447	0.584	0.751	0.908	0.931	0.656	0.611	0.623	0.552	0.466	0.418	0.443	0.275	0.199	0.146	0.104	0.094	0.106	0.074	0.070	0.095	0.064	0.042	0.030
23	0.949	0.852	0.711	0.559	0.586	0.701	0.855	0.907	0.737	0.731	0.801	0.956	0.934	0.881	0.934	0.380	0.239	0.158	0.127	0.125	0.138	0.126	0.124	0.138	0.118	0.093	0.063
24	0.939	0.835	0.723	0.622	0.716	0.791	0.908	1.031	0.894	0.948	1.036	1.079	0.948	0.897	0.983	0.386	0.260	0.189	0.150	0.160	0.199	0.155	0.163	0.219	0.145	0.101	0.070
25	1.055	0.989	0.894	0.681	0.663	0.726	0.837	0.914	0.813	0.840	0.916	0.892	0.837	0.785	0.846	0.359	0.237	0.179	0.151	0.150	0.175	0.145	0.128	0.147	0.116	0.086	0.058

Table 5.2 Concentrations of the 10 PAHs[a] in the data in Table 5.1.

Spectrum No.	PAH concentration mg l^{-1}									
	Py	Ace	Anth	Acy	Chry	Benz	Fluora	Fluore	Nap	Phen
1	0.456	0.120	0.168	0.120	0.336	1.620	0.120	0.600	0.120	0.564
2	0.456	0.040	0.280	0.200	0.448	2.700	0.120	0.400	0.160	0.752
3	0.152	0.200	0.280	0.160	0.560	1.620	0.080	0.800	0.160	0.188
4	0.760	0.200	0.224	0.200	0.336	1.080	0.160	0.800	0.040	0.752
5	0.760	0.160	0.280	0.120	0.224	2.160	0.160	0.200	0.160	0.564
6	0.608	0.200	0.168	0.080	0.448	2.160	0.040	0.800	0.120	0.940
7	0.760	0.120	0.112	0.160	0.448	0.540	0.160	0.600	0.200	0.188
8	0.456	0.080	0.224	0.160	0.112	2.160	0.120	1.000	0.040	0.188
9	0.304	0.160	0.224	0.040	0.448	1.620	0.200	0.200	0.040	0.376
10	0.608	0.160	0.056	0.160	0.336	2.700	0.040	0.200	0.080	0.188
11	0.608	0.040	0.224	0.120	0.560	0.540	0.040	0.400	0.040	0.564
12	0.152	0.160	0.168	0.200	0.112	0.540	0.080	0.200	0.120	0.752
13	0.608	0.120	0.280	0.040	0.112	1.080	0.040	0.600	0.160	0.376
14	0.456	0.200	0.056	0.040	0.224	0.540	0.120	0.800	0.080	0.376
15	0.760	0.040	0.056	0.080	0.112	1.620	0.160	0.400	0.080	0.940
16	0.152	0.040	0.112	0.040	0.336	2.160	0.080	0.400	0.200	0.376
17	0.152	0.080	0.056	0.120	0.448	1.080	0.080	1.000	0.080	0.564
18	0.304	0.040	0.168	0.160	0.224	1.080	0.200	0.400	0.120	0.188
19	0.152	0.120	0.224	0.080	0.224	2.700	0.080	0.600	0.040	0.940
20	0.456	0.160	0.112	0.080	0.560	1.080	0.120	0.200	0.200	0.940
21	0.608	0.080	0.112	0.200	0.224	1.620	0.040	1.000	0.200	0.752
22	0.304	0.080	0.280	0.080	0.336	0.540	0.200	1.000	0.160	0.940
23	0.304	0.200	0.112	0.120	0.112	2.700	0.200	0.800	0.200	0.564
24	0.760	0.080	0.168	0.040	0.560	2.700	0.160	1.000	0.120	0.376
25	0.304	0.120	0.056	0.200	0.560	2.160	0.200	0.600	0.080	0.752

[a] Abbreviations used in this chapter: Py = pyrene; Ace = acenaphthene; Anth = anthracene; Acy = acenaphthylene; Chry = chrysene; Benz = benzanthracene; Fluora = fluoranthene; Fluore = fluorene; Nap = naphthalene; Phen = phenanthracene.

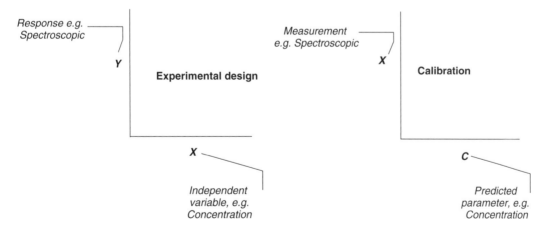

Figure 5.1
Different notations for calibration and experimental design as used in this book

5.2   Univariate Calibration

Univariate calibration involves relating two single variables to each other, and is often called linear regression. It is easy to perform using most data analysis packages.

5.2.1   Classical Calibration

One of the simplest problems is to determine the concentration of a single compound using the response at a single detector, for example a single spectroscopic wavelength or a chromatographic peak area.

Mathematically a series of experiments can be performed to relate the concentration to spectroscopic measurements as follows:

$$x \approx c.s$$

where, in the simplest case, x is a vector consisting, for example, of absorbances at one wavelength for a number of samples, and c is of the corresponding concentrations. Both vectors have length I, equal to the number of samples. The scalar s relates these parameters and is determined by regression. Classically, most regression packages try to find s.

A simple method for solving this equation is to use the pseudo-inverse (see Chapter 2, Section 2.2.2.3, for an introduction):

$$c'.x \approx (c'.c).s$$

so

$$(c'.c)^{-1}.c'.x \approx (c'.c)^{-1}.(c'.c).s$$

or

$$s \approx (c'.c)^{-1}.c'.x = \frac{\sum_{i=1}^{I} x_i c_i}{\sum_{i=1}^{I} c_i^2}$$

Many conventional texts express regression equations in the form of summations rather than matrices, but both approaches are equivalent; with modern spreadsheets and matrix oriented programming environments it is easier to build on the matrix based equations and the summations can become rather unwieldy if the problem is more complex. In Figure 5.2, the absorbance of the 25 spectra at 335 nm is plotted against the concentration of pyrene. The graph is approximately linear, and provides a best fit slope calculated by

$$\sum_{i=1}^{I} x_i c_i = 1.916$$

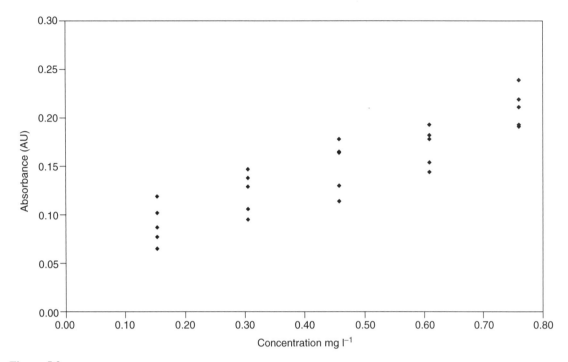

Figure 5.2
Absorbance at 335 nm for the PAH case study plotted against concentration of pyrene

and

$$\sum_{i=1}^{I} c_i^2 = 6.354$$

so that $\hat{x} = 0.301c$. The predictions are presented in Table 5.3. The spectra of the 10 pure standards are superimposed in Figure 5.3, with pyrene indicated in bold. It can be seen that pyrene has unique absorbances at higher wavelengths, so 335 nm will largely be characteristic of this compound. For most of the other compounds in these spectra, it would not be possible to obtain such good results from univariate calibration.

The quality of prediction can be determined by the residuals (or errors), i.e. the difference between the observed and predicted, i.e. $x - \hat{x}$, the smaller, the better. Generally, the root mean error is calculated:

$$E = \sqrt{\sum_{i=1}^{I} (x_i - \hat{x}_i)^2 / d}$$

where d is called the degrees of freedom. In the case of univariate calibration this equals the number of observations (N) minus the number of parameters in the model (P) or in this case, $25 - 1 = 24$ (see Chapter 2, Section 2.2.1), so that

$$E = \sqrt{0.0279/24} = 0.0341$$

Table 5.3 Concentration of pyrene, absorbance at 335 nm and predictions of absorbance, using single parameter classical calibration.

Concentration $(mg\,l^{-1})$	Absorbance at 335 nm	Predicted absorbance
0.456	0.165	0.137
0.456	0.178	0.137
0.152	0.102	0.046
0.760	0.191	0.229
0.760	0.239	0.229
0.608	0.178	0.183
0.760	0.193	0.229
0.456	0.164	0.137
0.304	0.129	0.092
0.608	0.193	0.183
0.608	0.154	0.183
0.152	0.065	0.046
0.608	0.144	0.183
0.456	0.114	0.137
0.760	0.211	0.229
0.152	0.087	0.046
0.152	0.077	0.046
0.304	0.106	0.092
0.152	0.119	0.046
0.456	0.130	0.137
0.608	0.182	0.183
0.304	0.095	0.092
0.304	0.138	0.092
0.760	0.219	0.229
0.304	0.147	0.092

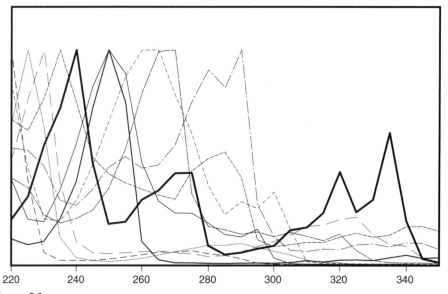

Figure 5.3
Spectra of pure standards, digitised at 5 nm intervals. Pyrene is indicated in bold

This error can be represented as a percentage of the mean, $E_\% = 100(E/\bar{x}) = 24.1\%$ in this case. Sometimes the percentage error is calculated relative to the standard deviation rather than mean: this is more appropriate if the data are mean centred (because the mean is 0), or if the data are all clustered at high values, in which case an apparently small error relative to the mean still may imply a fairly large deviation; there are no hard and fast rules and in this chapter we will calculate errors relative to the mean unless stated otherwise. It is always useful, however, to check the original graph (Figure 5.2) just to be sure, and this percentage appears reasonable. Provided that a consistent measure is used throughout, all percentage errors will be comparable.

This approach to calibration, although widely used throughout most branches of science, is nevertheless not always appropriate in all applications. We may want to answer the question 'can the absorbance in a spectrum be employed to determine the concentration of a compound?'. It is not the best approach to use an equation that predicts the absorbance from the concentration when our experimental aim is the reverse. In other areas of science the functional aim might be, for example, to predict an enzymic activity from its concentration. In the latter case univariate calibration as outlined in this section results in the correct functional model. Nevertheless, most chemists employ classical calibration and provided that the experimental errors are roughly normal and there are no significant outliers, all the different univariate methods should result in approximately similar conclusions.

For a new or unknown sample, however, the concentration can be estimated (approximately) by using the inverse of the slope or

$$\hat{c} = 3.32x$$

5.2.2 Inverse Calibration

Although classical calibration is widely used, it is not always the most appropriate approach in chemistry, for two main reasons. First, the ultimate aim is usually to predict the concentration (or independent variable) from the spectrum or chromatogram (response) rather than vice versa. The second relates to error distributions. The errors in the response are often due to instrumental performance. Over the years, instruments have become more reproducible. The independent variable (often concentration) is usually determined by weighings, dilutions and so on, and is often by far the largest source of errors. The quality of volumetric flasks, syringes and so on has not improved dramatically over the years, whereas the sensitivity and reproducibility of instruments has increased manyfold. Classical calibration fits a model so that all errors are in the response [Figure 5.4(a)], whereas a more appropriate assumption is that errors are primarily in the measurement of concentration [Figure 5.4(b)].

Calibration can be performed by the inverse method whereby

$$c \approx x \,.\, b$$

or

$$b \approx (x'.x)^{-1}.x'.c = \frac{\sum_{i=1}^{I} x_i c_i}{\sum_{i=1}^{I} x_i^2}$$

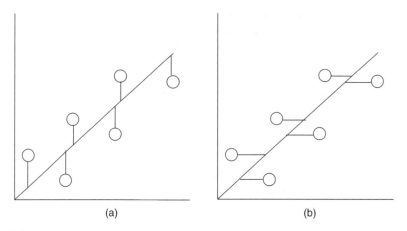

Figure 5.4
Difference between errors in (a) classical and (b) inverse calibration

giving for this example, $\hat{c} = 3.16x$, a root mean square error of 0.110 or 24.2 % relative to the mean. Note that b is only approximately the inverse of s (see above), because each model makes different assumptions about error distributions. The results are presented in Table 5.4. However, for good data, both models should provide fairly similar predictions, and if not there could be some other factor that influences the data, such as an intercept, nonlinearities, outliers or unexpected noise distributions. For heteroscedastic noise distributions there are a variety of enhancements to linear calibration. However, these are rarely taken into consideration when extending the principles to the multivariate calibration.

The best fit straight lines for both methods of calibration are given in Figure 5.5. At first it looks as if these are a poor fit to the data, but an important feature is that the intercept is assumed to be zero. The method of regression forces the line through the point (0,0). Because of other compounds absorbing in the spectrum, this is a poor approximation, so reducing the quality of regression. We look at how to improve this model below.

5.2.3   Intercept and Centring

In many situations it is appropriate to include extra terms in the calibration model. Most commonly an intercept (or baseline) term is included to give an inverse model of the form

$$c \approx b_0 + b_1 x$$

which can be expressed in matrix/vector notation by

$$c \approx X.b$$

for inverse calibration, where c is a column vector of concentrations and b is a column vector consisting of two numbers, the first equal to b_0 (the intercept) and the second to b_1 (the slope). X is now a matrix of two columns, the first of which is a column of ones and the second the absorbances.

Table 5.4 Concentration of pyrene, absorbance at 335 nm and predictions of concentration, using single parameter inverse calibration.

Concentration $(mg\,l^{-1})$	Absorbance at 335 nm	Predicted concentration $(mg\,l^{-1})$
0.456	0.165	0.522
0.456	0.178	0.563
0.152	0.102	0.323
0.760	0.191	0.604
0.760	0.239	0.756
0.608	0.178	0.563
0.760	0.193	0.611
0.456	0.164	0.519
0.304	0.129	0.408
0.608	0.193	0.611
0.608	0.154	0.487
0.152	0.065	0.206
0.608	0.144	0.456
0.456	0.114	0.361
0.760	0.211	0.668
0.152	0.087	0.275
0.152	0.077	0.244
0.304	0.106	0.335
0.152	0.119	0.377
0.456	0.130	0.411
0.608	0.182	0.576
0.304	0.095	0.301
0.304	0.138	0.437
0.760	0.219	0.693
0.304	0.147	0.465

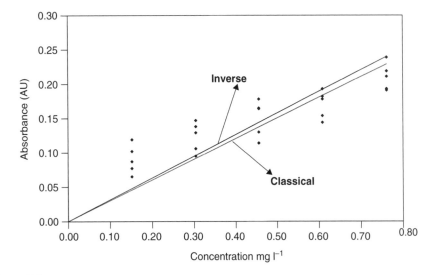

Figure 5.5
Best fit straight lines for classical and inverse calibration: data for pyrene at 335 nm, no intercept

Exactly the same principles can be employed for calculating the coefficients as in Section 2.1.2, but in this case b is a vector rather than scalar, and X is a matrix rather than a vector, so that

$$b \approx (X'.X)^{-1}.X'.c$$

or

$$\hat{c} = -0.173 + 4.227x$$

Note that the coefficients are different to those of Section 5.2.2. One reason is that there are still a number of interferents, from the other PAHs, in the spectrum at 335 nm, and these are modelled partly by the intercept term. The models of the previous sections force the best fit straight line to pass through the origin. A better fit can be obtained if this condition is not required. The new best fit straight line is presented in Figure 5.6 and results, visually, in a much better fit to the data.

The predicted concentrations are fairly easy to obtain, the easiest approach involving the use of matrix based methods, so that

$$\hat{c} = X.b$$

the root mean square error being given by

$$E = \sqrt{0.229/23} = 0.100 \text{ mg} \, l^{-1}$$

representing an $E_\%$ of 21.8 % relative to the mean. Note that the error term should be divided by 23 (number of degrees of freedom rather than 25) to reflect the *two* parameters used in the model.

One interesting and important consideration is that the apparent root mean square error in Sections 5.2.2 and 5.2.3 is only reduced by a small amount, yet the best fit straight line appears much worse if we neglect the intercept. The reason for this is that there is still a considerable replicate error, and this cannot readily be modelled using a

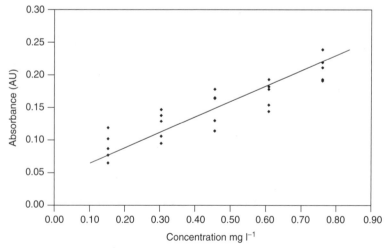

Figure 5.6
Best fit straight line using inverse calibration: data of Figure 5.5 and an intercept term

single compound model. If this contribution were removed the error would be reduced dramatically.

An alternative, and common, method for including the intercept is to mean centre both the x and the c variables to fit the equation

$$c - \bar{c} \approx (x - \bar{x})b$$

or

$$^{\text{cen}}c \approx {}^{\text{cen}}x\, b$$

or

$$b \approx ({}^{\text{cen}}x'.{}^{\text{cen}}x)^{-1}.{}^{\text{cen}}x'.{}^{\text{cen}}c = \frac{\displaystyle\sum_{i=1}^{I} (x_i - \bar{x})(c_i - \bar{c})}{\displaystyle\sum_{i=1}^{I} (x_i - \bar{x})^2}$$

It is easy to show algebraically that

- the value of b when both variables have been centred is identical with the value of b_1 obtained when the data are modelled including an intercept term (=4.227 in this example);
- the value of b_0 (intercept term for uncentred data) is given by $\bar{c} - b\bar{x} = 0.469 - 4.227 \times 0.149 = -0.173$, so the two methods are related.

It is common to centre both sets of variables for this reason, the calculations being mathematically simpler than including an intercept term. Note that both blocks must be centred, and the predictions are of the concentrations minus their mean, so the mean concentration must be added back to return to the original physical values.

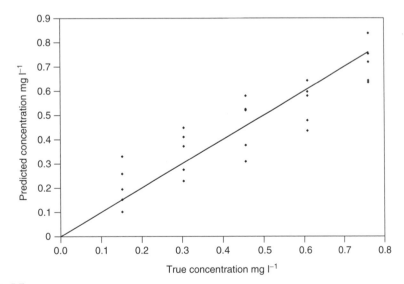

Figure 5.7
Predicted (vertical) versus known (horizontal) concentrations using the methods in Section 5.2.3

In calibration it is common to plot a graph of predicted versus observed concentrations as presented in Figure 5.7. This looks superficially similar to that in the previous figure, but the vertical scale is different and the graph goes through the origin (providing the data have been mean centred). There is a variety of potential graphical output and it is important not to be confused, but to distinguish each type of information carefully.

It is important to realise that the predictions for the method described in this section differ from those obtained for the uncentred data. It is also useful to realise that similar methods can be applied to classical calibration, the details being omitted for brevity, as it is recommended that inverse calibration is performed in normal circumstances.

5.3 Multiple Linear Regression

5.3.1 Multidetector Advantage

Multiple linear regression (MLR) is an extension when more than one response is employed. There are two principal reasons for this. The first is that there may be more than one component in a mixture. Under such circumstances it is usual to employ more than one response (the exception being if the concentrations of some of the components are known to be correlated): for N components, at least N wavelengths should normally be used. The second is that each detector contains extra, and often complementary, information: some individual wavelengths in a spectrum may be influenced by noise or unknown interferents. Using, for example, 100 wavelengths averages out the information, and will often provide a better result than relying on a single wavelength.

5.3.2 Multiwavelength Equations

In certain applications, equations can be developed that are used to predict the concentrations of compounds by monitoring at a finite number of wavelengths. A classical area is in pigment analysis by electronic absorption spectroscopy, for example in the area of chlorophyll chemistry. In order to determine the concentration of four pigments in a mixture, investigators recommend monitoring at four different wavelengths, and to use an equation that links absorbance at each wavelength to concentration of the pigments.

In the PAH case study, only certain compounds absorb above 330 nm, the main ones being pyrene, fluoranthene, acenaphthylene and benzanthracene (note that the small absorbance due to a fifth component may be regarded as an interferent, although adding this to the model will, of course, result in better predictions). It is possible to choose four wavelengths, preferably ones in which the absorbance ratios of these four compounds differ. The absorbance at wavelengths 330, 335, 340 and 345 nm are indicated in Figure 5.8. Of course, it is not necessary to select four sequential wavelengths; any four wavelengths would be sufficient, provided that the four compounds are the main ones represented by these variables to give an X matrix with four columns and 25 rows.

Calibration equations can be obtained, as follows, using inverse methods.

- First, select the absorbances of the 25 spectra at these four wavelengths.
- Second, obtain the corresponding C matrix consisting of the relevant concentrations. These new (reduced) matrices are presented in Table 5.5.
- The aim is to find coefficients B relating X and C by $C \approx X.B$, where B is a 4×4 matrix, each *column* representing a compound and each *row* a wavelength.

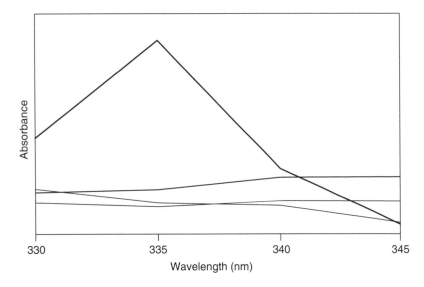

Figure 5.8
Absorbances of pure Pyr, Fluor, Benz and Ace between 330 and 345 nm

Table 5.5   Matrices for four components.

X				C			
330	335	340	345	Py	Ace	Benz	Fluora
0.127	0.165	0.110	0.075	0.456	0.120	1.620	0.120
0.150	0.178	0.140	0.105	0.456	0.040	2.700	0.120
0.095	0.102	0.089	0.068	0.152	0.200	1.620	0.080
0.134	0.191	0.107	0.060	0.760	0.200	1.080	0.160
0.170	0.239	0.146	0.094	0.760	0.160	2.160	0.160
0.135	0.178	0.115	0.078	0.608	0.200	2.160	0.040
0.129	0.193	0.089	0.041	0.760	0.120	0.540	0.160
0.127	0.164	0.113	0.078	0.456	0.080	2.160	0.120
0.104	0.129	0.098	0.074	0.304	0.160	1.620	0.200
0.157	0.193	0.134	0.093	0.608	0.160	2.700	0.040
0.100	0.154	0.071	0.030	0.608	0.040	0.540	0.040
0.056	0.065	0.053	0.036	0.152	0.160	0.540	0.080
0.094	0.144	0.078	0.043	0.608	0.120	1.080	0.040
0.079	0.114	0.064	0.040	0.456	0.200	0.540	0.120
0.143	0.211	0.114	0.067	0.760	0.040	1.620	0.160
0.081	0.087	0.081	0.069	0.152	0.040	2.160	0.080
0.071	0.077	0.061	0.045	0.152	0.080	1.080	0.080
0.081	0.106	0.072	0.047	0.304	0.040	1.080	0.200
0.114	0.119	0.115	0.096	0.152	0.120	2.700	0.080
0.098	0.130	0.080	0.051	0.456	0.160	1.080	0.120
0.133	0.182	0.105	0.059	0.608	0.080	1.620	0.040
0.070	0.095	0.064	0.042	0.304	0.080	0.540	0.200
0.124	0.138	0.118	0.093	0.304	0.200	2.700	0.200
0.163	0.219	0.145	0.101	0.760	0.080	2.700	0.160
0.128	0.147	0.116	0.086	0.304	0.120	2.160	0.200

Table 5.6 Matrix **B** for Section 5.3.2.

	Py	Ace	Benz	Fluor
330	−3.870	2.697	14.812	−4.192
335	8.609	−2.391	3.033	0.489
340	−5.098	4.594	−49.076	7.221
345	1.848	−4.404	65.255	−2.910

This equation can be solved using the regression methods in Section 5.2.2, changing vectors and scalars to matrices, so that $B = (X'.X)^{-1}.X'.C$, giving the matrix in Table 5.6.

- If desired, represent in equation form, for example, the first column of B suggests that

$$\text{estimated [pyrene]} = -3.870A_{330} + 8.609A_{335} - 5.098A_{340} + 1.848A_{345}$$

In many areas of optical spectroscopy, these types of equations are very common. Note, though, that changing the wavelengths can have a radical influence on the coefficients, and slight wavelength irreproducibility between spectrometers can lead to equations that are not easily transferred.

- Finally, estimate the concentrations by

$$\hat{C} = X.B$$

as indicated in Table 5.7.

The estimates by this approach are very much better than the univariate approaches in this particular example. Figure 5.9 shows the predicted versus known concentrations for pyrene. The root mean square error of prediction is now

$$E = \sqrt{\sum_{i=1}^{I} (c_i - \hat{c}_i)^2 / 21}$$

(note that the divisor is 21 not 25 as four degrees of freedom are lost because there are four compounds in the model), equal to 0.042 or 9.13 %, of the average concentration, a significant improvement. Further improvement could be obtained by including the intercept (usually performed by centring the data) and including the concentrations of more compounds. However, the number of wavelengths must be increased if the more compounds are used in the model.

It is possible also to employ classical methods. For the single detector, single wavelength model in Section 2.1.1,

$$\hat{c} = x(1/s)$$

where s is a scalar and x and c are vectors corresponding to the concentrations and absorbances for each of the I samples. Where there are several components in the mixture, this becomes

$$\hat{C} = X.S'.(S.S')^{-1}$$

Table 5.7 Estimated concentrations (mg l^{-1}) for four components as described in Section 5.3.2.

Py	Ace	Benz	Fluor
0.507	0.123	1.877	0.124
0.432	0.160	2.743	0.164
0.182	0.122	1.786	0.096
0.691	0.132	1.228	0.130
0.829	0.144	2.212	0.185
0.568	0.123	1.986	0.125
0.784	0.115	0.804	0.077
0.488	0.126	1.923	0.137
0.345	0.096	1.951	0.119
0.543	0.168	2.403	0.133
0.632	0.096	0.421	0.081
0.139	0.081	0.775	0.075
0.558	0.078	0.807	0.114
0.423	0.058	0.985	0.070
0.806	0.110	1.535	0.132
0.150	0.079	1.991	0.087
0.160	0.089	1.228	0.050
0.319	0.089	1.055	0.095
0.174	0.128	2.670	0.131
0.426	0.096	1.248	0.082
0.626	0.146	1.219	0.118
0.298	0.071	0.925	0.093
0.278	0.137	2.533	0.129
0.702	0.137	2.553	0.177
0.338	0.148	2.261	0.123

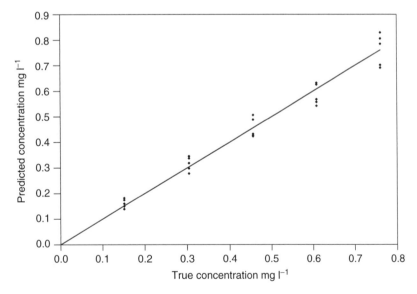

Figure 5.9
Predicted versus known concentration of pyrene, using a four component model and the wavelengths 330, 335, 340 and 345 nm (uncentred)

and the trick is to estimate S, which can be done in one of two ways: (1) by knowledge of the true spectra or (2) by regression since $C.S \approx X$, so $\hat{S} = (C'.C)^{-1}C'.X$. Note that

$$B \approx S'.(S.S')^{-1}$$

However, as in univariate calibration, the coefficients obtained using both approaches may not be exactly equal, as each method makes different assumptions about error structure.

Such equations make assumptions that the concentrations of the significant analytes are all known, and work well only if this is true. Application to mixtures where there are unknown interferents can result in serious estimation errors.

5.3.3 Multivariate Approaches

The methods in Section 5.3.2 could be extended to all 10 PAHs, and with appropriate choice of 10 wavelengths may give reasonable estimates of concentrations. However, all the wavelengths contain some information and there is no reason why most of the spectrum cannot be employed.

There is a fairly confusing literature on the use of multiple linear regression for calibration in chemometrics, primarily because many workers present their arguments in a very formalised manner. However, the choice and applicability of any method depends on three main factors:

1. the number of compounds in the mixture ($N = 10$ in this case) or responses to be estimated;
2. the number of experiments ($I = 25$ in this case), often spectra or chromatograms;
3. the number of variables ($J = 27$ wavelengths in this case).

In order to have a sensible model, the number of compounds must be less than or equal to the smaller of the number of experiments or number of variables. In certain specialised cases this limitation can be infringed if it is known that there are correlations between concentrations of different compounds. This may happen, for example, in environmental chemistry, where there could be tens or hundreds of compounds in a sample, but the presence of one (e.g. a homologous series) indicates the presence of another, so, in practice there are only a few independent factors or groups of compounds. Also, correlations can be built into the design. In most real world situations there definitely will be correlations in complex multicomponent mixtures. However, the methods described below are for the case where the number of compounds is smaller than the number of experiments or number of detectors.

The X data matrix is ideally related to the concentration and spectral matrices by

$$X \approx C.S$$

where X is a 25 $\times$ 27 matrix, C a 25 $\times$ 10 matrix and S a 10 $\times$ 27 matrix in the example discussed here. In calibration it is assumed that a series of experiments are performed in which C is known (e.g. a set of mixtures of compounds with known concentrations are recorded spectroscopically). An estimate of S can then be obtained by

$$\hat{S} = (C'.C)^{-1}.C'.X$$

and then the concentrations can be predicted using

$$\hat{C} = X.\hat{S}'.(\hat{S}.\hat{S}')^{-1}$$

exactly as above. This can be extended to estimating the concentrations in any unknown spectrum by

$$\hat{c} = x.\hat{S}'.(\hat{S}.\hat{S}')^{-1} = x.B$$

Unless the number of experiments is exactly equal to the number of compounds, the prediction will not be completely model the data. This approach works because the matrices $(C'.C)$ and $(\hat{S}.\hat{S}')$ are square matrices whose dimensions equal the number of compounds in the mixture (10×10) and have inverses, provided that experiments have been suitably designed and the concentrations of the compounds are not correlated. The predicted concentrations, using this approach, are given in Table 5.8, together with the percentage root mean square prediction error; note that there are only 15 degrees of freedom $(=25$ experiments -10 compounds). Had the data been centred, the number of degrees of freedom would be reduced further. The predicted concentrations are reasonably good for most compounds apart from acenaphthylene.

Table 5.8 Estimated concentrations for the case study using uncentred MLR and all wavelengths.

Spectrum No.	PAH concentration (mg l^{-1})									
	Py	Ace	Anth	Acy	Chry	Benz	Fluora	Fluore	Nap	Phen
1	0.509	0.092	0.200	0.151	0.369	1.731	0.121	0.654	0.090	0.433
2	0.438	0.100	0.297	0.095	0.488	2.688	0.148	0.276	0.151	0.744
3	0.177	0.150	0.303	0.217	0.540	1.667	0.068	0.896	0.174	0.128
4	0.685	0.177	0.234	0.150	0.369	1.099	0.128	0.691	0.026	0.728
5	0.836	0.137	0.304	0.155	0.224	2.146	0.159	0.272	0.194	0.453
6	0.593	0.232	0.154	0.042	0.435	2.185	0.071	0.883	0.146	1.030
7	0.777	0.164	0.107	0.129	0.497	0.439	0.189	0.390	0.158	0.206
8	0.419	0.040	0.198	0.284	0.044	2.251	0.143	1.280	0.088	0.299
9	0.323	0.141	0.247	0.037	0.462	1.621	0.196	0.101	−0.003	0.298
10	0.578	0.236	0.020	0.107	0.358	2.659	0.093	0.036	0.070	0.305
11	0.621	0.051	0.214	0.111	0.571	0.458	0.062	0.428	0.022	0.587
12	0.166	0.187	0.170	0.142	0.087	0.542	0.100	0.343	0.103	0.748
13	0.580	0.077	0.248	0.133	0.051	1.120	−0.042	0.689	0.176	0.447
14	0.468	0.248	0.057	−0.006	0.237	0.558	0.157	0.712	0.103	0.351
15	0.770	0.016	0.066	0.119	0.094	1.680	0.187	0.450	0.080	0.920
16	0.101	0.026	0.100	0.041	0.338	2.230	0.102	0.401	0.201	0.381
17	0.169	0.115	0.063	0.069	0.478	1.054	0.125	0.829	0.068	0.523
18	0.271	0.079	0.142	0.106	0.222	1.086	0.211	0.254	0.151	0.261
19	0.171	0.152	0.216	0.059	0.274	2.587	0.081	0.285	0.013	0.925
20	0.399	0.116	0.095	0.170	0.514	1.133	0.101	0.321	0.243	1.023
21	0.651	0.025	0.146	0.232	0.230	1.610	−0.013	0.940	0.184	0.616
22	0.295	0.135	0.256	0.052	0.349	0.502	0.237	0.970	0.161	1.037
23	0.296	0.214	0.116	0.069	0.144	2.589	0.202	0.785	0.162	0.588
24	0.774	0.085	0.187	−0.026	0.547	2.671	0.128	1.107	0.108	0.329
25	0.324	0.035	0.036	0.361	0.472	2.217	0.094	0.918	0.128	0.779
$E_\%$	9.79	44.87	15.58	69.43	13.67	4.71	40.82	31.38	29.22	16.26

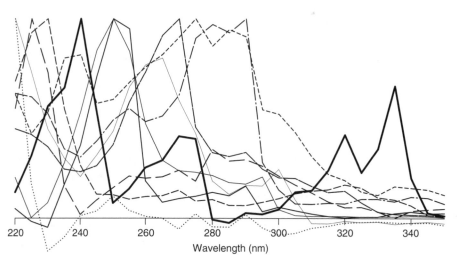

Figure 5.10
Normalised spectra of the 10 PAHs estimated by MLR, pyrene in bold

The predicted spectra are presented in Figure 5.10, and are not nearly as well predicted as the concentrations. In fact, it would be remarkable that for such a complex
mixture it is possible to reconstruct 10 spectra well, given that there is a great deal
of overlap. Pyrene, which is indicated in bold, exhibits most of the main peak maxima of the known pure data (compare with Figure 5.3). Often, other knowledge of the
system is required to produce better reconstructions of individual spectra. The reason
why concentration predictions appear to work significantly better than spectral reconstruction is that, for most compounds, there are characteristic regions of the spectrum
containing prominent features. These parts of the spectra for individual compounds
will be predicted well, and will disproportionately influence the effectiveness of the
method for determining concentrations. However, MLR as described in this section is
not an effective method for determining spectra in complex mixtures, and should be
employed primarily as a way of determining concentrations.

MLR predicts concentrations well in this case because all significant compounds
are included in the model, and so the data are almost completely modelled. If we
knew of only a few compounds, there would be much poorer predictions. Consider the
situation in which only pyrene, acenaphthene and anthracene are known. The C matrix
now has only three columns, and the predicted concentrations are given in Table 5.9.
The errors are, as expected, much larger than those in Table 5.8. The absorbances
of the remaining seven compounds are mixed up with those of the three modelled
components. This problem could be overcome if some characteristic wavelengths or
regions of the spectrum at which the selected compounds absorb most strongly are
identified, or if the experiments were designed so that there are correlations in the
data, or even by a number of methods for weighted regression, but the need to provide
information about all significant compounds is a major limitation of MLR.

The approach described above is a form of classical calibration, and it is also possible
to envisage an inverse calibration model since

$$\hat{C} = X.B$$

Table 5.9 Estimates for three PAHs using the full dataset and MLR but including only three compounds in the model.

Spectrum No.	PAH concentration (mg l^{-1})		
	Py	Ace	Anth
1	0.539	0.146	0.156
2	0.403	0.173	0.345
3	0.199	0.270	0.138
4	0.749	0.015	0.231
5	0.747	0.103	0.211
6	0.489	0.165	0.282
7	0.865	0.060	−0.004
8	0.459	0.259	0.080
9	0.362	0.121	0.211
10	0.512	0.351	−0.049
11	0.742	−0.082	0.230
12	0.209	0.023	0.218
13	0.441	0.006	0.202
14	0.419	0.095	0.051
15	0.822	0.010	0.192
16	0.040	0.255	0.151
17	0.259	0.162	0.122
18	0.323	0.117	0.104
19	0.122	0.179	0.346
20	0.502	0.085	0.219
21	0.639	0.109	0.130
22	0.375	−0.062	0.412
23	0.196	0.316	0.147
24	0.638	0.218	0.179
25	0.545	0.317	0.048
$E_\%$	22.04986	105.7827	52.40897

However, unlike in Section 2.2.2, there are now more wavelengths than samples or components in the mixture. The matrix B is given by

$$B = (X'.X)^{-1}.X'.C$$

as above. A problem with this approach is that the matrix $(X'X)$ is now a large matrix, with 27 rows and 27 columns, compared with the matrices used above which have 10 rows and 10 columns only. If there are only 10 components in a mixtures, in a noise free experiment, the matrix $X'X$ would only have 10 degrees of freedom and no inverse. In practice, a numerical inverse can be computed but it will be largely a function of noise, and often contain some very large (and meaningless) numbers, because many of the columns of the matrix will contain correlations, as the determinant of the matrix $X'.X$ will be very small. This use of the inverse is only practicable if

1. the number of experiments and wavelengths is at least equal to the number of components in the mixture, and
2. the number of experiments is at least equal to the number of wavelengths.

Condition 2 either requires a large number of extra experiments to be performed or a reduction to 25 wavelengths. There have been a number of algorithms developed for wavelength selection, so enabling inverse models to be produced, but there is no real advantage over classical least squares in these situations.

5.4 Principal Components Regression

MLR based methods have the disadvantage that all significant components must be known. PCA based methods do not require details about the spectra or concentrations of all the compounds in a mixture, although it is important to make a sensible estimate of how many significant components characterise a mixture, but not necessarily their characteristics.

Principal components are primarily abstract mathematical entities and further details are described in Chapter 4. In multivariate calibration the aim is to convert these to compound concentrations. PCR uses regression (sometimes also called transformation or rotation) to convert PC scores to concentrations. This process is often loosely called factor analysis, although terminology differs according to author and discipline. Note that although the chosen example in this chapter involves calibrating concentrations to spectral absorbances, it is equally possible, for example, to calibrate the property of a material to its structural features, or the activity of a drug to molecular parameters.

5.4.1 Regression

If c_n is a vector containing the known concentration of compound n in the spectra (25 in this instance), then the PC scores matrix, T, can be related as follows:

$$c_n \approx T.r_n$$

where r_n is a column vector whose length equals the number of PCs retained, sometimes called a rotation or transformation vector. Ideally, the length of r_n should be equal to the number of compounds in the mixture (= 10 in this case). However, noise, spectral similarities and correlations between concentrations often make it hard to provide an exact estimate of the number of significant components; this topic has been introduced in Section 4.3.3 of Chapter 4. We will assume, for the purpose of this section, that 10 PCs are employed in the model.

The scores of the first 10 PCs are presented in Table 5.10, using raw data.

The transformation vector can be obtained by using the pseudo-inverse of T:

$$r_n = (T'.T)^{-1}T'.c_n$$

Note that the matrix $(T'.T)$ is actually a diagonal matrix, whose elements consist of 10 eigenvalues of the PCs, and each element of r could be expressed as a summation:

$$r_{na} = \frac{\sum_{i=1}^{I} t_{ia}c_{in}}{\sum_{a=1}^{A} g_a}$$

Table 5.10 Scores of the first 10 PCs for the PAH case study.

2.757	0.008	0.038	0.008	0.026	0.016	0.012	−0.004	0.006	−0.006
3.652	−0.063	−0.238	−0.006	0.021	0.000	0.018	0.005	0.009	0.013
2.855	−0.022	0.113	0.049	−0.187	0.039	0.053	0.004	0.007	−0.003
2.666	0.267	0.040	−0.007	0.073	0.067	−0.002	−0.002	−0.013	−0.006
3.140	0.029	0.006	−0.153	0.111	−0.015	0.030	0.022	0.014	0.006
3.437	0.041	−0.090	0.034	−0.027	−0.018	−0.010	−0.014	−0.032	0.006
1.974	0.161	0.296	0.107	0.090	−0.010	0.003	0.037	0.004	0.008
2.966	−0.129	0.161	−0.147	−0.043	0.016	−0.006	−0.010	0.013	−0.035
2.545	−0.054	−0.143	0.080	0.074	0.073	0.013	0.008	0.025	−0.006
3.017	−0.425	0.159	0.002	0.096	0.049	−0.010	0.013	−0.018	0.022
2.005	0.371	0.003	0.120	0.093	0.032	0.015	−0.025	0.003	0.003
1.648	0.239	−0.020	−0.123	−0.090	0.051	−0.017	0.021	−0.009	0.007
1.884	0.215	0.020	−0.167	−0.024	−0.041	0.041	−0.007	−0.017	0.001
1.666	0.065	0.126	0.070	−0.007	−0.005	−0.016	0.036	−0.025	−0.009
2.572	0.085	−0.028	−0.095	0.184	−0.046	−0.045	−0.016	0.013	−0.006
2.532	−0.262	−0.126	0.047	−0.084	−0.076	0.004	0.005	0.017	0.010
2.171	0.014	0.028	0.166	−0.080	0.008	−0.018	−0.007	−0.003	−0.007
1.900	−0.020	0.027	−0.015	−0.006	−0.018	−0.005	0.030	0.029	−0.002
3.174	−0.114	−0.312	−0.059	−0.014	0.066	−0.009	−0.016	−0.011	0.007
2.610	0.204	0.037	0.036	−0.069	−0.041	−0.020	−0.005	0.012	0.033
2.567	0.119	0.155	−0.090	−0.017	−0.050	0.017	−0.023	−0.013	0.005
2.389	0.445	−0.190	0.045	−0.091	−0.026	−0.026	0.021	0.006	−0.015
3.201	−0.282	−0.043	−0.062	−0.066	−0.015	−0.026	0.032	−0.015	−0.009
3.537	−0.182	−0.071	0.166	0.094	−0.069	0.026	−0.013	−0.016	−0.021
3.343	−0.113	0.252	0.012	−0.086	0.019	−0.031	−0.048	0.018	0.004

Table 5.11 Vector r for pyrene.

0.166
0.470
0.624
−0.168
1.899
−1.307
1.121
0.964
−3.106
−0.020

or even as a product of vectors:

$$r_{na} = \frac{t'_a \cdot c_n}{\sum_{a=1}^{A} g_a}$$

We remind the reader of the main notation:

- n refers to compound number (e.g. pyrene = 1);
- a to PC number (e.g. 10 significant components, not necessarily equal to the number of compounds in a series of samples);
- i to sample number (=1–25 in this case).

This vector for pyrene using 10 PCs is presented in Table 5.11. If the concentrations of some or all the compounds are known, PCR can be extended simply by replacing the vector c_k with a matrix C, each column corresponding to a compound in the mixture, so that

$$C \approx T.R$$

and

$$R = (T'.T)^{-1}.T'.C$$

The number of PCs must be at least equal to the number of compounds of interest in the mixture. R has dimensions $A \times N$.

If the number of PCs and number of significant compounds are equal, so that, in this example, T and C are 25×10 matrices, then R is a square matrix of dimensions $N \times N$ and

$$\hat{X} = T.P = T.R.R^{-1}.P = \hat{C}.\hat{S}$$

Hence, by calculating $R^{-1}.P$, it is possible to determine the estimated spectra of each individual component without knowing this information in advance, and by calculating $T.R$ concentration estimates can be obtained. Table 5.12 provides the concentration estimates using PCR with 10 significant components. The percentage mean square

Table 5.12 Concentration estimates for the PAHs using PCR and 10 components (uncentred).

Spectrum No.	PAH concentration (mg l^{-1})									
	Py	Ace	Anth	Acy	Chry	Benz	Fluora	Fluore	Nap	Phen
1	0.505	0.113	0.198	0.131	0.375	1.716	0.128	0.618	0.094	0.445
2	0.467	0.120	0.286	0.113	0.455	2.686	0.137	0.381	0.168	0.782
3	0.161	0.178	0.296	0.174	0.558	1.647	0.094	0.836	0.162	0.161
4	0.682	0.177	0.231	0.165	0.354	1.119	0.123	0.720	0.049	0.740
5	0.810	0.128	0.297	0.156	0.221	2.154	0.159	0.316	0.189	0.482
6	0.575	0.170	0.159	0.107	0.428	2.240	0.072	0.942	0.146	1.000
7	0.782	0.152	0.104	0.152	0.470	0.454	0.162	0.477	0.169	0.220
8	0.401	0.111	0.192	0.170	0.097	2.153	0.182	1.014	0.062	0.322
9	0.284	0.084	0.237	0.106	0.429	1.668	0.166	0.241	0.022	0.331
10	0.578	0.197	0.023	0.157	0.321	2.700	0.077	0.194	0.090	0.300
11	0.609	0.075	0.194	0.103	0.550	0.460	0.080	0.472	0.038	0.656
12	0.185	0.172	0.183	0.147	0.083	0.558	0.086	0.381	0.101	0.701
13	0.555	0.103	0.241	0.092	0.084	1.104	0.007	0.576	0.156	0.475
14	0.461	0.167	0.067	0.089	0.212	0.624	0.111	0.812	0.114	0.304
15	0.770	0.019	0.076	0.089	0.115	1.669	0.178	0.393	0.068	0.884
16	0.109	0.033	0.101	0.040	0.349	2.189	0.108	0.352	0.190	0.376
17	0.178	0.102	0.073	0.086	0.481	1.057	0.112	0.805	0.073	0.486
18	0.271	0.067	0.145	0.104	0.221	1.077	0.183	0.273	0.142	0.250
19	0.186	0.135	0.217	0.101	0.253	2.618	0.071	0.369	0.036	0.919
20	0.406	0.109	0.111	0.145	0.534	1.126	0.111	0.306	0.220	0.973
21	0.665	0.110	0.152	0.130	0.284	1.541	0.044	0.720	0.165	0.614
22	0.315	0.112	0.258	0.092	0.336	0.501	0.205	0.981	0.162	1.009
23	0.327	0.179	0.126	0.115	0.126	2.610	0.160	0.847	0.161	0.537
24	0.766	0.075	0.168	0.029	0.525	2.692	0.121	1.139	0.124	0.383
25	0.333	0.110	0.053	0.210	0.539	2.151	0.135	0.709	0.086	0.738
$E_\%$	10.27	36.24	15.76	42.06	9.05	4.24	31.99	24.77	21.11	16.19

error of prediction (equalling the square root sum of squares of the errors of prediction divided by 15 minus the number of degrees of freedom which equals $25 - 10$ to account for the number of components in the model, and not by 25) for all 10 compounds is also presented. In most cases it is slightly better than using MLR; there are certainly fewer very large errors. However, the major advantage is that the prediction using PCR is the same if only one or all 10 compounds are included in the model. In this it differs radically from MLR; the estimates in Table 5.9 are much worse than those in Table 5.8, for example. The first main task when using PCR is to determine how many significant components are necessary to model the data.

5.4.2 Quality of Prediction

A key issue in calibration is to determine how well the data have been modelled. We have used only one indicator above, but it is important to appreciate that there are many other potential statistics.

5.4.2.1 Modelling the c Block

Most look at how well the concentration is predicted, or the c (or according to some authors y) block of data.

The simplest method is to determine the sum of square of residuals between the true and predicted concentrations:

$$S_c = \sum_{i=1}^{I} (c_{in} - \hat{c}_{in})^2$$

where

$$\hat{c}_{in} = \sum_{a=1}^{A} t_{ia} r_{an}$$

for compound n using a principal components. The larger this error, the worse is the prediction, so the error decreases as more components are calculated.

Often the error is reported as a root mean square error:

$$E = \sqrt{\frac{\sum_{i=1}^{I} (c_{in} - \hat{c}_{in})^2}{I - a}}$$

If the data are centred, a further degree of freedom is lost, so the sum of square residuals is divided by $I - a - 1$.

This error can also be presented as a percentage error:

$$E_{\%} = 100E/\bar{c}_n$$

where $\bar{c}_n$ is the mean concentration in the original units. Sometimes the percentage of the standard deviation is calculated instead, but in this text we will compute errors as a percentage of the mean unless specifically stated otherwise.

5.4.2.2   Modelling the x Block

It is also possible to report errors in terms of quality of modelling of spectra (or chromatograms), often called the x block error.

The quality of modelling of the spectra using PCA (the x variance) can likewise be calculated as follows:

$$S_x = \sum_{i=1}^{I} \sum_{j=1}^{J} (x_{ij} - \hat{x}_{ij})^2$$

where

$$\hat{x}_{ij} = \sum_{a=1}^{A} t_{ia} p_{aj}$$

However, this error also can be expressed in terms of eigenvalues or scores, so that

$$S_x = \sum_{i=1}^{I} \sum_{j=1}^{J} x_{ij}^2 - \sum_{a=1}^{A} g_a = \sum_{i=1}^{I} \sum_{j=1}^{J} x_{ij}^2 - \sum_{a=1}^{A} \sum_{i=1}^{I} t_{ia}^2$$

for A principal components. These can be converted to root mean square errors as above:

$$E = \sqrt{S_x / I.J}$$

Note that many people divide by $I.J$ ($= 25 \times 27 = 675$ in our case) rather than the more strictly correct $I.J - a$ (adjusting for degrees of freedom), because $I.J$ is very large relative to a, and we will adopt this convention.

The percentage root mean square error may be defined by (for uncentred data)

$$E_\% = 100E/\bar{x}$$

Note that if x is centred, the divisor is often given by

$$\sqrt{\sum_{i=1}^{I} \sum_{j=1}^{J} \frac{(x_{ij} - \bar{x}_j)^2}{I.J}}$$

where $\bar{x}_j$ is the average of all the measurements for the samples for variable j. Obviously there are several other ways of defining this error: if you try to follow a paper or a package, read very carefully the documents provided by the authors, and if there is no documentation, do not trust the answers.

Note that the x error depends only on the number of PCs, no matter how many compounds are being modelled, but the error in concentration estimates depends also on the specific compound, there being a different percentage error for each compound in the mixture. For 0 PCs, the estimates of the PCs and concentrations is simply 0 or, if mean-centred, the mean. The graphs of root mean square errors for both the concentration estimates of pyrene and spectra as increasing numbers of PCs are calculated are given in Figure 5.11, using a logarithmic scale for the error. Although the x error graph declines steeply, which might falsely suggest that only a small number of PCs

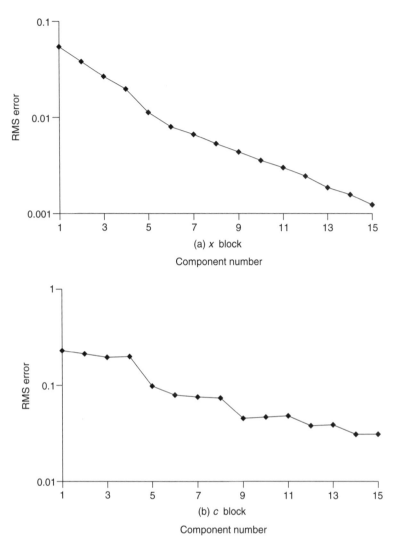

Figure 5.11
Root mean square errors of estimation of pyrene using uncentred PCR

are required for the model, the c error graph exhibits a much gentler decline. Sometimes these graphs are presented either as percentage variance remaining (or explained by each PC) or eigenvalues.

5.5 Partial Least Squares

PLS is often presented as the major regression technique for multivariate data. In fact its use is not always justified by the data, and the originators of the method were well aware of this, but, that being said, in some applications PLS has been spectacularly successful. In some areas such as QSAR, or even biometrics and psychometrics,

PLS is an invaluable tool, because the underlying factors have little or no physical meaning so a linearly additive model in which each underlying factor can be interpreted chemically is not anticipated. In spectroscopy of chromatography we usually expect linear additivity, and this is especially important for chemical instrumental data, and under such circumstances simpler methods such as MLR are often useful provided that there is a fairly full knowledge of the system. However, PLS is always an important tool when there is partial knowledge of the data, a well known example being the measurement of protein in wheat by NIR spectroscopy. A model can be obtained from a series of wheat samples, and PLS will use typical features in this dataset to establish a relationship to the known amount of protein. PLS models can be very robust provided that future samples contain similar features to the original data, but the predictions are essentially statistical. Another example is the determination of vitamin C in orange juices using spectroscopy: a very reliable PLS model could be obtained using orange juices from a particular region of Spain, but what if some Brazilian orange juice is included? There is no guarantee that the model will perform well on the new data, as there may be different spectral features, so it is always important to be aware of the limitations of the method, particularly to remember that the use of PLS cannot compensate for poorly designed experiments or inadequate experimental data.

An important feature of PLS is that it takes into account errors in both the concentration estimates and the spectra. A method such as PCR assumes that the concentration estimates are error free. Much traditional statistics rests on this assumption, that all errors are of the variables (spectra). If in medicine it is decided to determine the concentration of a compound in the urine of patients as a function of age, it is assumed that age can be estimated exactly, the statistical variation being in the concentration of a compound and the nature of the urine sample. Yet in chemistry there are often significant errors in sample preparation, for example accuracy of weighings and dilutions, and so the independent variable in itself also contains errors. Classical and inverse calibration force the user to choose which variable contains the error, whereas PLS assumes that it is equally distributed in both the x and c blocks.

5.5.1 PLS1

The most widespread approach is often called PLS1. Although there are several algorithms, the main ones due to Wold and Martens, the overall principles are fairly straightforward. Instead of modelling exclusively the x variables, two sets of models are obtained as follows:

$$X = T.P + E$$

$$c = T.q + f$$

where q has analogies to a loadings vector, although is not normalised. These matrices are represented in Figure 5.12. The product of T and P approximates to the spectral data and the product of T and q to the true concentrations; the common link is T. An important feature of PLS is that it is possible to obtain a scores matrix that is common to both the concentrations (c) and measurements (x). Note that T and P for PLS are different to T and P obtained in PCA, and unique sets of scores and loadings are obtained for each compound in the dataset. Hence if there are 10 compounds

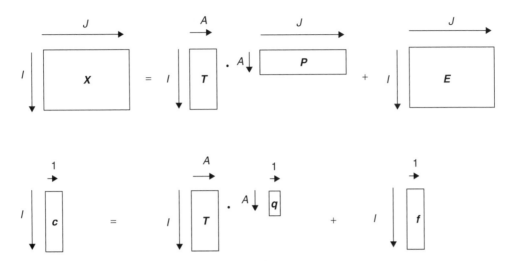

Figure 5.12
Principles of PLS1

of interest, there will be 10 sets of T, P and q. In this way PLS differs from PCR in which there is only one set of T and P, the PCA step taking no account of the c block. It is important to recognise that there are several algorithms for PLS available in the literature, and although the predictions of c are the same in each case, the scores and loadings are not. In this book and the associated Excel software, we use the algorithm of Appendix A.2.2. Although the scores are orthogonal (as in PCA), the loadings are not (which is an important difference to PCA), and, furthermore, the loadings are not normalised, so the sum of squares of each p vector does not equal one. If you are using a commercial software package, it is important to be check exactly what constraints and assumptions the authors make about the scores and loadings in PLS.

Additionally, the analogy to g_a or the eigenvalue of a PC involves multiplying the sum of squares of both t_a and p_a together, so we define the magnitude of a PLS component as

$$g_a = \left(\sum_{i=1}^{I} t_{ia}^2 \right) \left(\sum_{j=1}^{J} p_{aj}^2 \right)$$

This will have the property that the sum of values of g_a for all nonzero components add up to the sum of squares of the original (preprocessed) data. Note that in contrast to PCA, the size of successive values of g_a does not necessarily decrease as each component is calculated. This is because PLS does not only model the x data, and is a compromise between x and c block regression.

There are a number of alternative ways of presenting the PLS regression equations in the literature, all mathematically equivalent. In the models above, we obtain three arrays T, P and q. Some authors calculate a normalised vector, w, proportional to q, and the second equation becomes

$$c = T.B.w + f$$

where B is a diagonal matrix. Analogously, it is also possible to the define the PCA decomposition of x as a product of three arrays (the SVD method which is used in Matlab is a common alternative to NIPALS), but the models used in this chapter have the simplicity of using a single scores matrix for both blocks of data, and modelling each dataset using two matrices.

For a dataset consisting of 25 spectra observed at 27 wavelengths, for which eight PLS components are calculated, there will be

- a T matrix of dimensions 25×8;
- a P matrix of dimensions 8×27;
- an E matrix of dimensions 25×27;
- a q vector of dimensions 8×1;
- an f vector of dimensions 25×1.

Note that there will be 10 separate sets of these arrays, in the case discussed in this chapter, one for each compound in the mixture, and that the T matrix will be compound dependent, which differs from PCR.

Each successive PLS component approximates both the concentration and spectral data better. For each PLS component, there will be a

- spectral scores vector t;
- spectral loadings vector p;
- concentration loadings scalar q.

In most implementations of PLS it is conventional to centre both the x and c data, by subtracting the mean of each column, before analysis. In fact, there is no general scientific need to do this. Many spectroscopists and chromatographers perform PCA uncentred, but many early applications of PLS (e.g. outside chemistry) were of such a nature that centring the data was appropriate. Many of the historical developments in PLS as used for multivariate calibration in chemistry relate to applications in NIR spectroscopy, where there are specific spectroscopic problems, such as due to baselines, which, in turn would favour centring. However, as generally applied in many branches of chemistry, uncentred PLS is perfectly acceptable. Below, though, we use the most widespread implementation (involving centring) for the sake of compatibility with the most common computational implementations of the method.

For a given compound, the remaining percentage error in the x matrix for a PLS components can be expressed in a variety of ways as discussed in Section 5.4.2.2. Note that there are slight differences according to authors that take into account the number of degrees of freedom left in the model. The predicted measurements simply involve calculating $\hat{X} = T.P$ and adding on the column means where appropriate, and error indicators in the x block that can defined similarly to those used in PCR, see Section 5.4.2.2. The only difference is that each compound generates a separate scores matrix, unlike PCR where there is a single scores matrix for all compounds in the mixture and so there will be a different behaviour in the x block residuals according to compound.

The concentration of compound n is predicted by

$$\hat{c}_{in} = \sum_{a=1}^{A} t_{ian} q_{an} + \bar{c}_n$$

Table 5.13 Calculation of concentration estimates for pyrene using two PLS components.

Component 1 $q = 0.222$		Component 2 $q = 0.779$		Estimated concentration	
Scores	$t_{i1}q$	Scores	$t_{i2}q$	$t_{i1}q + t_{i2}q$	$t_{i1}q + t_{i2}q + \bar{c}$
0.088	0.020	0.052	0.050	0.058	0.514
0.532	0.118	−0.139	−0.133	0.014	0.470
0.041	0.009	−0.169	−0.162	−0.117	0.339
0.143	0.032	0.334	0.319	0.281	0.737
0.391	0.087	0.226	0.216	0.255	0.711
0.457	0.102	−0.002	−0.002	0.100	0.556
−0.232	−0.052	0.388	0.371	0.238	0.694
0.191	0.042	−0.008	−0.007	0.037	0.493
−0.117	−0.026	−0.148	−0.142	−0.137	0.319
0.189	0.042	−0.136	−0.130	−0.059	0.397
−0.250	−0.055	0.333	0.319	0.193	0.649
−0.621	−0.138	−0.046	−0.044	−0.173	0.283
−0.412	−0.092	0.105	0.101	−0.013	0.443
−0.575	−0.128	0.004	0.004	−0.125	0.331
0.076	0.017	0.264	0.253	0.214	0.670
−0.264	−0.059	−0.485	−0.464	−0.420	0.036
−0.358	−0.080	−0.173	−0.165	−0.209	0.247
−0.485	−0.108	−0.117	−0.112	−0.195	0.261
0.162	0.036	−0.356	−0.340	−0.229	0.227
0.008	0.002	0.105	0.100	0.080	0.536
0.038	0.008	0.209	0.200	0.164	0.620
−0.148	−0.033	0.080	0.076	0.026	0.482
0.197	0.044	−0.329	−0.315	−0.201	0.255
0.518	0.115	−0.041	−0.039	0.085	0.541
0.432	0.096	0.050	0.048	0.133	0.589

or, in matrix terms

$$c_n - \bar{c}_n = T_n . q_n$$

where $\bar{c}_n$ is a vector of the average concentration. Hence the scores of each PLS component are proportional to the contribution of the component to the concentration estimate. The method of the concentration estimation for two PLS components for pyrene is presented in Table 5.13.

The mean square error in the concentration estimate is defined just as in PCR. It is also possible to define this error in various different ways using t and q. In the case of the c block estimates, it is usual to divide the sum of squares by $I - A - 1$. These error terms have been discussed in greater detail in Section 5.4.2. The x block is usually mean centred and so to obtain a percentage error most people divide by the standard deviation, whereas for the c block the estimates are generally expressed in the original concentration units, so we will retain the convention of dividing by the mean concentration unless there is a specific reason for another approach. As in all areas of chemometrics, each group and software developer has their own favourite way of calculating parameters, so it is essential never to accept output from a package blindly.

The calculation of x block error is presented for the case of pyrene. Table 5.14 gives the magnitudes of the first 15 PLS components, defined as the product of the sum of squares for t and p of each component. The total sum of squares of the

Table 5.14 Magnitudes of first 15 PLS1 components (centred data) for pyrene.

Component	Magnitude	Component	Magnitude
1	7.944	9	0.004
2	1.178	10	0.007
3	0.484	11	0.001
4	0.405	12	0.002
5	0.048	13	0.002
6	0.158	14	0.003
7	0.066	15	0.001
8	0.01		

mean centred spectra is 10.313, hence the first two components account for $100 \times (7.944 + 1.178)/10.313 = 88.4\%$ of the overall variance, so the root mean square error after two PLS components have been calculated is $\sqrt{1.191/(27 \times 25)} = 0.042$ (since 1.191 is the residual error) or, expressed as a percentage of the mean centred data, $E_\% = 0.042/\sqrt{10.313/(27 \times 25)} = 40.0\%$. This could be expressed as a percentage of the mean of the raw data $= 0.042/0.430 = 9.76\%$. The latter appears much lower and is a consequence of the fact that the mean of the data is considerably higher than the standard deviation of the mean centred data. It is probably best simply to determine the percentage residual sum of square error ($= 100 - 88.4 = 11.6\%$) as more components are computed, but it is important to be aware that there are several approaches for the determination of errors.

The error in concentration predictions for pyrene using two PLS components can be computed from Table 5.13:

- the sum of squares of the errors is 0.385;
- dividing this by 22 and taking the square root leads to a root mean square error of 0.128 mg l^{-1};
- the average concentration of pyrene is 0.456 mg l^{-1};
- hence the percentage root mean square error (compared with the raw data) is 28.25 %.

Relative to the standard deviation of the centred data it is even higher. Hence the 'x' and 'c' blocks are modelled in different ways and it is important to recognise that the percentage error of prediction in concentration may diverge considerably from the percentage error of prediction of the spectra. It is sometimes possible to reconstruct spectral blocks fairly well but still not predict concentrations very effectively. It is best practice to look at errors in both blocks simultaneously to gain an understanding of the quality of predictions.

The root mean square errors for modelling both blocks of data as successive numbers of PLS components are calculated for pyrene are illustrated in Figure 5.13, and those for acenaphthene in Figure 5.14. Several observations can be made. First, the shape of the graph of residuals for the two blocks is often very different, see especially acenaphthene. Second, the graph of c residuals tends to change much more dramatically than that for x residuals, according to compound, as might be expected. Third, tests for numbers of significant PLS components might give different answers according to which block is used for the test.

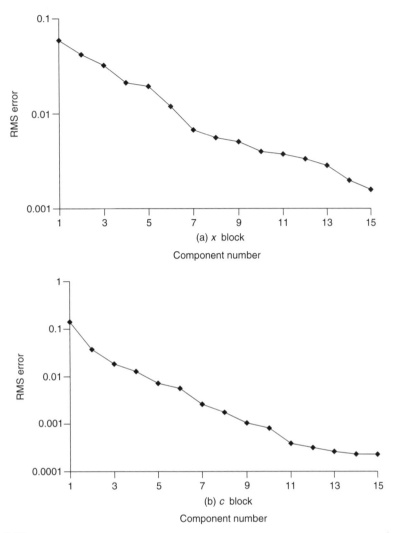

Figure 5.13
Root mean square errors in x and c blocks, PLS1 centred and pyrene

The errors using 10 PLS components are summarised in Table 5.15, and are better than PCR in this case. It is important, however, not to get too excited about the improved quality of predictions. The c or concentration variables may in themselves contain errors, and what has been shown is that PLS forces the solution to model the apparent c block better, but it does not necessarily imply that the other methods are worse at discovering the truth. If, however, we have a lot of confidence in the experimental procedure for determining c (e.g. weighing, dilution, etc.), PLS will result in a more faithful reconstruction.

5.5.2 PLS2

An extension to PLS1 was suggested some 15 years ago, often called PLS2. In fact there is little conceptual difference, except that the latter allows the use of

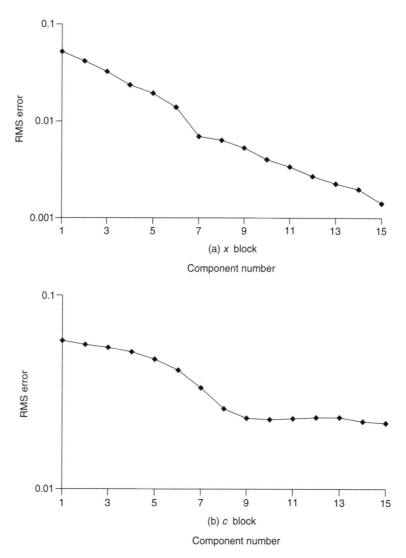

Figure 5.14
Root mean square errors in x and c blocks, PLS1 centred and acenaphthene

a concentration matrix, C, rather than concentration vectors for each individual compound in a mixture, and the algorithm is iterative. The equations above alter slightly in that Q is now a matrix not a vector, so that

$$X = T.P + E$$

$$C = T.Q + F$$

The number of columns in C and Q are equal to the number of compounds of interest. In PLS1 one compound is modelled at a time, whereas in PLS2 all known compounds can be included in the model simultaneously. This is illustrated in Figure 5.15.

Table 5.15 Concentration estimates of the PAHs using PLS1 and 10 components (centred).

Spectrum No.	PAH concentration (mg l^{-1})									
	Py	Ace	Anth	Acy	Chry	Benz	Fluora	Fluore	Nap	Phen
1	0.462	0.112	0.170	0.147	0.341	1.697	0.130	0.718	0.110	0.553
2	0.445	0.065	0.280	0.175	0.440	2.758	0.138	0.408	0.177	0.772
3	0.147	0.199	0.285	0.162	0.562	1.635	0.111	0.784	0.159	0.188
4	0.700	0.174	0.212	0.199	0.333	1.097	0.132	0.812	0.054	0.785
5	0.791	0.167	0.285	0.111	0.223	2.118	0.171	0.211	0.176	0.519
6	0.616	0.226	0.176	0.040	0.467	2.172	0.068	0.752	0.116	0.928
7	0.767	0.119	0.108	0.180	0.452	0.522	0.153	0.577	0.177	0.202
8	0.476	0.085	0.228	0.157	0.109	2.155	0.129	0.967	0.046	0.184
9	0.317	0.145	0.232	0.042	0.440	1.576	0.171	0.187	0.009	0.367
10	0.614	0.178	0.046	0.154	0.334	2.702	0.039	0.174	0.084	0.219
11	0.625	0.029	0.237	0.121	0.574	0.543	0.042	0.423	0.039	0.516
12	0.179	0.161	0.185	0.175	0.091	0.560	0.098	0.363	0.110	0.709
13	0.579	0.119	0.262	0.061	0.118	1.074	0.012	0.522	0.149	0.428
14	0.463	0.198	0.067	0.054	0.226	0.561	0.134	0.788	0.110	0.330
15	0.752	0.041	0.062	0.075	0.113	1.646	0.193	0.401	0.072	0.943
16	0.149	0.017	0.115	0.037	0.338	2.186	0.062	0.474	0.196	0.349
17	0.148	0.106	0.050	0.096	0.453	1.044	0.112	0.974	0.092	0.585
18	0.274	0.075	0.149	0.119	0.223	1.098	0.199	0.280	0.147	0.256
19	0.151	0.119	0.213	0.109	0.236	2.664	0.075	0.536	0.050	0.953
20	0.458	0.140	0.114	0.095	0.555	1.067	0.100	0.220	0.198	0.944
21	0.615	0.080	0.120	0.189	0.226	1.581	0.040	1.024	0.198	0.738
22	0.318	0.091	0.267	0.097	0.329	0.523	0.187	0.942	0.157	0.967
23	0.295	0.160	0.124	0.160	0.122	2.669	0.182	0.826	0.171	0.531
24	0.761	0.072	0.167	0.047	0.541	2.687	0.153	1.049	0.122	0.378
25	0.296	0.120	0.047	0.197	0.555	2.166	0.170	0.590	0.082	0.758
$E_\%$	5.47	19.06	7.85	22.48	3.55	2.46	21.96	12.96	16.48	7.02

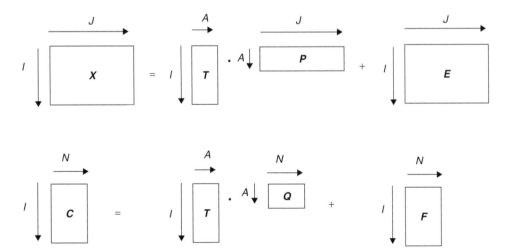

Figure 5.15
Principles of PLS2

Table 5.16 Concentration estimates of the PAHs using PLS2 and 10 components (centred).

Spectrum No.	PAH concentration mg l^{-1}									
	Py	Ace	Anth	Acy	Chry	Benz	Fluora	Fluore	Nap	Phen
1	0.505	0.110	0.193	0.132	0.365	1.725	0.125	0.665	0.089	0.459
2	0.460	0.116	0.285	0.105	0.453	2.693	0.144	0.363	0.150	0.760
3	0.162	0.180	0.294	0.173	0.563	1.647	0.094	0.787	0.161	0.157
4	0.679	0.173	0.224	0.164	0.343	1.134	0.123	0.752	0.038	0.748
5	0.811	0.135	0.294	0.149	0.230	2.152	0.162	0.221	0.183	0.475
6	0.575	0.182	0.156	0.108	0.442	2.228	0.077	0.827	0.153	1.002
7	0.779	0.151	0.107	0.143	0.469	0.453	0.167	0.484	0.156	0.199
8	0.397	0.100	0.198	0.183	0.093	2.165	0.181	1.035	0.070	0.306
9	0.295	0.089	0.238	0.108	0.433	1.665	0.158	0.238	0.032	0.341
10	0.581	0.203	0.029	0.148	0.327	2.690	0.079	0.191	0.088	0.287
11	0.609	0.070	0.207	0.108	0.559	0.453	0.079	0.484	0.049	0.636
12	0.190	0.176	0.186	0.144	0.086	0.549	0.083	0.411	0.105	0.709
13	0.565	0.107	0.249	0.095	0.092	1.088	0.000	0.595	0.173	0.478
14	0.468	0.173	0.067	0.089	0.214	0.610	0.108	0.830	0.124	0.322
15	0.771	0.018	0.073	0.096	0.112	1.668	0.175	0.415	0.077	0.906
16	0.119	0.030	0.110	0.037	0.345	2.189	0.101	0.442	0.192	0.369
17	0.181	0.098	0.070	0.090	0.468	1.061	0.106	0.903	0.076	0.510
18	0.278	0.067	0.151	0.102	0.226	1.073	0.178	0.292	0.147	0.249
19	0.184	0.131	0.218	0.102	0.245	2.617	0.071	0.434	0.034	0.925
20	0.410	0.120	0.111	0.134	0.543	1.117	0.115	0.243	0.215	0.963
21	0.663	0.100	0.147	0.129	0.262	1.558	0.040	0.845	0.152	0.630
22	0.308	0.108	0.257	0.093	0.335	0.509	0.209	0.954	0.156	0.998
23	0.320	0.179	0.123	0.114	0.129	2.610	0.164	0.817	0.157	0.537
24	0.763	0.072	0.165	0.038	0.524	2.696	0.120	1.123	0.130	0.390
25	0.327	0.110	0.049	0.216	0.544	2.150	0.139	0.650	0.091	0.746
$E_\%$	10.25	34.11	13.66	44.56	6.99	4.26	33.41	18.62	25.83	14.77

It is a simple extension to predict all the concentrations simultaneously, the PLS2 predictions, together with root mean square errors being given in Table 5.16. Note that there is now only one set of scores and loadings for the x (spectroscopic) dataset, and one set of g_a common to all 10 compounds. However, the concentration estimates are different when using PLS2 compared with PLS1. In this way PLS differs from PCR where it does not matter if each variable is modelled separately or all together. The reasons are rather complex but relate to the fact that for PCR the principal components are calculated independently of the c variables, whereas the PLS components are also influenced by both blocks of variables.

In some cases PLS2 is helpful, especially since it is easier to perform computationally if there are several c variables compared with PLS1. Instead of obtaining 10 independent models, one for each PAH, in this example, we can analyse all the data in one go. However, in many situations PLS2 concentration estimates are, in fact, worse than PLS1 estimates, so a good strategy might be to perform PLS2 as a first step, which could provide further information such as which wavelengths are significant and which concentrations can be determined with a high degree of confidence, and then perform PLS1 individually for the most appropriate compounds.

5.5.3 Multiway PLS

Two-way data such as HPLC–DAD, LC–MS and LC–NMR are increasingly common in chemistry, especially with the growth of coupled chromatography. Conventionally either a univariate parameter (e.g. a peak area at a given wavelength) (methods in Section 5.2) or a chromatographic elution profile at a single wavelength (methods in Sections 5.3 to 5.5.2) is used for calibration, allowing the use of normal regression techniques described above. However, additional information has been recorded for each sample, often involving both an elution profile and a spectrum. A series of two-way chromatograms are available, and can be organised into a three-way array, often visualised as a box, sometimes denoted by $\underline{X}$ where the line underneath the array name indicates a third dimension. Each level of the box consists of a single chromatogram. Sometimes these three-way arrays are called 'tensors', but tensors often have special properties in physics which are unnecessarily complex and confusing to the chemometrician. We will use the notation of tensors only where it helps in understanding the existing methods.

Enhancements of the standard methods for multivariate calibration are required. Although it is possible to use methods such as three-way MLR, most chemometricians have concentrated on developing approaches based on PLS, to which we will be restricted below. Theoreticians have extended these methods to cases where there are several dimensions in both the 'x' and 'c' blocks, but the most complex practical case is where there are three dimensions in the 'x' block, as happens for a series of coupled chromatograms or in fluorescence excitation–emission spectroscopy, for example. A simple simulated numerical example is presented in Table 5.17, in which the x block consists of four two-way chromatograms, each of dimensions 5×6. There are three components in the mixture, the c block consisting of a 4×3 matrix. We will restrict the discussion for the case where each column of c is to be estimated independently (analogous to PLS1) rather than all in one go. Note that although PLS is by far the most popular approach for multiway calibration, it is possible to envisage methods analogous to MLR or PCR, but they are rarely used.

5.5.3.1 Unfolding

One of the simplest methods is to create a single, long, data matrix from the original three-way tensor. In the case of Table 5.17, we have four samples, which could be arranged as a $4 \times 5 \times 6$ tensor (or 'box'). The three dimensions will be denoted I, J and K. It is possible to change the shape so that any binary combination of variables is converted to a new variable, for example, the intensity of the variable at $J = 2$ and $K = 3$, and the data can now be represented by $5 \times 6 = 30$ variables and is the unfolded form of the original data matrix. This operation is illustrated in Figure 5.16.

It is now a simple task to perform PLS (or indeed any other multivariate approach), as discussed above. The 30 variables are centred and the predictions of the concentrations performed when increasing number of components are used (note that three is the maximum permitted for column centred data in this case, so this example is somewhat simple). All the methods described above can be applied.

An important aspect of three-way calibration involves scaling, which can be rather complex. The are four fundamental ways in which the data can be treated:

1. no centring;
2. centre the columns in each $J \times K$ plane and then unfold with no further centring, so, for example, $x_{1,1,1}$ becomes $390 - (390 + 635 + 300 + 65 + 835)/5$;

Table 5.17   Three-way calibration dataset.

(a) X block, each of the 4 (=I) samples gives a two-way
$5 \times 6 \ (=J \times K)$ *matrix*

390	421	871	940	610	525
635	357	952	710	910	380
300	334	694	700	460	390
65	125	234	238	102	134
835	308	1003	630	1180	325
488	433	971	870	722	479
1015	633	1682	928	1382	484
564	538	1234	804	772	434
269	317	708	364	342	194
1041	380	1253	734	1460	375
186	276	540	546	288	306
420	396	930	498	552	264
328	396	860	552	440	300
228	264	594	294	288	156
222	120	330	216	312	114
205	231	479	481	314	268
400	282	713	427	548	226
240	264	576	424	336	232
120	150	327	189	156	102
385	153	482	298	542	154

*(b) C block, concentrations of three compounds in each
of the four samples*

1	9	10
7	11	8
6	2	6
3	4	5

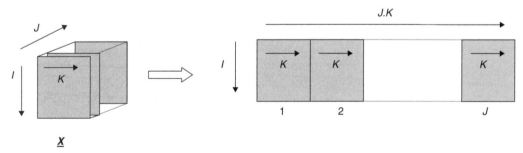

Figure 5.16
Unfolding a data matrix

3. unfold the raw data and centre afterwards, so, for example, $x_{1,1,1}$ becomes $390-(390 + 488 + 186 + 205)/4 = 72.75$;

4. combine methods 2 and 3, start with centring as in step 2, then unfold and recentre a second time.

These four methods are illustrated in Table 5.18 for the case of the $x_{i,1,1}$, the variables in the top left-hand corner of each of the four two-way datasets. Note that methods 3

Table 5.18 Four methods of mean centring the data in Table 5.17, illustrated by the variable $x_{i,1,1}$ as discussed in Section 5.5.3.1.

Sample	Method 1	Method 2	Method 3	Method 4
1	390	−55	72.75	44.55
2	488	−187.4	170.75	−87.85
3	186	−90.8	−131.25	8.75
4	205	−65	−112.25	34.55

and 4 provide radically different answers; for example, sample 2 has the highest value (=170.75) using method 3, but the lowest using method 4 (= −87.85).

Standardisation is also sometimes employed, but must be done before unfolding for meaningful results; an example might be in the GC–MS of a series of samples, each mass being of different absolute intensity. A sensible strategy might be as follows:

1. standardise each mass in each individual chromatogram, to provide I standardised matrices of dimensions $J \times K$;
2. unfold;
3. centre each of the variables.

Standardising at the wrong stage of the analysis can result in meaningless data so it is always essential to think carefully of the physical (and numerical) consequences of any preprocessing which is far more complex and has far more options than for simple two-way data.

After this preprocessing, all the normal multivariate calibration methods can be employed.

5.5.3.2 Trilinear PLS1

Some of the most interesting theoretical developments in chemometrics over the past few years have been in so-called 'multiway' or 'multimode' data analysis. Many such methods have been available for some years, especially in the area of psychometrics, and a few do have relevance to chemistry. It is important, though, not to get too carried away with the excitement of these novel theoretical approaches. We will restrict the discussion here to trilinear PLS1, involving a three-way x block and a single c variable. If there are several known calibrants, the simplest approach is to perform trilinear PLS1 individually on each variable.

Since centring can be fairly complex for three-way data, and there is no inherent reason to do this, for simplicity it is assumed that data are not centred, so raw concentrations and chromatographic/spectroscopic measurements are employed. The data in Table 5.17 can be considered to be arranged in the form of a cube, with three dimensions, I for the number of samples and J and K for the measurements.

Trilinear PLS1 attempts to model both the 'x' and 'c' blocks simultaneously. Here we will illustrate the use with the algorithm of Appendix A.2.4, based on methods proposed by de Jong and Bro.

Superficially, trilinear PLS1 has many of the same objectives as normal PLS1, and the method as applied to the x block is often represented diagrammatically as in Figure 5.17, replacing 'squares' or matrices by 'boxes' or tensors, and replacing, where necessary, the dot product ('.') by something called a tensor product ('⊗'). The 'c'

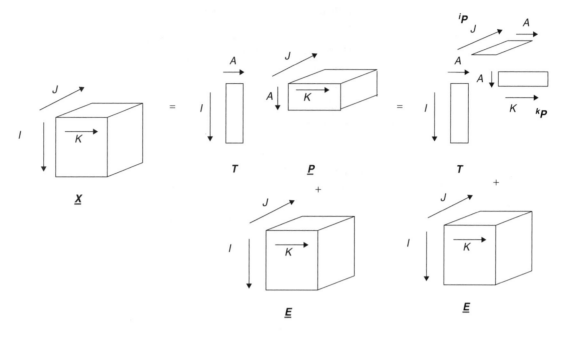

Figure 5.17
Representation of trilinear PLS1

block decomposition can be represented as per PLS1 and is omitted from the diagram for brevity. In fact, as we shall see, this is an oversimplification, and is not an entirely accurate description of the method.

In trilinear PLS1, for each component it is possible to determine

- a scores vector (t), of length I or 4 in this example;
- a weight vector, which has analogy to a loadings vector ($^j p$) of length J or 5 in this example, referring to one of the dimensions (e.g. time), whose sum of squares equals 1;
- another weight vector, which has analogy to a loadings vector ($^k p$) of length K or 6 in this example, referring to the other one of the dimensions (e.g. wavelength) whose sum of squares also equals 1.

Superficially these vectors are related to scores and loadings in normal PLS, but in practice they are completely different, a key reason being that these vectors are not orthogonal in trilinear PLS1 influencing the additivity of successive components. Here, we keep the notation scores and loadings, simply for the purpose of retaining familiarity with terminology usually used in two-way data analysis.

In addition, a vector q is determined after each new component, by

$$q = (T'.T)^{-1}.T'.c$$

so that

$$\hat{c} = T.q$$

or

$$c = T.q + f$$

where T is the scores matrix, the columns of which consist of the individual scores vectors for each component and has dimensions $I \times A$ or 4×3 in this example if three PLS components are computed, and q is a column vector of dimensions $A \times 1$ or 3×1 in our example.

A key difference from bilinear PLS1 as described in Section 5.5.1 is that the elements of q have to be recalculated afresh as new components are computed, whereas for two-way PLS, the first element of q, for example, is the same no matter how many components are calculated. This limitation is a consequence of nonorthogonality of individual columns of matrix T.

The x block residuals after each component are often computed conventionally by

$$^{resid,a}x_{ijk} = {}^{resid,a-1}x - t_i{}^jp^kp$$

where $^{resid,a}x_{ijk}$ is the residual after a components are calculated, which would lead to a model

$$\hat{x}_{ijk} = \sum_{a=1}^{A} t_i{}^jp_j{}^kp_k$$

Sometimes these equations are written as tensor products, but there are numerous ways of multiplying tensors together, so this notation can be confusing and it is often conceptually more convenient to deal directly with vectors and matrices, just as in Section 5.5.3.1 by unfolding the data. This procedure can be called *matricisation*.

In mathematical terms, we can state that

$$^{unfolded}\hat{X} = \sum_{a=1}^{A} t_a.{}^{unfolded}p_a$$

where $^{unfolded}p_a$ is simply a row vector of length $J.K$. Where trilinear PLS1 differs from unfolded PLS described in Section 5.5.3.1 is that a matrix P_a of dimensions $J \times K$ can be obtained for each PLS component given by

$$P_a = {}^jp_a.{}^kp_a$$

and P_a is unfolded to give $^{unfolded}p_a$.

Figure 5.18 represents this procedure, avoiding tensor multiplication, using conventional matrices and vectors together with unfolding. A key problem with the common implementation of trilinear PLS1 is that, since the scores and loadings of successive components are not orthogonal, the methods for determining residual errors are simply an approximation. Hence the x block residual is not modelled very well, and the error matrices (or tensors) do not have an easily understood physical meaning. It also means that there are no obvious analogies to eigenvalues. This means that it is not easy to determine the size of the components or the modelling power using the x scores and loadings, but, nevertheless, the main aim is to predict the concentration (or c block),

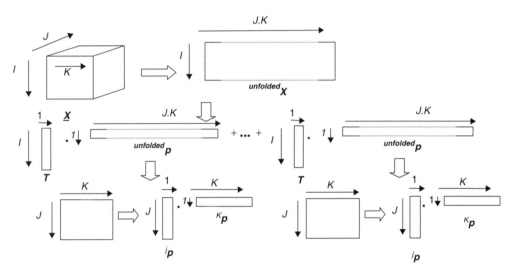

Figure 5.18
Matricisation in three-way calibration

Table 5.19 Calculation of three trilinear PLS1 components for the data in Table 5.17.

Component	t	^jp	^kp	q	$\hat{c}$	RMS concentration residuals	RMS of x 'residuals'
1	3135.35	0.398	0.339	0.00140	4.38	20.79	2.35×10^6
	4427.31	0.601	0.253		6.19		
	2194.17	0.461	0.624		3.07		
	1930.02	0.250	0.405		2.70		
		0.452	0.470				
			0.216				
2	−757.35	−0.252	0.381	0.00177	1.65	1.33	1.41×10^6
	−313.41	0.211	0.259	0.00513	6.21		
	511.73	0.392	0.692		6.50		
	−45.268	0.549	0.243		3.18		
		−0.661	0.485				
			0.119				
3	−480.37	−0.875	−0.070	0.00201	1	0.00	1.01×10^6
	−107.11	−0.073	0.263	0.00508	7		
	−335.17	−0.467	0.302	0.00305	6		
	−215.76	−0.087	0.789		3		
		0.058	0.004				
			0.461				

so we are not always worried about this limitation, so trilinear PLS1 is an acceptable method, provided that care is taken to interpret the output and not to expect the
residuals to have a physical meaning.

In order to understand this method further, trilinear PLS1 is performed on the first
compound. The main results are given in Table 5.19 (using uncentred data). It can be
seen that three components provide an exact model of the concentration, but there is

still an apparent residual error in the x matrix, representing 2.51 % of the overall sum of squares of the data: this error has no real physical or statistical meaning, except that it is fairly small. Despite this problem, it is essential to recognise that the concentration has been modelled correctly. Analytical chemists who expect to relate errors directly to physical properties often find this hard to appreciate.

A beauty of multimode methods is that the dimensions of c (or indeed $\underline{X}$) can be changed, for example, a matrix C can be employed consisting of several different compounds, exactly as in PLS2, or even a tensor. It is possible to define the number of dimensions in both the x and c blocks, for example, a three-way x block and a two-way c block may consist of a series of two-way chromatograms each containing several compounds. However, unless one has a good grip of the theory or there is a real need from the nature of the data, it is best to reduce the problem to one of trilinear PLS1: for example, a concentration matrix C can be treated as several concentration vectors, in the same way that a calibration problem that might appear to need PLS2 can be reduced to several calculations using PLS1.

Whereas there has been a huge interest in multimode calibration in the theoretical chemometrics literature, it is important to recognise that there are limitations to the applicability of such techniques. Good, very high order, data are rare in chemistry. Even three-way calibration, such as in HPLC–DAD, has to be used cautiously as there are frequent experimental difficulties with exact alignments of chromatograms in addition to interpretation of the numerical results. However, there have been some significant successes in areas such as sensory research and psychometrics and certain techniques such as fluorescence excitation–emission spectroscopy where the wavelengths are very stable show promise for the future.

Using genuine three-way methods (even when redefined in terms of matrices) differs from unfolding in that the connection between different variables is retained; in an unfolded data matrix there is no indication of whether two variables share the same elution time or spectroscopic wavelength.

5.6 Model Validation

Unquestionably one of the most important aspects of all calibration methods is model validation. Several questions need to be answered:

- how many significant components are needed to characterise a dataset?;
- how well is an unknown predicted?;
- how representative are the data used to produce a model?

It is possible to obtain as close a fit as desired using more and more PLS or PCA components, until the raw data are fitted exactly; however, the later terms are unlikely to be physically meaningful. There is a large literature on how to decide what model to adapt, which requires an appreciation of model validation, experimental design and how to measure errors. Most methods aim to guide the experimenter as to how many significant components to retain. The methods are illustrated below with reference to PLS1, but similar principles apply to all calibration methods, including MLR, PCR, PLS2 and trilinear PLS1.

5.6.1 Autoprediction

The simplest approach to determining number of significant components is by measuring the autoprediction error. This is the also called the root mean square error of calibration.

Usually (but not exclusively) the error is calculated on the concentration data matrix (c), and we will restrict the discussion below to errors in concentration for brevity: it is important to understand that similar equations can be obtained for the 'x' data block.

As more components are calculated, the residual error decreases. There are two ways of calculating this error:

$$^1E_{cal} = \sqrt{\frac{\sum_{i=1}^{I}(c_i - \hat{c}_i)^2}{I - a - 1}}$$

where a PLS components have been calculated and the data have been centred, or

$$^2E_{cal} = \sqrt{\frac{\sum_{i=1}^{I}(c_i - \hat{c}_i)^2}{I}}$$

Note that these errors can easily be converted to a percentage variance or mean square errors as described in Sections 5.4 and 5.5.

The value of $^2E_{cal}$ will always decline in value as more components are calculated, whereas that of $^1E_{cal}$ has the possibility of increasing slightly in size although, in most well behaved cases, it will also decrease with increase in the number of components and if it does increase against component number it is indicative that there may be problems with the data. These two autopredictive errors for acenaphthylene are presented in Figure 5.19, using PLS1. Notice how $^1E_{cal}$ increases slightly at the end. In this book we will use the first type of calibration error by default.

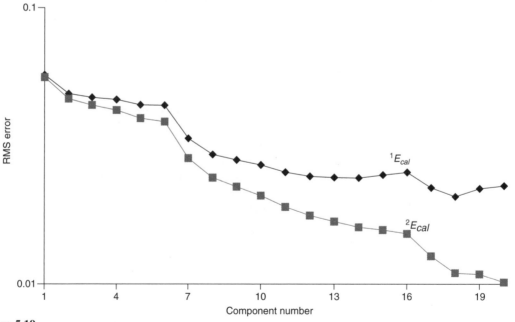

Figure 5.19
Autopredictive errors for acenaphthylene using PLS1

The autopredictive error can be used to determine how many PLS components to use in the model, in a number of ways:

1. A standard cut-off percentage error can be used, for example, 1 %. Once the error has reduced to this cut-off, ignore later PLS (or PCA) components.
2. Sometimes an independent measure of the noise level is possible. Once the error has declined to the noise level, ignore later PLS (or PCA) components.
3. Occasionally the error can reach a plateau. Take the PLS components up to this plateau.

By plotting the magnitudes of each successive components (or errors in modelling the 'x' block), it is also possible to determine prediction errors for the 'x' block. However, the main aim of calibration is to predict concentrations rather than spectra, so this information, although useful, is less frequently employed in calibration. More details have been discussed in the context of PCA in Chapter 4, Section 4.3.3, and also Section 5.4.2; the ideas for PLS are similar.

Many statistically oriented chemometricians do not like to use autoprediction for the determination of the number of significant PCs (sometimes called rank) as it is always possible to fit data perfectly simply by increasing the number of terms (or components) in the model. There is, though, a difference between statistical and chemical thinking. A chemist might know (or have a good intuitive feel for) parameters such as noise levels, and, therefore, in some circumstances be able to successfully interpret the autopredictive errors in a perfectly legitimate physically meaningful manner.

5.6.2 Cross-validation

An important chemometric tool is called cross-validation. The basis of the method is that the predictive ability of a model formed on part of a dataset can be tested by how well it predicts the remainder of the data, and has been introduced previously in other contexts (see Chapter 4, Sections 4.3.3.2 and 4.5.1.2).

It is possible to determine a model using $I - 1$ (=24) samples leaving out one sample (i). How well does this model fit the original data? Below we describe the method when the data are centred, the most common approach.

The following steps are used to determine cross-validation errors for PCR.

1. If centring the data prior to PCA, centre both the $I - 1$ (=24 in this example) concentrations and spectra each time, remembering to calculate the means $\bar{c}_i$ and $\bar{x}_i$ involved, removing sample i and subtracting these means from the original data. This process needs repeating I times.
2. Decide how many PCs are in the model, which determines the size of the matrices. Normally the procedure is repeated using successively more PCs, and a cross-validation error is obtained each time.
3. Perform PCA to give loadings and scores matrices T and P for the x data, then obtain a vector r for the c data using standard regression techniques. Note that these arrays will differ according to which sample is removed from the analysis.

Predicting the concentration of an unknown sample is fairly straightforward.

1. Call the spectrum of sample i x_i (a row vector).
2. Subtract the mean of the $I - 1$ samples from this to give $x_i - \bar{x}_i$, where $\bar{x}_i$ is the mean spectrum *excluding* sample i, if mean centring.

3. Estimate $\hat{t}_i = (x_i - \bar{x}_i).P'$, where P are the loadings obtained from the PCA model using $I - 1$ samples *excluding* sample i or $\hat{t}_i = x_i.P'$ if not mean-centred.

4. Then calculate $^{cv}\hat{c}_i = \hat{t}_i.r + \bar{c}_i$(centred) or $^{cv}\hat{c}_i = \hat{t}_i.r_i$ (uncentred) which is the estimated concentration of sample i using the model based on the remaining $(I - 1)$ (=24 samples).

Most methods of cross-validation then repeat the calculation leaving another spectrum out, and so on, until the entire procedure has been repeated I (=25 in our case) times over. The root mean square of these errors is then calculated as follows:

$$E_{cv} = \sqrt{\frac{\sum_{i=1}^{I}(c_i - {}^{cv}\hat{c}_i)^2}{I}}$$

Note that unlike the autoprediction error, this term is always divided by I because each sample in the original dataset represents an additional degree of freedom, however many PCA components have been calculated and however the data have been preprocessed. Note that it is conventional to calculate this error on the 'c' block of the data.

Cross-validation in PLS is slightly more complicated. The reason is that the scores are obtained using both the 'c' and 'x' blocks simultaneously, so steps 3 and 4 of the prediction above are slightly more elaborate. The product of T and P is no longer the best least squares approximation to the x block so it is not possible to obtain an estimate $\hat{t}_i$ using just P and x_i. Many people use what is called a weights vector which can be employed in prediction. The method is illustrated in more detail in Problem 5.8, but the interested reader should first look at Appendix A.2.2 for a description of the PLS1 algorithm. Below we will apply cross-validation to PLS predictions. The same comments about calculating E_{cv} from $\hat{c}$ apply to PLS as they do to PCR.

For acenaphthylene using PLS1, the cross-validated error is presented in Figure 5.20. An immediate difference between autoprediction and cross-validation is evident. In the former case the data will always be better modelled as more components are employed in the calculation, so the error will always decrease (with occasional rare exceptions in the case of $^1E_{cal}$ when a large number of PLS components have been computed). However, cross-validated errors normally reach a minimum as the correct number of components are found and then increase afterwards. This is because later components really represent noise and not systematic information in the data. In this case the cross-validation error has a minimum after nine components are used and then increases steadily afterwards; the value for seven PLS components is probably due to noise, although some people might conclude that there are only six components. Note that there will be a similar type of graph for the x block, but the minimum may be reached at a different stage. Also, the interpretation of cross-validated errors is somewhat different in calibration to pattern recognition.

Cross-validation has two main purposes.

1. In can be employed as a method for determining how many components characterise the data. From Figure 5.20, it appears that nine components are necessary to obtain an optimum model for acenaphthylene. This number will rarely equal the number of chemicals in the mixture, as spectral similarity and correlations between

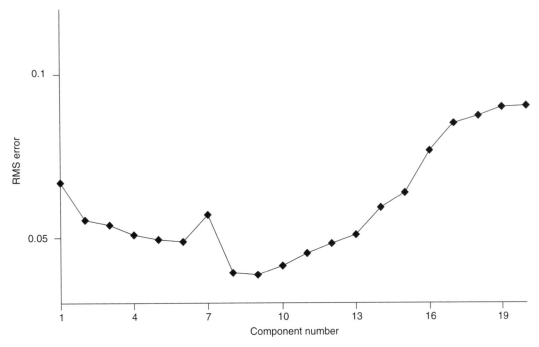

Figure 5.20
Cross-validated error for acenaphthylene

concentrations will often reduce it whereas impurities or interferents may increase it. Later components probably model noise and it would be dangerous to include them if used to predict concentrations of unknowns.
2. It can be employed as a fairly realistic error estimate for predictive ability. The minimum cross-validated prediction error for acenaphthylene of 0.0493 mg l^{-1}equals 41.1 %. This compares with an autopredictive error of 0.0269 mg l^{-1}or 22.48 % using 10 components and PLS1, which is an over-optimistic estimate.

Many refinements to cross-validation have been proposed in the literature which have been discussed in Chapter 4, and it is equally possibly to apply these to calibration in addition to pattern recognition.

5.6.3 Independent Test Sets

A significant weakness of cross-validation is that it depends on the design and scope of the original dataset used to form the model. This dataset is often called a 'training' set (see Chapter 4, Section 4.5.1.1). Consider a situation in which a series of mixture spectra are recorded, but it happens that the concentrations of two compounds are correlated, so that the concentration of compound A is high when compound B likewise is high, and vice versa. A calibration model can be obtained from analytical data, which predicts both concentrations fairly well. Even cross-validation might suggest that the model is good. However, if asked to predict a spectrum where compound A is in a high concentration and compound B in a low concentration it is likely to give very poor results, as it has not been trained to cope with this new situation. Cross-validation is very useful for removing the influence of internal factors such as instrumental noise or dilution errors or instrumental noise, but cannot help very much if there are correlations in the concentration of compounds in the training set.

In some cases there will inevitably be correlations in the concentration data, because it is not easy to find samples without this. Examples routinely occur in environmental monitoring. Several compounds often arise from a single source. For example, pol-yaromatic hydrocarbons (PAHs) are well known pollutants, so if one or two PAHs are present in high concentrations it is a fair bet that others will be too. There may be some correlations, for example, in the occurrence of compounds of different molecular weights if a homologous series occurs, e.g. as the byproduct of a biological pathway, there may be an optimum chain length which is most abundant in samples from a certain source. It would be hard to find a series of field samples in which there are no correlations between the concentrations of compounds in the samples. Consider, for example, setting up a model of PAHs coming from rivers close to several spe-cific sources of pollution. The model may behave well on this training set, but can it be safely used to predict the concentrations of PAHs in an unknown sample from a very different source? Another serious problem occurs in process control. Consider trying to set up a calibration model using NIR spectroscopy to determine the concen-tration of chemicals in a manufacturing process. If the process is behaving well, the predictions may be fairly good, but it is precisely to detect problems in the process that the calibration model is effective: is it possible to rely on the predictions if data have a completely different structure? Some chemometricians do look at the structure of the data and samples that do not fit into the structure of the calibration set are often called outliers. It is beyond the scope of this text to provide extensive discussion about how to spot outliers, as this depends on the software package and often the type of data, but it is important to understand how the design of training sets influences model validation.

Instead of validating the predictions internally, it is possible to test the predictions against an independent data set, often called a 'test' set. Computationally the procedure is similar to cross-validation. For example, a model is obtained using I samples, and then the predictions are calculated using an independent test set of L samples, to give

$$E_{test} = \sqrt{\frac{\sum_{l=1}^{L}(c_l - {}^{test}\hat{c}_l)^2}{L}}$$

The value of $\hat{c}_l$ is determined in exactly the same way as per cross-validation (see Section 5.6.2), but only one calibration model is formed, from the training set.

We will use the data in Table 5.1 for the training set, but test the predictions using the spectra obtained from a new dataset presented in Table 5.20. In this case, each dataset has the same number of samples, but this is not a requirement. The graph of E_{test} for acenaphthylene is presented in Figure 5.21 and shows similar trends to that of E_{cv} although the increase in error when a large number of components are calculated is not so extreme. The minimum error is 37.4 %, closely comparable to that for cross-validation. Normally the minimum test set error is higher than that for cross-validation, but if the structure of the test set is encompassed in the training set, these two errors will be very similar.

Table 5.20 Independent test set.

(a) Spectra ('x' block)

	220	225	230	235	240	245	250	255	260	265	270	275	280	285	290	295	300	305	310	315	320	325	330	335	340	345	350
1	0.687	0.706	0.660	0.476	0.490	0.591	0.723	0.783	0.589	0.611	0.670	0.663	0.646	0.624	0.678	0.280	0.185	0.144	0.127	0.122	0.139	0.120	0.102	0.106	0.097	0.075	0.053
2	0.710	0.644	0.603	0.483	0.552	0.646	0.808	0.885	0.722	0.759	0.821	0.808	0.737	0.698	0.771	0.286	0.199	0.143	0.117	0.127	0.161	0.133	0.130	0.160	0.119	0.087	0.062
3	0.682	0.698	0.655	0.466	0.477	0.575	0.707	0.768	0.578	0.601	0.661	0.654	0.636	0.616	0.666	0.273	0.178	0.138	0.120	0.115	0.131	0.113	0.095	0.099	0.091	0.069	0.047
4	0.790	0.803	0.713	0.545	0.608	0.705	0.829	0.917	0.705	0.738	0.852	0.944	0.922	0.873	0.943	0.363	0.233	0.177	0.152	0.156	0.185	0.154	0.156	0.196	0.142	0.100	0.070
5	0.521	0.522	0.507	0.434	0.539	0.593	0.661	0.718	0.558	0.571	0.624	0.476	0.364	0.335	0.353	0.218	0.159	0.140	0.112	0.115	0.150	0.093	0.097	0.151	0.068	0.030	0.018
6	0.677	0.631	0.606	0.464	0.507	0.470	0.486	0.478	0.320	0.298	0.337	0.574	0.259	0.236	0.261	0.119	0.079	0.076	0.072	0.091	0.129	0.090	0.103	0.161	0.072	0.030	0.017
7	0.609	0.536	0.522	0.486	0.556	0.748	0.847	0.844	0.555	0.504	0.539	0.479	0.489	0.448	0.500	0.241	0.155	0.118	0.095	0.105	0.133	0.096	0.111	0.167	0.088	0.049	0.034
8	0.787	0.612	0.541	0.464	0.556	0.604	0.659	0.655	0.518	0.486	0.510	0.479	0.392	0.355	0.370	0.224	0.147	0.116	0.088	0.090	0.106	0.070	0.079	0.118	0.056	0.029	0.019
9	0.812	0.634	0.613	0.563	0.671	0.704	0.752	0.794	0.675	0.656	0.703	0.777	0.665	0.611	0.683	0.284	0.207	0.154	0.122	0.133	0.168	0.124	0.140	0.206	0.109	0.062	0.040
10	0.682	0.509	0.420	0.415	0.519	0.657	0.820	0.847	0.657	0.639	0.641	0.612	0.558	0.513	0.564	0.278	0.184	0.126	0.092	0.086	0.091	0.076	0.070	0.078	0.068	0.056	0.042
11	0.920	0.795	0.703	0.559	0.594	0.679	0.792	0.858	0.749	0.768	0.831	0.940	0.902	0.869	0.918	0.368	0.233	0.161	0.134	0.140	0.164	0.147	0.146	0.170	0.134	0.103	0.073
12	0.955	0.781	0.681	0.587	0.594	0.848	1.064	1.128	0.924	0.941	1.006	1.058	0.980	0.926	0.984	0.394	0.256	0.175	0.137	0.135	0.153	0.135	0.128	0.144	0.125	0.101	0.073
13	0.825	0.777	0.672	0.471	0.484	0.584	0.725	0.817	0.720	0.765	0.831	0.821	0.798	0.767	0.820	0.305	0.220	0.161	0.133	0.128	0.145	0.107	0.107	0.144	0.102	0.081	0.055
14	0.842	0.792	0.722	0.573	0.631	0.775	0.949	1.039	0.812	0.844	0.924	0.984	0.929	0.888	0.936	0.361	0.244	0.173	0.144	0.149	0.176	0.150	0.146	0.174	0.136	0.099	0.070
15	0.731	0.718	0.632	0.474	0.527	0.621	0.776	0.841	0.666	0.711	0.790	0.742	0.695	0.665	0.707	0.306	0.183	0.183	0.145	0.136	0.152	0.117	0.105	0.126	0.094	0.068	0.048
16	0.744	0.778	0.721	0.560	0.629	0.685	0.766	0.819	0.620	0.626	0.710	0.766	0.698	0.664	0.742	0.265	0.166	0.129	0.116	0.135	0.182	0.144	0.152	0.205	0.128	0.081	0.056
17	0.610	0.618	0.620	0.528	0.629	0.687	0.760	0.796	0.588	0.598	0.667	0.602	0.492	0.472	0.506	0.251	0.177	0.158	0.136	0.145	0.192	0.134	0.139	0.207	0.109	0.057	0.040
18	0.606	0.591	0.547	0.400	0.460	0.474	0.522	0.563	0.449	0.463	0.527	0.529	0.453	0.426	0.476	0.218	0.156	0.128	0.103	0.110	0.145	0.100	0.108	0.162	0.082	0.040	0.024
19	0.520	0.490	0.514	0.491	0.618	0.651	0.701	0.709	0.510	0.482	0.519	0.481	0.364	0.332	0.352	0.211	0.154	0.134	0.107	0.119	0.158	0.104	0.120	0.187	0.083	0.038	0.026
20	0.718	0.625	0.560	0.452	0.529	0.595	0.662	0.658	0.487	0.453	0.497	0.549	0.504	0.472	0.519	0.240	0.146	0.111	0.091	0.093	0.114	0.087	0.090	0.123	0.075	0.046	0.030
21	0.761	0.624	0.579	0.526	0.646	0.736	0.834	0.864	0.694	0.677	0.729	0.770	0.665	0.622	0.690	0.293	0.205	0.152	0.117	0.123	0.148	0.110	0.119	0.173	0.097	0.058	0.038
22	0.646	0.471	0.373	0.361	0.442	0.527	0.619	0.647	0.546	0.534	0.545	0.477	0.397	0.362	0.379	0.232	0.164	0.124	0.090	0.084	0.094	0.068	0.063	0.084	0.053	0.035	0.023
23	0.801	0.650	0.604	0.488	0.551	0.659	0.810	0.840	0.687	0.689	0.729	0.837	0.800	0.766	0.811	0.319	0.229	0.156	0.120	0.118	0.122	0.113	0.109	0.125	0.101	0.077	0.052
24	0.826	0.657	0.589	0.521	0.610	0.809	1.008	1.063	0.832	0.830	0.875	0.905	0.866	0.827	0.879	0.374	0.236	0.159	0.122	0.113	0.117	0.105	0.095	0.100	0.098	0.084	0.060
25	0.986	0.887	0.783	0.548	0.536	0.582	0.685	0.785	0.745	0.798	0.890	0.933	0.911	0.861	0.890	0.325	0.234	0.172	0.144	0.145	0.162	0.145	0.135	0.151	0.122	0.094	0.064

(continued overleaf)

Table 5.20 (continued)

(b) Concentrations ("c" block)

Spectrum No.	PAH concentration mg l^{-1}									
	Py	Ace	Anth	Acy	Chry	Benz	Fluora	Fluore	Nap	Phen
1	0.456	0.120	0.168	0.120	0.336	1.620	0.120	0.600	0.120	0.564
2	0.456	0.040	0.224	0.160	0.560	2.160	0.120	1.000	0.040	0.188
3	0.152	0.160	0.224	0.200	0.448	1.620	0.200	0.200	0.040	0.376
4	0.608	0.160	0.280	0.160	0.336	2.700	0.040	0.200	0.080	0.188
5	0.608	0.200	0.224	0.120	0.560	0.540	0.040	0.400	0.040	0.564
6	0.760	0.160	0.168	0.200	0.112	0.540	0.080	0.200	0.120	0.376
7	0.608	0.120	0.280	0.040	0.112	1.080	0.040	0.600	0.080	0.940
8	0.456	0.200	0.056	0.040	0.224	0.540	0.120	0.400	0.200	0.940
9	0.760	0.040	0.056	0.080	0.112	1.620	0.080	1.000	0.200	0.752
10	0.152	0.040	0.112	0.040	0.336	1.080	0.200	1.000	0.160	0.940
11	0.152	0.080	0.056	0.120	0.224	2.700	0.200	0.800	0.200	0.564
12	0.304	0.040	0.168	0.080	0.560	2.700	0.160	1.000	0.120	0.752
13	0.152	0.120	0.112	0.200	0.560	2.160	0.200	0.600	0.160	0.376
14	0.456	0.080	0.280	0.200	0.448	2.700	0.120	0.800	0.080	0.376
15	0.304	0.200	0.280	0.160	0.560	1.620	0.160	0.400	0.080	0.188
16	0.760	0.200	0.224	0.200	0.336	2.160	0.080	0.400	0.040	0.376
17	0.760	0.160	0.280	0.120	0.448	1.080	0.080	0.200	0.080	0.564
18	0.608	0.200	0.168	0.160	0.224	1.080	0.040	0.400	0.120	0.188
19	0.760	0.120	0.224	0.080	0.224	0.540	0.080	0.600	0.040	0.752
20	0.456	0.160	0.112	0.080	0.112	1.080	0.120	0.200	0.160	0.752
21	0.608	0.080	0.112	0.040	0.224	1.620	0.040	0.800	0.160	0.940
22	0.304	0.080	0.056	0.080	0.336	0.540	0.160	0.800	0.200	0.752
23	0.304	0.040	0.112	0.120	0.112	2.160	0.160	1.000	0.160	0.564
24	0.152	0.080	0.168	0.040	0.448	2.160	0.200	0.800	0.120	0.940
25	0.304	0.120	0.056	0.160	0.448	2.700	0.160	0.600	0.200	0.188

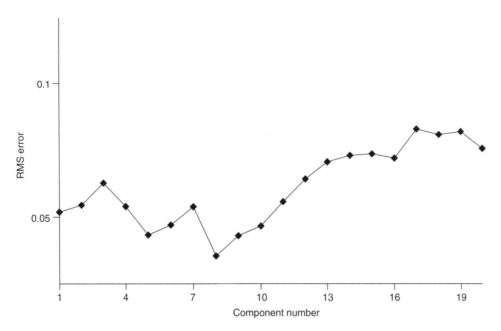

Figure 5.21
Root mean square error using data in Table 5.1 as a training set and data in Table 5.20 as a test set, PLS1 (centred) and acenaphthylene

If, however, we use the data in Table 5.20 as the training set and those in Table 5.1 as the test set, a very different story emerges, as shown in Figure 5.22 for acenaphthylene. The minimum error is 53.96 % and the trends are very different. Despite this, the values of $^1E_{cal}$ using 10 PLS components are very similar for both datasets. In fact, neither autoprediction nor cross-validation will distinguish the behaviour of either dataset especially. However, the results when the test and training sets are swapped around differ considerably, and give us a clue as to what is going wrong. It appears that the data in Table 5.1 provide a good training set whereas those in Table 5.20 are not so useful. This suggests that the data in Table 5.1 encompass the features of those in Table 5.20, but not vice versa. The reason for this problem relates to the experimental design. In Chapter 2, Section 2.3.4, we discuss experimental design for multivariate calibration, and it can be shown that the concentrations the first dataset are orthogonal, but not the second one, explaining this apparent anomaly. The orthogonal training set predicts both itself and a nonorthogonal test set well, but a nonorthogonal training set, although is very able to produce a model that predicts itself well, cannot predict an orthogonal (and more representative) test set as well.

It is important to recognise that cross-validation can therefore sometimes give a misleading and over-optimistic answer. However, whether cross-validation is useful depends in part on the practical aim of the measurements. If, for example, data containing all the possible features of Table 5.1 are unlikely ever to occur, it may be safe to use the model obtained from Table 5.20 for future predictions. For example, if it is desired to determine the amount of vitamin C in orange juices from a specific region of Spain, it might be sufficient to develop a calibration method only on these juices. It could be expensive and time consuming to find a more representative calibration set.

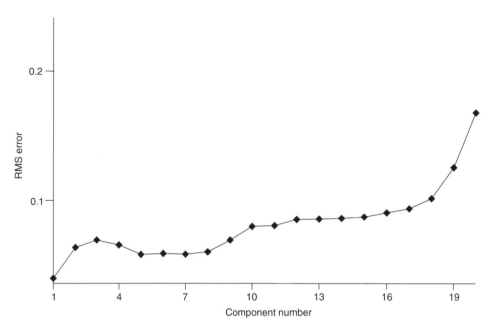

Figure 5.22
Root mean square error using data in Table 5.20 as a training set and data in Table 5.1 as a test set, PLS1 (centred) and acenaphthylene

Brazilian orange juice may exhibit different features, but is it necessary to go to all the trouble of setting up a calibration experiment that includes Brazilian orange juice? Is it really necessary or practicable to develop a method to measure vitamin C in all conceivable orange juices or foodstuffs? The answer is no, and so, in some circumstances, living within the limitations of the original dataset is entirely acceptable and finding a test set that is wildly unlikely to occur in real situations represents an artificial experiment; remember we want to save time (and money) when setting up the calibration model. If at some future date extra orange juice from a new region is to be analysed, the first step is to set up a dataset from this new source of information as a test set and so determine whether the new data fit into the structure of the existing database or whether the calibration method must be developed afresh. It is very important, though, to recognise the limitations and calibration models especially if they are to be applied to situations that are wider than those represented by the initial training sets.

There are a number of variations on the theme of test sets, one being simply to take a few samples from a large training set and assign them to a test set, for example, to take five out of the 25 samples from the case study and assign them to a test set, using the remaining 20 samples for the training set. Alternatively, the two could be combined, and 40 out of the 50 used for determining the model, the remaining 10 for independent testing.

The optimum size and representativeness of training and test sets for calibration modelling are a big subject. Some chemometricians recommend using hundreds or thousands of samples, but this can be expensive and time consuming. In some cases a completely orthogonal dataset is unlikely ever to occur and field samples with these features cannot be found. Hence there is no 'perfect' way of validating calibration models, but it is essential to have some insight into the aims of an experiment before deciding how important the structure of the training set is to the success of the model in foreseeable future circumstances. Sometimes it can take so long to produce good calibration models that new instruments and measurement techniques become available, superseding the original datasets.

Problems

Problem 5.1 Quantitative Retention Property Relationships of Benzodiazepines

Section 5.5   Section 5.6.2

QSAR, QSPR and similar techniques are used to relate one block of properties to another. In this dataset, six chromatographic retention parameters of 13 compounds are used to predict a biological property (A1).

Compound	Property A1	Retention parameters					
		C18	Ph	CN-R	NH2	CN-N	Si
1	−0.39	2.90	2.19	1.49	0.58	−0.76	−0.41
2	−1.58	3.17	2.67	1.62	0.11	−0.82	−0.52
3	−1.13	3.20	2.69	1.55	−0.31	−0.96	−0.33
4	−1.18	3.25	2.78	1.78	−0.56	−0.99	−0.55
5	−0.71	3.26	2.77	1.83	−0.53	−0.91	−0.45
6	−1.58	3.16	2.71	1.66	0.10	−0.80	−0.51
7	−0.43	3.26	2.74	1.68	0.62	−0.71	−0.39
8	−2.79	3.29	2.96	1.67	−0.35	−1.19	−0.71
9	−1.15	3.59	3.12	1.97	−0.62	−0.93	−0.56
10	−0.39	3.68	3.16	1.93	−0.54	−0.82	−0.50
11	−0.64	4.17	3.46	2.12	−0.56	−0.97	−0.55
12	−2.14	4.77	3.72	2.29	−0.82	−1.37	−0.80
13	−3.57	5.04	4.04	2.44	−1.14	−1.40	−0.86

1. Perform standardised cross-validated PLS on the data (note that the property parameter should be mean centred but there is no need to standardise), calculating five PLS components. If you are not using the Excel Add-in the following steps may be required, as described in more detail in Problem 5.8:
   - Remove one sample.
   - Calculate the standard deviation and mean of the c and x block parameters for the remaining 12 samples (note that it is not strictly necessary to standardise the c block parameter).
   - Standardise these according to the parameters calculated above and then perform PLS.
   - Use this model to predict the property of the 13th sample. Remember to correct this for the standard deviation and mean of the 12 samples used to form the model. This step is a tricky one as it is necessary to use a weight vector.
   - Continue for all 13 samples.

- When calculating the room mean square error, divide by the number of samples (13) rather than the number of degrees of freedom.

  If this is your first experience of cross-validation you are recommended to use the Excel Add-in, or first to attempt Problem 5.8.

  Produce a graph of root mean square error against component number. What appears to be the optimum number of components in the model?

2. Using the optimum number of PLS components obtained in question 1, perform PLS (standardising the retention parameters) on the overall dataset, and obtain a table of predictions for the parameter A1. What is the root mean square error?

3. Plot a graph of predicted versus observed values of parameter A1 from the model in question 2.

Problem 5.2 Classical and Inverse Univariate Regression

Section 5.2

The following are some data: a response (x) is recorded at a number of concentrations (c).

c	x
1	0.082
2	0.174
3	0.320
4	0.412
5	0.531
6	0.588
7	0.732
8	0.792
9	0.891
10	0.975

1. There are two possible regression models, namely the inverse $\hat{c} = b_0 + b_1 x$ and the classical $\hat{x} = a_0 + a_1 c$. Show how the coefficients on the right of each equation would relate to each other algebraically if there were no errors.

2. The regression models can be expressed in matrix form, $\hat{c} = X.b$ and $\hat{x} = C.a$. What are the dimensions of the six matrices/vectors in these equations? Using this approach, calculate the four coefficients from question 1.

3. Show that the coefficients for each model are approximately related as in question 1 and explain why this is not exact.

4. What different assumptions are made by both models?

Problem 5.3 Multivariate Calibration with Several x and c Variables, Factors that Influence the Taste and Quality of Cocoa

Section 4.3   Section 5.5.1   Section 5.5.2   Section 5.6.1

Sometimes there area several variables in both the 'x' and 'c' blocks, and in many applications outside analytical chemistry of mixtures it is difficult even to define which variables belong to which block.

The following data represent eight blends of cocoa, together with scores obtained from a taste panel of various qualities.

Sample No.	Ingredients			Assessments					
	Cocoa (%)	Sugar (%)	Milk (%)	Lightness	Colour	Cocoa odour	Smooth texture	Milk taste	Sweetness
1	20.00	30.00	50.00	44.89	1.67	6.06	8.59	6.89	8.48
2	20.00	43.30	36.70	42.77	3.22	6.30	9.09	5.17	9.76
3	20.00	50.00	30.00	41.64	4.82	7.09	8.61	4.62	10.50
4	26.70	30.00	43.30	42.37	4.90	7.57	5.96	3.26	6.69
5	26.70	36.70	36.70	41.04	7.20	8.25	6.09	2.94	7.05
6	26.70	36.70	36.70	41.04	6.86	7.66	6.74	2.58	7.04
7	33.30	36.70	30.00	39.14	10.60	10.24	4.55	1.51	5.48
8	40.00	30.00	30.00	38.31	11.11	11.31	3.42	0.86	3.91

Can we predict the assessments from the ingredients? To translate into the terminology of this chapter, refer to the ingredients as x and the assessments as c.

1. Standardise, using the population standard deviation, all the variables. Perform PCA separately on the $8 \times 3 \, X$ matrix and on the $8 \times 6 \, C$ matrix. Retain the first two PCs, plot the scores and loadings of PC2 versus PC1 of each of the blocks, labelling the points, to give four graphs. Comment.
2. After standardising both the C and X matrices, perform PLS1 on the six c block variables against the three x block variables. Retain two PLS components for each variable (note that it is not necessary to retain the same number of components in each case), and calculate the predictions using this model. Convert this matrix (which is standardised) back to the original nonstandardised matrix and present these predictions as a table.
3. Calculate the percentage root mean square prediction errors for each of the six variables as follows. (i) Calculate residuals between predicted and observed. (ii) Calculate the root mean square of these residuals, taking care to divide by 5 rather than 8 to account for the loss of three degrees of freedom due to the PLS components and the centring. (iii) Divide by the sample standard deviation for each parameter and multiply by 100 (note that it is probably more relevant to use the standard deviation than the average in this case).
4. Calculate the six correlation coefficients between the observed and predicted variables and plot a graph of the percentage root mean square error obtained in question 3 against the correlation coefficient, and comment.
5. It is possible to perform PLS2 rather than PLS1. To do this you must either produce an algorithm as presented in Appendix A.2.3 in Matlab or use the VBA Add-in. Use a six variable C block, and calculate the percentage root mean square errors as in question 3. Are they better or worse than for PLS1?
6. Instead of using the ingredients to predict taste/texture, it is possible to use the sensory variables to predict ingredients. How would you do this (you are not required to perform the calculations in full)?

Problem 5.4 Univariate and Multivariate Regression

Section 5.2   Section 5.3.3   Section 5.5.1

The following represent 10 spectra, recorded at eight wavelengths, of two compounds A and B, whose concentrations are given by the two vectors.

Spectrum No.								
1	0.227	0.206	0.217	0.221	0.242	0.226	0.323	0.175
2	0.221	0.412	0.45	0.333	0.426	0.595	0.639	0.465
3	0.11	0.166	0.315	0.341	0.51	0.602	0.537	0.246
4	0.194	0.36	0.494	0.588	0.7	0.831	0.703	0.411
5	0.254	0.384	0.419	0.288	0.257	0.52	0.412	0.35
6	0.203	0.246	0.432	0.425	0.483	0.597	0.553	0.272
7	0.255	0.326	0.378	0.451	0.556	0.628	0.462	0.339
8	0.47	0.72	0.888	0.785	1.029	1.233	1.17	0.702
9	0.238	0.255	0.318	0.289	0.294	0.41	0.444	0.299
10	0.238	0.305	0.394	0.415	0.537	0.585	0.566	0.253

Concentration vectors:

Spectrum No.	Conc. A	Conc. B
1	1	3
2	3	5
3	5	1
4	7	2
5	2	5
6	4	3
7	6	1
8	9	6
9	2	4
10	5	2

Most calculations will be on compound A, although in certain cases we may utilise the information about compound B.

1. After centring the data matrix down the columns, perform univariate calibration at each wavelength for compound A (only), using a method that assumes all errors are in the response (spectral) direction. Eight slopes should be obtained, one for each wavelength.

2. Predict the concentration vector for compound A using the results of regression at each wavelength, giving eight predicted concentration vectors. From these vectors, calculate the root mean square error of predicted minus true concentrations at each wavelength, and indicate which wavelengths are best for prediction.

3. Perform multilinear regression using all the wavelengths at the same time for compound A, as follows. (i) On the uncentred data, find $\hat{s}$, assuming $X \approx c.s$, where

c is the concentration vector for compound A. (ii) Predict the concentrations since $\hat{c} = X.\hat{s}'.(\hat{s}.\hat{s}')^{-1}$. What is the root mean square error of prediction?

4. Repeat the calculations of question 3, but this time include both concentration vectors in the calculations, replacing the vector c by the matrix C. Comment on the errors.

5. After centring the data both in the concentration and spectral directions, calculate the first and second components for the data and compound A using PLS1.

6. Calculate the estimated concentration vector for component A, (i) using one and (ii) using two PLS components. What is the root mean square error for prediction in each case?

7. Explain why information only on compound A is necessary for good prediction using PLS, but both information on both compounds is needed for good prediction using MLR.

Problem 5.5 Principal Components Regression

Section 4.3   Section 5.4

A series of 10 spectra of two components are recorded at eight wavelengths. The following are the data.

Spectrum No.								
1	0.070	0.124	0.164	0.171	0.184	0.208	0.211	0.193
2	0.349	0.418	0.449	0.485	0.514	0.482	0.519	0.584
3	0.630	0.732	0.826	0.835	0.852	0.848	0.877	0.947
4	0.225	0.316	0.417	0.525	0.586	0.614	0.649	0.598
5	0.533	0.714	0.750	0.835	0.884	0.930	0.965	0.988
6	0.806	0.979	1.077	1.159	1.249	1.238	1.344	1.322
7	0.448	0.545	0.725	0.874	1.005	1.023	1.064	1.041
8	0.548	0.684	0.883	0.992	1.166	1.258	1.239	1.203
9	0.800	0.973	1.209	1.369	1.477	1.589	1.623	1.593
10	0.763	1.019	1.233	1.384	1.523	1.628	1.661	1.625

Concentration vectors:

Spectrum No.	Conc. A	Conc. B
1	1	1
2	2	5
3	3	9
4	4	2
5	5	6
6	6	10
7	7	3
8	8	4
9	9	8
10	10	7

1. Perform PCA on the centred data, calculating loadings and scores for the first two PCs. How many PCs are needed to model the data almost perfectly (you are not asked to do cross-validation)?
2. Perform principal components regression as follows. Take each compound in turn, and regress the concentrations on to the scores of the two PCs; you should centre the concentration matrices first. You should obtain two coefficients of the form $\hat{c} = t_1 r_1 + t_2 r_2$ for each compound. Verify that the results of PCR provide a good estimate of the concentration for each compound. Note that you will have to add the mean concentration back to the results.
3. Using the coefficients obtained in question 2, give the 2×2 rotation or transformation matrix, that rotates the scores of the first two PCs on to the centred estimates of concentrations.
4. Calculate the inverse of the matrix obtained in question 3 and, hence, determine the spectra of both components in the mixture from the loadings.
5. A different design can be used as follows:

Spectrum No.	Conc. A	Conc. B
1	1	10
2	2	9
3	3	8
4	4	7
5	5	6
6	6	5
7	7	4
8	8	3
9	9	2
10	10	1

Using the spectra obtained in question 4, and assuming no noise, calculate the data matrix that would be obtained. Perform PCA on this matrix and explain why there are considerable differences between the results using this design and the earlier design, and hence why this design is not satisfactory.

Problem 5.6 Multivariate Calibration and Prediction in Spectroscopic Monitoring of Reactions

Section 4.3   Section 5.3.3   Section 5.5.1

The aim is to monitor reactions using a technique called flow injection analysis (FIA) which is used to record a UV/vis spectrum of a sample. These spectra are reported as summed intensities over the FIA trace below. The reaction is of the form $A + B \longrightarrow C$ and so the aim is to quantitate the three compounds and produce a profile with time. A series of 25 three component mixtures are available; their spectra and concentrations of each component are presented below, with spectral intensities and wavelengths (nm) recorded.

	234.39	240.29	246.20	252.12	258.03	263.96	269.89	275.82	281.76	287.70	293.65	299.60	305.56	311.52	317.48	323.45	329.43	335.41	341.39	347.37	353.36	358.86
1	12.268	14.149	12.321	6.863	2.479	1.559	1.695	2.261	3.262	4.622	6.283	8.199	10.199	12.041	13.617	14.453	14.068	13.014	11.167	8.496	6.139	4.456
2	7.624	9.817	8.933	4.888	1.346	0.621	0.738	1.101	1.683	2.452	3.372	4.409	5.417	6.292	6.857	6.976	6.487	5.575	4.396	3.005	1.829	1.113
3	10.259	11.537	9.759	5.170	1.914	1.317	1.447	1.848	2.544	3.516	4.770	6.155	7.820	9.289	10.953	12.315	12.990	13.131	12.597	11.171	9.428	7.551
4	8.057	8.464	7.115	4.239	2.056	1.534	1.587	2.033	2.922	4.145	5.610	7.391	9.198	11.007	12.527	13.257	12.597	11.428	9.346	6.380	4.086	2.749
5	18.465	21.359	18.855	10.408	3.448	2.012	2.208	2.957	4.235	5.981	8.165	10.601	13.244	15.543	17.655	18.938	18.925	17.988	16.159	13.274	10.336	7.820
6	11.687	12.672	10.783	6.238	2.739	1.943	2.020	2.584	3.678	5.187	7.023	9.207	11.469	13.649	15.530	16.517	15.939	14.688	12.395	9.012	6.225	4.407
7	14.424	17.271	15.422	8.688	2.886	1.619	1.753	2.386	3.497	4.996	6.816	8.912	11.003	12.896	14.285	14.784	13.965	12.361	10.026	7.018	4.496	2.927
8	10.096	12.184	10.821	6.017	2.001	1.152	1.256	1.701	2.477	3.533	4.822	6.304	7.820	9.196	10.285	10.782	10.351	9.365	7.840	5.797	4.020	2.816
9	11.158	12.751	10.981	6.012	2.229	1.463	1.573	2.027	2.844	3.977	5.405	7.024	8.833	10.472	12.079	13.172	13.329	12.889	11.718	9.715	7.700	5.948
10	14.251	15.483	13.098	7.348	3.112	2.207	2.310	2.914	4.066	5.675	7.684	10.009	12.567	14.950	17.251	18.727	18.759	18.015	16.180	13.035	10.071	7.652
11	16.518	18.466	16.008	9.102	3.572	2.343	2.460	3.185	4.545	6.417	8.709	11.376	14.161	16.741	18.919	20.045	19.508	18.069	15.515	11.681	8.338	5.998
12	16.513	19.051	16.833	9.579	3.459	2.103	2.227	2.935	4.224	5.990	8.144	10.635	13.176	15.491	17.325	18.131	17.368	15.745	13.166	9.591	6.523	4.501
13	17.159	19.509	16.987	9.452	3.451	2.187	2.333	3.024	4.265	5.983	8.131	10.566	13.214	15.593	17.782	19.153	19.080	18.155	16.233	13.157	10.150	7.670
14	16.913	18.425	15.741	8.950	3.759	2.614	2.717	3.432	4.804	6.713	9.080	11.827	14.780	17.523	19.979	21.421	21.154	20.033	17.698	13.915	10.441	7.795
15	14.437	16.934	15.067	8.745	3.224	1.940	2.027	2.700	3.948	5.651	7.680	10.087	12.406	14.592	16.101	16.467	16.101	15.159	13.023	9.955	3.100	1.555
16	16.117	19.285	17.162	9.433	2.960	1.623	1.788	2.401	3.430	4.837	6.602	8.558	10.665	12.468	14.060	14.994	14.949	14.082	12.572	10.341	8.033	6.058
17	5.991	6.551	5.665	3.476	1.684	1.233	1.248	1.566	2.225	3.139	4.228	5.574	6.900	8.244	9.291	9.696	9.006	7.927	6.177	3.837	2.105	1.215
18	12.738	15.538	13.856	7.603	2.386	1.307	1.436	1.928	2.754	3.886	5.298	6.873	8.547	9.990	11.210	11.873	11.710	10.889	9.556	7.676	5.820	4.320
19	10.106	10.967	9.247	5.196	2.273	1.654	1.723	2.138	2.952	4.096	5.534	7.205	9.075	10.832	12.582	13.774	13.873	13.427	12.118	9.853	7.705	5.928
20	14.632	16.343	14.095	7.981	3.205	2.158	2.254	2.869	4.030	5.644	7.641	9.961	12.433	14.737	16.796	17.974	17.675	16.590	14.485	11.230	8.311	6.137
21	13.294	15.525	13.701	7.851	2.932	1.819	1.907	2.494	3.585	5.081	6.893	9.023	11.163	13.166	14.698	15.292	14.439	12.847	10.395	7.150	4.512	2.930
22	10.969	13.628	12.320	6.932	2.199	1.160	1.254	1.720	2.533	3.637	4.965	6.503	7.994	9.336	10.222	10.415	9.628	8.244	6.388	4.166	2.357	1.327
23	7.356	9.084	8.112	4.465	1.461	0.844	0.921	1.221	1.739	2.450	3.332	4.336	5.397	6.340	7.139	7.577	7.442	6.913	6.023	4.772	3.584	2.663
24	7.748	8.738	7.523	4.228	1.742	1.224	1.278	1.586	2.174	3.007	4.050	5.263	6.602	7.852	9.060	9.863	9.896	9.512	8.551	6.946	5.430	4.187
25	7.501	8.829	7.837	4.589	1.822	1.181	1.220	1.568	2.243	3.173	4.289	5.629	6.949	8.223	9.164	9.485	8.808	7.691	6.010	3.843	2.176	1.272

The concentrations of the three compounds in the 25 calibration samples are as follows.

	Conc. A (mM)	Conc. B (mM)	Conc. C (mM)
1	0.276	0.090	0.069
2	0.276	0.026	0.013
3	0.128	0.026	0.126
4	0.128	0.153	0.041
5	0.434	0.058	0.126
6	0.200	0.153	0.069
7	0.434	0.090	0.041
8	0.276	0.058	0.041
9	0.200	0.058	0.098
10	0.200	0.121	0.126
11	0.357	0.153	0.098
12	0.434	0.121	0.069
13	0.357	0.090	0.126
14	0.276	0.153	0.126
15	0.434	0.153	0.013
16	0.434	0.026	0.098
17	0.128	0.121	0.013
18	0.357	0.026	0.069
19	0.128	0.090	0.098
20	0.276	0.121	0.098
21	0.357	0.121	0.041
22	0.357	0.058	0.013
23	0.200	0.026	0.041
24	0.128	0.058	0.069
25	0.200	0.090	0.013

The spectra recorded with time (minutes) along the first column are presented on page 331. The aim is to estimate the concentrations of each compound in the mixture.

1. Perform PCA (uncentred) on the 25 calibration spectra, calculating the first two PCs. Plot the scores of PC2 versus PC1 and label the points. Perform PCA (uncentred) on the 25×3 concentration matrix of A, B and C, calculating the first two PCs, likewise plotting the scores of PC2 versus PC1 and labelling the points. Comment.
2. Predict the concentrations of A in the calibration set using MLR and assuming only compound A can be calibrated, as follows. (i) Determine the vector $\hat{s} = c'.X/\sum c^2$, where X is the 25×22 spectral calibration matrix and c a 25×1 vector. (ii) Determine the predicted concentration vector $\hat{c} = X.\hat{s}'/\sum \hat{s}^2$ (note that the denominator is simply the sum of squares when only one compound is used).
3. In the model of question 2, plot a graph of predicted against true concentrations. Determine the root mean square error both in mM and as a percentage of the average. Comment.
4. Repeat the predictions using MLR but this time for all three compounds simultaneously as follows. (i) Determine the matrix $\hat{S} = (C'.C)^{-1}.C'.X$ where X is

	234.39	240.29	246.20	252.12	258.03	263.96	269.89	275.82	281.76	287.70	293.65	299.60	305.56	311.52	317.48	323.45	329.43	335.41	341.39	347.37	353.36	358.86
1.08	11.657	12.457	10.788	6.372	2.609	1.700	1.747	2.278	3.312	4.721	6.391	8.417	10.358	12.262	13.583	13.905	12.656	10.765	7.999	4.499	1.927	0.722
3.67	11.821	12.617	10.921	6.461	2.664	1.747	1.793	2.330	3.377	4.805	6.500	8.554	10.526	12.462	13.812	14.142	12.873	10.962	8.157	4.602	1.993	0.765
7.22	11.922	12.725	11.015	6.518	2.692	1.767	1.815	2.355	3.407	4.842	6.547	8.614	10.600	12.541	13.903	14.241	12.977	11.068	8.254	4.686	2.065	0.822
9.48	11.886	12.688	10.986	6.497	2.685	1.768	1.817	2.356	3.401	4.828	6.520	8.571	10.541	12.466	13.816	14.160	12.927	11.042	8.268	4.745	2.140	0.896
13.07	11.755	12.561	10.866	6.401	2.633	1.735	1.792	2.327	3.355	4.757	6.420	8.429	10.373	12.261	13.605	13.969	12.790	10.980	8.290	4.864	2.315	1.071
15.45	11.478	12.260	10.596	6.218	2.553	1.694	1.759	2.281	3.273	4.621	6.222	8.149	10.023	11.832	13.141	13.521	12.441	10.750	8.226	4.984	2.541	1.303
19.17	11.769	12.561	10.823	6.309	2.588	1.738	1.824	2.352	3.339	4.675	6.268	8.162	10.029	11.815	13.148	13.603	12.664	11.112	8.746	5.647	3.242	1.920
22.06	11.459	12.232	10.520	6.076	2.465	1.664	1.760	2.272	3.207	4.474	5.990	7.777	9.564	11.253	12.560	13.076	12.303	10.945	8.819	5.983	3.721	2.387
25.01	11.848	12.632	10.831	6.220	2.524	1.719	1.827	2.344	3.282	4.553	6.081	7.865	9.680	11.378	12.737	13.354	12.706	11.471	9.470	6.723	4.456	3.012
28.02	12.151	12.936	11.063	6.310	2.551	1.747	1.862	2.379	3.312	4.576	6.109	7.888	9.722	11.424	12.848	13.539	13.034	11.922	10.038	7.389	5.127	3.592
31.04	11.844	12.609	10.764	6.108	2.463	1.692	1.807	2.305	3.196	4.409	5.884	7.586	9.365	11.001	12.406	13.131	12.737	11.746	10.028	7.545	5.380	3.843
34.09	11.882	12.640	10.770	6.077	2.427	1.663	1.783	2.278	3.165	4.369	5.840	7.526	9.304	10.934	12.360	13.144	12.815	11.902	10.254	7.835	5.686	4.109
37.15	12.065	12.813	10.901	6.126	2.443	1.680	1.805	2.301	3.185	4.387	5.860	7.544	9.337	10.972	12.429	13.263	13.022	12.173	10.604	8.241	6.092	4.460
40.36	12.152	12.902	10.960	6.140	2.451	1.695	1.819	2.313	3.187	4.376	5.842	7.506	9.298	10.920	12.400	13.291	13.128	12.372	10.885	8.604	6.474	4.794
43.49	12.055	12.799	10.855	6.052	2.406	1.669	1.798	2.278	3.131	4.290	5.725	7.347	9.114	10.702	12.189	13.113	13.043	12.387	11.011	8.834	6.761	5.056
47.18	12.113	12.843	10.872	6.030	2.379	1.648	1.778	2.260	3.102	4.254	5.684	7.292	9.059	10.641	12.148	13.124	13.122	12.535	11.235	9.118	7.057	5.315
50.22	11.908	12.622	10.677	5.907	2.331	1.620	1.752	2.223	3.045	4.169	5.566	7.133	8.866	10.411	11.907	12.896	12.950	12.426	11.201	9.168	7.154	5.409
55.2	11.655	12.358	10.441	5.762	2.272	1.586	1.716	2.173	2.969	4.056	5.411	6.927	8.616	10.114	11.590	12.583	12.690	12.228	11.091	9.155	7.206	5.482
60.06	12.194	12.913	10.892	5.982	2.341	1.630	1.768	2.242	3.067	4.191	5.601	7.168	8.928	10.486	12.041	13.117	13.290	12.876	11.747	9.783	7.764	5.935
64.59	12.662	13.389	11.277	6.179	2.417	1.687	1.833	2.319	3.164	4.315	5.762	7.365	9.179	10.775	12.385	13.528	13.762	13.391	12.289	10.319	8.251	6.333
70.34	12.372	13.083	11.008	6.014	2.353	1.648	1.794	2.265	3.078	4.190	5.592	7.136	8.899	10.445	12.035	13.185	13.481	13.192	12.193	10.331	8.331	6.425
75.41	12.389	13.095	11.007	5.995	2.332	1.634	1.781	2.250	3.059	4.165	5.562	7.097	8.862	10.400	12.001	13.176	13.524	13.276	12.327	10.508	8.524	6.599
81.09	11.939	12.629	10.611	5.767	2.239	1.570	1.715	2.169	2.946	4.012	5.355	6.830	8.532	10.011	11.566	12.714	13.081	12.859	11.968	10.233	8.316	6.442
86.49	12.120	12.809	10.758	5.839	2.269	1.595	1.742	2.201	2.985	4.059	5.418	6.905	8.628	10.122	11.704	12.880	13.264	13.075	12.194	10.465	8.525	6.614
92.43	11.463	12.129	10.185	5.523	2.147	1.514	1.654	2.087	2.825	3.836	5.117	6.518	8.145	9.554	11.057	12.193	12.589	12.434	11.628	10.009	8.182	6.362
99.09	12.277	12.969	10.871	5.876	2.269	1.596	1.747	2.210	2.998	4.077	5.443	6.935	8.673	10.177	11.794	13.023	13.472	13.345	12.516	10.818	8.869	6.904
105.32	12.211	12.894	10.808	5.837	2.259	1.593	1.745	2.202	2.978	4.046	5.398	6.871	8.594	10.083	11.692	12.924	13.409	13.308	12.519	10.864	8.943	6.976
110.39	11.263	11.923	9.997	5.396	2.090	1.479	1.622	2.044	2.759	3.740	4.988	6.346	7.938	9.311	10.804	11.967	12.431	12.361	11.651	10.131	8.357	6.526
123.41	11.438	12.089	10.123	5.451	2.111	1.498	1.644	2.066	2.786	3.771	5.028	6.386	7.996	9.375	10.899	12.098	12.615	12.585	11.919	10.430	8.644	6.763
144.31	11.871	12.527	10.472	5.606	2.153	1.526	1.681	2.121	2.861	3.875	5.172	6.565	8.233	9.653	11.249	12.525	13.131	13.165	12.537	11.047	9.215	7.237

the 25×22 spectral calibration matrix and c a 25×1 vector. (ii) Determine the predicted concentration matrix $\hat{C} = X.\hat{S}'.(\hat{S}.\hat{S}')^{-1}$.

5. In the model of question 3, plot a graph of predicted against true concentration for compound A. Determine the root mean square error both in mM and as a percentage of the average for all three compounds. Comment.

6. Repeat questions 2 and 3, but instead of MLR use PLS1 (centred) for the prediction of the concentration of A retaining the first three PLS components. Note that to obtain a root mean square error it is best to divide by 21 rather than 25 if three components are retained. You are not asked to cross-validate the models. Why are the predictions much better?

7. Use the 25×22 calibration set as a training set, obtain a PLS1 (centred) model for all three compounds retaining three components in each case, and centring the spectroscopic data. Use this model to predict the concentrations of compounds A–C in the 30 reaction spectra. Plot a graph of estimated concentrations of each compound against time.

Problem 5.7 PLS1 Algorithm

Section 5.5.1   Section A.2.2

The PLS1 algorithm is fairly simple and described in detail in Appendix A.2.2. However, it can be easily set up in Excel or programmed into Matlab in a few lines, and the aim of this problem is to set up the matrix based calculations for PLS.

The following is a description of the steps you are required to perform.

(a) Centre both the x and c blocks by subtracting the column means.
(b) Calculate the scores of the first PLS component by

$$h = X'.c$$

and then

$$t = \frac{X.h}{\sqrt{\sum h^2}}$$

(c) Calculate the x loadings of the first PLS component by

$$p = t'.X \Big/ \sum t^2$$

Note that the denominator is simply the sum of squares of the scores.

(d) Calculate the c loadings (a scalar in PLS1) by

$$q = c'.t \Big/ \sum t^2$$

(e) Calculate the contribution to the concentration estimate by $t.q$ and the contribution to the x estimate by $t.p$

(f) Subtract the contributions in step (e) from the current c vector and X matrix, and use these residuals for the calculation of the next PLS component by returning to step (b).

(g) To obtain the overall concentration estimate simply multiply $T.q$, where T is a scores matrix with A columns corresponding to the PLS components and q a column vector of size A. Add back the mean value of the concentrations to produce real estimates.

The method will be illustrated by a small simulated dataset, consisting of four samples, five measurements and one c parameter which is exactly characterised by three PLS components.

X					c
10.1	6.6	8.9	8.2	3.8	0.5
12.6	6.3	7.1	10.9	5.3	0.2
11.3	6.7	10.0	9.3	2.9	0.5
15.1	8.7	7.8	12.9	9.3	0.3

1. Calculate the loadings and scores of the first three PLS components, laying out the calculations in full.
2. What are the residual sum of squares for the 'x' and 'c' blocks as each successive component is computed (hint: start from the centred data matrix and simply sum the squares of each block, repeat for the residuals)? What percentage of the overall variance is accounted for by each component?
3. How many components are needed to describe the data exactly? Why does this answer not say much about the underlying structure of the data?
4. Provide a table of true concentrations, and of predicted concentrations as one, two and three PLS components are calculated.
5. If only two PLS components are used, what is the root mean square error of prediction of concentrations over all four samples? Remember to divide by 1 and not 4 (why is this?).

Problem 5.8 Cross-validation in PLS

Section 5.5.1   Section 5.6.2   Section A.2.2

The following consists of 10 samples, whose spectra are recorded at six wavelengths. The concentration of a component in the samples is given by a c vector. This dataset has been simulated to give an exact fit for two components as an example of how cross-validation works.

Sample	Spectra						c
1	0.10	0.22	0.20	0.06	0.29	0.10	1
2	0.20	0.60	0.40	0.20	0.75	0.30	5
3	0.12	0.68	0.24	0.28	0.79	0.38	9
4	0.27	0.61	0.54	0.17	0.80	0.28	3
5	0.33	0.87	0.66	0.27	1.11	0.42	6
6	0.14	0.66	0.28	0.26	0.78	0.36	8
7	0.14	0.34	0.28	0.10	0.44	0.16	2
8	0.25	0.79	0.50	0.27	0.98	0.40	7
9	0.10	0.22	0.20	0.06	0.29	0.10	1
10	0.19	0.53	0.38	0.17	0.67	0.26	4

1. Select samples 1–9, and calculate their means. Mean centre both the x and c variables over these samples.

2. Perform PLS1, calculate two components, on the first nine samples, centred as in question 1. Calculate t, p, h and the contribution to the c values for each PLS component (given by $q.t$), and verify that the samples can be exactly modelled using two PLS components (note that you will have to add on the mean of samples 1–9 to c after prediction). You should use the algorithm of Problem 5.7 or Appendix A.2.2 and you will need to find the vector h to answer question 3.

3. Cross-validation is to be performed on sample 10, using the model of samples 1–9 as follows.
    (a) Subtract the means of samples 1–9 from sample 10 to produce a new x vector, and similarly for the c value.
    (b) Then calculate the predicted score for the first PLS component and sample 10 by $\hat{t}_{10,1} = x_{10}.h_1/\sqrt{\Sigma h_1^2}$, where h_1 has been calculated above on samples 1–9 for the first PLS component, and calculate the new residual spectral vector $x_{10} - \hat{t}_{10,1}.p_1$.
    (c) Calculate the contribution to the mean centred concentration for sample 10 as $\hat{t}_{10,1}.q_1$, where q_1 is the value of q for the first PLS component using samples 1–9, and calculate the residual concentration $c_{10} - \hat{t}_{10,1}q_1$.
    (d) Find $\hat{t}_{10,2}$ for the second component using the residual vectors above using the vector h determined for the second component using the prediction set of nine samples.
    (e) Calculate the contribution to predicting c and x from the second component.

4. Demonstrate that, for this particular set, cross-validation results in an exact prediction of concentration for sample 10; remember to add the mean of samples 1–9 back after prediction.

5. Unlike for PCA, it is not possible to determine the predicted scores by $x.p'$ but it is necessary to use a vector h. Why is this?

Problem 5.9 Multivariate Calibration in Three-way Diode Array HPLC

The aim of this problem is to perform a variety of methods of calibration on a three-way dataset. Ten chromatograms are recorded of 3-hydroxypyridine impurity within a main peak of 2-hydroxypyridine. The aim is to employ PLS to determine the concentration of the minor component.

For each concentration a 20×10 chromatogram is presented, taken over 20 s in time (1 s digital resolution), and in this dataset, for simplicity, absorbances every 12 nm starting at 230 nm are presented.

Five concentrations are used, replicated twice. The 10 concentrations (mM) in the following table are presented in the arrangement on the following pages.

0.0158	0.0158
0.0315	0.0315
0.0473	0.0473
0.0631	0.0631
0.0789	0.0789

1. One approach to calibration is to use one-way PLS. This can be in either the spectroscopic or time direction. In fact, the spectroscopic dimension is often more

Upper block

1	2	3	4	5	6	7	8	9	10	11	12	13	14	15	16	17	18	19	20
0.089	0.011	0.020	0.048	0.097	0.132	0.116	0.055	0.005	0.000	0.087	0.011	0.020	0.048	0.096	0.130	0.114	0.054	0.005	0.000
0.150	0.015	0.032	0.080	0.161	0.223	0.198	0.092	0.008	0.001	0.148	0.015	0.031	0.079	0.159	0.220	0.196	0.090	0.008	0.001
0.224	0.020	0.046	0.118	0.239	0.334	0.299	0.137	0.011	0.001	0.221	0.019	0.045	0.117	0.236	0.329	0.295	0.135	0.010	0.001
0.300	0.024	0.060	0.158	0.320	0.449	0.402	0.184	0.013	0.001	0.296	0.024	0.059	0.156	0.315	0.442	0.396	0.181	0.013	0.001
0.366	0.028	0.073	0.192	0.389	0.548	0.492	0.224	0.016	0.001	0.360	0.027	0.071	0.189	0.383	0.539	0.484	0.220	0.015	0.001
0.412	0.030	0.081	0.216	0.438	0.618	0.555	0.252	0.017	0.001	0.405	0.030	0.080	0.213	0.430	0.607	0.546	0.247	0.017	0.001
0.436	0.032	0.085	0.229	0.463	0.654	0.588	0.267	0.018	0.001	0.428	0.031	0.084	0.225	0.455	0.642	0.578	0.262	0.017	0.001
0.439	0.031	0.086	0.231	0.467	0.659	0.594	0.269	0.018	0.001	0.431	0.031	0.084	0.226	0.458	0.647	0.582	0.264	0.017	0.001
0.428	0.030	0.083	0.224	0.453	0.641	0.577	0.261	0.017	0.001	0.419	0.030	0.082	0.220	0.445	0.629	0.566	0.256	0.017	0.001
0.405	0.028	0.079	0.212	0.429	0.607	0.547	0.247	0.016	0.001	0.397	0.028	0.077	0.208	0.421	0.596	0.536	0.243	0.016	0.001
0.377	0.026	0.073	0.197	0.398	0.564	0.508	0.230	0.015	0.001	0.370	0.026	0.072	0.193	0.391	0.554	0.499	0.225	0.015	0.001
0.346	0.024	0.067	0.180	0.365	0.517	0.466	0.211	0.014	0.001	0.340	0.023	0.066	0.177	0.359	0.508	0.458	0.207	0.013	0.001
0.315	0.022	0.061	0.164	0.332	0.470	0.423	0.191	0.013	0.001	0.309	0.021	0.060	0.161	0.326	0.462	0.416	0.188	0.012	0.001
0.285	0.020	0.055	0.148	0.300	0.424	0.382	0.173	0.011	0.000	0.280	0.019	0.054	0.146	0.295	0.418	0.376	0.170	0.011	0.001
0.257	0.018	0.049	0.133	0.270	0.382	0.344	0.156	0.010	0.000	0.253	0.017	0.049	0.131	0.266	0.377	0.339	0.153	0.010	0.000
0.231	0.016	0.044	0.120	0.243	0.344	0.310	0.140	0.009	0.000	0.228	0.016	0.044	0.118	0.240	0.339	0.306	0.138	0.009	0.000
0.208	0.014	0.040	0.108	0.219	0.309	0.279	0.126	0.008	0.000	0.206	0.014	0.039	0.107	0.216	0.306	0.275	0.124	0.008	0.000
0.188	0.013	0.036	0.097	0.197	0.279	0.251	0.113	0.007	0.000	0.186	0.013	0.036	0.096	0.195	0.276	0.248	0.112	0.007	0.000
0.170	0.012	0.033	0.088	0.178	0.252	0.227	0.102	0.007	0.000	0.168	0.012	0.032	0.087	0.176	0.249	0.224	0.101	0.006	0.000
0.154	0.010	0.029	0.080	0.161	0.228	0.205	0.093	0.006	0.000	0.152	0.010	0.029	0.079	0.159	0.226	0.203	0.092	0.006	0.000

Lower block

1	2	3	4	5	6	7	8	9	10	11	12	13	14	15	16	17	18	19	20
0.055	0.011	0.015	0.032	0.063	0.081	0.069	0.035	0.005	0.001	0.067	0.013	0.018	0.038	0.076	0.099	0.085	0.042	0.006	0.001
0.100	0.014	0.024	0.056	0.111	0.149	0.130	0.062	0.007	0.001	0.120	0.016	0.028	0.066	0.132	0.178	0.156	0.074	0.008	0.001
0.163	0.019	0.036	0.088	0.177	0.243	0.215	0.101	0.010	0.001	0.191	0.021	0.042	0.103	0.207	0.284	0.252	0.118	0.011	0.001
0.237	0.023	0.050	0.126	0.255	0.353	0.315	0.145	0.012	0.001	0.271	0.026	0.057	0.144	0.291	0.404	0.360	0.166	0.014	0.001
0.309	0.027	0.063	0.164	0.331	0.462	0.413	0.190	0.015	0.001	0.347	0.030	0.071	0.183	0.370	0.518	0.464	0.213	0.016	0.001
0.369	0.030	0.074	0.195	0.394	0.552	0.495	0.226	0.017	0.001	0.406	0.033	0.082	0.214	0.433	0.608	0.545	0.249	0.018	0.001
0.409	0.032	0.081	0.215	0.435	0.612	0.550	0.250	0.018	0.001	0.443	0.034	0.088	0.233	0.472	0.664	0.596	0.272	0.019	0.001
0.426	0.032	0.084	0.224	0.454	0.639	0.575	0.261	0.018	0.001	0.457	0.035	0.091	0.240	0.486	0.685	0.616	0.280	0.020	0.001
0.426	0.032	0.084	0.224	0.452	0.638	0.574	0.261	0.018	0.001	0.452	0.034	0.089	0.237	0.480	0.677	0.609	0.277	0.019	0.001
0.411	0.030	0.081	0.216	0.436	0.615	0.554	0.251	0.017	0.001	0.433	0.032	0.085	0.227	0.460	0.649	0.584	0.265	0.018	0.001
0.387	0.028	0.076	0.203	0.410	0.579	0.522	0.237	0.016	0.001	0.406	0.029	0.079	0.213	0.430	0.607	0.547	0.248	0.017	0.001
0.359	0.026	0.070	0.188	0.379	0.536	0.483	0.219	0.015	0.001	0.375	0.027	0.073	0.196	0.396	0.560	0.504	0.229	0.015	0.001
0.328	0.023	0.064	0.171	0.347	0.490	0.441	0.200	0.013	0.001	0.342	0.024	0.066	0.178	0.361	0.510	0.459	0.208	0.014	0.001
0.298	0.021	0.058	0.155	0.314	0.444	0.400	0.181	0.012	0.001	0.310	0.022	0.060	0.161	0.326	0.461	0.415	0.188	0.013	0.001
0.269	0.019	0.052	0.140	0.284	0.401	0.361	0.164	0.011	0.001	0.279	0.020	0.054	0.145	0.294	0.416	0.374	0.170	0.011	0.000
0.243	0.017	0.047	0.126	0.255	0.361	0.325	0.147	0.010	0.001	0.252	0.018	0.049	0.131	0.264	0.374	0.336	0.152	0.010	0.000
0.218	0.015	0.042	0.114	0.230	0.324	0.292	0.132	0.009	0.000	0.226	0.016	0.044	0.117	0.238	0.336	0.302	0.137	0.009	0.000
0.197	0.014	0.038	0.102	0.207	0.292	0.263	0.119	0.008	0.000	0.204	0.014	0.039	0.106	0.214	0.302	0.272	0.123	0.008	0.000
0.177	0.013	0.034	0.092	0.186	0.263	0.237	0.107	0.007	0.000	0.184	0.013	0.036	0.095	0.193	0.272	0.245	0.111	0.007	0.000
0.160	0.011	0.031	0.083	0.168	0.237	0.214	0.097	0.006	0.000	0.166	0.012	0.032	0.086	0.174	0.246	0.221	0.100	0.007	0.000

(continued overleaf)

0.074	0.016	0.021	0.044	0.086	0.110	0.093	0.047	0.008	0.001	0.080	0.017	0.023	0.047	0.093	0.119	0.102	0.051	0.008	0.001
0.130	0.020	0.032	0.073	0.145	0.193	0.168	0.081	0.010	0.001	0.137	0.021	0.033	0.077	0.153	0.204	0.178	0.086	0.010	0.001
0.203	0.025	0.046	0.111	0.222	0.302	0.267	0.126	0.013	0.001	0.210	0.025	0.047	0.114	0.229	0.313	0.276	0.130	0.013	0.001
0.283	0.029	0.061	0.152	0.306	0.422	0.376	0.175	0.016	0.002	0.288	0.029	0.061	0.155	0.311	0.430	0.383	0.177	0.015	0.001
0.358	0.033	0.074	0.190	0.384	0.534	0.477	0.220	0.018	0.002	0.358	0.033	0.074	0.191	0.385	0.536	0.479	0.220	0.019	0.001
0.414	0.036	0.085	0.220	0.443	0.619	0.555	0.254	0.020	0.002	0.411	0.035	0.084	0.218	0.440	0.616	0.552	0.252	0.020	0.001
0.447	0.037	0.090	0.237	0.478	0.670	0.601	0.275	0.020	0.001	0.441	0.036	0.089	0.233	0.471	0.662	0.593	0.271	0.020	0.001
0.457	0.036	0.092	0.242	0.488	0.686	0.616	0.281	0.020	0.001	0.449	0.035	0.090	0.237	0.480	0.674	0.605	0.275	0.019	0.001
0.450	0.035	0.090	0.237	0.479	0.674	0.606	0.276	0.018	0.001	0.440	0.034	0.088	0.232	0.469	0.661	0.594	0.270	0.018	0.001
0.429	0.033	0.085	0.226	0.456	0.642	0.578	0.263	0.017	0.001	0.419	0.032	0.083	0.221	0.446	0.629	0.565	0.257	0.016	0.001
0.401	0.030	0.079	0.210	0.425	0.599	0.539	0.245	0.016	0.001	0.391	0.029	0.077	0.206	0.416	0.586	0.527	0.239	0.015	0.001
0.369	0.027	0.072	0.193	0.390	0.550	0.495	0.225	0.014	0.001	0.360	0.027	0.071	0.189	0.382	0.538	0.484	0.220	0.014	0.001
0.336	0.025	0.066	0.175	0.355	0.500	0.450	0.205	0.013	0.001	0.328	0.024	0.064	0.172	0.347	0.490	0.440	0.200	0.012	0.001
0.304	0.022	0.059	0.158	0.320	0.452	0.407	0.185	0.011	0.001	0.297	0.022	0.058	0.155	0.314	0.442	0.398	0.180	0.011	0.001
0.273	0.020	0.053	0.143	0.288	0.407	0.366	0.166	0.010	0.001	0.267	0.020	0.052	0.140	0.282	0.398	0.358	0.162	0.010	0.001
0.246	0.018	0.048	0.128	0.259	0.365	0.329	0.149	0.009	0.000	0.240	0.018	0.047	0.126	0.254	0.358	0.322	0.146	0.009	0.000
0.221	0.016	0.043	0.115	0.233	0.328	0.295	0.134	0.008	0.000	0.216	0.016	0.042	0.113	0.228	0.321	0.289	0.131	0.008	0.000
0.199	0.015	0.039	0.104	0.209	0.295	0.265	0.120	0.007	0.000	0.195	0.014	0.038	0.102	0.205	0.289	0.260	0.118	0.007	0.000
0.179	0.013	0.035	0.093	0.188	0.265	0.239	0.108	0.007	0.000	0.175	0.013	0.035	0.092	0.185	0.260	0.234	0.106	0.006	0.000
0.162	0.012	0.032	0.085	0.170	0.240	0.215	0.098	0.006	0.000	0.159	0.012	0.031	0.083	0.167	0.235	0.211	0.096	0.006	0.000

0.061	0.018	0.021	0.038	0.075	0.090	0.074	0.040	0.008	0.001	0.067	0.019	0.022	0.042	0.081	0.099	0.082	0.043	0.009	0.001
0.108	0.022	0.030	0.063	0.125	0.161	0.137	0.068	0.010	0.001	0.118	0.023	0.032	0.068	0.135	0.175	0.151	0.075	0.011	0.002
0.173	0.026	0.042	0.097	0.193	0.258	0.225	0.108	0.013	0.002	0.188	0.027	0.045	0.104	0.208	0.278	0.244	0.117	0.014	0.002
0.249	0.030	0.056	0.136	0.273	0.371	0.328	0.154	0.015	0.002	0.266	0.031	0.059	0.144	0.290	0.396	0.351	0.164	0.016	0.002
0.324	0.034	0.070	0.174	0.351	0.484	0.430	0.200	0.018	0.002	0.341	0.034	0.073	0.183	0.368	0.508	0.453	0.210	0.018	0.002
0.385	0.036	0.081	0.206	0.415	0.576	0.515	0.237	0.019	0.002	0.399	0.037	0.083	0.213	0.428	0.596	0.533	0.245	0.020	0.002
0.425	0.037	0.087	0.226	0.456	0.637	0.570	0.261	0.020	0.002	0.435	0.038	0.089	0.231	0.465	0.651	0.583	0.267	0.021	0.002
0.443	0.037	0.090	0.235	0.474	0.664	0.595	0.272	0.020	0.002	0.448	0.037	0.091	0.237	0.478	0.670	0.601	0.275	0.021	0.002
0.441	0.036	0.089	0.233	0.471	0.661	0.594	0.271	0.020	0.002	0.442	0.036	0.089	0.233	0.471	0.662	0.594	0.271	0.020	0.002
0.425	0.034	0.085	0.225	0.453	0.637	0.572	0.261	0.019	0.001	0.423	0.033	0.085	0.223	0.450	0.633	0.569	0.259	0.019	0.001
0.400	0.031	0.080	0.211	0.426	0.599	0.538	0.245	0.017	0.001	0.396	0.031	0.079	0.208	0.421	0.592	0.532	0.242	0.017	0.001
0.370	0.029	0.073	0.194	0.393	0.553	0.497	0.226	0.016	0.001	0.365	0.028	0.072	0.191	0.387	0.544	0.489	0.223	0.016	0.001
0.338	0.026	0.067	0.177	0.358	0.504	0.453	0.206	0.015	0.001	0.332	0.025	0.066	0.174	0.352	0.495	0.445	0.203	0.014	0.001

0.306	0.023	0.060	0.160	0.324	0.457	0.410	0.187	0.013	0.001
0.276	0.021	0.054	0.145	0.292	0.411	0.370	0.168	0.012	0.001
0.249	0.019	0.049	0.130	0.262	0.370	0.332	0.151	0.011	0.001
0.223	0.017	0.044	0.117	0.236	0.332	0.298	0.135	0.009	0.000
0.201	0.015	0.040	0.105	0.212	0.298	0.268	0.122	0.008	0.000
0.181	0.014	0.036	0.095	0.191	0.268	0.241	0.109	0.008	0.000
0.163	0.013	0.033	0.086	0.172	0.242	0.217	0.099	0.007	0.000
0.082	0.023	0.027	0.051	0.100	0.122	0.102	0.054	0.011	0.002
0.140	0.028	0.038	0.081	0.160	0.207	0.178	0.088	0.013	0.002
0.213	0.032	0.052	0.119	0.237	0.317	0.277	0.133	0.016	0.002
0.293	0.036	0.066	0.160	0.320	0.436	0.386	0.182	0.019	0.002
0.366	0.039	0.079	0.197	0.396	0.546	0.485	0.226	0.021	0.002
0.420	0.040	0.089	0.225	0.452	0.628	0.561	0.259	0.022	0.002
0.451	0.041	0.093	0.240	0.484	0.675	0.604	0.278	0.022	0.002
0.459	0.040	0.094	0.244	0.492	0.688	0.617	0.283	0.022	0.002
0.450	0.038	0.091	0.238	0.481	0.674	0.604	0.276	0.021	0.002
0.428	0.035	0.086	0.226	0.457	0.640	0.575	0.263	0.020	0.002
0.399	0.032	0.080	0.210	0.425	0.596	0.535	0.244	0.018	0.001
0.366	0.029	0.073	0.193	0.389	0.547	0.491	0.224	0.016	0.001
0.333	0.026	0.066	0.175	0.353	0.496	0.446	0.203	0.015	0.001
0.301	0.024	0.060	0.158	0.319	0.448	0.402	0.183	0.013	0.001
0.271	0.021	0.054	0.142	0.286	0.402	0.362	0.165	0.012	0.001
0.243	0.019	0.048	0.128	0.257	0.361	0.324	0.148	0.011	0.001
0.218	0.017	0.044	0.115	0.231	0.324	0.291	0.132	0.010	0.001
0.196	0.016	0.040	0.103	0.207	0.291	0.261	0.119	0.009	0.001
0.177	0.014	0.036	0.093	0.187	0.262	0.235	0.107	0.008	0.000
0.160	0.013	0.033	0.084	0.169	0.236	0.212	0.096	0.007	0.000

0.301	0.023	0.059	0.157	0.318	0.448	0.403	0.183	0.013	0.001
0.271	0.021	0.053	0.142	0.286	0.403	0.362	0.165	0.012	0.001
0.244	0.018	0.048	0.127	0.257	0.362	0.325	0.148	0.010	0.001
0.219	0.017	0.043	0.114	0.231	0.325	0.292	0.133	0.009	0.000
0.197	0.015	0.039	0.103	0.207	0.292	0.262	0.119	0.008	0.000
0.177	0.014	0.035	0.093	0.187	0.263	0.236	0.107	0.007	0.000
0.160	0.012	0.032	0.084	0.169	0.237	0.213	0.097	0.007	0.000
0.081	0.023	0.027	0.051	0.098	0.120	0.100	0.053	0.011	0.002
0.137	0.027	0.038	0.080	0.158	0.204	0.175	0.087	0.013	0.002
0.210	0.032	0.051	0.118	0.234	0.313	0.273	0.131	0.016	0.002
0.290	0.035	0.066	0.158	0.317	0.432	0.382	0.180	0.018	0.002
0.363	0.038	0.079	0.196	0.393	0.542	0.482	0.224	0.020	0.002
0.418	0.040	0.088	0.224	0.451	0.626	0.558	0.258	0.022	0.002
0.450	0.040	0.093	0.240	0.483	0.674	0.603	0.277	0.022	0.002
0.459	0.039	0.094	0.244	0.492	0.688	0.617	0.282	0.022	0.002
0.450	0.038	0.091	0.239	0.482	0.675	0.605	0.277	0.021	0.002
0.429	0.035	0.086	0.227	0.458	0.642	0.576	0.263	0.019	0.002
0.400	0.032	0.080	0.211	0.426	0.598	0.537	0.245	0.018	0.001
0.367	0.029	0.073	0.193	0.390	0.548	0.493	0.224	0.016	0.001
0.334	0.026	0.066	0.175	0.354	0.498	0.447	0.204	0.015	0.001
0.301	0.024	0.060	0.158	0.319	0.449	0.403	0.184	0.013	0.001
0.271	0.021	0.054	0.142	0.287	0.403	0.362	0.165	0.012	0.001
0.243	0.019	0.048	0.128	0.257	0.362	0.325	0.148	0.010	0.001
0.219	0.017	0.044	0.115	0.231	0.324	0.291	0.133	0.009	0.000
0.196	0.016	0.039	0.103	0.208	0.291	0.261	0.119	0.008	0.000
0.177	0.014	0.036	0.093	0.187	0.262	0.235	0.107	0.007	0.000
0.159	0.013	0.032	0.084	0.169	0.236	0.212	0.096	0.007	0.000

useful. Produce a 10×10 table of summed intensities over the 20 chromatographic points in time at each wavelength for each sample.

2. Standardise the data, and perform autopredictive PLS1, calculating three PLS components. Why is it useful to standardise the measurements?

3. Plot graphs of predicted versus known concentrations for one, two and three PLS components, and calculate the root mean square errors in mM.

4. Perform PLS1 cross-validation on the c values for the first eight components and plot a graph of cross-validated error against component number.

5. Unfold the original datamatrix to give a 10×200 data matrix.

6. It is desired to perform PLS calibration on this dataset, but first to standardise the data. Explain why there may be problems with this approach. Why is it desirable to reduce the number of variables from 200, and why was this variable selection less important in the PLS1 calculations?

7. Why is the standard deviation a good measure of variable significance? Reduce the dataset to 100 significant variables with the highest standard deviations to give a 10×100 data matrix.

8. Perform autopredictive PLS1 on the standardised reduced unfolded data of question 7 and calculate the errors as one, two and three components are computed.

9. How might you improve the model of question 8 still further?

6 Evolutionary Signals

6.1 Introduction

Some of the classical applications of chemometrics are to evolutionary data. Such a type of information is increasingly common, and normally involves simultaneously recording spectra whilst a physical parameter such as time or pH is changed, and signals evolve during the change of this parameter.

In the modern laboratory, one of the most widespread applications is in the area of coupled chromatography, such as HPLC–DAD (high-performance liquid chromatography–diode array detector), LC–MS (liquid chromatography–mass spectrometry) and LC–NMR (liquid chromatography–nuclear magnetic resonance). A chromatogram is recorded whilst a UV/vis, mass or NMR spectrum is recorded. The information can be presented in matrix form, with time along the rows and wavelength, mass number or frequency along the columns, as already introduced in Chapter 4. Multivariate approaches can be employed to analyse these data. However, there are a wide variety of other applications, ranging from pH titrations to processes that change in a systematic way with time to spectroscopy of mixtures. Many of the approaches in this chapter have wide applicability, for example, baseline correction, data scaling and 3D PC plots, but for brevity we illustrate the chapter primarily with case studies from coupled chromatography, as this has been the source of a huge literature over the past two decades.

With modern laboratory computers it is possible to obtain huge quantities of information very rapidly. For example, spectra sampled at 1 nm intervals over a 200 nm region can be obtained every second using modern chromatography, hence in an hour 3600 spectra in time × 200 spectral frequencies or 720 000 pieces of information can be produced from a single chromatogram. A typical medium to large industrial site may contain 100 or more coupled chromatographs, meaning the potential of acquiring 72 million data-points per hour of this type of information. Add on all the other instruments, and it is not difficult to see how billions of numbers can be generated on a daily basis.

In Chapters 4 and 5, we discussed a number of methods for multivariate data analysis, but the methods described did not take into account the sequential nature of the information. When performing PCA on a data matrix, the order of the rows and columns is irrelevant. Figure 6.1 represents three cross-sections through a data matrix. The first could correspond to a chromatographic peak, the others not. However, since PCA and most other classical methods for pattern recognition would not distinguish these sequences, clearly other approaches are useful.

In many cases, underlying factors corresponding to individual compounds in a mixture are unimodal in time, that this, they have one maximum. The aim is to deconvolute the experimentally observed sum into individual components and determine the features of each component. The change in spectral characteristics across the chromatographic peak can be used to provide this information.

Figure 6.1
Three possible sequential patterns that would be treated identically using standard multivariate techniques

To the practising chemist, there are three main questions that can be answered by applied chemometric techniques to coupled chromatography, of increasing difficulty.

1. *How many peaks in a cluster?* Can we detect small impurities? Can we detect metabolites against a background? Can we determine whether there are embedded peaks?
2. *What are the characteristics of each pure compound?* What are the spectra? Can we obtain mass spectra and NMR spectra of embedded chromatographic peaks at low levels of sufficient quality that we can be confident of their identities?

3. *What are the quantities of each component?* Can we quantitate small impurities? Could we use chromatography of mixtures for reaction monitoring and kinetics? Can we say with certainty the level of a dopant or a potential environmental pollutant when it is detected in low concentrations buried within a major peak?

There are a large number of 'named' methods in the literature, but they are based mainly around certain main principles of evolutionary factor analysis, whereby factors corresponding to individual compounds evolve in time (or any other sequential parameter such as pH).

Such methods are applicable not only to coupled chromatography but also in areas such as pH dependence of equilibria, whereby the spectra of a mixture of chemical species can be followed with change of pH. It would be possible to record 20 spectra and then treat each independently. Sometimes this can lead to good quantification, but including the information that each component will be unimodal or monotonic over the course of a pH titration results in further insight. Another important application is in industrial process control where concentrations of compounds or levels of various factors may have a specific evolution over time.

Below we will illustrate the main methods of resolution of two-way data, primarily as applied to HPLC–DAD, but also comment on the specific enhancements required for other instrumental techniques and applications. Some techniques have already been introduced in Chapters 4 and 5, but we elaborate on them in this chapter.

A few of the methods discussed in this chapter, such as 3D PC plots and variable selection, have significant roles in most applications of chemometrics, so the interest in the techniques is by no means restricted to chromatographic applications, but in order to reduce excessive repetition the methods are introduced in one main context.

6.2 Exploratory Data Analysis And Preprocessing

6.2.1 Baseline Correction

A preliminary first step before applying most methods in this chapter is often baseline correction, especially when using older instruments. The reason for this is that most chemometric approaches look at variation above a baseline, so if baseline correction is not done artefacts can be introduced.

Baseline correction is best performed on each variable (such as mass or wavelength) independently. There are many ways of doing this, but it is first important to identify regions of baseline and of peaks, as in Figure 6.2 which is for an LC–MS dataset. Note that the right-hand side of this tailing peak is not used: we take only regions that we are confident in. Then normally a function is fitted to the baseline regions. This can simply involve the average or else a linear or polynomial best fit. Sometimes the baseline both before and after a peak cluster is useful, but if the cluster is fairly sharp, this is not essential, and in the case illustrated it would be hard. Sometimes the baseline is calculated over the entire chromatogram, in other cases separately for each pack cluster. After that, we obtain a simple mathematical model, and then subtract the baseline from the entire region of interest, separately for each variable. In the examples in this chapter it is assumed that either there are no baseline problems or that correction has already been performed.

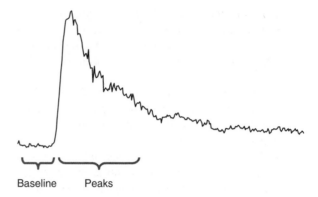

Figure 6.2
Dividing data into regions prior to baseline correction

6.2.2 Principal Component Based Plots

Scores and loadings plots have been introduced in Chapter 4 (Section 4.3.5). In this chapter we will explore some further properties, especially useful where one or both of the variables are related in sequence. Table 6.1 represents a two-way dataset, corresponding to HPLC–DAD, each elution time being represented by a row and each measurement (such as successive wavelengths) by a column, giving a 25×12 data matrix, which will be called dataset A. The data represent two partially overlapping chromatographic peaks. The profile (sum of intensity over the spectrum at each elution time) is presented in Figure 6.3.

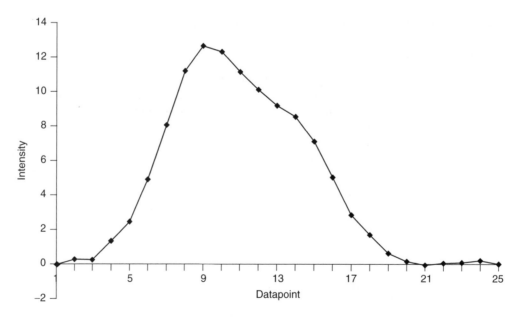

Figure 6.3
Profile of data in Table 6.1

Table 6.1 Dataset A.

	A	B	C	D	E	F	G	H	I	J	K	L
1	0.1102	−0.0694	−0.0886	0.0622	−0.0079	−0.0336	0.0518	−0.0459	−0.032	0.0645	0.0174	−0.0558
2	−0.0487	0.0001	0.0507	−0.0014	0.072	−0.0377	0.0123	0.1377	−0.0034	−0.0015	0.0355	0.0608
3	0.036	0.0277	0.1005	−0.0009	0.0386	−0.0528	−0.0612	−0.0259	0.0293	−0.0246	0.0283	0.048
4	0.2104	0.1564	0.1828	0.1073	0.0185	0.1912	0.0499	−0.0587	0.0669	0.1275	0.1371	0.1521
5	0.1713	0.3206	0.4304	0.3531	0.2383	0.1575	−0.0015	0.1367	0.1024	0.1143	0.2227	0.2164
6	0.497	0.6192	0.7367	0.7042	0.3234	0.293	0.1919	0.1325	0.341	0.4269	0.4225	0.2212
7	0.6753	1.1198	1.3239	1.0167	0.6054	0.3783	0.3703	0.3343	0.529	0.5496	0.5986	0.5579
8	1.0412	1.5129	1.6344	1.3823	1.0843	0.6825	0.3584	0.334	0.5212	0.8334	0.9435	0.8741
9	1.0946	1.5543	1.9253	1.5951	1.1767	0.7215	0.5764	0.5695	0.7138	0.8645	0.9545	0.9038
10	0.9955	1.4794	1.5299	1.5679	1.1986	0.793	0.7043	0.5333	0.7661	0.9224	0.9744	0.8434
11	0.672	1.1315	1.2793	1.254	1.0619	0.9552	0.907	0.7855	0.78	0.7912	0.8432	0.6739
12	0.469	0.7531	0.8139	1.0496	1.094	1.1321	1.1164	1.0237	0.7796	0.6313	0.6549	0.5869
13	0.3113	0.3894	0.5844	0.7349	0.9656	1.2339	1.3362	1.2283	0.959	0.5641	0.4393	0.4386
14	0.0891	0.2121	0.3344	0.5837	0.9758	1.3175	1.3713	1.2238	0.9459	0.6646	0.4327	0.3938
15	0.0567	0.1408	0.169	0.4609	0.7807	1.1592	1.3094	1.1237	0.7724	0.4457	0.3217	0.3639
16	0.0391	−0.0211	0.1684	0.3332	0.5427	0.8509	0.9616	0.7876	0.5951	0.3343	0.2212	0.2178
17	0.0895	−0.0086	0.079	0.1721	0.2747	0.4634	0.582	0.5677	0.3231	0.1546	0.0379	0.1021
18	0.007	−0.024	0.0842	0.1622	0.1922	0.2974	0.3571	0.2925	0.1289	0.0491	0.0518	0.0979
19	0.0146	−0.0567	−0.0672	−0.0239	−0.0113	0.2454	0.1721	0.1047	0.1577	0.0129	0.0458	0.0307
20	0.0012	−0.0043	−0.0362	−0.0564	0.0693	0.0468	0.0213	0.1182	0.0152	−0.0342	−0.014	0.0308
21	−0.0937	0.0324	0.0371	−0.0405	−0.0648	−0.0053	0.0218	0.0975	−0.0222	−0.0138	−0.0065	0.0017
22	−0.0031	0.0127	0.0323	−0.0533	0.067	0.0716	0.0479	−0.0383	0.0038	−0.0186	−0.0026	−0.0653
23	−0.0387	−0.0041	0.0175	0.0052	0.0199	−0.0507	0.0263	0.0342	0.0072	0.0242	0.0579	−0.0072
24	−0.0449	0.0076	−0.0191	0.0046	0.0572	0.0946	−0.0018	0.0182	−0.0368	−0.0236	0.0619	0.0853
25	−0.0986	0.0244	0.0185	0.0395	−0.0291	0.0236	−0.0137	−0.0263	0.0156	0.003	0.0237	0.027

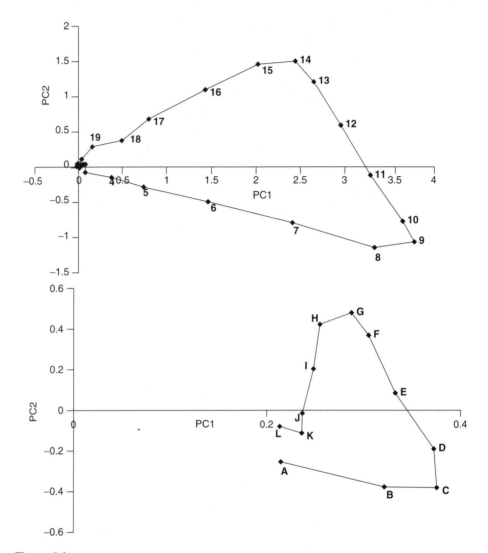

Figure 6.4
Scores and loadings plots of PC2 versus PC1 of the raw data in Table 6.1

The simplest plots are the scores and loadings plots of the first two PCs of the raw data (see Figure 6.4). These would suggest that there are two components, with a region of overlap between times 9 and 14, with wavelengths H and G most strongly associated with the slowest eluting compound and wavelengths A, B, C, L and K with the fastest eluting compound. For further discussion of the interpretation of these types of graph, see Section 4.3.5.

The dataset in Table 6.2 is of the same size but represents three partially overlapping peaks. The profile (Figure 6.5) appears to be slightly more complex than that for dataset A, and the PC scores plot presented in Figure 6.6 definitely appears to contain more features. Each turning point represents a pure compound, so it appears that there are three compounds, centred at times 9, 13 and 17. In addition, the spectral characteristics

Table 6.2 Dataset B.

	A	B	C	D	E	F	G	H	I	J	K	L
1	-0.1214	0.0097	-0.0059	0.0136	0.0399	0.0404	-0.0530	-0.0066	0.0078	-0.0257	0.0641	0.0784
2	0.0750	-0.0200	0.0183	-0.0251	-0.1072	0.0218	0.0074	0.0876	-0.0341	-0.0295	-0.0715	-0.0746
3	-0.0256	0.1103	-0.0246	-0.0229	-0.0347	0.0102	0.0365	-0.1114	-0.0079	-0.0271	-0.0654	-0.0437
4	0.0838	0.0486	-0.0155	-0.0142	0.0045	-0.0213	0.0396	-0.0499	0.0421	0.0396	0.1128	0.0614
5	0.1956	0.2059	0.1601	0.1567	0.1594	0.0886	0.0310	0.0657	0.0898	0.0472	0.1688	0.1335
6	0.4605	0.5753	0.5696	0.3477	0.1596	0.0725	0.1147	0.0382	0.1049	0.3195	0.3498	0.2801
7	0.9441	1.1101	0.9926	0.6515	0.3413	0.2480	0.1340	0.1802	0.3163	0.4346	0.6187	0.7223
8	1.3161	1.6053	1.3641	0.9179	0.5933	0.3124	0.2991	0.3650	0.6141	0.8600	0.8351	0.8289
9	1.5698	1.8485	1.5372	1.1441	0.6591	0.3880	0.4163	0.4519	0.7866	1.0155	1.0880	0.8584
10	1.3576	1.6975	1.5099	1.1872	0.7743	0.6088	0.6598	0.6486	0.7726	1.0264	0.9573	0.8324
11	1.0215	1.1341	1.0579	0.9695	0.8587	0.9031	0.9594	0.9789	0.9984	0.9810	0.7305	0.8146
12	0.5267	0.6154	0.7074	0.8352	0.9716	1.2316	1.3408	1.1761	0.9874	0.7335	0.6364	0.5917
13	0.3936	0.3650	0.5143	0.6098	1.0398	1.2065	1.3799	1.3346	0.9871	0.7738	0.7354	0.6865
14	0.4351	0.3077	0.3630	0.5386	0.8882	1.1305	1.2870	1.2373	0.9289	0.8656	0.8967	0.9762
15	0.7120	0.4754	0.2429	0.3367	0.5563	0.7994	0.9244	0.8403	0.9415	0.9667	1.2267	1.2778
16	1.0076	0.5493	0.3146	0.2574	0.2568	0.4061	0.5775	0.6494	0.8331	1.1937	1.5703	1.7106
17	1.2155	0.5669	0.3203	0.1255	0.1279	0.1431	0.2725	0.3421	0.6828	1.1376	1.6209	1.8048
18	1.1392	0.4750	0.1603	0.0936	0.1297	0.0917	0.1154	0.2421	0.5582	0.9945	1.4607	1.6280
19	0.6988	0.4000	0.1668	0.0071	-0.0290	0.0012	0.0062	0.1631	0.3016	0.6982	1.0083	1.1998
20	0.3291	0.1766	0.0825	0.0714	-0.0180	-0.0078	-0.0121	0.0176	0.2330	0.2982	0.4253	0.6602
21	0.2183	0.1892	0.0436	-0.0689	0.0239	0.0410	0.0367	0.0322	0.1284	0.2204	0.2760	0.1539
22	0.1135	0.0517	0.0418	-0.0307	0.0017	0.0023	0.1068	-0.0517	0.0411	0.0370	0.0782	0.0798
23	-0.0442	0.0156	0.0520	-0.0867	0.0020	-0.1177	0.0374	-0.0282	-0.0036	0.0217	-0.0521	0.0518
24	-0.0013	-0.1103	-0.0536	-0.0875	-0.0212	-0.0066	0.0199	-0.0258	-0.0119	-0.1066	0.0664	-0.0489
25	0.0697	0.0827	0.0093	0.0298	-0.0511	0.0637	-0.1094	0.0358	-0.0279	0.0319	0.0480	-0.0383

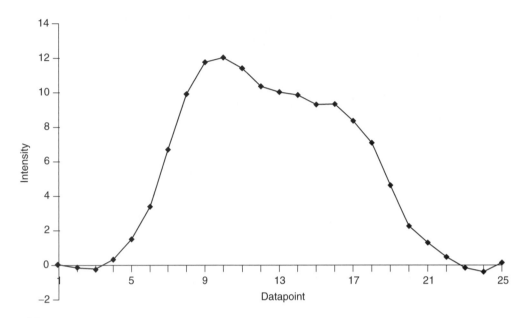

Figure 6.5
Profile of data in Table 6.2

of the compounds centred at times 9 and 17 are probably similar compared with that centred at time 13. Comparing the loadings plot suggests that wavelengths F, G and H are strongly associated with the middle eluting compound, whereas A, B, J, K and L are associated with the other two compounds. There is some distinction, in that wavelengths A, K and L appear most associated with the slowest eluting compound (centred at time 17) and B and J with the fastest. The loadings and scores could be combined into a biplot (Chapter 4, Section 4.3.7.1).

It is sometimes clearer to present these graphs in three dimensions as in Figures 6.7 and 6.8 adding a third PC. Note that the three-dimensional scores plot for dataset A is not particularly informative and the two-dimensional plot shows the main trends more clearly. The reason for this is that there are only two main components in the system, so the third dimension consists primarily of noise and thus degrades the information. If the three dimensions were scaled according to the size of the PCs (or the eigenvalues), the graphs in Figure 6.7 would be flat. However for dataset B, the directions are much clearer than in the two-dimensional projections, so adding an extra PC can be beneficial if there are more than two significant components.

A useful trick is to normalise the scores. This involves calculating

$$^{norm}t_{ia} = \frac{t_{ia}}{\sqrt{\sum_{a=1}^{A} t_{ia}^2}}$$

Note that there is often confusing and conflicting terminology in the literature, some authors called this summing to a constant total normalisation, but we will adopt only one convention in this book; however, if you read the original literature be very careful

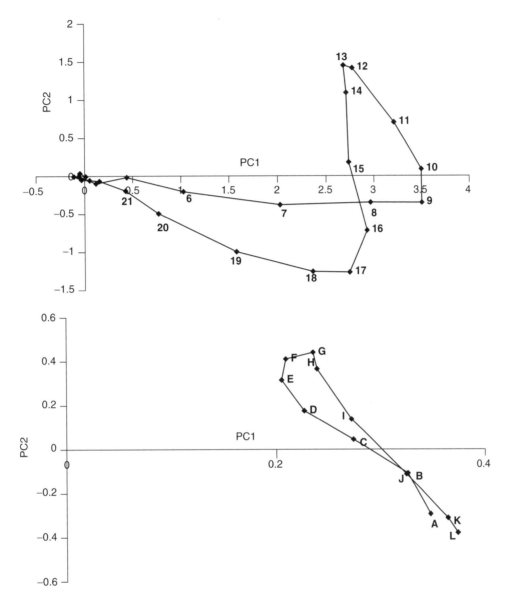

Figure 6.6
Scores and loadings plots of PC2 versus PC1 of the raw data in Table 6.1

about terminology. If only two PCs are used this will project the scores on to a circle, whereas if three PCs are used the projection will be on to a sphere. It is best to set *A* according to the number of compounds in the region of the chromatogram being studied.

Figure 6.9 illustrates the scores of dataset A normalised over two PCs. Between times 3 and 21, the points in the chromatogram are in sequence on the arc of a circle. The extremes (3 and 21) could represent the purest elution times, but points influenced primarily by noise might lie anywhere on the circle. Hence time 25, which is clearly

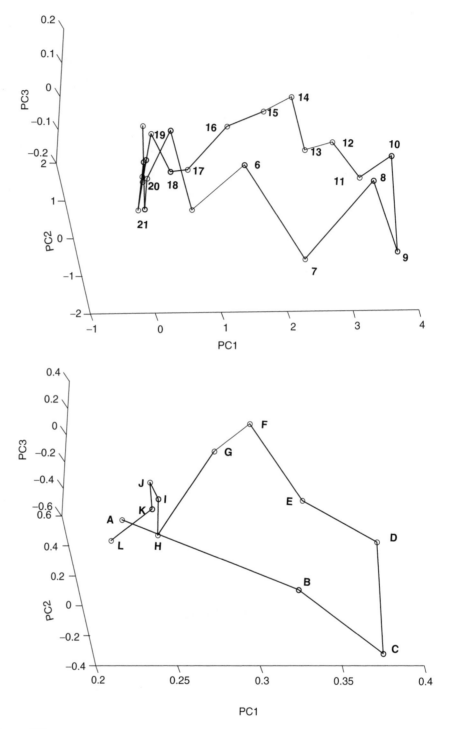

Figure 6.7
Three-dimensional projections of scores (top) and loadings (bottom) for dataset A

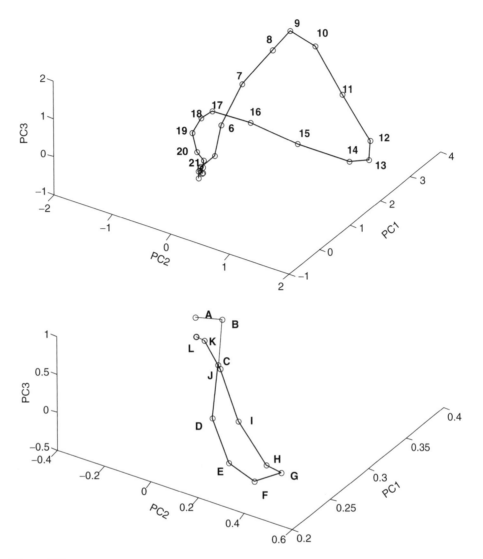

Figure 6.8
Three-dimensional projections of scores (top) and loadings (bottom) for dataset B

not representative of the fastest eluting component, is close to time 3 (this is entirely fortuitous and depends on the noise distribution). Because elution times 4–9 are closely clustered, they probably better represent the faster eluting compound. Note how points on a straight line (Figure 6.4) in the raw scores plot project on to clusters in the normalised scores plot.

The normalised scores of dataset B [Figure 6.10(a)] show a clearer pattern. The figure suggests the following:

- points 1–4 are mainly noise as they form a fairly random pattern;
- the purest points for the fastest eluting peak are 6 and 7, because these correspond to a turning point;

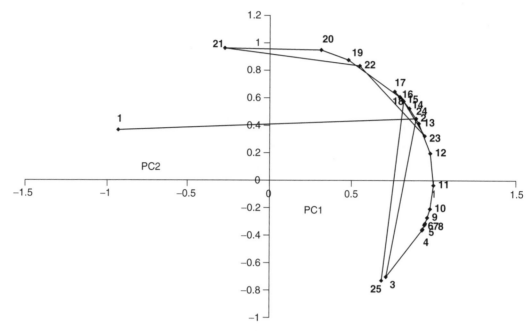

Figure 6.9
Scores of dataset A normalised over the first two principal components

- the purest points for the middle eluting peak are 12 and 13, again a turning point;
- the purest points for the slowest eluting peak are 18–20;
- points 23–25 are mainly dominated by noise.

It is probably best to remove the noise points 1–4 and 23–15, and show the normalised scores plot as in Figure 6.10(b). Notice that we come to a slightly different conclusion from Figure 6.6 as to which are the most representative elution times (or spectra) for each component. This is mainly because the ends of each limb in the raw scores plot correspond to the peak maxima, which are not necessarily the purest regions. For the fastest and slowest eluting components the purest regions will be at more extreme elution times before noise dominates: if the noise levels are low they may be at the base rather than top of the peak clusters. For the central peak the purest region is still at the same position, probably because this peak does not have a selective or pure region. The data could also be normalised over three dimensions with pure points falling on the surface of a sphere; the clustering becomes more obvious (see Figure 6.11). Note that similar calculations can be performed on the loadings plots and it is possible to normalise the loadings instead.

6.2.3 Scaling the Data

It is also possible to scale the raw data *prior* to performing PCA.

6.2.3.1 Scaling the Rows

Each successive row in a data matrix formed from a coupled chromatogram corresponds to a spectrum taken at a given elution time. One of the simplest methods of scaling

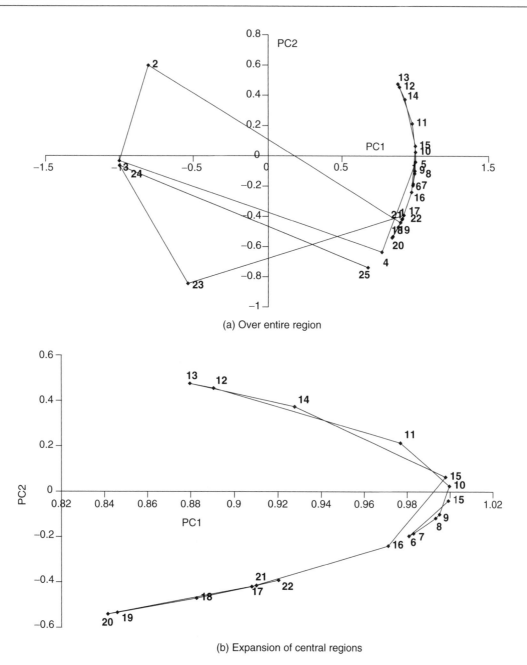

(a) Over entire region

(b) Expansion of central regions

Figure 6.10
Scores of dataset B normalised over the first two principal components (a) Over entire region (b) Expansion of central regions

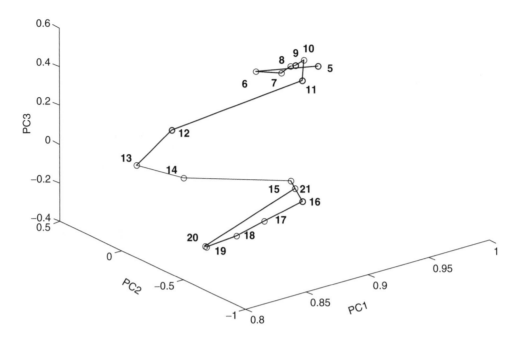

Figure 6.11
Scores corresponding to Figure 6.10(b) but normalised over three PCs and presented in three dimensions

involves summing each row to a constant total. Put mathematically:

$$^{rs}x_{ij} = \frac{x_{ij}}{\sum\limits_{j=1}^{J} x_{ij}}$$

Note that some people call this normalisation, but we will avoid that terminology, as this method is distinct from that in Section 6.2.2. The influence on PC scores plots has already been introduced (Chapter 4, Section 4.3.6.2) but will be examined in more detail in this chapter.

Figure 6.12(a) shows what happens if the rows of dataset A are first scaled to a constant total and then PCA performed on this data. At first glance this appears rather discouraging, but that is because the noise points have a disproportionate influence. These points contain largely nonsensical data, which is emphasised when scaling each point in time to the same total. An expansion of points 5–19 is slightly more encouraging [Figure 6.12(b)], but still not very good. Performing PCA only on points 5–19 (after scaling the rows as described above), however, provides a very clear picture of what is happening; all the points fall roughly on a straight line, with the purest points at the end [Figure 6.12(c)]. Unlike normalising the scores after PCA (Section 6.2.2), where the data must fall exactly on a geometric figure such as a circle or sphere (dependent on the number of PCs chosen), the straight line is only approximate and depends on there being two components in the region of the data that have been chosen.

The corresponding scores plot for the first two PCs of dataset B, using points 5–20, is presented in Figure 6.13(a). There are now two linear regions, one between compounds

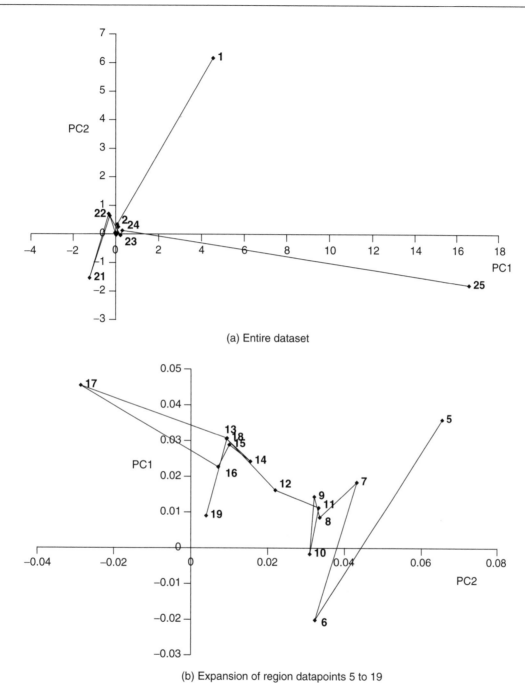

(a) Entire dataset

(b) Expansion of region datapoints 5 to 19

Figure 6.12
Scores plot of dataset A, each row summed to a constant total, PC2 versus PC1

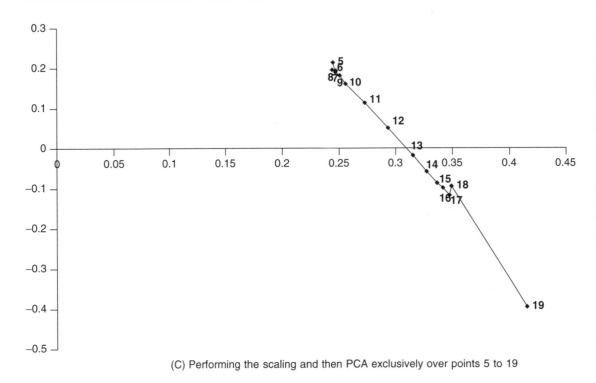

(C) Performing the scaling and then PCA exclusively over points 5 to 19

Figure 6.12
(*continued*)

A (fastest) and B, and another between compounds B and C (slowest). Some important features are of interest. The first is that there are now three main directions in the graph, but the direction due to B is unlikely to represent the pure compound, and probably the line would need to be extended further along the top right-hand corner. However, it looks likely that there is only a small or negligible region where the three components co-elute, otherwise the graph could not easily be characterised by two straight lines. The trends are clearer in three dimensions [Figure 6.13(b)]. Note that the point at time 5 is probably influenced by noise.

Summing each row to a constant total is not the only method of dealing with individual rows or spectra. Two variations below can be employed.

1. *Selective summation to constant total.* This allows each portion of a row to be scaled to a constant total, for example it might be interesting to scale the wavelengths 200–300, 400–500 and 500–600 nm each to 1. Or perhaps the wavelengths 200–300 nm are more diagnostic than the others, so why not scale these to a total of 5, and the others to a total of 1? Sometimes more than one type of measurement can be used to study an evolutionary process, such as UV/vis and MS, and each data block could be scaled to a constant total. When doing selective summation it is important to consider very carefully the consequences of preprocessing.

2. *Scaling to a base peak.* In some forms of measurement, such as mass spectrometry (e.g. LC–MS or GC–MS), it is possible to select a base peak and scale to this; for

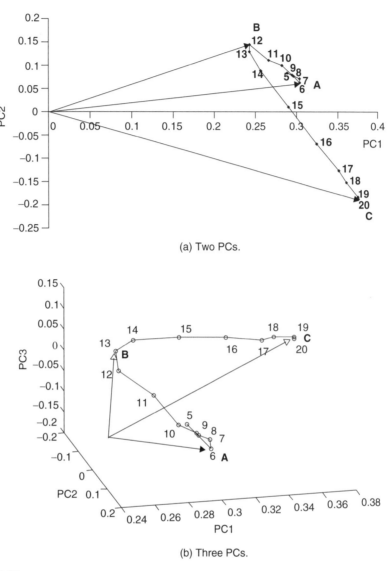

Figure 6.13
Scores plot of dataset B with rows summed to a constant total between times 5 and 20 and three main directions indicated (a) Two PCs (b) Three PCs

example, if the aim is to analyse the LC–MS results for two isomers, ratioing to the molecular ion can be performed, so that

$$^{scaled}x_{ij} = \frac{x_{ij}}{x_{i(\text{molecular ion})}}$$

In certain cases the molecular ion can then be discarded. This method of preprocessing can be used to investigate how the ratio of fragment ions varies across a cluster.

6.2.3.2 Scaling the Columns

In many cases it is useful to scale along the columns, e.g. each wavelength or mass number or spectral frequency. This can be used to put all the variables on a similar scale.

Mean centring, involving subtracting the mean of each column, is the simplest method. Many PC packages do this automatically, but in the case of signal analysis is often inappropriate, because the interest is about variability above the baseline rather that around an average.

Standardisation is a common technique that has already been discussed (Chapter 4, Section 4.3.6.4) and is sometimes called *autoscaling*. It can be mathematically described by

$$^{stand}x_{ij} = \frac{x_{ij} - \overline{x}_j}{\sqrt{\sum_{i=1}^{I}(x_{ij} - \overline{x}_j)^2 / I}}$$

where there are I points in time and $\overline{x}_j$ is the average of variable j. Note that it is conventional to divide by I rather than $I - 1$ in this application, if doing the calculations check whether the package defaults to the 'population' rather than 'sample' standard deviation. Matlab users should be careful when performing this scaling. This can be useful, for example, in mass spectrometry where the variation of an intense peak (such as a molecular ion of isomers) is no more significant than that of a much less intense peak, such as a significant fragment ion. However, standardisation will also emphasize variables that are pure noise, and if there are, for example, 200 mass numbers of which 180 correspond to noise, this could substantially degrade the analysis.

The most dramatic change is normally to the loadings plot. Figure 6.14 illustrates this for dataset B. The scores plot hardly changes in appearance. The loadings plot however, has changed considerably in appearance, however, and is much clearer and more spread out than in Figure 6.6.

Standardisation is most useful if the magnitudes of the variables are very different, as might occur in LC–MS. Table 6.3 is of dataset C, which consists of 25 points in time and eight measurements, making a 25×8 data matrix. As can be seen, the magnitude of the measurements is different, with variable H having a maximum of 100, but others being much smaller. We assume that the variables are not in a particular sequence, or are not best represented sequentially, so the loadings graphs will consist of a series of points that are not joined up. Figure 6.15 is of the raw profile together with scores and loadings plots. The scores plot suggests that there are two components in the mixture, but the loadings are not very well distinguished and are dominated by variable H. Standardisation (Figure 6.16) largely retains the pattern in the scores plot but the loadings change radically in appearance, and in this case fall approximately on a circle because there are two main components in the mixture. The variables corresponding most to each pure component fall at the ends of the circle. It is important to recognise that this pattern is an approximation and will only happen if there are two main components, otherwise the loadings will fall on to the surface of a sphere (if three PCs are employed and there are three compounds in the mixture) and so on. However, standardisation can have a remarkable influence on the appearance of loadings plots.

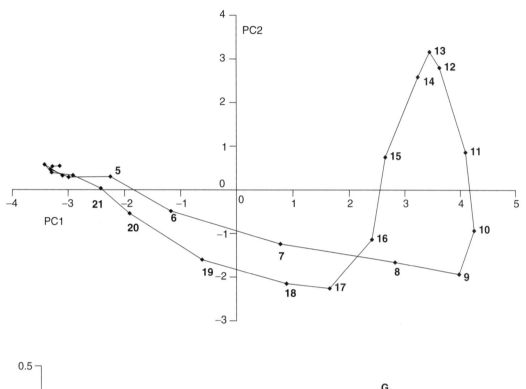

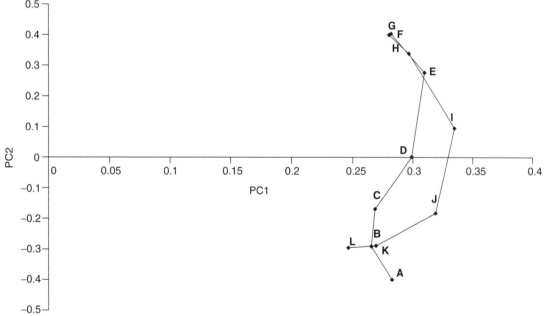

Figure 6.14
Scores and loadings of PC2 versus PC1 after dataset B has been standardised

Table 6.3 Two-way dataset C.

	A	B	C	D	E	F	G	H
1	0.407	0.149	0.121	0.552	−0.464	0.970	0.389	−0.629
2	0.093	−0.062	0.084	−0.015	−0.049	0.178	0.478	1.073
3	0.044	0.809	0.874	0.138	0.529	−1.180	0.040	1.454
4	−0.073	0.307	−0.205	0.518	1.314	2.053	0.658	7.371
5	1.461	1.359	−0.272	1.087	2.801	0.321	0.080	20.763
6	1.591	4.580	0.207	2.381	5.736	3.334	2.155	41.393
7	4.058	7.030	0.280	2.016	9.001	4.651	3.663	67.949
8	4.082	8.492	0.304	4.180	11.916	5.705	4.360	92.152
9	5.839	10.469	0.529	3.764	12.184	6.808	3.739	105.228
10	5.688	10.525	1.573	5.193	12.100	5.720	5.621	106.111
11	3.883	10.111	2.936	4.802	10.026	5.292	7.061	99.404
12	3.630	9.139	2.356	4.739	9.257	4.478	7.530	92.409
13	2.279	8.052	3.196	3.777	9.926	3.228	10.012	92.727
14	2.206	7.952	4.229	5.118	8.629	1.869	9.403	86.828
15	1.403	5.906	2.867	4.229	7.804	1.234	8.774	73.230
16	1.380	5.523	1.720	2.529	4.845	2.249	6.621	52.831
17	0.991	2.820	0.825	1.986	2.790	1.229	3.571	31.438
18	0.160	0.993	0.715	0.591	1.594	0.880	1.662	15.701
19	0.562	−0.018	−0.348	−0.290	0.567	0.070	1.257	6.528
20	0.590	−0.308	−0.715	0.490	0.384	0.595	0.409	2.657
21	0.309	0.371	−0.394	0.077	−0.517	0.434	−0.250	0.551
22	−0.132	−0.081	−0.861	−0.279	−0.622	−0.640	1.166	0.079
23	0.371	0.342	−0.226	0.374	−0.284	0.177	−0.751	−0.197
24	−0.215	−0.577	−0.297	0.834	0.720	−0.248	0.470	−1.053
25	−0.051	0.608	−0.070	−0.087	−0.068	−0.537	−0.208	0.601

Sometimes weighting by the standard deviation can be performed without centring, so that

$$^{scaled}x_{ij} = \frac{x_{ij}}{\sqrt{\sum_{i=1}^{I}(x_{ij} - \bar{x}_j)^2/I}}$$

It is, of course, possible to use any weighting criterion for the columns, so that

$$^{scaled}x_{ij} = {}^{j}w.x_{ij}$$

where w is a weighting factor. The weights may relate to noise content or standard deviations or significance of a variable. Fairly complex criteria can be employed. In the extreme if $w = 0$, this becomes a form of variable selection, which will be discussed in Section 6.2.4.

In rare and interesting cases it is possible to rank the size of the variables along each column. The suitability depends on the type of preprocessing performed first on the rows. However, a common method is to give the most intense reading in any column a value of I and the least intense 1. If the absolute values of each variable are not very meaningful, this procedure is an alternative that takes into account relative intensities. This procedure is exemplified by reference to the dataset C, and illustrated in Table 6.4.

1. Choose a region where the peaks elute, in this case from time 4 to 19 as suggested by the scores plot in Figure 6.15.

2. Scale the data in this region, so that each row is of a constant total.
3. Rank the data in each column, from 1 (low) to 16 (high).

The PC scores and loadings plots are presented in Figure 6.17. Many similar conclusions can be deduced as in Figure 6.16. For example, the loadings arising from measurement C are close to the slowest eluting peak centred on times 14–16, whereas measurements A–F correspond mainly to the fastest eluting peak. When ranking variables it is unlikely that the resultant scores and loadings plots will fall on to a smooth geometric figure such as a circle or a line. However, this procedure can be useful for

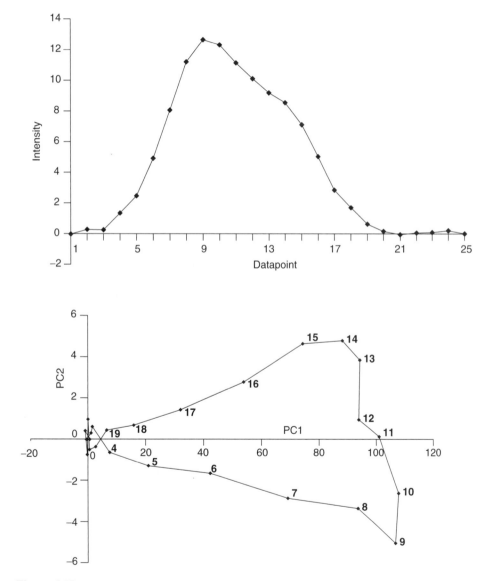

Figure 6.15
Intensity profile and unscaled scores and loadings of PC2 versus PC1 from dataset in Table 6.3

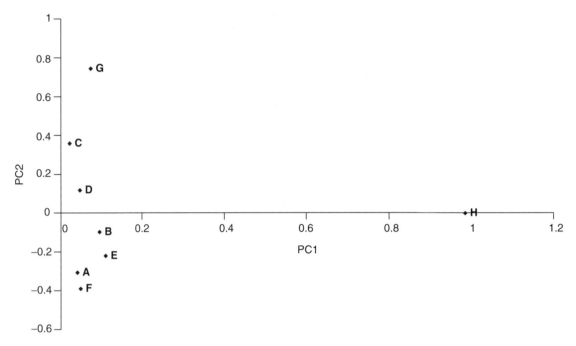

Figure 6.15
(*continued*)

exploratory graphical analysis, especially if the dataset is fairly complex with several different compounds and also many measurements on different intensity scales.

It is, of course, possible to scale both the rows and columns simultaneously, first by scaling the rows and then the columns. Note that the reverse (scaling the columns first) is rarely useful and standardisation followed by summing to a constant total has no physical meaning.

6.2.4 Variable Selection

Variable selection has an important role throughout chemometrics, but will be described below in the context of coupled chromatography. This involves keeping only a portion of the original measurements, selecting only those such as wavelengths or masses that are most relevant to the underlying problem. There are a huge number of combinations of approaches limited only by the imagination of the chromatographer or spectroscopist. In this section we give only a brief summary of some of the main methods. Often several steps are combined.

Variable selection is particularly important in LC–MS and GC–MS. Raw data form what is sometimes called a *sparse* data matrix, in which the majority of data points are zero or represent noise. In fact, only a small percentage (perhaps 5 % or less) of the measurements are of any interest. The trouble with this is that if multivariate methods are applied to the raw data, often the results are nonsense, dominated by noise. Consider the case of performing LC–MS on two closely eluting isomers, whose fragment ions are of principal interest. The most intense peak might be the molecular

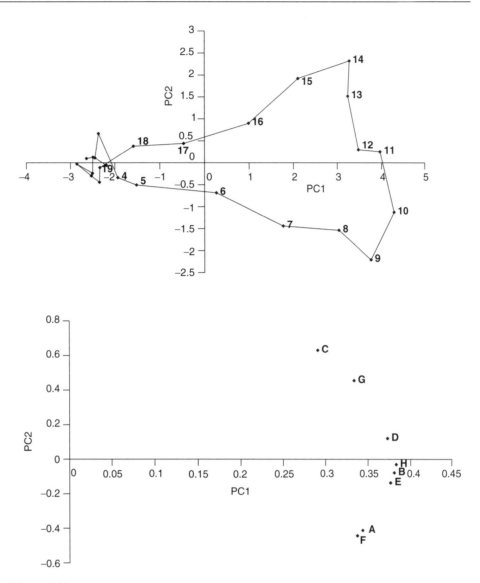

Figure 6.16
Scores and loadings of PC2 versus PC1 after the data in Table 6.3 have been standardised

ion, but in order to study the fragmentation ions, a method such as standardisation described above is required to place equal significance on all the ions. Unfortunately, not only are perhaps 20 or so fragment ions increased in importance, but so are 200 or so ions that represent pure noise, so the data become worse, not better. Typically, out of 200–300 masses, there may be around 20 significant ones, and the aim of variable selection is to find these key measurements. However, too much variable reduction has the disadvantage that the dimensions of the multivariate matrices are reduced. It is important to find an optimum size as illustrated in Figure 6.18. What tricks can we use to remove irrelevant variables?

Table 6.4 Method for ranking variables using the dataset in Table 6.4.

(a) Data between times 4 and 19 each row summed to a total of 1

Time	A	B	C	D	E	F	G	H
4	−0.006	0.026	−0.017	0.043	0.110	0.172	0.055	0.617
5	0.053	0.049	−0.010	0.039	0.101	0.012	0.003	0.752
6	0.026	0.075	0.003	0.039	0.093	0.054	0.035	0.674
7	0.041	0.071	0.003	0.020	0.091	0.047	0.037	0.689
8	0.031	0.065	0.002	0.032	0.091	0.043	0.033	0.702
9	0.039	0.070	0.004	0.025	0.082	0.046	0.025	0.708
10	0.037	0.069	0.010	0.034	0.079	0.038	0.037	0.696
11	0.027	0.070	0.020	0.033	0.070	0.037	0.049	0.693
12	0.027	0.068	0.018	0.035	0.069	0.034	0.056	0.692
13	0.017	0.060	0.024	0.028	0.075	0.024	0.075	0.696
14	0.017	0.063	0.034	0.041	0.068	0.015	0.074	0.688
15	0.013	0.056	0.027	0.040	0.074	0.012	0.083	0.694
16	0.018	0.071	0.022	0.033	0.062	0.029	0.085	0.680
17	0.022	0.062	0.018	0.044	0.061	0.027	0.078	0.689
18	0.007	0.045	0.032	0.027	0.071	0.039	0.075	0.704
19	0.067	−0.002	−0.042	−0.035	0.068	0.008	0.151	0.784

(b) Ranked data over these times

Time	A	B	C	D	E	F	G
4	1	2	2	15	16	16	8
5	15	4	3	12	15	2	1
6	8	16	6	11	14	15	4
7	14	15	5	2	13	14	6
8	11	9	4	6	12	12	3
9	13	13	7	3	11	13	2
10	12	11	8	9	10	10	5
11	9	12	11	8	6	9	7
12	10	10	9	10	5	8	9
13	4	6	13	5	9	5	12
14	5	8	16	14	4	4	10
15	3	5	14	13	8	3	14
16	6	14	12	7	2	7	15
17	7	7	10	16	1	6	13
18	2	3	15	4	7	11	11
19	16	1	1	1	3	1	16

Some simple methods, often used as an initial filter of irrelevant variables, are as follows; note that it is often important first to have performed baseline correction (Section 6.2.1).

1. Remove variables outside a given region, e.g. in mass spectrometry these may be at low or high m/z values, in UV/vis spectroscopy there may be a significant wavelength range where there is no absorbance.
2. Sometimes is possible to measure the noise content of the chromatograms for each variable simply by looking at the standard deviation of the noise region. The higher the noise, the less significant is the mass. This technique is useful in combination with other methods often as a first step.

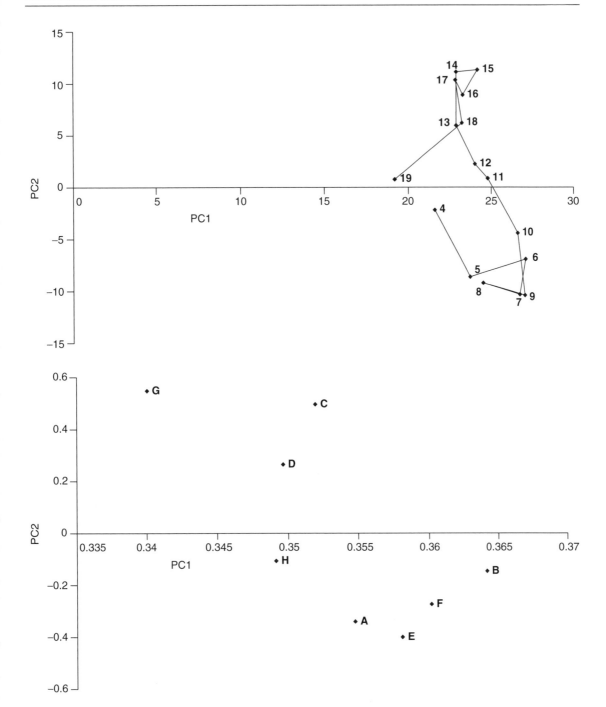

Figure 6.17
Scores and loadings of PC2 versus PC1 of the ranked data in Table 6.4

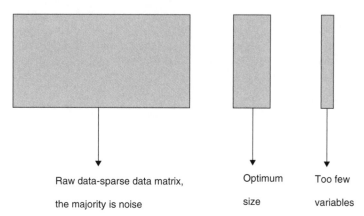

Raw data-sparse data matrix,   Optimum   Too few

the majority is noise   size   variables

Figure 6.18
Optimum size for variable reduction

Many methods then use simple functions of the data, choosing the variables with the highest values.

1. The very simplest is to order the variables according to their mean, e.g. $\overline{x}_j$, which is the average of column j. If all the measurements are on approximately the same scale such as in many forms of spectroscopic detection, this is a good approach, but is less useful if there remain significant background peaks or if there are dominant high intensity peaks such as molecular ions that are not necessarily very diagnostic.

2. A variant is to employ the variance v_j (or standard deviation). Large peaks that do not vary much may have a small standard deviation. However, this depends crucially on determining a region of the data where compounds are present, and if noise regions are included, this method will often fail.

3. A compromise is to select peaks according to a criterion of variance over mean, $v_j/\overline{x}_j$. This may pick some less intense measurements that vary through interesting regions of the data. Intense peaks may still have a large variance but this might not be particularly significant relative to the average intensity. The problem with this approach, however, is that some measurements that are primarily noise could have a mean close to zero, so the ratio becomes large and they will be accidentally selected. To prevent this, first remove the noisy variables by another method and then from the remaining select those with highest relative variance. Variables can have low noise but still be uninteresting if they correspond, for example, to solvent or base peaks.

4. A modification of the method in point 3 is to select peaks using a criterion of $v_j/(\overline{x}_j + e)$, where e relates to noise level. The advantage of this is that variables with low means are not accidentally selected. Of course, the value of e must be carefully chosen.

There are no general guidelines as to how many variables should be selected; some people use statistical tests, others cut off the selection according to what appears sensible or manageable. The optimum method depends on the technique employed and the general features of a particular source of data.

There are numerous other approaches, for example to look at smoothness of variables, correlations between successive points, and so on. In some cases after selecting variables, contiguous variables can then be combined into a smaller number of very significant (and less noisy) measurements; this could be valuable in the case of LC–NMR or GC–IR, where neighbouring variables often correspond to peaks in a spectrum of a single compound in a mixture, but is unlikely be valuable in HPLC–DAD, where there are often contiguous regions of a spectrum that correspond to different compounds. Sometimes features of variables, such as the change in relative intensity over a peak cluster, can also be taken into account; variables diagnostic for an individual compound are likely to vary in a smooth and predictable way, whereas those due to noise will vary in a random manner.

For each type of coupled chromatography (and indeed for any technique where chemometric methods are employed) there are specific methods for variable selection. In some cases such as LC–MS this is a crucial first step prior to further analysis, whereas in the case of HPLC–DAD it is often less essential, and omitted.

6.3 Determining Composition

After exploring data via PC plots, baseline correction, preprocessing, scaling and variable selection, as required, the next step is normally to look at the composition of different regions of the chromatograms. Most chemometric techniques try to identify pure variables that are associated with one specific component in a mixture. In chromatography these are usually regions in time where a single compound elutes, although they can also be measurements such as an m/z value characteristic of a single compound or a peak in IR spectroscopy. Below we will concentrate primarily on methods for determining pure variables in the chromatographic direction, but many can be modified fairly easily for spectroscopy. There is an enormous battery of techniques but below we summarise the main groups of approaches.

6.3.1 Composition

The concept of composition is an important one. There are many alternative ways of expressing the same idea, that of rank being popular also, which derives from matrices: ideally the rank of a matrix equals the number of independent components or nonzero eigenvectors.

A region of composition 0 contains no compounds, one of composition 1 one compound, and so on. Composition 1 regions are also selective or pure regions. A complexity arises in that because of noise, a matrix over a region of composition 1 will not necessarily be described by only one PC, and it is important to try to identify how many PCs are significant and correspond to real information.

There are many cases of varying difficulty. Figure 6.19 illustrates four cases. Case (a) is the most studied and easiest, in which each peak has a composition 1 or selective region. Although not the hardest of problems, there is often considerable value in the application of chemometrics techniques in such a situation. For example, there may be a requirement for quantification in which the complete peak profile is required, including the area of each peak in the region of overlap. The spectra of the compounds might not be very clear and chemometrics can improve the quality. In complex peak clusters it might simply be important to identify how many compounds are present, which regions are pure, what the spectra in the selective regions are and whether it is necessary to improve the chromatography. Finally, this has potential in the area of automation of pulling out spectra from chromatograms containing several compounds where there is some overlap. Case (b) involves a peak cluster where one or more do not have a selective region. Case (c) is of an embedded impurity peak and is surprisingly common. Many modern separations involve asymmetric peaks such as in case (d), and many conventional chemometrics methods fail under such circumstances.

To understand the problem, it is possible to produce a graph of ratios of intensities between the various components. For ideal noise free peakshapes corresponding to the four cases above, these are presented in Figure 6.20. Note the use of a logarithmic scale, as the ideal ratios will vary over a large range. Case (a) corresponds to two Gaussian peaks (for more information about peakshapes, see Chapter 3, Section 3.2.1) and is straightforward. In case (b), the ratios of the first to second and of the second to third peaks are superimposed, and it can be seen that the rate of change is different for each pair of peaks; this relates to the different separation of the elution time maxima. Note that there is a huge dynamic range, which is due to noise-free simulations being used. Case (c) is typical of an embedded peak, showing a purity maximum for the smaller component in the centre of its elution. Finally, the ratio arising from case (d) is typical of tailing peaks; many multivariate methods cannot cope easily with this type of data. However, these graphs are of ideal situations and, in practice, it is only

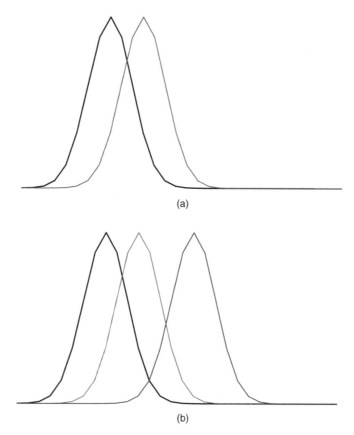

(a)

(b)

Figure 6.19
Different types of problems in chromatography

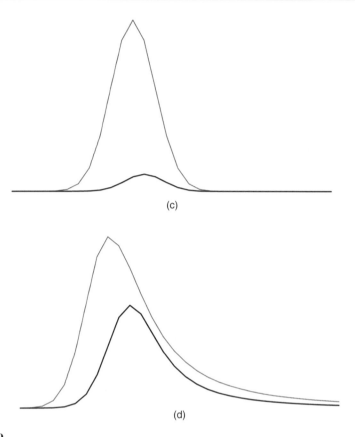

(c)

(d)

Figure 6.19
(*continued*)

practicable to observe small effects if the data are of an appropriate quality. In reality, measurement noise and spectral similarity limit data quality. In practice, it is only realistic to detect two (or more components) if the ratios of intensities of the two peaks are within a certain range, for example no more than 50:1, as indicated by region a in Figure 6.21. Outside these limits, it is unlikely that a second component will be detected. In addition, when the intensity of signal is sufficiently low (say 1 % of the maximum, outside region b in Figure 6.21), the signal may be swamped by noise, and so no signal detected. Region a would appear to be of composition 2, the overlap between regions a and b composition 1 and the chromatogram outside region b composition 0. If noise levels are higher, these regions become narrower.

Below we indicate a number of approaches for the determination of composition.

6.3.2 Univariate Methods

By far the simplest are univariate approaches. It is important not to overcomplicate a problem if not justified by the data. Most conventional chromatography software contains methods for estimating ratios between peak intensities. If two spectra are sufficiently dissimilar then this method can work well. The measurements most diagnostic for each compound can be chosen by a number of means. For the data in Table 6.1 we can look at the loadings plot of Figure 6.4. At first glance it may appear that measurements C and G are most appropriate, but this is not so. The problem is that the most diagnostic wavelengths for one compound may correspond to zero or very low intensity for the other one. This would mean that there will be regions of the chromatogram where one number is close to zero or even negative (because of noise), leading to very large or negative ratios. Measurements that are characteristic of both compounds but exhibit distinctly different features in each case, are better. Figure 6.22(a) plots the ratio of intensities of variables D to F. Initially this plot looks slightly discouraging, but that is because there are noise regions where almost any ratio could be obtained.

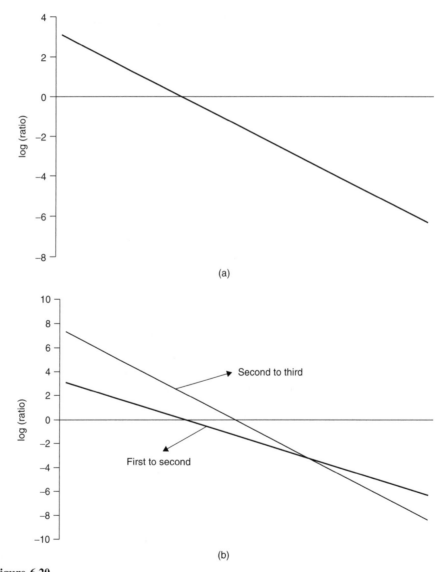

Figure 6.20
Ratios of peak intensities for the case studies (a)–(d) assuming ideal peakshapes and peaks detectable over an indefinite region

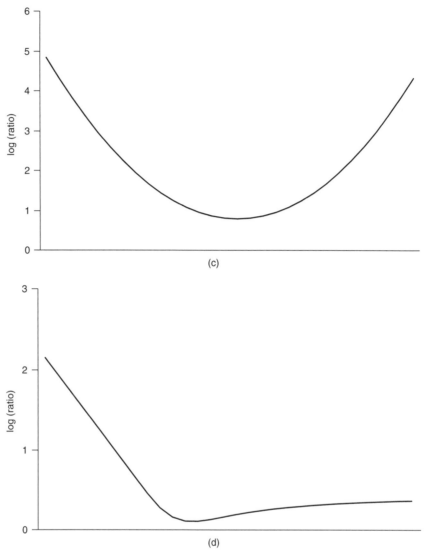

Figure 6.20
(*continued*)

Cutting the region down to times 5–18 (within which range the intensities of both variables are positive) improves the situation. It is also helpful to consider the best way to display the graph, the reason being that a ratio of 2:1 is no more significant to a ratio of 1:2, yet using a linear scale there is an arbitrary asymmetry, for example, moving from 0 to 50% of compound one may change the ratio from 0.1 to 1 but moving from 50% it could change from 1 to 10. To overcome this, either use a logarithmic scale [Figure 6.22(b)] or take the smaller of the ratios D:F and F:D [Figure 6.22(c)].

The peak ratio plots suggest that there is a composition 2 region starting between times 9 and 10 and finishing between times 14 and 15. There is some ambiguity about the exact start and end, largely because there is noise imposed upon the data. In some cases peak

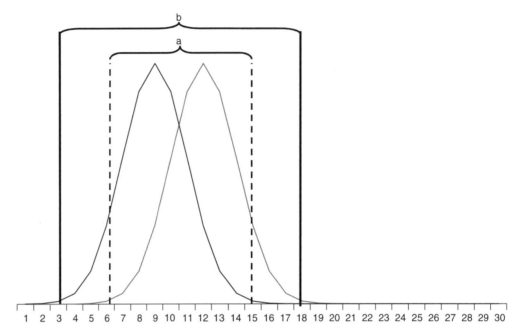

Figure 6.21
Regions of chromatogram (a) in Figure 6.19. Region a is where the ratio of the two components is between 50:1 and 1:50 and region b where the overall intensity is more than 1 % of the maximum intensity

ratio plots are very helpful, but they do depend on having adequate signal to noise ratios and finding suitable variables. If a spectrum is monitored over 200 wavelengths this may not be so easy, and approaches that use all the wavelengths may be more successful. In addition, good diagnostic measurements are required, noise regions have to be eliminated and also the graphs can become complicated if there are several compounds in a portion of the chromatogram. An ideal situation would be to calculate several peak ratios simultaneously, but this then suggests that multivariate methods, as described below, have an important role to play.

Another simple trick is to sum the data to constant total at each point in time, as described above, so as to obtain values of

$$^{rs}x_{ij} = \frac{x_{ij}}{\sum\limits_{j=1}^{J} x_{ij}}$$

Provided that noise regions are discarded, the relative intensities of diagnostic wavelengths should change according to the composition of the data. Unlike using ratios of intensities, we are able to choose strongly associated wavelengths such as C and G, as an intensity of zero (or even a small negative number) will not unduly influence the appearance of the graph, given in Figure 6.23. The regions of composition 1 are somewhat flatter but influenced by noise, but where the relative intensity changes most is in the composition 2 region.

This approach is not ideal, but the graphs of Figure 6.23 and 6.22 are intuitively easy for the practising chromatographer (or spectroscopist) and result in the creation of a form of purity curve. The appearance of the curves can be enhanced by selecting variables that are least noisy and then calculating the relative intensity curves as a function of time for several (rather than just two) variables. Some will increase and others decrease according to whether the variable is most associated with the fastest or slowest eluting compound. Reversing the curve for one set of variables results in several superimposed purity curves, which can be averaged to give a good picture of changes over the chromatogram.

These methods can be extended to cases of embedded peaks, in which the purest point for the embedded peak does not correspond to a selective region; a weakness of using this method of ratios is that it is not always possible to determine whether a maximum (or minimum) in the purity curve is genuinely a consequence of a composition 1 region or simply the portion of the chromatogram where the concentration of one analyte is highest.

Such simple approaches can become rather messy when there are several compounds in a cluster, especially if the spectra are similar, but in favourable cases they are very effective.

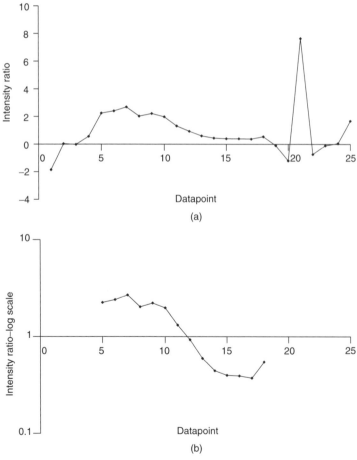

Figure 6.22
Ratio of intensity of measurements D–F for the data in Table 6.1. (a) Raw information; (b) logarithmic scale between points 5 and 18; (c) the minimum of the ratio of intensity D: F and F: D between points 5 and 18

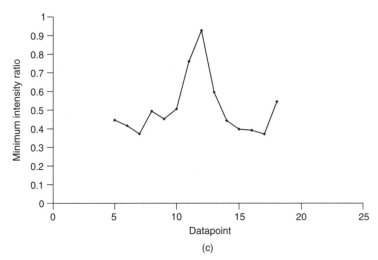

(c)

Figure 6.22
(*continued*)

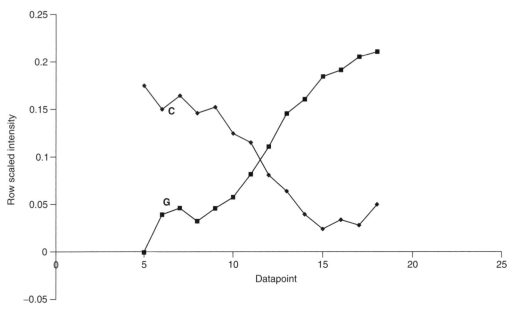

Figure 6.23
Intensities for wavelengths C and G using the data in Table 6.1, summing the measurements at each successive
point to constant total of 1

6.3.3 Correlation and Similarity Based Methods

Another set of methods is based on correlation coefficients. The principle is that the
correlation coefficient between two successive points in time defined by

$$r_{i-1,i} = \sum_{j=1}^{J} \frac{(x_{ij} - \overline{x}_i)(x_{i-1j} - \overline{x}_{i-1})}{s_i s_{i-1}}$$

where $\bar{x}_i$ is the mean measurement at time i and s_i the corresponding standard deviation, will have the following characteristics:

1. it will be close to 1 in regions of composition 1;
2. it will be close to 0 in noise regions;
3. it will be below 1 in regions of composition 2.

Table 6.5 gives of the correlation coefficients between successive points in time for the data in Table 6.1. The data are plotted in Figure 6.24, and suggest that

- points 7–9 are composition 1;
- points 10–13 are composition 2;
- points 14–18 are composition 1.

Note that because the correlation coefficient is between two successive points, the three point dip in Figure 6.24 actually suggests four composition 2 points.

The principles can be further extended to finding the points in time corresponding to the purest regions of the data. This is sometimes useful, for example to obtain the spectrum of each compound that has a composition 1 region. The highest correlation is between points 15 and 16, so one of these is the purest point in the chromatogram. The correlation between points 14 and 15 is higher than that between points 16 and 17, so point 15 is chosen as the elution time best representative of slowest eluting compound.

Table 6.5 Correlation coefficients for the data in Table 6.1.

Time	$r_{i,i-1}$	$r_{i,15}$
1		−0.045
2	−0.480	0.227
3	−0.084	−0.515
4	0.579	−0.632
5	0.372	−0.651
6	0.802	−0.728
7	0.927	−0.714
8	0.939	−0.780
9	0.973	−0.696
10	0.968	−0.643
11	0.817	−0.123
12	0.489	0.783
13	0.858	0.974
14	0.976	0.990
15	0.990	1.000
16	0.991	0.991
17	0.968	0.967
18	0.942	0.950
19	0.708	0.809
20	0.472	0.633
21	0.332	0.326
22	−0.123	0.360
23	−0.170	0.072
24	−0.070	0.276
25	0.380	0.015

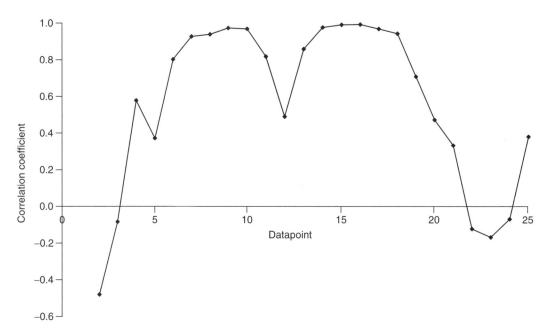

Figure 6.24
Graph of correlation between successive points in the data in Table 6.1

The next step is to calculate the correlation coefficient $r_{15,i}$ between this selected point in time and all other points. The lowest correlation is likely to belong to the second component. The data are presented in Table 6.5 and Figure 6.25. The trends are very clear. The most negative correlation occurs at point 8, being the purity maximum for the fastest eluting peak. The composition 2 region is somewhat smaller than estimated in the previous section, but probably more accurate. One reason why these graphs are an improvement over the univariate ones is that they take all the data into account rather than single measurements. Where there are 100 or more spectral frequencies these approaches can have significant advantages, but it is important to ensure that most of the variables are meaningful. In the case of NMR or MS, 95 % or more of the measurements may simply arise from noise, so a careful choice of variables using the methods in Section 6.2.4 is a prior necessity.

This method can be extended to fairly complex peak clusters, as presented in Figure 6.26 for the data in Table 6.2. Ignoring the noise at the beginning, it is fairly clear that there are three components in the data. Note that the central component eluting approximately between times 10 and 15 does not have a true composition 1 region because the correlation coefficient only reaches approximately 0.9, whereas the other two compounds have well established selective areas. It is possible to determine the purest point for each component in the mixture successively by extending the approach illustrated above.

Another related aspect involves using these graphs to select pure variables. This is often useful in spectroscopy, where certain masses or frequencies are most diagnostic of different components in a mixture, and finding these helps in the later stages of the analysis.

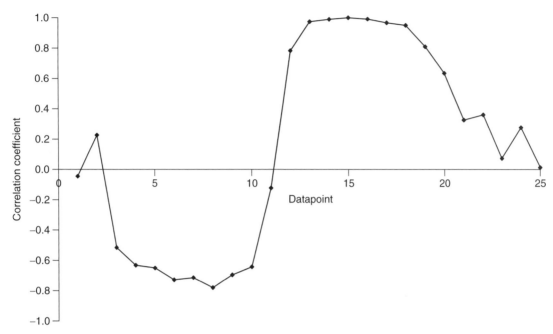

Figure 6.25
Correlation between point 15 and the data in Table 6.1

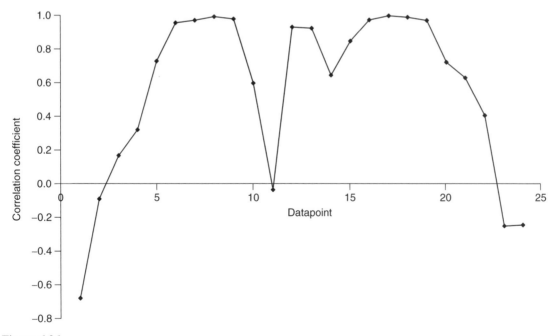

Figure 6.26
Graph corresponding to that in Figure 6.24 for the data in Table 6.2

It is not essential to use correlation coefficients as a method for determining similarity, and other measures, even of dissimilarity, can be proposed, for example Euclidean distances between normalised spectra – the larger the distance, the less is the similarity. It is also possible to start with an average or most intense spectrum in the chromatogram and determine how the spectrum at each point in time differs from this. The variations on this theme are endless and most papers and software packages contain some differences reflecting the authors' favourite approaches. When using correlation or similarity based methods it is important to work carefully through the details and not accept the results as a 'black box'. There is insufficient space in this text to itemise each published method, but the general principles should be clear for any user.

6.3.4 Eigenvalue Based Methods

Many chemometricians like multivariate approaches, most of which are based on PCA. A large number of methods are available in the literature, such as evolving factor analysis (EFA), fixed sized window factor analysis (WFA) and heuristic evolving latent projections (HELP), among many, that contain one step that involves calculating eigenvalues to determine the composition of regions of evolutionary two-way data. The principle is that the more components there are in a particular region of a chromatogram, the greater is the number of significant eigenvalues.

There are two fundamental groups of approaches. The first involves performing PCA on an expanding window which we will refer to as EFA. There are several variants on this theme, but a popular one is indicated below.

1. Perform uncentred PCA on the first few datapoints of the series, e.g. points 1–4. In the case of Table 6.1, this will involve starting with a 4×12 matrix.
2. Record the first few eigenvalues of this matrix, which should be more than the number of components expected in the mixture and cannot be more than the smallest dimension of the starting matrix. We will keep four eigenvalues.
3. Extend the matrix by an extra point in time to a matrix of points 1–5 in this example and repeat PCA, keeping the same number of eigenvalues as in step 2.
4. Continue until the entire data matrix is employed, so the final step involves performing PCA on a data matrix of dimensions 25×12 and keeping four eigenvalues.
5. Produce a table of the eigenvalues against matrix size. In this example, there will be $22 (= 25 - 4 + 1)$ rows and four columns. This procedure is called *forward expanding factor analysis*.
6. Next, take a matrix at the opposite end of the dataset, from points 21–25, to give another 4×12 matrix, and perform steps 1 and 2 on this matrix.
7. Expand the matrix backwards, so the second calculation is from points 20–25, the third from points 19–25, and so on.
8. Produce a table similar to that in step 5. This procedure is called *backward expanding factor analysis*.

The results are given in Table 6.6. Note that the first columns of the two datasets should be properly aligned. Normally the eigenvalues are plotted on a logarithmic scale. The

Table 6.6 Results of forward and backward EFA for the data in Table 6.1.

Time	Forward				Backward			
1	n/a	n/a	n/a	n/a	86.886	12.168	0.178	0.154
2	n/a	n/a	n/a	n/a	86.885	12.168	0.167	0.147
3	n/a	n/a	n/a	n/a	86.879	12.166	0.165	0.128
4	0.231	0.062	0.024	0.006	86.873	12.160	0.164	0.128
5	0.843	0.088	0.036	0.012	86.729	12.139	0.161	0.111
6	3.240	0.101	0.054	0.015	86.174	12.057	0.161	0.100
7	9.708	0.118	0.055	0.026	84.045	11.802	0.156	0.067
8	22.078	0.118	0.104	0.047	78.347	11.064	0.107	0.057
9	37.444	0.132	0.106	0.057	67.895	9.094	0.101	0.056
10	51.219	0.182	0.131	0.105	55.350	6.278	0.069	0.056
11	61.501	0.678	0.141	0.105	44.609	3.113	0.069	0.051
12	69.141	2.129	0.141	0.109	35.796	1.123	0.069	0.046
13	75.034	4.700	0.146	0.119	27.559	0.272	0.067	0.044
14	80.223	7.734	0.148	0.119	19.260	0.112	0.050	0.042
15	83.989	10.184	0.148	0.119	11.080	0.057	0.049	0.038
16	85.974	11.453	0.148	0.123	4.878	0.051	0.045	0.038
17	86.611	11.924	0.155	0.127	1.634	0.051	0.039	0.032
18	86.855	12.067	0.156	0.127	0.514	0.050	0.035	0.025
19	86.879	12.150	0.160	0.134	0.155	0.037	0.025	0.025
20	86.881	12.162	0.163	0.139	0.046	0.034	0.025	0.018
21	86.881	12.164	0.176	0.139	0.037	0.030	0.019	0.010
22	86.882	12.166	0.177	0.139	0.032	0.024	0.012	0.009
23	86.882	12.166	0.178	0.139	n/a	n/a	n/a	n/a
24	86.886	12.167	0.178	0.154	n/a	n/a	n/a	n/a
25	86.886	12.168	0.178	0.154	n/a	n/a	n/a	n/a

results for the first three eigenvalues (the last is omitted in order not to complicate the graph) are illustrated in Figure 6.27. What can we tell from this?

1. Although the third eigenvalue increases slightly, this is largely due to the data matrix increasing in size and does not indicate a third component.
2. In the forward plot, it is clear that the fastest eluting component has started to become significant in the matrix by time 4, so the elution window starts at time 4.
3. In the forward plot, the slowest component starts to become significant by time 10.
4. In the backward plot, the slowest eluting component starts to become significant at time 19.
5. In the backward plot, the fastest eluting component starts to become significant at time 13.

Hence,

- there are two significant components in the mixture;
- the elution region of the fastest is between times 4 and 13;
- the elution region of the slowest between times 10 and 19;
- so between times 10 and 13 the chromatogram is composition 2, consisting of overlapping elution.

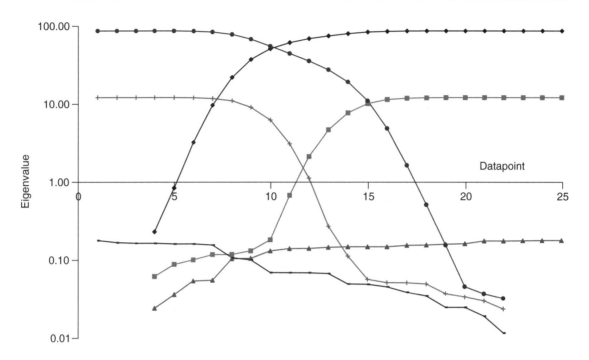

Figure 6.27
Forward and backward EFA plots of the first three eigenvalues from the data in Table 6.1

We have interpreted the graphs visually but, of course, some people like to use statistical methods, which can be useful for automation but rely on noise behaving in a specific manner, and it is normally better to produce a graph to see that the conclusions are sensible.

The second approach, which we will call WFA, involves using a fixed sized window as follows.

1. Choose a window size, usually a small odd number such as 3 or 5. This window should be at least the maximum composition expected in the chromatogram, preferably one point wider. It does not need to be as large as the number of components expected in the system, only the maximum overlap anticipated. We will use a window size of 3.

2. Perform uncentred PCA on the first points of the chromatogram corresponding to the window size, in this case points 1–3, resulting in a matrix of size 3×12.

3. Record the first few eigenvalues of this matrix, which should be no more than the highest composition expected in the mixture and cannot be more than the smallest dimension of the starting matrix. We will keep three eigenvalues.

4. Move the window successively along the chromatogram, so that the next window will consist of points 2–4, and the final one of points 23–25. In most implementations, the window is not changed in size.

5. Produce a table of the eigenvalues against matrix centre. In this example, there will be 23 rows and three columns.

Table 6.7 Fixed sized window factor
analysis applied to the data in Table 6.1
using a three point window.

Centre			
1	n/a	n/a	n/a
2	0.063	0.027	0.014
3	0.231	0.036	0.013
4	0.838	0.057	0.012
5	3.228	0.058	0.035
6	9.530	0.054	0.025
7	21.260	0.098	0.045
8	34.248	0.094	0.033
9	41.607	0.133	0.051
10	39.718	0.391	0.046
11	32.917	0.912	0.018
12	27.375	1.028	0.024
13	25.241	0.587	0.028
14	22.788	0.169	0.022
15	17.680	0.049	0.014
16	10.584	0.031	0.013
17	4.770	0.024	0.013
18	1.617	0.047	0.013
19	0.505	0.046	0.018
20	0.147	0.031	0.016
21	0.036	0.027	0.015
22	0.031	0.021	0.009
23	0.030	0.021	0.010
24	0.032	0.014	0.009
25	n/a	n/a	n/a

The results for the data in Table 6.1 are given in Table 6.7, with the graph, again presented on a logarithmic axis, shown in Figure 6.28. What can we tell from this?

1. It appears fairly clear that there are no regions where more than two components elute.
2. The second eigenvalue appears to become significant between points 10 and 14. However, since a three point window has been employed, this suggests that the chromatogram is composition 2 between points 9 and 15. This is a slightly larger region than expanding factor analysis finds. One problem about using a fixed sized window in this case is that the dataset is rather small, each matrix having a size of size 3×12, and so can be sensitive to noise. If more measurements are not available, a solution is to use a larger window size, but then the accuracy in time may be less. However, it is often not possible to predict elution windows to within one point in time, and the overall conclusions of the two methods are fairly similar in nature.
3. The first eigenvalue mainly reflects the overall intensity of the chromatogram (see Figure 6.28).

The regions of elution for each component can be similarly defined as for EFA above. There are numerous variations on fixed sized window factor analysis, such as changing the window size across the chromatogram, and the results can change

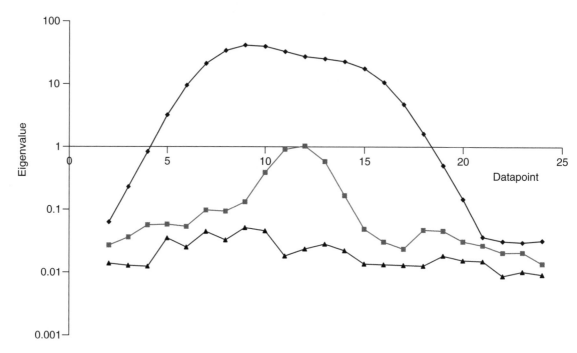

Figure 6.28
Three-point FSW graph for the data in Table 6.1

dramatically when using different forms of data scaling. However, this is a fairly simple visual technique that is popular. The region where the second eigenvalue is significant in Figure 6.28 can be compared with the dip in Figure 6.24, the ascent in Figure 6.25, the peak in Figure 6.22(c) and various features of the scores plots. In most cases similar regions are predicted within a datapoint.

Eigenvalue based methods are effective in many cases but may break down for unusual peakshapes. They normally depend on peakshapes being symmetrical with roughly equal peak widths for each compound in a mixture. The interpretation of eigenvalue plots for tailing peakshapes (Chapter 3, Section 3.2.1.3) is difficult. They also depend on a suitable selection of variables. If, as in the case of raw mass spectral data, the majority of variables are noise or consist mainly of baseline, they will not always give clear answers; however, reducing these will improve the appearance of the eigenvalue plots significantly.

6.3.5 Derivatives

Finally, there are a number of approaches based on calculating derivatives. The principle is that a spectrum will not change significantly in nature during a selective or composition 1 region. Derivatives measure change, hence we can exploit this.

There are a large number of approaches to incorporating information about derivatives into methods for determining composition. However the method below, illustrated by reference to the dataset in Table 6.1 is effective.

1. Scale the spectra at each point in time as described in Section 6.2.3.1. For pure regions, the spectra should not change in appearance.
2. Calculate the first derivative at each wavelength and each point in time. Normally the Savitsky–Golay method, described in Chapter 3, Section 3.3.2, can be employed. The simplest case is a five point quadratic first derivative (see Table 3.6), so that

$$\Delta_{ij} = \frac{dx_{ij}}{di} \approx \left[-2x_{(i-2)j} - x_{(i-1)j} + x_{(i+1)j} + 2x_{(i+2)j}\right]/10$$

Note that it is not always appropriate to choose a five point window – this depends very much on the nature of the raw data.
3. The closer the magnitude of this is to zero, the more likely the point represents a pure region, hence it is easier to convert to the absolute value of the derivative. If there are not too many variables and these variables are on a similar scale it is possible to superimpose the graphs from either step 2 or 3 to have a preliminary look at the data. Sometimes there are points at the end of a region that represent noise and will dominate the overall average derivative calculation; these points may be discarded, and often this can simply be done by removing points whose average intensity is below a given threshold.
4. To obtain an overall consensus, average the absolute value of the derivatives at each variable in time. If the variables are of fairly different is size, it is first useful to scale each variable to a similar magnitude, but setting the sum of each column (or variable) to a constant total:

$$^{const}\Delta_{ij} = |\Delta_{ij}| \left/ \sum_{i=w-1}^{I-w+1} |\Delta_{ij}| \right.$$

where | indicates an absolute value and the window size for the derivatives equals w. This step is optional.
5. The final step involves averaging the values obtained in step 3 or 4 above as follows:

$$d_i = \sum_{j=1}^{J} {}^{const}\Delta_{ij} \left/ J \right. \quad \text{or} \quad d_i = \sum_{j=1}^{J} |\Delta_{ij}| \left/ J \right.$$

The calculation is illustrated for the data in Table 6.1. Table 6.8(a) shows the data where each row is summed to a constant total. Note that the first and last rows contain some large numbers: this is because the absolute intensity is low, with several negative as well as positive numbers that are similar in magnitude. Table 6.8(b) gives the absolute value of the first derivatives. Note the very large, and not very meaningful, numbers at times 3 and 23, a consequence of the five point window encompassing rows 1 and 25. Row 22 also contains a fairly large number, and so is not very diagnostic. In Table 6.8(c), points 3, 22 and 23 have been rejected, and the columns have now been set to a constant total of 1. Finally, the consensus absolute value of the derivative is presented in Table 6.8(d).

The resultant value of d_i is best presented on a logarithmic scale as in Figure 6.29. The regions of highest purity can be pinpointed fairly well as minima at times 7 and 15. These graphs are most useful for determining the purest points in time rather than

Table 6.8 Derivative calculation for determining the purity of regions in the data in Table 6.1.

(a) Scaling the rows to constant total

1	-4.066	2.561	3.269	-2.295	0.292	1.240	-1.911	1.694	1.181	-2.380	-0.642	2.059
2	-0.176	0.000	0.183	-0.005	0.260	-0.136	0.045	0.498	-0.012	-0.005	0.128	0.220
3	0.145	0.111	0.404	-0.004	0.155	0.212	-0.246	-0.104	0.118	-0.099	0.114	0.193
4	0.157	0.117	0.136	0.080	0.014	0.143	0.037	-0.044	0.050	0.095	0.102	0.113
5	0.070	0.130	0.175	0.143	0.097	0.064	-0.001	0.056	0.042	0.046	0.090	0.088
6	0.101	0.126	0.150	0.143	0.066	0.060	0.039	0.027	0.069	0.087	0.086	0.045
7	0.084	0.139	0.164	0.126	0.075	0.047	0.046	0.041	0.066	0.068	0.074	0.069
8	0.093	0.135	0.146	0.123	0.097	0.061	0.032	0.030	0.047	0.074	0.084	0.078
9	0.087	0.123	0.152	0.126	0.093	0.057	0.046	0.045	0.056	0.068	0.075	0.071
10	0.081	0.120	0.124	0.127	0.097	0.064	0.057	0.043	0.062	0.075	0.079	0.069
11	0.060	0.102	0.115	0.113	0.095	0.086	0.081	0.071	0.070	0.071	0.076	0.061
12	0.046	0.075	0.081	0.104	0.108	0.112	0.110	0.101	0.077	0.062	0.065	0.058
13	0.034	0.042	0.064	0.080	0.105	0.134	0.145	0.134	0.104	0.061	0.048	0.048
14	0.010	0.025	0.039	0.068	0.114	0.154	0.160	0.143	0.111	0.078	0.051	0.046
15	0.008	0.020	0.024	0.065	0.110	0.163	0.184	0.158	0.109	0.063	0.045	0.051
16	0.008	-0.004	0.033	0.066	0.108	0.169	0.191	0.157	0.118	0.066	0.044	0.043
17	0.032	-0.003	0.028	0.061	0.097	0.163	0.205	0.200	0.114	0.054	0.013	0.036
18	0.004	-0.014	0.050	0.096	0.113	0.175	0.211	0.172	0.076	0.029	0.031	0.058
19	0.023	-0.091	-0.108	-0.038	-0.018	0.393	0.275	0.168	0.252	0.021	0.073	0.049
20	0.008	-0.027	-0.230	-0.358	0.439	0.297	0.135	0.750	0.096	-0.217	-0.089	0.195
21	1.664	-0.575	-0.659	0.719	1.151	0.094	-0.387	-1.732	0.394	0.245	0.115	-0.030
22	-0.057	0.235	0.597	-0.985	1.238	1.323	0.885	-0.708	0.070	-0.344	-0.048	-1.207
23	-0.422	-0.045	0.191	0.057	0.217	-0.553	0.287	0.373	0.079	0.264	0.631	-0.079
24	-0.221	0.037	-0.094	0.023	0.281	0.466	-0.009	0.090	-0.181	-0.116	0.305	0.420
25	-12.974	3.211	2.434	5.197	-3.829	3.105	-1.803	-3.461	2.053	0.395	3.118	3.553

(b) *Absolute value of first derivative using a five-point Savitsky–Golay quadratic smoothing function*

1												
2												
3	0.8605	0.4745	0.6236	0.4962	0.0636	0.2073	0.3814	0.3818	0.2216	0.4953	0.1439	0.4049
4	0.0480	0.0270	0.0296	0.0444	0.0448	0.0244	0.0235	0.0783	0.0087	0.0330	0.0108	0.0455
5	0.0178	0.0065	0.0466	0.0323	0.0108	0.0414	0.0586	0.0362	0.0085	0.0326	0.0095	0.0316
6	0.0114	0.0046	0.0009	0.0070	0.0144	0.0180	0.0036	0.0133	0.0017	0.0020	0.0052	0.0089
7	0.0026	0.0006	0.0049	0.0055	0.0023	0.0013	0.0085	0.0018	0.0007	0.0031	0.0032	0.0000
8	0.0038	0.0028	0.0064	0.0032	0.0081	0.0020	0.0036	0.0036	0.0024	0.0024	0.0013	0.0049
9	0.0059	0.0090	0.0120	0.0023	0.0041	0.0081	0.0096	0.0072	0.0025	0.0006	0.0002	0.0027
10	0.0119	0.0142	0.0168	0.0053	0.0025	0.0131	0.0193	0.0169	0.0075	0.0021	0.0039	0.0051
11	0.0140	0.0207	0.0221	0.0116	0.0035	0.0202	0.0253	0.0235	0.0111	0.0026	0.0070	0.0058
12	0.0167	0.0250	0.0222	0.0151	0.0043	0.0228	0.0271	0.0263	0.0131	0.0004	0.0085	0.0058
13	0.0141	0.0213	0.0224	0.0131	0.0035	0.0197	0.0256	0.0217	0.0111	0.0001	0.0075	0.0031
14	0.0103	0.0180	0.0134	0.0090	0.0004	0.0143	0.0200	0.0135	0.0087	0.0009	0.0044	0.0026
15	0.0007	0.0120	0.0077	0.0041	0.0023	0.0073	0.0150	0.0146	0.0027	0.0025	0.0076	0.0026
16	0.0011	0.0101	0.0025	0.0050	0.0015	0.0042	0.0121	0.0100	0.0064	0.0106	0.0072	0.0008
17	0.0027	0.0231	0.0247	0.0177	0.0251	0.0465	0.0202	0.0035	0.0245	0.0122	0.0043	0.0010
18	0.0008	0.0134	0.0661	0.0947	0.0548	0.0485	0.0042	0.1153	0.0095	0.0600	0.0206	0.0317
19	0.3269	0.1158	0.1653	0.0864	0.2434	0.0017	0.1260	0.3287	0.0581	0.0135	0.0085	0.0005
20	0.1518	0.0013	0.0543	0.1404	0.3419	0.1998	0.0687	0.3660	0.0130	0.0521	0.0115	0.2609
21	0.0956	0.0354	0.1423	0.0438	0.1269	0.0865	0.0773	0.1047	0.0374	0.0360	0.1157	0.1658
22	0.2543	0.0660	0.1121	0.0098	0.1250	0.0309	0.0386	0.0785	0.0871	0.0220	0.1303	0.0401
23	2.9440	0.7375	0.5495	0.9964	1.0917	0.5164	0.3725	0.2660	0.3065	0.0527	0.6359	0.8792
24												
25												

(continued overleaf)

Table 6.8 (continued)

(c) Rejecting points 3, 22 and 23 and putting the measurements on a common scale

1											
2											
3											
4	0.065	0.045	0.082	0.050	0.042	0.043	0.066	0.038	0.124	0.046	0.079
5	0.024	0.071	0.060	0.012	0.071	0.107	0.031	0.037	0.122	0.040	0.055
6	0.015	0.001	0.013	0.016	0.031	0.007	0.011	0.008	0.007	0.022	0.015
7	0.003	0.007	0.010	0.003	0.002	0.016	0.002	0.003	0.012	0.013	0.000
8	0.005	0.010	0.006	0.009	0.003	0.007	0.003	0.010	0.009	0.005	0.008
9	0.008	0.018	0.004	0.005	0.014	0.018	0.006	0.011	0.002	0.001	0.005
10	0.016	0.025	0.010	0.003	0.023	0.035	0.014	0.033	0.008	0.016	0.009
11	0.019	0.033	0.021	0.004	0.035	0.046	0.020	0.049	0.010	0.029	0.010
12	0.023	0.034	0.028	0.005	0.039	0.049	0.022	0.058	0.001	0.036	0.010
13	0.019	0.034	0.024	0.004	0.034	0.047	0.018	0.049	0.001	0.032	0.005
14	0.014	0.020	0.017	0.000	0.025	0.037	0.011	0.038	0.003	0.019	0.005
15	0.001	0.012	0.008	0.003	0.013	0.027	0.012	0.012	0.009	0.032	0.005
16	0.001	0.004	0.009	0.002	0.007	0.022	0.008	0.028	0.040	0.030	0.001
17	0.004	0.037	0.033	0.028	0.080	0.037	0.003	0.108	0.046	0.018	0.002
18	0.001	0.100	0.175	0.061	0.084	0.008	0.097	0.042	0.225	0.087	0.055
19	0.444	0.250	0.160	0.272	0.003	0.230	0.277	0.256	0.051	0.036	0.001
20	0.206	0.082	0.260	0.382	0.345	0.125	0.309	0.057	0.195	0.049	0.450
21	0.130	0.216	0.081	0.142	0.149	0.141	0.088	0.164	0.135	0.489	0.286
22											
23											
24											
25											

(d) Calculating the final consensus derivative

i	d_i
1	
2	
3	
4	0.063
5	0.054
6	0.013
7	0.006
8	0.007
9	0.010
10	0.019
11	0.028
12	0.031
13	0.027
14	0.020
15	0.014
16	0.015
17	0.038
18	0.081
19	0.192
20	0.205
21	0.177
22	
23	
24	
25	

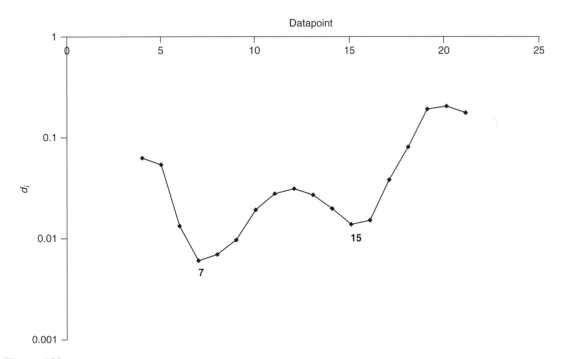

Figure 6.29
Derivative purity plot for the data in Table 6.1 with the purest points indicated

regions of differing composition, but the visual display is often very informative and can cope well with unusual peakshapes.

6.4 Resolution

Resolution or deconvolution of two-way chromatograms or mixture spectra involves converting a cluster of peaks into its constituent parts, each ideally representing a component of the signal from a single compound. The number of named methods in the literature is enormous, and it would be completely outside the scope of this text to discuss each approach in detail. In areas such as chemical pattern recognition or calibration, certain generic approaches are accepted as part of an overall strategy and the data preprocessing, variable selection, etc., are regarded as extra steps. In the field of resolution of evolutionary data, there is a fondness for packaging a series of steps into a named method, so there are probably 20 or more named methods, and maybe as many unnamed approaches reported in the literature. However, most are based on a number of generic principles, which are described in this chapter.

There are several aims for resolution.

1. Obtaining the profiles for each resolved compound. These might be the elution profiles (in chromatography) or the concentration distribution in a series of compounds (in spectroscopy of mixtures) or the pH profiles of different chemical species.

2. Obtaining the spectra of each pure compound. This allows identification or library searching. In some cases, this procedure merely uses the multivariate signals to improve on the quality of the individual spectra, which may be noisy, but in other cases, such as an embedded peak, genuinely difficult information can be gleaned. This is particularly useful in impurity monitoring.
3. Obtaining quantitative information. This involves using the resolved two-way data to provide concentrations (or relative concentrations when pure standards are not available).
4. Automation. Complex chromatograms may consist of 50 or more peaks, some of which will be noisy and overlapping. Speeding up procedures, for example, using rapid chromatography in a matter of minutes resulting in considerable overlap, rather than taking 30 min per chromatogram, also results in embedded peaks. Chemometrics can ideally pull out the constituents' spectra and profiles.

The methods in this chapter differ from those in Chapter 5 in that pure standards are not required for the model.

Whereas some datasets can be very complicated, it is normal to divide the data into small regions where there are signals from only a few components. Even in the spectroscopy of mixtures, in many cases such as MIR or NMR it is normally easy to find regions of the spectra where only two or three compounds at the most absorb, so this process of finding windows rather than analysing an entire dataset in one go is normal. Hence we will limit the discussion to three peak clusters in this section. Naturally the methods in Section 6.3 would usually first be applied to the entire dataset to identify these regions. We will illustrate the discussion below primarily in the context of coupled chromatography.

6.4.1 Selectivity for All Components

These methods involve first finding some pure or selective (composition 1) region in the chromatogram or selective spectral measurement such as an m/z value for each compound in a mixture.

6.4.1.1 Pure Spectra and Selective Variables

The most straightforward situation is when each compound has a composition 1 region. The simplest approach is to estimate the pure spectrum in such a region. There are several methods.

1. Take the spectrum at the point of maximum purity for each compound.
2. Average the spectra for each compound over each composition 1 region.
3. Perform PCA over each composition 1 region separately (so if there are three compounds, perform three PCA calculations) and then take the loadings of the first PC as an estimate of the pure spectrum. PCA is used as a smoothing technique, the idea being that the noise is banished to later PCs.

Some rather elaborate multivariate methods are also available that, instead of using the spectra in the composition 1 regions, use the elution profiles. In the case of Table 6.1 we might guess that the fastest eluting compound A has a composition 1 region between points 4 and 8, and the slowest eluting B between points 15 and 19. Hence we could divide up the chromatogram as follows.

1. points 1–3: no compounds elute;
2. points 4–8: compound A elutes selectively;
3. points 9–14: co-elution;
4. points 15–19: compound B elutes selectively;
5. points 20–25: no compounds elute.

As discussed above, there can be slight variations on this theme. This is represented in Figure 6.30. Chemometrics is used to fill in the remaining pieces of the jigsaw. The only unknowns are the elution profiles in the composition 2 regions. The profiles in the composition 1 regions can be estimated either by using the summed profiles or by performing PCA in these regions and taking the scores of the first PC.

An alternative is to find pure variables rather than composition 1 regions. These methods are popular when using various types of spectroscopy such as in LC–MS or in the MIR of mixtures. Wavelengths, frequencies or masses belonging to single compounds can often be identified. In the case of Table 6.3, we suspect that variables C and F are diagnostic of the two compounds (see Figure 6.16), and their profiles are presented in Figure 6.31. Note that these profiles are somewhat noisy. This is fairly common in techniques such as mass spectrometry. It is possible to improve the quality of the profiles by using methods for smoothing as described in Chapter 3, or to average profiles from several pure variables. The latter technique is useful in NMR or IR spectroscopy where a peak might be defined by several datapoints, or where there could be a number of selective regions in the spectrum.

The result of this section will be to produce either a first guess of all or part of the concentration profiles, represented by the matrix $\hat{C}$ or of the spectra $\hat{S}$.

6.4.1.2   Multiple Linear Regression

If pure profiles can be obtained from all components, the next step in deconvolution is straightforward.

In the case of Table 6.1, we can guess the pure spectra for A as the average of the data between times 4 and 8, and for B as the average between times 15 and 19. These

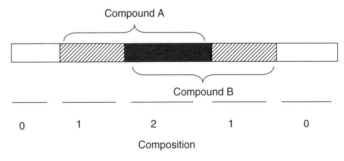

Figure 6.30
Composition of regions in chromatogram deriving from Table 6.1

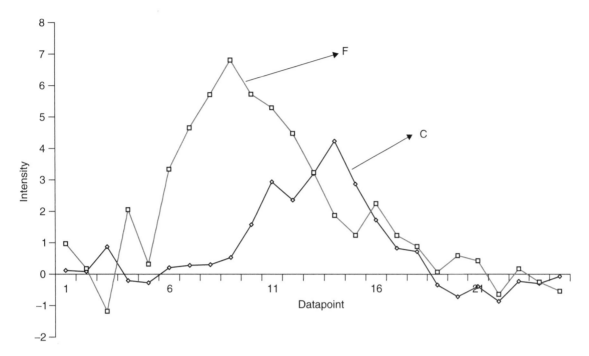

Figure 6.31
Profiles of variables C and F in Table 6.3

make up a 2×12 data matrix $\hat{S}$. Since

$$X \approx \hat{C}.\hat{S}$$

therefore

$$\hat{C} = X.\hat{S}'.(\hat{S}.\hat{S}')^{-1}$$

as discussed in Chapter 5 (Section 5.3). The estimated spectra are listed in Table 6.9 and the resultant profiles are presented in Figure 6.32. Note that the vertical scale in fact has no direct physical meaning: intensity data can only be reconstructed by multiplying the profiles by the spectra. However, MLR has provided a very satisfactory estimate, and provided that pure regions are available for each significant component, is probably entirely adequate as a tool in many cases.

If pure variables such as spectral frequencies or m/z values can be determined, even if there are embedded peaks, it is also possible to use these to obtain first estimates of

Table 6.9 Estimated spectra obtained from the composition 1 regions in the example of Table 6.1.

A	B	C	D	E	F	G	H	I	J	K	L
0.519	0.746	0.862	0.713	0.454	0.341	0.194	0.176	0.312	0.410	0.465	0.404
0.041	0.006	0.087	0.221	0.356	0.603	0.676	0.575	0.395	0.199	0.136	0.162

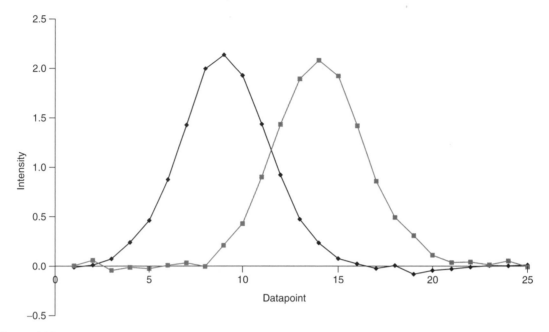

Figure 6.32
Reconstructed profiles for the data in Table 6.1 using MLR

elution profiles, $\hat{C}$, then the spectra can be obtain using all (or a great proportion of) the variables by

$$\hat{S} = (\hat{C}'.\hat{C})^{-1}.\hat{C}'.X$$

The concentration profile can be improved by increasing the variables; so, for example, the first guess might involve using one variable per compound, the next 20 significant variables and the final 100 or more. This approach is also useful in spectroscopy of mixtures, if pure frequencies can be identified for each compound. Using these for initial estimates of the concentrations of each compound in each spectrum, the full spectra can be reconstructed even when there are overlapping regions. Such approaches are useful in MIR, but not so valuable in NIR or UV/vis spectroscopy where it is often hard to find selective wavelengths and the effectiveness depends on the type of spectroscopy employed.

6.4.1.3 Principal Components Regression

PCR is an alternative to MLR (Section 5.4) and can be used in signal analysis just as in calibration. There are a number of ways of employing PCA, but a simple approach is to note that the scores and loadings can be related to the concentration profile and spectra by

$$X \approx \hat{C}.\hat{S} = T.R.R^{-1}.P$$

hence

$$\hat{C} = T.R$$

and

$$\hat{S} = R^{-1}.P$$

If we perform PCA on the dataset, and know the pure spectra, it is possible to find the matrix R^{-1} simply by regression since

$$R^{-1} = \hat{S}.P'$$

[because the loadings are orthonormal (Chapter 4, Section 4.3.2) this equation is simple]. It is then easy to obtain $\hat{C}$. This procedure is illustrated in Table 6.10 using the spectra as obtained from Table 6.9. The profiles are very similar to those presented in Figure 6.32 and so are not presented graphically for brevity.

PCR can be employed in more elaborate ways using the known profiles in the composition 1 (and sometimes composition 0) region for each compound. These methods were the basis of some of the earliest approaches to resolution of two-way chromatographic data. There are several variants, and one is as follows.

1. Choose only those regions where one component elutes. In our example in Table 6.1, we will use the regions between times 4–8 and 15–19 inclusive, which involves 10 points.
2. For each compound, use either the estimated profiles if the region is composition 1 or 0 if another compound elutes in this region. A matrix is obtained of size $Z \times 2$ whose columns correspond to each component, where Z equals the total number of composition 1 datapoints. In our example, the matrix is of size 10×2, half of the values being 0 and half consisting of the profile in the composition 1 region. Call this matrix Z.
3. Perform PCA on the overall matrix.
4. Find a matrix R such that $Z \approx T.R$ using the known profiles obtained in step 2, simply by using regression so that $R = (T'.T)^{-1}.T'.Z$ but including the scores *only* of the composition 1 region.
5. Knowing R, it is a simple matter to reconstruct the concentration profiles by including the scores over the entire data matrix as above, and similarly the spectra.

The key steps in the calculation are presented in Table 6.11. Note that the magnitude of the numbers in the matrix R differ from those presented in Table 6.10. This is simply because the magnitudes of the estimates of the spectra and profiles are different, and have no physical significance. The resultant profiles obtained by the multiplication $\hat{C} = T.R$ on the entire dataset are illustrated in Figure 6.33.

In straightforward cases, PCR is unnecessary and if not carefully controlled may provide worse results than MLR. However, for more complex systems it can be very useful.

Table 6.10 Estimation of profiles using PCA for the data in Table 6.9.

Loadings

0.215	0.321	0.375	0.372	0.333	0.305	0.288	0.255	0.248	0.237	0.236	0.214
−0.254	−0.377	−0.381	−0.191	0.085	0.369	0.479	0.422	0.203	−0.015	−0.113	−0.080

Matrix R^{-1}

1.667	−0.573
0.982	0.781

Matrix R

0.419	0.307
−0.527	0.894

Scores		*T.R*	
−0.019	0.007	−0.012	0.001
0.079	0.040	0.012	0.060
0.077	−0.076	0.072	−0.044
0.380	−0.147	0.237	−0.015
0.746	−0.284	0.462	−0.024
1.464	−0.493	0.873	0.009
2.412	−0.795	1.429	0.031
3.332	−1.147	2.000	−0.001
3.775	−1.066	2.143	0.208
3.646	−0.770	1.933	0.432
3.286	−0.116	1.438	0.906
2.954	0.593	0.925	1.438
2.650	1.210	0.473	1.896
2.442	1.504	0.231	2.095
2.020	1.461	0.077	1.927
1.432	1.098	0.022	1.421
0.803	0.682	−0.022	0.856
0.495	0.377	0.009	0.489
0.158	0.288	−0.086	0.306
0.037	0.112	−0.043	0.111
−0.013	0.046	−0.029	0.037
0.026	0.039	−0.010	0.043
0.026	0.009	0.006	0.016
0.057	0.041	0.003	0.054
0.011	−0.012	0.011	−0.007

6.4.2 Partial Selectivity

More difficult situations occur when only some components exhibit selectivity. A common example is a completely embedded peak in HPLC–DAD. In the case of LC–MS or LC–NMR, this problem is often solved by finding pure variables, but because UV/vis spectra are often completely overlapping it is not always possible to treat data in this manner.

Fortunately, PCA comes to the rescue. In Chapter 5 we discussed the different applicabilities of PCR and MLR. Using the former method, we stated that it was not necessary to have information about the concentration of every component in the mixture, simply a good idea of how many significant components there are. So in the case of resolution of two-way data these approaches can easily be extended. We

Table 6.11 Key steps in the calculation of the rotation matrix for the data in Table 6.1 using scores in composition 1 regions.

Time	Z		T	
	Compound A	Compound B	PC 1	PC 2
1				
2				
3				
4	1.341	0.000	0.380	−0.147
5	2.462	0.000	0.746	−0.284
6	4.910	0.000	1.464	−0.493
7	8.059	0.000	2.412	−0.795
8	11.202	0.000	3.332	−1.147
9				
10				
11				
12				
13				
14				
15	0.000	7.104	2.020	1.461
16	0.000	5.031	1.432	1.098
17	0.000	2.838	0.803	0.682
18	0.000	1.696	0.495	0.377
19	0.000	0.625	0.158	0.288
20				
21				
22				
23				
24				
25				

Matrix R

2.311	1.094
−3.050	3.197

will illustrate this using the data in Table 6.2 which correspond to three peaks, the middle one being completely embedded in the others. There are several different ways of exploiting this.

One approach uses the idea of a zero concentration window. The first step is to identify compounds that we know have selective regions, and determine where they do *not* elute. This information may be obtained from a variety of approaches such as eigenvalue or PC plots. In this region we expect the intensity of the data to be zero, so it is possible to find a vector r for each compound so that

$$0 = T_0.r$$

where 0 is a vector of zeros, and T_0 stands for the portion of the scores in this region; normally one excludes the region where no peaks elute, and then finds the zero component region for each component. For example, if we record a cluster over 50 datapoints and we suspect that there are three peaks eluting between points 10 and 25, 20 and 35 and 30 and 45, then there are three T_0 matrices, and for the fastest eluting

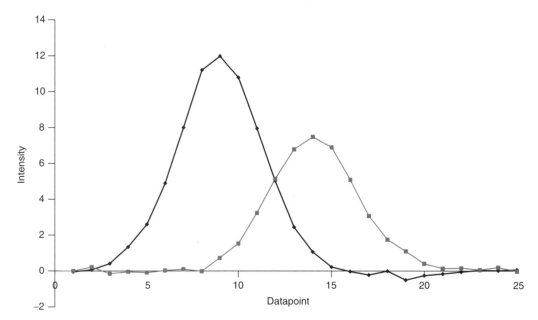

Figure 6.33
Profiles obtained as described in Section 6.4.1.3

component this is between points 26 and 45. Note that the number of PCs should be made equal to the number of compounds in the overall mixture. There is one small problem in that one value of the vector r must be set to an arbitrary number; usually the first coefficient is set to 1, but this does not have a serious effect on the algorithm. The equation can then be solved as follows. Separate out the contribution from the first PC to that from all the others so, setting r_1 to 1,

$$T_{0(2:K)}.r_{2:K} \approx -t_{01}$$

so that

$$r_{2:K} = \left[T'_{0(2:K)}.T_{0(2:K)} \right]^{-1}.T'_{0(2:K)}.t_{01}$$

where $r_{2:K}$ is a column vector of length $K - 1$ where there are K PCs, t_{01} is the scores of the first PC over the zero concentration window and $T_{0(2:K)}$ the scores of the remaining PCs. It is important to ensure that K equals the number of components suspected to elute within the cluster of peaks. It is possible to perform this operation on any embedded peaks because these also exhibit zero composition regions.

The profiles of all the compounds can now be obtained over the entire region by

$$\hat{C} = T.R$$

and the spectra by

$$\hat{S} = R^{-1}.P$$

We will illustrate this with the example of Table 6.2. From inspecting the data we might conclude that compound A elutes between times 4 and 13, B between times

9 and 17 and C between times 13 and 22. This information could be obtained by a variety of methods, as discussion in Section 6.3. Hence the zero composition regions are as follows:

- compound A: points 14–22;
- compound B: points 4–8 and 18–22;
- compound C: points 4–12.

Obviously different approaches may identify slightly different regions. The calculation is presented in Table 6.12 for the data in Table 6.2, and the resultant profiles and spectra are presented in Figure 6.34.

Table 6.12 Determing spectrum and elution profiles of an embedded peak.

(a) Choosing matrices T_0: the composition 0 regions are used to identify the portions of the overall scores matrix for compounds A, B and C

Time	T			Composition 0 regions		
				A	B	C
1	0.011	−0.006	−0.052			
2	−0.049	0.036	0.035			
3	−0.059	−0.002	0.084			
4	0.120	−0.099	−0.033		0	0
5	0.439	−0.018	0.129		0	0
6	1.029	−0.205	0.476		0	0
7	2.025	−0.379	0.808		0	0
8	2.962	−0.348	1.133		0	0
9	3.505	−0.351	1.287			0
10	3.501	0.088	1.108			0
11	3.213	0.704	0.353			0
12	2.774	1.417	−0.224			0
13	2.683	1.451	−0.646			
14	2.710	1.091	−0.885	0		
15	2.735	0.178	−0.918	0		
16	2.923	−0.718	−0.950	0		
17	2.742	−1.265	−0.816	0		
18	2.359	−1.256	−0.761	0	0	
19	1.578	−0.995	−0.495	0	0	
20	0.768	−0.493	−0.231	0	0	
21	0.428	−0.195	−0.065	0	0	
22	0.156	−0.066	−0.016	0	0	
23	−0.031	−0.049	0.005			
24	−0.110	−0.007	−0.095			
25	0.052	−0.057	0.074			

(b) Determining a matrix R

A	B	C
1	1	1
0.078	2.603	−2.476
3.076	−1.560	−3.333

(continued overleaf)

Table 6.12   *(continued)*

(c) Determining the concentration profiles
using $\hat{C} = T.R$

	A	B	C
1	−0.150	0.079	0.200
2	0.062	−0.009	−0.255
3	0.200	−0.196	−0.335
4	0.009	−0.085	0.475
5	0.835	0.192	0.052
6	2.476	−0.245	−0.050
7	4.480	−0.222	0.272
8	6.419	0.288	0.049
9	7.436	0.584	0.086
10	6.916	2.001	−0.410
11	4.354	4.494	0.295
12	2.194	6.813	0.012
13	0.809	7.466	1.244
14	0.072	6.931	2.959
15	−0.075	4.630	5.353
16	−0.055	2.536	7.869
17	0.133	0.722	8.596
18	−0.079	0.277	8.004
19	−0.022	−0.239	5.691
20	0.020	−0.155	2.757
21	0.213	0.022	1.126
22	0.100	0.009	0.374
23	−0.021	−0.167	0.076
24	−0.401	0.019	0.223
25	0.274	−0.211	−0.053

Many papers and theses have been written about this problem, and there are a large number of modifications to this approach, but in this text we illustrate using one of the best established approaches.

6.4.3   Incorporating Constraints

Finally, it is important to mention another class of methods. In many cases it is not possible to obtain a unique mathematical solution to the multivariate resolution of complex mixtures, and the problem of embedded peaks without selectivity, which may occur, for example, in impurity monitoring, causes difficulties when using many conventional approaches.

There is a huge literature on algorithm development under such circumstances, which cannot be fully reviewed in this book. However, many modern methods attempt to incorporate chemical knowledge or constraints about a system. For example, underlying chromatographic profiles should be unimodal, and spectra and chromatographic profiles positive, so the reconstructions of Figure 6.34, whilst providing a good starting point, suggest that there is still some way to go before the embedded peak is modelled correctly. In many cases there exist a large number of equally good statistical solutions that fit a dataset, but many are unrealistic in chemical terms. Most algorithms try to narrow down the possible solutions to those that obey constraints. Often this is done

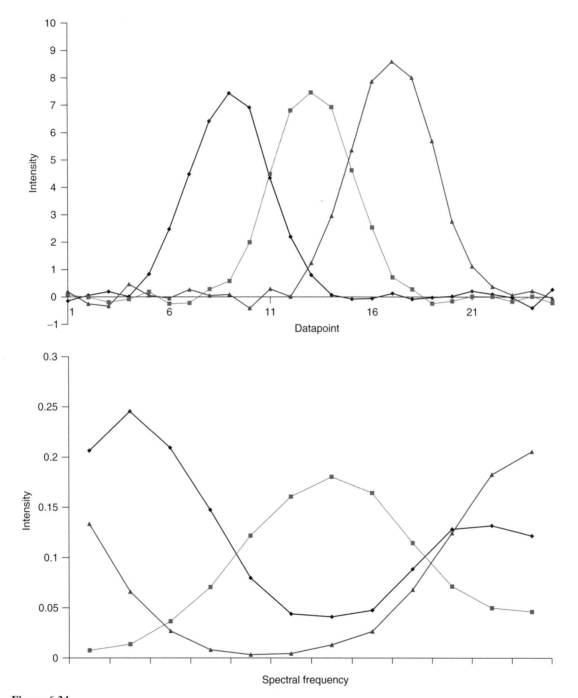

Figure 6.34
Profiles and spectra of three peaks obtained as in Section 6.4.2

in an iterative manner, improving the fit to the data at the same time as ensuring that the solutions are physically meaningful.

These approaches are fairly complex and the enthusiast should either develop their own methods or code in the approaches from source literature. There are very few public domain software packages in this area, although some have been specially commissioned for industry or instrument manufacturers. One of the difficulties is that different problems occur according to instrumental technique and, with rapid changes in technology, new types of measurement come into vogue. A good example is the movement away from HPLC–DAD towards LC–MS and LC–NMR. The majority of the chemometrics literature in the area of resolution of two-way chromatograms still involves HPLC–DAD, where spectra are often overlapping, so that there is often no selectivity in the spectral dimension. However, in many other types of coupled chromatography there are often some selective variables, but many new difficulties relating to preprocessing, variable selection and preparation of the data arise. For example, in LC–NMR, Fourier transformation, spectral smoothing, alignment, baseline correction and variable selection play an important role, but it is often easier to find selective variables compared with HPLC–DAD, so the effort is concentrated in other areas. Also in MS and NMR there will be different sorts of spectral information that can be exploited, so sophisticated knowledge can be incorporated into an algorithm. When developing methods for other applications such as infrared spectroscopy, reaction monitoring or equilibria studies, very specific and technique dependent knowledge must be introduced.

Problems

Problem 6.1 Determining of Purity Within a Two-component Cluster: Derivatives, Correlation Coefficients and PC plots

Section 6.2.2   Section 6.3.3   Section 6.3.5

The table on page 399 represents an HPLC–DAD chromatogram recorded at 27 wavelengths (the low digital resolution is used for illustrative purposes) and 30 points in time. The wavelengths in nanometres are presented at the top.

1. Calculate the 29 correlation coefficients between successive points in time, and plot a graph of these. Remove the first correlation coefficients and replot the graph. Comment on these graphs. How might you improve the graph still further?
2. Use first derivatives to look at purity as follows.
   a. Sum the spectrum (to a total of 1) at each point in time.
   b. At each wavelength and points 3–28 in time, calculate the absolute value of the five point quadratic Savitsky–Golay derivative (see Chapter 3, Table 3.6). You should produce a matrix of size 26×27.
   c. Average these over all wavelengths and plot this graph against time.
   d. Improve this graph by using a logarithmic scale for the parameter calculated in step c.
   Comment on what you observe.
3. Perform PCA on the raw uncentred data, retaining the first two PCs. Plot a graph of the scores of PC2 versus 1, labelling the points. What do you observe from this graph and how does it compare to the plots in questions 1 and 2?

Problem 6.1

206.33	213.35	220.38	227.41	234.45	241.50	248.56	255.62	262.70	269.77	276.86	283.95	291.05	298.15	305.27	312.38	319.51	326.64	333.78	340.92	348.07	355.22	361.48	368.05	375.24	382.43	389.63
0.0204	0.0153	0.0127	0.0073	0.0051	0.0026	0.0016	-0.0020	0.0002	-0.0004	0.0111	0.0092	0.0006	0.0015	-0.0028	-0.0004	0.0008	-0.0011	0.0021	0.0026	0.0042	0.0035	-0.0045	0.0052	0.0026	0.0019	0.0062
0.0258	0.0211	0.0197	0.0141	0.0093	0.0078	0.0088	0.0058	0.0070	0.0081	0.0258	0.0260	0.0052	0.0032	-0.0012	-0.0008	0.0022	0.0009	0.0048	0.0040	0.0038	0.0044	-0.0026	0.0053	0.0022	0.0033	0.0058
0.0361	0.0327	0.0347	0.0273	0.0169	0.0183	0.0228	0.0209	0.0214	0.0234	0.0545	0.0605	0.0136	0.0063	0.0011	0.0032	0.0048	0.0036	0.0077	0.0071	0.0050	0.0056	-0.0006	0.0081	0.0003	0.0050	0.0057
0.0522	0.0518	0.0620	0.0513	0.0294	0.0356	0.0468	0.0469	0.0463	0.0501	0.1046	0.1207	0.0277	0.0119	0.0036	0.0068	0.0087	0.0077	0.0111	0.0121	0.0086	0.0077	0.0015	0.0096	0.0003	0.0050	0.0056
0.0769	0.0812	0.1040	0.0883	0.0487	0.0611	0.0833	0.0871	0.0851	0.0922	0.1811	0.2120	0.0488	0.0209	0.0068	0.0117	0.0134	0.0145	0.0168	0.0198	0.0136	0.0116	0.0038	0.0088	0.0020	0.0049	0.0050
0.1098	0.1194	0.1589	0.1371	0.0744	0.0944	0.1313	0.1402	0.1371	0.1480	0.2813	0.3323	0.0756	0.0335	0.0114	0.0184	0.0199	0.0242	0.0254	0.0303	0.0203	0.0177	0.0071	0.0084	0.0049	0.0064	0.0044
0.1465	0.1617	0.2194	0.1910	0.1033	0.1316	0.1844	0.1984	0.1957	0.2103	0.3921	0.4654	0.1051	0.0477	0.0173	0.0265	0.0281	0.0351	0.0351	0.0405	0.0285	0.0250	0.0106	0.0130	0.0062	0.0079	0.0047
0.1795	0.1997	0.2742	0.2389	0.1291	0.1653	0.2323	0.2506	0.2485	0.2653	0.4917	0.5843	0.1315	0.0589	0.0233	0.0334	0.0359	0.0443	0.0432	0.0494	0.0360	0.0327	0.0150	0.0194	0.0065	0.0086	0.0065
0.1994	0.2243	0.3102	0.2697	0.1456	0.1872	0.2638	0.2850	0.2830	0.3015	0.5574	0.6613	0.1491	0.0660	0.0275	0.0371	0.0404	0.0506	0.0478	0.0553	0.0414	0.0381	0.0182	0.0215	0.0062	0.0091	0.0070
0.2030	0.2307	0.3194	0.2771	0.1498	0.1926	0.2724	0.2945	0.2919	0.3115	0.5748	0.6810	0.1540	0.0684	0.0292	0.0379	0.0404	0.0528	0.0490	0.0579	0.0423	0.0404	0.0195	0.0202	0.0051	0.0108	0.0064
0.1925	0.2193	0.3027	0.2622	0.1420	0.1822	0.2579	0.2797	0.2771	0.2956	0.5437	0.6451	0.1458	0.0656	0.0278	0.0365	0.0377	0.0505	0.0472	0.0552	0.0392	0.0393	0.0192	0.0197	0.0040	0.0108	0.0061
0.1724	0.1954	0.2688	0.2321	0.1261	0.1608	0.2277	0.2482	0.2469	0.2615	0.4789	0.5683	0.1295	0.0585	0.0247	0.0325	0.0337	0.0441	0.0424	0.0481	0.0348	0.0352	0.0166	0.0191	0.0028	0.0100	0.0064
0.1484	0.1672	0.2283	0.1961	0.1071	0.1356	0.1917	0.2114	0.2123	0.2207	0.4005	0.4732	0.1094	0.0492	0.0214	0.0271	0.0299	0.0355	0.0360	0.0395	0.0294	0.0292	0.0120	0.0174	0.0025	0.0093	0.0069
0.1253	0.1414	0.1903	0.1615	0.0893	0.1124	0.1589	0.1793	0.1837	0.1836	0.3249	0.3806	0.0899	0.0409	0.0190	0.0225	0.0269	0.0278	0.0292	0.0317	0.0241	0.0238	0.0066	0.0148	0.0026	0.0089	0.0069
0.1067	0.1213	0.1601	0.1326	0.0752	0.0944	0.1338	0.1577	0.1666	0.1550	0.2611	0.3032	0.0730	0.0349	0.0178	0.0201	0.0242	0.0215	0.0237	0.0253	0.0188	0.0198	0.0041	0.0120	0.0028	0.0086	0.0064
0.0931	0.1068	0.1382	0.1110	0.0647	0.0814	0.1164	0.1470	0.1612	0.1354	0.2104	0.2429	0.0603	0.0305	0.0176	0.0188	0.0222	0.0163	0.0194	0.0201	0.0145	0.0158	0.0041	0.0093	0.0026	0.0062	0.0064
0.0829	0.0963	0.1225	0.0950	0.0574	0.0718	0.1046	0.1430	0.1628	0.1227	0.1710	0.1964	0.0512	0.0272	0.0175	0.0181	0.0212	0.0126	0.0160	0.0167	0.0117	0.0121	0.0044	0.0081	0.0027	0.0060	0.0067
0.0750	0.0888	0.1105	0.0825	0.0519	0.0639	0.0954	0.1406	0.1650	0.1133	0.1389	0.1586	0.0440	0.0244	0.0169	0.0177	0.0212	0.0099	0.0142	0.0135	0.0094	0.0091	0.0039	0.0075	0.0026	0.0072	0.0069
0.0681	0.0815	0.0998	0.0710	0.0463	0.0561	0.0862	0.1354	0.1620	0.1032	0.1108	0.1261	0.0377	0.0219	0.0165	0.0169	0.0206	0.0076	0.0126	0.0104	0.0083	0.0071	0.0033	0.0066	0.0028	0.0079	0.0068
0.0609	0.0729	0.0882	0.0599	0.0400	0.0479	0.0756	0.1247	0.1515	0.0904	0.0866	0.0977	0.0319	0.0192	0.0161	0.0150	0.0190	0.0056	0.0116	0.0082	0.0075	0.0057	0.0013	0.0072	0.0034	0.0069	0.0069
0.0529	0.0630	0.0750	0.0488	0.0334	0.0397	0.0636	0.1083	0.1345	0.0754	0.0666	0.0735	0.0263	0.0158	0.0149	0.0119	0.0169	0.0037	0.0098	0.0059	0.0063	0.0051	-0.0004	0.0088	0.0034	0.0041	0.0073
0.0449	0.0531	0.0613	0.0387	0.0275	0.0319	0.0515	0.0887	0.1132	0.0601	0.0501	0.0540	0.0207	0.0122	0.0126	0.0087	0.0148	0.0025	0.0079	0.0039	0.0049	0.0048	-0.0015	0.0097	0.0033	0.0035	0.0073
0.0384	0.0447	0.0488	0.0308	0.0223	0.0252	0.0400	0.0699	0.0908	0.0462	0.0372	0.0387	0.0154	0.0094	0.0101	0.0061	0.0121	0.0017	0.0062	0.0023	0.0036	0.0050	-0.0024	0.0084	0.0021	0.0037	0.0074
0.0336	0.0375	0.0387	0.0247	0.0179	0.0198	0.0300	0.0546	0.0700	0.0345	0.0272	0.0273	0.0104	0.0074	0.0075	0.0050	0.0087	0.0008	0.0049	0.0013	0.0033	0.0046	-0.0032	0.0087	0.0018	0.0041	0.0069
0.0301	0.0316	0.0312	0.0195	0.0145	0.0157	0.0227	0.0425	0.0531	0.0252	0.0204	0.0201	0.0068	0.0060	0.0051	0.0048	0.0064	0.0005	0.0043	0.0008	0.0031	0.0033	-0.0044	0.0089	0.0020	0.0044	0.0061
0.0270	0.0272	0.0258	0.0149	0.0117	0.0124	0.0176	0.0324	0.0405	0.0176	0.0157	0.0160	0.0044	0.0045	0.0030	0.0043	0.0052	-0.0001	0.0033	-0.0002	0.0026	0.0017	-0.0050	0.0078	0.0023	0.0041	0.0055
0.0245	0.0232	0.0219	0.0115	0.0093	0.0094	0.0136	0.0238	0.0312	0.0118	0.0119	0.0131	0.0030	0.0028	0.0014	0.0033	0.0042	0.0000	0.0023	-0.0004	0.0026	0.0013	-0.0057	0.0069	0.0019	0.0029	0.0049
0.0228	0.0199	0.0188	0.0095	0.0073	0.0067	0.0096	0.0167	0.0239	0.0077	0.0097	0.0105	0.0017	0.0010	0.0005	0.0021	0.0026	-0.0003	0.0020	-0.0008	0.0027	0.0016	-0.0060	0.0079	0.0001	0.0012	0.0054
0.0215	0.0172	0.0164	0.0085	0.0059	0.0046	0.0061	0.0114	0.0172	0.0048	0.0085	0.0080	0.0006	-0.0002	-0.0005	0.0008	0.0006	-0.0009	0.0025	-0.0012	0.0025	0.0018	-0.0051	0.0092	-0.0009	0.0006	0.0070
0.0205	0.0152	0.0147	0.0077	0.0052	0.0033	0.0035	0.0073	0.0113	0.0020	0.0081	0.0067	0.0002	-0.0004	-0.0015	-0.0006	-0.0004	-0.0018	0.0038	-0.0010	0.0018	0.0023	-0.0042	0.0076	-0.0003	0.0007	0.0088

Problem 6.2 Evolutionary and Window Factor Analysis in the Detection of an Embedded Peak

Section 6.3.4   Section 6.4.1.3   Section 6.4.2

The following small dataset represents an evolutionary process consisting of two peaks, one embedded, recorded at 16 points in time and over six variables.

0.156	0.187	0.131	0.119	0.073	0.028
0.217	0.275	0.229	0.157	0.096	0.047
0.378	0.456	0.385	0.215	0.121	0.024
0.522	0.667	0.517	0.266	0.178	0.065
0.690	0.792	0.705	0.424	0.186	0.060
0.792	0.981	0.824	0.541	0.291	0.147
0.841	1.078	0.901	0.689	0.400	0.242
0.832	1.144	0.992	0.779	0.568	0.308
0.776	1.029	0.969	0.800	0.650	0.345
0.552	0.797	0.749	0.644	0.489	0.291
0.377	0.567	0.522	0.375	0.292	0.156
0.259	0.330	0.305	0.202	0.158	0.068
0.132	0.163	0.179	0.101	0.043	0.029
0.081	0.066	0.028	0.047	0.006	0.019
0.009	0.054	0.056	0.013	−0.042	−0.031
0.042	−0.005	0.038	−0.029	−0.013	−0.057

1. Perform EFA on the data (using uncentred PCA) as follows.
   - For forward EFA, perform PCA on the 3×6 matrix consisting of the first three spectra, and retain the three eigenvalues.
   - Then perform PCA on the 4×6 matrix consisting of the first four spectra, retaining the first three eigenvalues.
   - Continue this procedure, increasing the matrix by one row at a time, until a 14×3 matrix is obtained whose rows correspond to the ends of each window (from 3 to 16) and columns to the eigenvalues.
   - Repeat the same process but for backward EFA, the first matrix consisting of the three last spectra (14–16) to give another 14×3 matrix.
   If you are able to program in Matlab or VBA, it is easiest to automate this, but it is possible simply to use repeatedly the PCA add-in for each calculation.
2. Produce EFA plots, first converting the eigenvalues to a logarithmic scale, superimposing six graphs; always plot the eigenvalues against the extreme rather than middle or starting value of each window. Comment.
3. Perform WFA, again using an uncentred data matrix, with a window size of 3. To do this, simply perform PCA on spectra 1–3, and retain the first three eigenvalues. Repeat this for spectra 2–4, 3–5, and so on. Plot the logarithms of the first three eigenvalues against window centre and comment.
4. There are clearly two components in this mixture. Show how you could distinguish the situation of an embedded peak from that of two peaks with a central region of co-elution, and demonstrate that we are dealing with an embedded peak in this situation.
5. From the EFA plot it is possible to identify the composition 1 regions for the main peak. What are these? Calculate the average spectrum over these regions, and use this as an estimate of the spectrum of the main component.

Problem 6.3 Variable Selection and PC plots in LCMS

Section 6.2.2   Section 6.2.3.2   Section 6.2.4   Section 6.2.3.1

The table on page 402 represents the intensity of 49 masses in LC–MS of a peak cluster recorded at 25 points in time. The aim of this exercise is to look at variable selection and the influence on PC plots. The data have been transposed to fit on a page, with the first column representing the mass numbers. Some preprocessing has already been performed with the original masses reduced slightly and the ion current at each mass set to a minimum of 0. You will probably wish to transpose the matrix so that the columns represent different masses.

1. Plot the total ion current (using the masses listed) against time. This is done by summing the intensity over all masses at each point in time.
2. Perform PCA on the dataset, but standardise the intensities at each mass, and retain two PCs. Present the scores plot of PC2 versus PC1, labelling all the points in time, starting from 1 the lowest to 25 the highest. Produce a similar loadings plot, also labelling the points and comment on the correspondence between these graphs.
3. Repeat this but sum the intensities at each point in time to 1 prior to standardising and performing PCA and produce scores plot of PC2 versus PC1, and comment. Why might it be desirable to remove points 1–3 in time? Repeat the procedure, this time using only points 4–25 in time. Produce PC2 versus PC1 scores and loadings plots and comment.
4. A very simple approach to variable selection involves sorting according to standard deviation. Take the standard deviations of the 49 masses using the raw data, and list the 10 masses with highest standard deviations.
5. Perform PCA, standardised, again on the reduced 25×10 dataset consisting of the best 10 masses according to the criterion of question 4, and present the labelled scores and loadings plots. Comment. Can you assign m/z values to the components in the mixture?

Problem 6.4 Use of Derivatives, MLR and PCR in Signal Analysis

Section 6.3.5   Section 6.4.1.2   Section 6.4.1.3

The following data represent HPLC data recorded at 30 points in time and 10 wavelengths.

0.042	0.076	0.043	0.089	0.105	−0.004	0.014	0.030	0.059	0.112
0.009	0.110	0.127	0.179	0.180	0.050	0.015	0.168	0.197	0.177
−0.019	0.118	0.182	0.264	0.362	0.048	0.147	0.222	0.375	0.403
0.176	0.222	0.329	0.426	0.537	0.115	0.210	0.328	0.436	0.598
0.118	0.304	0.494	0.639	0.750	0.185	0.267	0.512	0.590	0.774
0.182	0.364	0.554	0.825	0.910	0.138	0.343	0.610	0.810	0.935
0.189	0.405	0.580	0.807	1.005	0.209	0.404	0.623	0.811	1.019
0.193	0.358	0.550	0.779	0.945	0.258	0.392	0.531	0.716	0.964
0.156	0.302	0.440	0.677	0.715	0.234	0.331	0.456	0.662	0.806
0.106	0.368	0.485	0.452	0.666	0.189	0.220	0.521	0.470	0.603
0.058	0.262	0.346	0.444	0.493	0.188	0.184	0.336	0.367	0.437

(*continued on p. 403*)

Problem 6.3

ID		334	905	1286	1917	1620	1831	1694	1423	1860	1678	1770	1668	1148	1492	1571	1402	1181	1207	1173	1146	919	527	1146	1002
41	0	334	905	1286	1917	1620	1831	1694	1423	1860	1678	1770	1668	1148	1492	1571	1402	1181	1207	1173	1146	919	527	1146	1002
42	530	578	386	245	229	119	69	164	162	176	34	286	283	106	72	67	315	239	284	335	191	288	157	638	236
44	242	567	353	188	178	0	158	328	341	455	325	517	162	262	265	180	137	110	102	91	141	158	64	354	219
50	66	111	60	73	39	0	3	39	52	83	88	59	75	63	7	73	91	68	46	107	24	34	208	76	67
54	165	83	107	88	11	110	30	35	45	39	13	98	69	99	132	134	77	158	110	558	65	135	208	128	104
59	1111	1143	593	706	43	558	19	139	499	130	379	222	0	240	537	299	234	803	375	106	41	299	658	623	739
61	360	278	515	375	14	131	0	162	206	661	180	363	224	322	225	252	98	310	70	324	179	194	364	173	463
68	594	430	770	919	1626	1530	1238	1260	722	108	550	523	648	594	749	771	600	632	494	514	612	364	479	568	650
69	287	562	440	367	550	259	233	124	0	104	57	131	292	125	258	234	438	263	425	212	430	436	320	403	385
72	962	135	427	153	314	278	16	118	109	405	173	124	105	160	268	196	72	84	180	882	152	199	103	265	163
76	1225	1256	114	589	498	453	41	130	398	294	592	325	56	506	574	128	608	1136	580	811	584	726	132	1028	754
77	35	1150	622	1094	0	54	1713	289	758	4479	3830	323	301	906	665	449	473	759	639	4122	125	340	901	123	925
78	246	256	73	0	271	671	81	2733	2633	46	36	4564	4476	4396	3225	4688	5054	4824	4836	135	3714	3703	4279	2996	3313
82	131	123	98	120	107	126	72	47	40	5	9	46	15	111	110	128	102	134	113	86	139	139	158	100	194
86	188	59	89	94	68	48	45	33	2	25	0	15	5	9	61	42	92	50	143	173	121	131	115	66	96
94	0	108	139	46	40	92	0	23	5	4578	4817	17	61	104	115	121	117	59	125	198	198	187	40	38	92
95	0	721	2172	3216	3545	5536	4551	5060	4589	12223	9457	5014	4648	4550	4515	4203	5025	5535	3103	4273	2273	3468	2416	3073	2926
96	0	2246	5742	9371	13004	13709	11352	14185	12855	1613	9457	13212	12832	12833	11816	11459	11006	9069	9768	11437	7258	7432	7375	10484	8036
97	132	631	377	1123	1489	1309	1170	1437	1515	59	1507	1426	1717	1717	1766	1624	1191	1201	1250	1322	1050	704	828	704	872
104	221	113	121	112	83	104	67	56	57	29	36	138	46	85	81	69	102	84	130	117	132	116	99	116	92
108	470	289	198	97	70	105	203	47	18	396	687	318	96	7	43	69	6	0	49	85	76	10	76	128	112
109	274	109	448	212	0	184	114	407	380	61	61	54	333	317	496	335	282	160	337	280	89	233	170	443	249
110	527	212	247	108	191	90	214	70	86	91	91	43	129	91	133	127	117	131	187	205	149	130	125	184	166
118	484	284	295	339	122	231	213	122	50	294	552	348	98	231	254	238	263	292	250	310	164	300	350	192	300
119	263	855	357	206	0	110	213	416	0	16	18	35	296	430	302	340	345	28	431	291	124	149	163	525	467
126	663	177	168	155	166	134	115	117	710	884	507	868	44	127	109	199	161	269	172	211	149	299	235	261	281
127	217	861	675	669	793	538	766	906	36	11	0	9	770	673	645	740	849	914	766	331	1050	758	0	880	1113
128	174	223	102	135	57	85	25	53	5	27	0	48	98	83	9	122	149	184	138	89	147	135	154	168	160
131	441	76	91	44	82	90	43	40	436	524	0	372	65	57	89	40	93	43	69	120	116	104	103	83	113
135	2452	598	555	249	235	103	0	359	892	2608	422	1531	512	548	377	504	209	296	491	301	254	228	325	712	524
136	1118	3865	4081	0	1256	1492	485	3372	485	190	1527	352	2492	2470	1717	3113	2322	1753	2687	4391	3904	3323	2472	4982	2813
137	216	234	222	68	123	72	557	412	21	14	606	27	0	151	278	74	176	133	437	574	458	303	768	268	783
138	286	152	15	184	71	57	145	22	40	66	0	32	54	111	42	66	138	86	166	146	172	184	143	87	221
140	283	262	149	189	237	139	151	74	96	95	71	57	47	90	85	57	50	128	148	185	177	177	213	186	219
141	553	295	194	107	45	13	435	9	51	0	0	151	28	91	13	10	217	76	27	19	19	39	25	0	108
142	360	304	187	332	256	243	269	38	0	19	13	11	53	178	110	123	100	343	320	304	351	363	262	245	378
144	204	190	191	124	158	138	104	62	33	1737	19	35	11	65	81	52	129	110	149	197	179	196	164	157	173
149	0	80	133	104	73	104	119	27	33	2259	1678	1478	43	86	60	104	103	112	103	106	87	196	133	141	115
154	0	497	1225	1782	2516	2625	2111	2027	1806	28	1929	2232	1444	1358	1566	1200	1224	1046	907	1044	796	779	967	765	874
155	101	423	1179	1823	1670	1904	1900	1988	2013	68	43	2232	1886	2031	1819	2036	1909	1720	1708	1427	1408	1375	1250	1434	1388
158	547	19	45	54	0	41	160	5	46	1575	0	89	42	74	63	52	69	91	61	50	54	83	51	11	88
167	269	396	231	196	286	282	85	77	75	6	1567	89	180	112	147	155	172	314	238	301	377	390	320	337	259
168	334	146	132	207	163	93	43	73	21	145	51	35	58	19	98	124	137	184	161	101	220	189	112	129	147
169	58	207	206	229	92	132	141	141	148	206	115	105	123	179	139	173	202	218	183	253	180	180	140	163	206
172	380	0	137	345	633	876	1141	1261	1321	45	257	1328	1390	1657	1467	1447	1183	1310	1122	1005	815	838	920	679	773
195	200	187	184	195	164	159	0	37	20	0	0	124	64	95	180	63	98	221	204	302	206	318	223	217	305
196	109	415	160	125	82	28	71	202	172	0	0	211	15	184	246	146	147	204	166	166	45	57	88	11	54
197	92	178	250	115	77	50	0	151	236	0	0	17	140	35	46	122	48	70	37	29	20	140	220	249	159
198		97	54	39	44	33	63	26	5				8							37	64	38	25	29	49

(*continued from p. 401*)

0.159	0.281	0.431	0.192	0.488	0.335	0.196	0.404	0.356	0.265
0.076	0.341	0.629	0.294	0.507	0.442	0.252	0.592	0.352	0.196
0.138	0.581	0.883	0.351	0.771	0.714	0.366	0.805	0.548	0.220
0.223	0.794	1.198	0.543	0.968	0.993	0.494	1.239	0.766	0.216
0.367	0.865	1.439	0.562	1.118	1.130	0.578	1.488	0.837	0.220
0.310	0.995	1.505	0.572	1.188	1.222	0.558	1.550	0.958	0.276
0.355	0.895	1.413	0.509	1.113	1.108	0.664	1.423	0.914	0.308
0.284	0.723	1.255	0.501	0.957	0.951	0.520	1.194	0.778	0.219
0.350	0.593	0.948	0.478	0.738	0.793	0.459	0.904	0.648	0.177
0.383	0.409	0.674	0.454	0.555	0.629	0.469	0.684	0.573	0.126
0.488	0.220	0.620	0.509	0.494	0.554	0.580	0.528	0.574	0.165
0.695	0.200	0.492	0.551	0.346	0.454	0.695	0.426	0.584	0.177
0.877	0.220	0.569	0.565	0.477	0.582	0.747	0.346	0.685	0.168
0.785	0.230	0.486	0.724	0.346	0.601	0.810	0.370	0.748	0.147
0.773	0.204	0.435	0.544	0.321	0.442	0.764	0.239	0.587	0.152
0.604	0.141	0.417	0.504	0.373	0.458	0.540	0.183	0.504	0.073
0.493	0.083	0.302	0.359	0.151	0.246	0.449	0.218	0.392	0.110
0.291	0.050	0.096	0.257	0.034	0.199	0.238	0.142	0.271	0.018
0.204	0.034	0.126	0.097	0.092	0.095	0.215	0.050	0.145	0.034

The aim of this problem is to explore different approaches to signal resolution using a variety of common chemometric methods.

1. Plot a graph of the sum of intensities at each point in time. Verify that it looks as if there are three peaks in the data.
2. Calculate the derivative of the spectrum, scaled at each point in time to a constant sum, and at each wavelength as follows.
   a. Rescale the spectrum at each point in time by dividing by the total intensity at that point in time so that the total intensity at each point in time equals 1.
   b. Then calculate the smoothed five point quadratic Savitsky–Golay first derivatives as presented in Chapter 3, Table 3.6, independently for each of the 10 wavelengths. A table consisting of derivatives at 26 times and 10 wavelengths should be obtained.
   c. Superimpose the 10 graphs of derivatives at each wavelength.
3. Summarise the change in derivative with time by calculating the mean of the absolute value of the derivative over all 10 wavelengths at each point in time. Plot a graph of this, and explain why a value close to zero indicates a good pure or composition 1 point in time. Show that this suggests that points 6, 17 and 26 are good estimates of pure spectra for each component.
4. The concentration profiles of each component can be estimated using MLR as follows.
   a. Obtain estimates of the spectra of each pure component at the three points of highest purity, to give an estimated spectral matrix $\hat{S}$.
   b. Using MLR calculate $\hat{C} = X.\hat{S}'.(\hat{S}.\hat{S}')^{-1}$.
   c. Plot a graph of the predicted concentration profiles
5. An alternative method is PCR. Perform *uncentred* PCA on the raw data matrix X and verify that there are approximately three components.

6. Using estimates of each pure component given in question 4(a), perform PCR as follows.
   a. Using regression find the matrix R for which $\hat{S} = R^{-1}.P$ where P is the loadings matrix obtained in question 5; keep three PCs only.
   b. Estimate the elution profiles of all three peaks since $\hat{C} \approx T.R$.
   c. Plot these graphically.

Problem 6.5 Titration of Three Spectroscopically Active Compounds with pH

Section 6.2.2   Section 6.3.3   Section 6.4.1.2

The data in the table on page 405 represent the spectra of a mixture of three spectroscopically active species recorded at 25 wavelengths over 36 different values of pH.

1. Perform PCA on the raw uncentred data, and obtain the scores and loadings for the first three PCs.
2. Plot a graph of the loadings of the first PC and superimpose this on the graph of the average spectrum over all the observed pHs, scaling the two graphs so that they are of approximately similar size. Comment on why the first PC is not very useful for discriminating between the compounds.
3. Calculate the logarithm of the correlation coefficient between each successive spectrum, and plot this against pH (there will be 35 numbers; plot the logarithm of the correlation between the spectra at pH 2.15 and 2.24 against the lower pH). Show how this is consistent with there being three different spectroscopic species in the mixture. On the basis of three components, are there pure regions for each components, and over which pH ranges are these?
4. Centre the data and produce three scores plots, those of PC2 vs PC1, PC3 vs PC1 and PC3 vs PC2. Label each point with pH (Excel users will have to adapt the macro provided). Comment on these plots, especially in the light of the correlation graph in question 3.
5. Normalise the scores of the first two PCs obtained in question 4 by dividing by the square root of the sum of squares at each pH. Plot the graph of the normalised scores of PC2 vs PC1, labelling each point as in question 4, and comment.
6. Using the information above, choose one pH which best represents the spectra for each of the three compounds (there may be several answers to this, but they should not differ by a great deal). Plot the spectra of each pure compound, superimposed on one another.
7. Using the guesses of the spectra for each compound in question 7, perform MLR to obtain estimated profiles for each species by $\hat{C} = X.S'.(S.S')^{-1}$. Plot a graph of the pH profiles of each species.

Problem 6.6 Resolution of Mid-infrared Spectra of a Three-component Mixture

Section 6.2.2   Section 6.2.3.1   Section 6.4.1.2

The table on page 406 represents seven spectra consisting of different mixtures of three compounds, 1,2,3-trimethylbenzene, 1,3,5-trimethylbenzene and toluene, whose mid-infrared spectra have been recorded at 16 cm^{-1} intervals between 528 and 2000 nm, which you will need to reorganise as a matrix of dimensions 7×93.

Problem 6.5

pH	Wavelength (nm)																								
---	240	244	248	252	256	260	264	268	272	276	280	284	288	292	296	300	304	308	312	316	320	324	328	332	336
2.15	0.382	0.479	0.571	0.638	0.673	0.663	0.632	0.584	0.520	0.449	0.379	0.306	0.234	0.166	0.104	0.056	0.026	0.013	0.006	0.005	0.002	0.001	0.000	0.000	0.000
2.24	0.386	0.482	0.573	0.639	0.672	0.661	0.630	0.582	0.518	0.448	0.376	0.302	0.229	0.162	0.102	0.054	0.025	0.012	0.006	0.005	0.002	0.001	0.001	0.000	0.000
2.44	0.391	0.488	0.576	0.640	0.671	0.657	0.626	0.578	0.516	0.444	0.372	0.296	0.223	0.156	0.098	0.051	0.023	0.011	0.005	0.004	0.002	0.001	0.001	0.001	0.000
2.68	0.402	0.496	0.582	0.642	0.669	0.655	0.623	0.574	0.508	0.437	0.364	0.287	0.214	0.147	0.091	0.048	0.023	0.012	0.006	0.005	0.003	0.002	0.001	0.001	0.000
3.00	0.409	0.503	0.586	0.643	0.668	0.650	0.616	0.567	0.501	0.428	0.352	0.274	0.202	0.137	0.084	0.045	0.021	0.011	0.006	0.004	0.002	0.001	0.001	0.001	0.001
3.25	0.424	0.515	0.597	0.651	0.674	0.653	0.617	0.566	0.496	0.421	0.343	0.263	0.192	0.129	0.079	0.042	0.021	0.011	0.007	0.005	0.005	0.003	0.003	0.002	0.002
3.47	0.430	0.519	0.599	0.652	0.673	0.652	0.613	0.559	0.487	0.410	0.331	0.253	0.180	0.119	0.072	0.039	0.019	0.011	0.007	0.005	0.005	0.003	0.003	0.002	0.002
3.72	0.435	0.522	0.599	0.651	0.670	0.650	0.607	0.550	0.477	0.397	0.317	0.238	0.167	0.107	0.064	0.034	0.018	0.010	0.007	0.005	0.004	0.003	0.002	0.002	0.002
4.04	0.444	0.528	0.604	0.653	0.672	0.648	0.602	0.542	0.464	0.382	0.300	0.218	0.149	0.092	0.053	0.028	0.015	0.009	0.005	0.004	0.003	0.003	0.002	0.002	0.002
4.40	0.454	0.537	0.611	0.658	0.675	0.650	0.601	0.537	0.457	0.371	0.285	0.203	0.135	0.079	0.044	0.023	0.013	0.008	0.005	0.004	0.003	0.003	0.002	0.001	0.002
4.77	0.462	0.544	0.615	0.661	0.676	0.649	0.599	0.531	0.449	0.361	0.274	0.192	0.124	0.069	0.037	0.018	0.010	0.007	0.005	0.004	0.003	0.002	0.002	0.002	0.002
5.06	0.470	0.550	0.623	0.666	0.680	0.651	0.601	0.532	0.447	0.356	0.269	0.185	0.116	0.063	0.032	0.016	0.009	0.007	0.005	0.004	0.003	0.003	0.002	0.002	0.002
5.40	0.476	0.555	0.626	0.670	0.683	0.654	0.601	0.531	0.445	0.354	0.264	0.180	0.110	0.058	0.028	0.014	0.009	0.007	0.005	0.004	0.004	0.003	0.002	0.002	0.002
5.68	0.480	0.559	0.628	0.671	0.685	0.656	0.602	0.532	0.446	0.353	0.262	0.179	0.109	0.056	0.026	0.013	0.009	0.007	0.004	0.005	0.004	0.003	0.003	0.002	0.002
5.98	0.483	0.562	0.631	0.675	0.687	0.658	0.605	0.533	0.446	0.352	0.263	0.177	0.109	0.055	0.026	0.013	0.008	0.007	0.004	0.005	0.003	0.003	0.003	0.002	0.002
6.25	0.484	0.566	0.635	0.677	0.690	0.660	0.606	0.534	0.445	0.353	0.262	0.177	0.107	0.055	0.025	0.012	0.009	0.006	0.004	0.004	0.004	0.003	0.003	0.001	0.002
6.49	0.487	0.566	0.636	0.680	0.692	0.662	0.608	0.536	0.447	0.354	0.263	0.178	0.107	0.054	0.025	0.013	0.008	0.006	0.005	0.004	0.004	0.003	0.003	0.003	0.003
6.85	0.490	0.570	0.639	0.683	0.696	0.665	0.611	0.538	0.449	0.355	0.264	0.178	0.108	0.055	0.026	0.013	0.008	0.007	0.006	0.005	0.004	0.003	0.003	0.003	0.003
7.00	0.492	0.571	0.641	0.684	0.697	0.667	0.611	0.538	0.450	0.356	0.264	0.179	0.108	0.055	0.025	0.013	0.009	0.007	0.005	0.005	0.005	0.004	0.004	0.003	0.003
7.47	0.494	0.572	0.642	0.685	0.698	0.668	0.613	0.539	0.451	0.358	0.266	0.179	0.108	0.058	0.026	0.013	0.009	0.006	0.005	0.005	0.005	0.004	0.004	0.003	0.004
7.75	0.500	0.578	0.647	0.691	0.703	0.673	0.618	0.545	0.456	0.361	0.268	0.181	0.109	0.056	0.027	0.015	0.010	0.008	0.006	0.006	0.006	0.006	0.005	0.003	0.004
7.96	0.501	0.580	0.648	0.692	0.704	0.676	0.619	0.546	0.456	0.362	0.269	0.181	0.110	0.057	0.027	0.014	0.008	0.006	0.006	0.006	0.005	0.005	0.004	0.005	0.005
8.12	0.503	0.580	0.649	0.692	0.705	0.674	0.620	0.547	0.457	0.362	0.270	0.183	0.110	0.057	0.027	0.015	0.010	0.008	0.007	0.006	0.005	0.005	0.005	0.004	0.004
8.51	0.503	0.579	0.645	0.689	0.702	0.673	0.622	0.548	0.458	0.364	0.270	0.181	0.110	0.056	0.028	0.015	0.011	0.009	0.007	0.006	0.006	0.005	0.005	0.005	0.005
8.82	0.506	0.578	0.644	0.688	0.702	0.677	0.628	0.553	0.462	0.366	0.270	0.179	0.108	0.055	0.027	0.015	0.011	0.009	0.007	0.006	0.006	0.006	0.006	0.004	0.005
9.11	0.509	0.578	0.642	0.685	0.702	0.677	0.627	0.557	0.469	0.368	0.270	0.178	0.107	0.055	0.027	0.016	0.011	0.009	0.007	0.006	0.007	0.006	0.005	0.005	0.005
9.38	0.511	0.576	0.639	0.684	0.701	0.680	0.633	0.563	0.471	0.371	0.269	0.175	0.104	0.054	0.028	0.017	0.012	0.010	0.008	0.007	0.007	0.006	0.006	0.007	0.007
9.61	0.515	0.575	0.638	0.682	0.702	0.684	0.640	0.570	0.478	0.375	0.270	0.172	0.100	0.050	0.026	0.016	0.013	0.010	0.007	0.007	0.007	0.007	0.006	0.006	0.006
9.89	0.516	0.573	0.634	0.680	0.702	0.688	0.649	0.580	0.487	0.380	0.270	0.169	0.094	0.047	0.025	0.016	0.012	0.010	0.008	0.007	0.007	0.007	0.006	0.006	0.006
10.38	0.520	0.573	0.632	0.680	0.707	0.699	0.664	0.598	0.503	0.390	0.270	0.163	0.085	0.040	0.021	0.015	0.013	0.011	0.010	0.008	0.008	0.007	0.007	0.006	0.006
10.57	0.522	0.573	0.631	0.681	0.708	0.701	0.670	0.604	0.508	0.392	0.270	0.160	0.081	0.037	0.020	0.016	0.013	0.010	0.009	0.008	0.008	0.007	0.007	0.007	0.006
10.74	0.523	0.573	0.631	0.680	0.709	0.704	0.674	0.608	0.510	0.395	0.270	0.158	0.081	0.037	0.021	0.015	0.013	0.012	0.009	0.008	0.008	0.008	0.007	0.007	0.007
11.01	0.524	0.573	0.631	0.680	0.710	0.707	0.678	0.612	0.514	0.396	0.271	0.157	0.079	0.035	0.020	0.016	0.013	0.012	0.009	0.008	0.009	0.009	0.007	0.007	0.007
11.27	0.526	0.573	0.631	0.680	0.712	0.708	0.679	0.614	0.516	0.399	0.270	0.156	0.078	0.035	0.017	0.016	0.014	0.012	0.009	0.008	0.008	0.008	0.008	0.008	0.007
11.47	0.527	0.575	0.631	0.681	0.712	0.709	0.680	0.614	0.517	0.399	0.270	0.156	0.077	0.036	0.021	0.016	0.015	0.013	0.012	0.009	0.009	0.008	0.008	0.008	0.008
11.64	0.529	0.576	0.631	0.680	0.710	0.708	0.680	0.615	0.517	0.399	0.271	0.157	0.079	0.036	0.022	0.017	0.015	0.012	0.010	0.010	0.009	0.009	0.008	0.008	0.008

Problem 6.6 Samples along rows, wavelength in cm^{-1}

Sample	2000	1984	1968	1952	1936	1920	1904	1888	1872	1856	1840	1824	1808	1792	1776	1760	1744	1728	1712	1696	1680	1664	1648	1632	1616
1	211	214	256	381	422	531	456	354	382	421	447	312	398	426	425	657	617	472	647	359	343	378	488	840	5859
2	143	145	167	241	262	301	259	246	247	224	237	197	218	222	215	581	514	318	521	252	221	224	271	688	6503
3	164	169	258	489	484	374	285	265	424	412	368	272	472	495	336	475	487	379	461	303	285	265	325	738	5062
4	184	189	252	427	440	416	339	307	395	384	371	282	409	428	341	618	588	418	582	330	304	300	370	834	6563
5	211	217	310	552	574	578	469	340	495	566	549	355	590	639	524	459	502	485	513	368	374	399	525	756	3027
6	272	277	336	508	556	673	576	458	503	542	570	404	521	556	539	866	812	611	844	468	444	481	617	1110	7995
7	190	197	305	587	581	455	347	304	504	508	455	324	579	612	421	491	524	440	497	349	337	319	399	811	4825

Sample	1600	1584	1568	1552	1536	1520	1504	1488	1472	1456	1440	1424	1408	1392	1376	1360	1344	1328	1312	1296	1280	1264	1248	1232	1216
1	4251	1936	1507	1617	1956	2318	3339	5298	5282	4086	2429	1090	1153	1810	1453	771	585	457	434	382	356	397	432	339	363
2	4053	1407	1282	1412	1739	2051	2599	3195	2844	2635	1515	720	853	1245	1117	605	432	294	272	248	214	223	225	181	201
3	3587	1355	1163	1281	1662	2006	3724	4358	3345	2862	1755	879	866	1377	1141	600	432	332	326	284	248	267	319	270	316
4	4408	1657	1433	1573	1976	2356	3669	4573	3874	3343	1992	961	1025	1575	1337	713	516	380	363	322	283	304	338	277	314
5	2983	1707	1173	1237	1543	1869	3785	6011	5666	4036	2505	1143	1042	1771	1299	669	524	461	456	387	371	426	507	406	448
6	5693	2501	1985	2139	2602	3085	4468	6806	6642	5229	3106	1409	1497	2338	1898	1008	759	586	557	490	454	503	546	431	465
7	3678	1520	1236	1352	1759	2134	4243	5233	4140	3354	2080	1029	971	1586	1268	661	484	390	387	333	298	326	397	333	387

Sample	1200	1184	1168	1152	1136	1120	1104	1088	1072	1056	1040	1024	1008	992	976	960	944	928	912	896	880	864	848	832	816
1	372	352	384	356	329	390	1091	1462	972	839	988	887	936	959	718	410	343	553	420	382	290	274	1556	2919	684
2	218	194	216	213	207	223	434	579	506	650	751	507	595	558	353	215	228	466	316	286	186	186	1802	3472	663
3	314	270	290	271	258	334	696	1151	994	830	1143	982	585	578	447	300	289	468	362	351	260	229	1180	2182	502
4	322	285	310	295	282	338	732	1094	904	865	1099	885	725	707	507	320	310	546	401	375	267	248	1691	3198	674
5	437	407	435	389	353	468	1371	2007	1355	873	1174	1235	881	952	808	479	367	468	408	391	331	287	582	951	409
6	478	448	488	455	424	500	1349	1828	1255	1112	1323	1159	1190	1209	898	519	444	732	551	503	376	355	2120	3987	910
7	381	331	354	326	306	406	927	1509	1232	938	1313	1196	686	698	568	374	339	506	408	399	308	266	1049	1891	495

Sample	800	784	768	752	736	720	704	688	672	656	640	624	608	592	576	560	544	528
1	296	1238	4468	3372	2742	2709	2229	2593	1091	264	240	230	258	287	347	554	625	598
2	171	452	1404	1130	1084	1089	1071	2363	972	169	155	134	163	149	210	265	322	366
3	220	625	1924	2592	4946	5042	3565	3207	1376	252	222	216	239	288	357	451	538	658
4	240	725	5384	2396	3520	3568	2710	3175	1340	251	225	211	240	264	338	444	527	606
5	327	1482	5384	4846	5896	5921	4201	2737	1200	313	277	287	304	395	447	713	797	798
6	377	1510	5384	4180	3686	3659	2982	3550	1493	344	312	297	335	370	453	701	797	784
7	267	867	2821	3492	6235	6342	4432	3533	1529	299	263	262	285	357	430	571	670	791

1. Scale the data so that the sum of the spectral intensities at each wavelength equals 1 (note that this differs from the usual method which is along the rows, and is a way of putting equal weight on each wavelength). Perform PCA, without further preprocessing, and produce a plot of the loadings of PC2 vs PC1.
2. Many wavelengths are not very useful if they are low intensity. Identify those wavelengths for which the sum over all seven spectra is greater than 10 % of the wavelength that has the maximum sum, and label these in the graph in question 1.
3. Comment on the appearance of the graph in question 2, and suggest three wavelengths that are typical of each of the compounds.
4. Using the three wavelengths selected in question 3, obtain a 7×3 matrix of relative concentrations in each of the spectra and call this $\hat{C}$.
5. Calling the original data X, obtain the estimated spectra for each compound by $S = (\hat{C}'.\hat{C})^{-1}.\hat{C}'.X$ and plot these graphically.

Appendices

A.1 Vectors and Matrices

A.1.1 Notation and Definitions

A single number is often called a scalar, and is represented by italics, e.g. x.

A vector consists of a row or column of numbers and is represented by bold lower case italics, e.g. $\boldsymbol{x}$. For example, $\boldsymbol{x} = \begin{pmatrix} 3 & -11 & 9 & 0 \end{pmatrix}$ is a row vector and

$$\boldsymbol{y} = \begin{pmatrix} 5.6 \\ 2.8 \\ 1.9 \end{pmatrix}$$

is a column vector.

A matrix is a two-dimensional array of numbers and is represented by bold upper case italics e.g. $\boldsymbol{X}$. For example,

$$\boldsymbol{X} = \begin{pmatrix} 12 & 3 & 8 \\ -2 & 14 & 1 \end{pmatrix}$$

is a matrix.

The dimensions of a matrix are normally presented with the number of rows first and the number of columns second, and vectors can be considered as matrices with one dimension equal to 1, so that $\boldsymbol{x}$ above has dimensions 1×4 and $\boldsymbol{X}$ has dimensions 2×3.

A square matrix is one where the number of columns equals the number of rows. For example,

$$\boldsymbol{Y} = \begin{pmatrix} -7 & 4 & -1 \\ 11 & -3 & 6 \\ 2 & 4 & -12 \end{pmatrix}$$

is a square matrix.

An identity matrix is a square matrix whose elements are equal to 1 in the diagonal and 0 elsewhere, and is often denoted by $\boldsymbol{I}$. For example,

$$\boldsymbol{I} = \begin{pmatrix} 1 & 0 \\ 0 & 1 \end{pmatrix}$$

is an identity matrix.

The individual elements of a matrix are often referenced as scalars, with subscripts referring to the row and column; hence, in the matrix above, $y_{21} = 11$, which is the element in row 2 and column 1. Optionally, a comma can be placed between the subscripts for clarity; this is useful if one of the dimensions exceeds 9.

A.1.2 Matrix and Vector Operations

A.1.2.1 Addition and Subtraction

Addition and subtraction is the most straightforward operation. Each matrix (or vector) must have the same dimensions, and simply involves performing the operation element by element. Hence

$$
\begin{pmatrix} 9 & 7 \\ -8 & 4 \\ -2 & 4 \end{pmatrix} + \begin{pmatrix} 0 & -7 \\ -11 & 3 \\ 5 & 6 \end{pmatrix} = \begin{pmatrix} 9 & 0 \\ -19 & 7 \\ 3 & 10 \end{pmatrix}
$$

A.1.2.2 Transpose

Transposing a matrix involves swapping the columns and rows around, and may be denoted by a right-hand-side superscript ($'$). For example, if

$$
Z = \begin{pmatrix} 3.1 & 0.2 & 6.1 & 4.8 \\ 9.2 & 3.8 & 2.0 & 5.1 \end{pmatrix}
$$

then

$$
Z' = \begin{pmatrix} 3.1 & 9.2 \\ 0.2 & 3.8 \\ 6.1 & 2.0 \\ 4.8 & 5.1 \end{pmatrix}
$$

Some authors used a superscript ^{T} instead.

A.1.2.3 Multiplication

Matrix and vector multiplication using the 'dot' product is denoted by the symbol '.' between matrices. It is only possible to multiply two matrices together if the number of columns of the first matrix equals the number of rows of the second matrix. The number of rows of the product will equal the number of rows of the first matrix, and the number of columns equal the number of columns of the second matrix. Hence a 3×2 matrix when multiplied by a 2×4 matrix will give a 3×4 matrix.

Multiplication of matrices is not commutative, that is, generally $A.B \neq B.A$ even if the second product is allowable. Matrix multiplication can be expressed in the form of summations. For arrays with more than two dimensions (e.g. tensors), conventional symbolism can be awkard and it is probably easier to think in terms of summations.

If matrix A has dimensions $I \times J$ and matrix B has dimensions $J \times K$, then the product C of dimensions $I \times K$ has elements defined by

$$
c_{ik} = \sum_{j=1}^{J} a_{ij} b_{jk}
$$

Hence

$$
\begin{pmatrix} 1 & 7 \\ 9 & 3 \\ 2 & 5 \end{pmatrix} \cdot \begin{pmatrix} 6 & 10 & 11 & 3 \\ 0 & 1 & 8 & 5 \end{pmatrix} = \begin{pmatrix} 6 & 17 & 67 & 38 \\ 54 & 93 & 123 & 42 \\ 12 & 25 & 62 & 31 \end{pmatrix}
$$

To illustrate this, the element of the first row and second column of the product is given by $17 = 1 \times 10 + 7 \times 1$.

When several matrices are multiplied together it is normal to take any two neighbouring matrices, multiply them together and then multiply this product with another neighbouring matrix. It does not matter in what order this is done, hence $A.B.C = (A.B).C = A.(B.C)$. Hence matrix multiplication is associative. Matrix multiplication is also distributive, that is, $A.(B + C) = A.B + A.C$.

A.1.2.4 Inverse

Most square matrices have inverses, defined by the matrix which when multiplied with the original matrix gives the identity matrix, and is represented by a $^{-1}$ as a right-hand-side superscript, so that $D.D^{-1} = I$. Note that some square matrices do not have inverses: this is caused by there being correlations in the original matrix; such matrices are called *singular* matrices.

A.1.2.5 Pseudo-inverse

In several sections of this text we use the idea of a pseudo-inverse. If matrices are not square, it is not possible to calculate an inverse, but the concept of a pseudo-inverse exists and is employed in regression analysis.

If $A = B.C$ then $B'.A = B'.B.C$, so $(B'.B)^{-1}.B'.A = C$ and $(B'.B)^{-1}.B'$ is said to be the left pseudo-inverse of B.

Equivalently, $A.C' = B.C.C'$, so $A.C'.(C.C')^{-1} = B$ and $C'.(C.C')^{-1}$ is said to be the right pseudo-inverse of C.

In regression, the equation $A \approx B.C$ is an approximation; for example, A may represent a series of spectra that are approximately equal to the product of two matrices such as scores and loadings matrices, hence this approach is important to obtain the best fit model for C knowing A and B or for B knowing A and C.

A.1.2.6 Trace and Determinant

Other properties of square matrices sometimes encountered are the trace, which is the sum of the diagonal elements, and the determinant, which relates to the size of the matrix. A determinant of 0 indicates a matrix without an inverse. A very small determinant often suggests that the data are fairly correlated or a poor experimental design resulting in fairly unreliable predictions. If the dimensions of matrices are large and the magnitudes of the measurements are small, e.g. 10^{-3}, it is sometimes possible to obtain a determinant close to zero even though the matrix has an inverse; a solution to this problem is to multiply each measurement by a number such as 10^3 and then remember to readjust the magnitude of the numbers in resultant calculations to take account of this later.

A.1.2.7 Vector length

An interesting property that chemometricians sometimes use is that the product of the transpose of a column vector with itself equals the sum of square of elements of the vector, so that $x'.x = \Sigma x^2$. The length of a vector is given by $\sqrt{(x'.x)} = \sqrt{\Sigma x^2}$ or the square root of the sum of its elements. This can be visualised in geometry as the length of the line from the origin to the point in space indicated by the vector.

A.2 Algorithms

There are many different descriptions of the various algorithms in the literature. This Appendix describes one algorithm for each of four regression methods.

A.2.1 Principal components analysis

NIPALS is a common, iterative algorithm often used for PCA. Some authors use another method called SVD (singular value decomposition). The main difference is that NIPALS extracts components one at a time, and can be stopped after the desired number of PCs has been obtained. In the case of large datasets with, for example, 200 variables (e.g. in spectroscopy), this can be very useful and reduce the amount of effort required. The steps are as follows.

Initialisation

1. Take a matrix Z and, if required, preprocess (e.g. mean centre or standardise) to give the matrix X which is used for PCA.

New Principal Component

2. Take a column of this matrix (often the column with greatest sum of squares) as the first guess of the scores first principal component; call it $^{initial}\hat{t}$.

Iteration for each Principal Component

3. Calculate

$$^{unnorm}\hat{p} = \frac{^{initial}\hat{t}'.X}{\sum \hat{t}^2}$$

4. Normalise the guess of the loadings, so

$$\hat{p} = \frac{^{unnorm}\hat{p}}{\sqrt{\sum {^{unnorm}\hat{p}^2}}}$$

5. Now calculate a new guess of the scores:

$$^{new}\hat{t} = X.\hat{p}'$$

Check for Convergence

6. Check if this new guess differs from the first guess; a simple approach is to look at the size of the sum of square difference in the old and new scores, i.e. $\sum (^{initial}\hat{t} - {^{new}\hat{t}})^2$. If this is small the PC has been extracted, set the PC scores (t) and loadings (p) for the current PC to $\hat{t}$ and $\hat{p}$. Otherwise, return to step 3, substituting the initial scores by the new scores.

Compute the Component and Calculate Residuals

7. Subtract the effect of the new PC from the datamatrix to obtain a residual data matrix:

$$^{resid}X = X - t.p$$

Further PCs

8. If it is desired to compute further PCs, substitute the residual data matrix for X and go to step 2.

A.2.2 PLS1

There are several implementations; the one below is noniterative.

Initialisation

1. Take a matrix Z and, if required, preprocess (e.g. mean centre or standardise) to give the matrix X which is used for PLS.
2. Take the concentration vector k and preprocess it to give the vector c which is used for PLS. Note that if the data matrix Z is centred down the columns, the concentration vector must also be centred. Generally, centring is the only form of preprocessing useful for PLS1. Start with an estimate of $\hat{c}$ that is a vector of 0s (equal to the mean concentration if the vector is already centred).

New PLS Component

3. Calculate the vector

$$h = X'.c$$

4. Calculate the scores, which are simply given by

$$t = \frac{X.h}{\sqrt{\sum h^2}}$$

5. Calculate the x loadings by

$$p = \frac{t'.X}{\sum t^2}$$

6. Calculate the c loading (a scalar) by

$$q = \frac{c'.t}{\sum t^2}$$

Compute the Component and Calculate Residuals

7. Subtract the effect of the new PLS component from the data matrix to get a residual data matrix:

$$^{resid}X = X - t.p$$

8. Determine the new concentration estimate by

$$^{new}\hat{c} = {}^{initial}\hat{c} + t.q$$

and sum the contribution of all components calculated to give an estimated $\hat{c}$. Note that the initial concentration estimate is 0 (or the mean) before the first component has been computed. Calculate

$$^{resid}c = {}^{true}c - {}^{new}\hat{c}$$

where ^{true}c is, like all values of c, after the data have been preprocessed (such as centring).

Further PLS Components

9. If further components are required, replace both X and c by the residuals and return to step 3.

Note that in the implementation used in this text the PLS loadings are neither normalised nor orthogonal. There are several different PLS1 algorithms, so it is useful to check exactly what method a particular package uses, although the resultant concentration estimates should be identical for each method (unless there is a problem with convergence in iterative approaches).

A.2.3 PLS2

This is a straightforward, iterative, extension of PLS1. Only small variations are required. Instead of c being a vector it is now a matrix C and instead of q being a scalar it is now a vector q.

Initialisation

1. Take a matrix Z and, if required, preprocess (e.g. mean centre or standardise) to give the matrix X which is used for PLS.
2. Take the concentration matrix K and preprocess it to give the vector c which is used for PLS. Note that if the data matrix is centred down the columns, the concentration vector must also be centred. Generally, centring is the only form of preprocessing useful for PLS2. Start with an estimate of $\hat{C}$ that is a vector of 0s (equal to the mean concentration if the vector is already centred).

New PLS Component

3. An extra step is required to identify a vector u which can be a guess (as in PCA), but can be chosen as one of the columns in the initial preprocessed concentration matrix, C.
4. Calculate the vector

$$h = X'.u$$

5. Calculate the guessed scores by

$$^{new}\hat{t} = \frac{X.h}{\sqrt{\sum h^2}}$$

6. Calculate the guessed x loadings by

$$\hat{p} = \frac{\hat{t}'.X}{\sum \hat{t}^2}$$

7. Calculate the c loadings (a vector rather than scalar in PLS2) by

$$\hat{q} = \frac{C'.\hat{t}}{\sum \hat{t}^2}$$

8. If this is the first iteration, remember the scores, and call them $^{initial}t$, then produce a new vector u by

$$u = \frac{C.\hat{q}}{\sum q^2}$$

and return to step 4.

Check for Convergence

9. If this is the second time round, compare the new and old scores vectors for example, by looking at the size of the sum of square difference in the old and new scores, i.e. $\sum (^{initial}\hat{t} - ^{new}\hat{t})^2$. If this is small the PLS component has been adequately modelled, set the PLS scores (t) and both types of loadings (p and c) for the current PC to $\hat{t}$, $\hat{p}$, and $\hat{q}$. Otherwise, calculate a new value of u as in step 8 and return to step 4.

Compute the Component and Calculate Residuals

10. Subtract the effect of the new PLS component from the data matrix to obtain a residual data matrix:

$$^{resid}X = X - t.p$$

11. Determine the new concentration estimate by

$$^{new}\hat{C} = {}^{initial}\hat{C} + t.q$$

and sum the contribution of all components calculated to give an estimated $\hat{c}$. Calculate

$$^{resid}C = {}^{true}C - \hat{C}$$

Further PLS Components

12. If further components are required, replace both X and C by the residuals and return to step 3.

A.2.4   Tri-linear PLS1

The algorithm below is based closely on PLS1 and is suitable when there is only one column in the c vector.

Initialisation

1. Take a three-way tensor $\underline{Z}$ and, if required, preprocess (e.g. mean centre or standardise) to give the tensor $\underline{X}$ which is used for PLS. Perform all preprocessing on this tensor. The tensor has dimensions $I \times J \times K$.
2. Preprocess the concentrations if appropriate to give a vector c.

New PLS Component

3. From the original tensor, create a new matrix H with dimensions $J \times K$ which is the sum of each of the I matrices for each of the samples multiplied by the concentration of the analyte for the relevant sample, i.e.

$$H = X_1 c_1 + X_2 c_2 + \cdots + X_I c_I$$

or, as a summation

$$h_{jk} = \sum_{i=1}^{I} c_i x_{ijk}$$

4. Perform PCA on H to obtain the scores and loadings, $^h t$ and $^h p$ for the first PC of H. Note that only the first PC is retained, and for each new PLS component a fresh H matrix is obtained.
5. Calculate the two x loadings for the current PLS component of the overall dataset by normalising the scores and loadings of H, i.e.

$$^j p = \frac{^h t'}{\sqrt{\sum {}^h t^2}}$$

$$^k p = \frac{^h p}{\sqrt{\sum {}^h p^2}}$$

(the second step is generally not necessary for most PCA algorithm as $^h p$ is usually normalised).

6. Calculate the overall scores by

$$t_i = \sum_{j=1}^{J} \sum_{k=1}^{K} x_{ijk} {}^j p_j {}^k p_k$$

7. Calculate the c loadings vector

$$q = (T'.T)^{-1}.T'.c$$

where T is the scores matrix each column consisting of one component (a vector for the first PLS component).

Compute the Component and Calculate Residuals

8. Subtract the effect of the new PLS component from the original data matrix to obtain a residual data matrix (for each sample i):

$$^{resid}X_i = X_i - t_i \cdot {}^jp \cdot {}^kp$$

9. Determine the new concentration estimates by

$$\hat{c} = T \cdot q$$

Calculate

$$^{resid}c = {}^{true}c - \hat{c}$$

Further PLS Components

10. If further components are required, replace both X and c by the residuals and return to step 3.

A.3 Basic Statistical Concepts

There are numerous texts on basic statistics, some of them oriented towards chemists. It is not the aim of this section to provide a comprehensive background, but simply to provide the main definitions and tables that are helpful for using this text.

A.3.1 Descriptive Statistics

A.3.1.1 Mean

The mean of a series of measurements is defined by

$$\bar{x} = \sum_{i=1}^{I} x_i / I$$

Conventionally a bar is placed above the letter. Sometimes the letter m is used, but in this text we will avoid this, as m is often used to denote an index. Hence the mean of the measurements

$$4 \quad 8 \quad 5 \quad -6 \quad 2 \quad -5 \quad 6 \quad 0$$

is $\bar{x} = (4 + 8 + 5 - 6 + 2 - 5 + 6 + 0)/8 = 1.75$.

Statistically, this sample mean is often considered an estimate of the true population mean sometimes denoted by μ. The population involves all possible samples, whereas only a selection are observed. In some cases in chemometrics this distinction is not so clear; for example, the mean intensity at a given wavelength over a chromatogram is a purely experimental variable.

A.3.1.2 Variance and Standard Deviation

The estimated or sample variance of a series of measurements is defined by

$$v = \sum_{i=1}^{I} (x_i - \bar{x})^2/(I-1)$$

which can also be calculated using the equation

$$v = \sum_{i=1}^{I} x_i^2/(I-1) - \bar{x}^2 \times I/(I-1)$$

So the variance of the data in Section A.3.1.1 is

$$v = (4^2 + 8^2 + 5^2 + 6^2 + 2^2 + 5^2 + 6^2 + 0^2)/7 - 1.75^2 \times 8/7 = 25.928$$

This equation is useful when it is required to estimate the variance from a series of samples. However, the true population variance is defined by

$$v = \sum_{i=1}^{I} (x_i - \bar{x})^2/I = \sum_{i=1}^{I} x_i^2/I - \bar{x}^2$$

The reason why there is a factor of $I-1$ when using measurements in a number of samples to estimate statistics is because one degree of freedom is lost when determining variance experimentally. For example, if we record one sample, the sum of squares $\sum_{i=1}^{I} (x_i - \bar{x})^2$ must be equal to 0, but this does not imply that the variance of the parent population is 0. As the number of samples increases, this small correction is not very important, and sometimes ignored.

The standard deviation, s, is simply the square root of the variance. The population standard deviation is sometimes denoted by σ.

In chemometrics it is usual to use the population and not the sample standard deviation for standardising a data matrix. The reason is that we are not trying to estimate parameters in this case, but just to put different variables on a similar scale.

A.3.1.3 Covariance and Correlation Coefficient

The covariance between two variables is a method for determining how closely they follow similar trends. It will never exceed in magnitude the geometric mean of the variance of the two variables; the lower is the value, the less close are the trends. Both variables must be measured for an identical number of samples, I in this case. The sample or estimated covariance between variables x and y is defined by

$$\text{cov}_{xy} = \sum_{i=1}^{I} (x_i - \bar{x})(y_i - \bar{y})/(I-1)$$

whereas the population statistic is given by

$$\text{cov}_{xy} = \sum_{i=1}^{I} (x_i - \overline{x})(y_i - \overline{y})/I$$

Unlike the variance, it is perfectly possible for a covariance to take on negative values. Many chemometricians prefer to use the correlation coefficient, given by

$$r_{xy} = \frac{\text{cov}_{xy}}{s_x \cdot s_y} = \frac{\displaystyle\sum_{i=1}^{I} (x_i - \overline{x})(y_i - \overline{y})}{\sqrt{\displaystyle\sum_{i=1}^{I} (x_i - \overline{x})^2 \sum_{i=1}^{I} (y_i - \overline{y})^2}}$$

Note that the definition of the correlation coefficient is identical both for samples and populations.

The correlation coefficient has a value between -1 and $+1$. If close to $+1$, the two variables are perfectly correlated. In many applications, correlation coefficients of -1 also indicate a perfect relationship. Under such circumstances, the value of y can be exactly predicted if we know x. The closer the correlation coefficients are to zero, the harder it is to use one variable to predict another. Some people prefer to use the square of the correlation coefficient which varies between 0 and 1.

If two columns of a matrix have a correlation coefficient of ± 1, the matrix is said to be rank deficient and has a determinant of 0, and so no inverse; this has consequences both in experimental design and in regression. There are various ways around this, such as by removing selected variables.

In some areas of chemometrics we used a *variance–covariance* matrix. This is a square matrix, whose dimensions usually equal the number of variables in a dataset, for example, if there are 20 variables the matrix has dimensions 20×20. The diagonal elements equal the variance of each variable and the off-diagonal elements the covariances. This matrix is symmetric about the diagonal. It is usual to employ population rather than sample statistics for this calculation.

A.3.2 Normal Distribution

The normal distribution is an important statistical concept. There are many ways of introducing such distributions. Many texts use a probability density function

$$f(x) = \frac{1}{\sigma\sqrt{2\pi}} \exp\left[-\frac{1}{2}\left(\frac{x-\mu}{\sigma}\right)^2\right]$$

This rather complicated equation can be interpreted as follows. The function $f(x)$ is proportional to the probability that a measurement has a value x for a normally distributed population of mean μ and standard deviation σ. The function is scaled so that the area under the normal distribution curve is 1.

Table A.1 Cumulative standardised normal distribution.

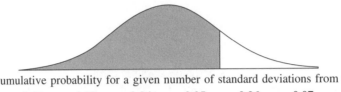

Values of cumulative probability for a given number of standard deviations from the mean.

	0.00	0.01	0.02	0.03	0.04	0.05	0.06	0.07	0.08	0.09
0.0	0.500 00	0.503 99	0.507 98	0.511 97	0.515 95	0.519 94	0.523 92	0.527 90	0.531 88	0.535 86
0.1	0.539 83	0.543 80	0.547 76	0.551 72	0.555 67	0.559 62	0.563 56	0.567 49	0.571 42	0.575 35
0.2	0.579 26	0.583 17	0.587 06	0.590 95	0.594 83	0.598 71	0.602 57	0.606 42	0.610 26	0.614 09
0.3	0.617 91	0.621 72	0.625 52	0.629 30	0.633 07	0.636 83	0.640 58	0.644 31	0.648 03	0.651 73
0.4	0.655 42	0.659 10	0.662 76	0.666 40	0.670 03	0.673 64	0.677 24	0.680 82	0.684 39	0.687 93
0.5	0.691 46	0.694 97	0.698 47	0.701 94	0.705 40	0.708 84	0.712 26	0.715 66	0.719 04	0.722 40
0.6	0.725 75	0.729 07	0.732 37	0.735 65	0.738 91	0.742 15	0.745 37	0.748 57	0.751 75	0.754 90
0.7	0.758 04	0.761 15	0.764 24	0.767 30	0.770 35	0.773 37	0.776 37	0.779 35	0.782 30	0.785 24
0.8	0.788 14	0.791 03	0.793 89	0.796 73	0.799 55	0.802 34	0.805 11	0.807 85	0.810 57	0.813 27
0.9	0.815 94	0.818 59	0.821 21	0.823 81	0.826 39	0.828 94	0.831 47	0.833 98	0.836 46	0.838 91
1.0	0.841 34	0.843 75	0.846 14	0.848 49	0.850 83	0.853 14	0.855 43	0.857 69	0.859 93	0.862 14
1.1	0.864 33	0.866 50	0.868 64	0.870 76	0.872 86	0.874 93	0.876 98	0.879 00	0.881 00	0.882 98
1.2	0.884 93	0.886 86	0.888 77	0.890 65	0.892 51	0.894 35	0.896 17	0.897 96	0.899 73	0.901 47
1.3	0.903 20	0.904 90	0.906 58	0.908 24	0.909 88	0.911 49	0.913 08	0.914 66	0.916 21	0.917 74
1.4	0.919 24	0.920 73	0.922 20	0.923 64	0.925 07	0.926 47	0.927 85	0.929 22	0.930 56	0.931 89
1.5	0.933 19	0.934 48	0.935 74	0.936 99	0.938 22	0.939 43	0.940 62	0.941 79	0.942 95	0.944 08
1.6	0.945 20	0.946 30	0.947 38	0.948 45	0.949 50	0.950 53	0.951 54	0.952 54	0.953 52	0.954 49
1.7	0.955 43	0.956 37	0.957 28	0.958 18	0.959 07	0.959 94	0.960 80	0.961 64	0.962 46	0.963 27
1.8	0.964 07	0.964 85	0.965 62	0.966 38	0.967 12	0.967 84	0.968 56	0.969 26	0.969 95	0.970 62
1.9	0.971 28	0.971 93	0.972 57	0.973 20	0.973 81	0.974 41	0.975 00	0.975 58	0.976 15	0.976 70
2.0	0.977 25	0.977 78	0.978 31	0.978 82	0.979 32	0.979 82	0.980 30	0.980 77	0.981 24	0.981 69
2.1	0.982 14	0.982 57	0.983 00	0.983 41	0.983 82	0.984 22	0.984 61	0.985 00	0.985 37	0.985 74
2.2	0.986 10	0.986 45	0.986 79	0.987 13	0.987 45	0.987 78	0.988 09	0.988 40	0.988 70	0.988 99
2.3	0.989 28	0.989 56	0.989 83	0.990 10	0.990 36	0.990 61	0.990 86	0.991 11	0.991 34	0.991 58
2.4	0.991 80	0.992 02	0.992 24	0.992 45	0.992 66	0.992 86	0.993 05	0.993 24	0.993 43	0.993 61
2.5	0.993 79	0.993 96	0.994 13	0.994 30	0.994 46	0.994 61	0.994 77	0.994 92	0.995 06	0.995 20
2.6	0.995 34	0.995 47	0.995 60	0.995 73	0.995 85	0.995 98	0.996 09	0.996 21	0.996 32	0.996 43
2.7	0.996 53	0.996 64	0.996 74	0.996 83	0.996 93	0.997 02	0.997 11	0.997 20	0.997 28	0.997 36
2.8	0.997 44	0.997 52	0.997 60	0.997 67	0.997 74	0.997 81	0.997 88	0.997 95	0.998 01	0.998 07
2.9	0.998 13	0.998 19	0.998 25	0.998 31	0.998 36	0.998 41	0.998 46	0.998 51	0.998 56	0.998 61

	0.0	0.1	0.2	0.3	0.4	0.5	0.6	0.7	0.8	0.9
3.0	0.998 65	0.999 03	0.999 31	0.999 52	0.999 66	0.999 77	0.999 84	0.999 89	0.999 93	0.999 95
4.0	0.999 968	0.999 979	0.999 987	0.999 991	0.999 995	0.999 997	0.999 998	0.999 999	0.999 999	1.000 000

Most tables deal with the *standardised* normal distribution. This involves first standardising the raw data, to give a new value z, and the equation simplifies to

$$f(z) = \frac{1}{\sqrt{2\pi}} \exp\left(-\frac{z^2}{2}\right)$$

Instead of calculating $f(z)$, most people look at the area under the normal distribution curve. This is proportional to the probability that a measurement is between certain limits. For example the probability that a measurement is between one and two standard deviations can be calculated by taking the proportion of the overall area for which $1 \leq z \leq 2$.

These numbers can be obtained using simple functions, e.g. in a spreadsheet, but are often conventionally presented in tabular form. There are a surprisingly large number of types of tables, but Table A.1 allows the reader to calculate relevant information. This table is of the cumulative normal distribution, and represents the area to the left of the curve for a specified number of standard deviations from the mean. The number of standard deviations equals the sum of the left-hand column and the top row, so, for example, the area for 1.17 standard deviations equals 0.879 00.

Using this table it is then possible to determine the probability of a measurement between any specific limits.

- The probability that a measurement is above 1 standard deviation from the mean is equal to $1 - 0.841\,34 = 0.158\,66$.
- The probability that a measurement is more than 1 standard deviation from the mean will be twice this, because both positive and negative deviations are possible and the curve is symmetrical, and is equal to 0.317 32. Put another way, around a third of all measurements will fall outside 1 standard deviation from the mean.
- The probability that a measurement falls between -2 and $+1$ standard deviations from the mean can be calculated as follows:
  — the probability that a measurement falls between 0 and -2 standard deviations is the same as the probability it falls between 0 and $+2$ standard deviations and is equal to $0.977\,25 - 0.5 = 0.477\,25$;
  — the probability that a measurement falls between 0 and $+1$ standard deviations is equal to $0.841\,34 - 0.5 = 0.341\,34$;
  — therefore the total probability is $0.477\,25 + 0.341\,34 = 0.818\,59$.

The normal distribution curve is not only a probability distribution but is also used to describe peakshapes in spectroscopy and chromatography.

A.3.3 F Distribution

The F-test is normally used to compare two variances or errors and ask either whether one variance is significantly greater than the other (one-tailed) or whether it differs significantly (two-tailed). In this book we use only the one-tailed F-test, mainly to see whether one error (e.g. lack-of-fit) is significantly greater than a second one (e.g. experimental or analytical).

The F statistic is the ratio between these two variances, normally presented as a number greater than 1, i.e. the largest over the smallest. The F distribution depends on the number of degrees of freedom of each variable, so, if the highest variance is obtained from 10 samples, and the lowest from seven samples, the two variables have nine and six degrees of freedom, respectively. The F distribution differs according to the number of degrees of freedom, and it would be theoretically possible to produce an F distribution table for every possible combination of degrees of freedom, similar to the normal distribution table. However, this would mean an enormous number of tables (in theory an infinite number), and it is more usual simply to calculate the F statistic at certain well defined probability levels.

A one-tailed F statistic at the 1 % probability level is the value of the F ratio above which only 1 % of measurements would fall if the two variances were not significantly

Table A.2 One-tailed F distribution at the 1 % level.

df	1	2	3	4	5	6	7	8	9	10	15	20	25	30	50	100	∞
1	4052.18	4999.34	5403.53	5624.26	5763.96	5858.95	5928.33	5980.95	6022.40	6055.93	6156.97	6208.66	6239.86	6260.35	6302.26	6333.92	6365.59
2	98.5019	99.0003	99.1640	99.2513	99.3023	99.3314	99.3568	99.3750	99.3896	99.3969	99.4332	99.4478	99.4587	99.4660	99.4769	99.4914	99.4987
3	34.1161	30.8164	29.4567	28.7100	28.2371	27.9106	27.6714	27.4895	27.3449	27.2285	26.8719	26.6900	26.5791	26.5045	26.3544	26.2407	26.1252
4	21.1976	17.9998	16.6942	15.9771	15.5219	15.2068	14.9757	14.7988	14.6592	14.5460	14.1981	14.0194	13.9107	13.8375	13.6897	13.5769	13.4633
5	16.2581	13.2741	12.0599	11.3919	10.9671	10.6722	10.4556	10.2893	10.1577	10.0511	9.7223	9.5527	9.4492	9.3794	9.2377	9.1300	9.0204
6	13.7452	10.9249	9.7796	9.1484	8.7459	8.4660	8.2600	8.1017	7.9760	7.8742	7.5590	7.3958	7.2960	7.2286	7.0914	6.9867	6.8801
7	12.2463	9.5465	8.4513	7.8467	7.4604	7.1914	6.9929	6.8401	6.7188	6.6201	6.3144	6.1555	6.0579	5.9920	5.8577	5.7546	5.6496
8	11.2586	8.6491	7.5910	7.0061	6.6318	6.3707	6.1776	6.0288	5.9106	5.8143	5.5152	5.3591	5.2631	5.1981	5.0654	4.9633	4.8588
9	10.5615	8.0215	6.9920	6.4221	6.0569	5.8018	5.6128	5.4671	5.3511	5.2565	4.9621	4.8080	4.7130	4.6486	4.5167	4.4150	4.3106
10	10.0442	7.5595	6.5523	5.9944	5.6364	5.3858	5.2001	5.0567	4.9424	4.8491	4.5582	4.4054	4.3111	4.2469	4.1155	4.0137	3.9090
11	9.6461	7.2057	6.2167	5.6683	5.3160	5.0692	4.8860	4.7445	4.6315	4.5393	4.2509	4.0990	4.0051	3.9411	3.8097	3.7077	3.6025
12	9.3303	6.9266	5.9525	5.4119	5.0644	4.8205	4.6395	4.4994	4.3875	4.2961	4.0096	3.8584	3.7647	3.7008	3.5692	3.4668	3.3608
13	9.0738	6.7009	5.7394	5.2053	4.8616	4.6203	4.4410	4.3021	4.1911	4.1003	3.8154	3.6646	3.5710	3.5070	3.3752	3.2723	3.1654
14	8.8617	6.5149	5.5639	5.0354	4.6950	4.4558	4.2779	4.1400	4.0297	3.9394	3.6557	3.5052	3.4116	3.3476	3.2153	3.1118	3.0040
15	8.6832	6.3588	5.4170	4.8932	4.5556	4.3183	4.1416	4.0044	3.8948	3.8049	3.5222	3.3719	3.2782	3.2141	3.0814	2.9772	2.8684
16	8.5309	6.2263	5.2922	4.7726	4.4374	4.2016	4.0259	3.8896	3.7804	3.6909	3.4090	3.2587	3.1650	3.1007	2.9675	2.8627	2.7528
17	8.3998	6.1121	5.1850	4.6689	4.3360	4.1015	3.9267	3.7909	3.6823	3.5931	3.3117	3.1615	3.0676	3.0032	2.8694	2.7639	2.6531
18	8.2855	6.0129	5.0919	4.5790	4.2479	4.0146	3.8406	3.7054	3.5971	3.5081	3.2273	3.0771	2.9831	2.9185	2.7841	2.6779	2.5660
19	8.1850	5.9259	5.0103	4.5002	4.1708	3.9386	3.7653	3.6305	3.5225	3.4338	3.1533	3.0031	2.9089	2.8442	2.7092	2.6023	2.4893
20	8.0960	5.8490	4.9382	4.4307	4.1027	3.8714	3.6987	3.5644	3.4567	3.3682	3.0880	2.9377	2.8434	2.7785	2.6430	2.5353	2.4212
25	7.7698	5.5680	4.6755	4.1774	3.8550	3.6272	3.4568	3.3239	3.2172	3.1294	2.8502	2.6993	2.6041	2.5383	2.3999	2.2888	2.1694
30	7.5624	5.3903	4.5097	4.0179	3.6990	3.4735	3.3045	3.1726	3.0665	2.9791	2.7002	2.5487	2.4526	2.3860	2.2450	2.1307	2.0062
35	7.4191	5.2679	4.3958	3.9082	3.5919	3.3679	3.1999	3.0687	2.9630	2.8758	2.5970	2.4448	2.3480	2.2806	2.1374	2.0202	1.8910
40	7.3142	5.1785	4.3126	3.8283	3.5138	3.2910	3.1238	2.9930	2.8876	2.8005	2.5216	2.3689	2.2714	2.2034	2.0581	1.9383	1.8047
45	7.2339	5.1103	4.2492	3.7674	3.4544	3.2325	3.0658	2.9353	2.8301	2.7432	2.4642	2.3109	2.2129	2.1443	1.9972	1.8751	1.7374
50	7.1706	5.0566	4.1994	3.7195	3.4077	3.1864	3.0202	2.8900	2.7850	2.6981	2.4190	2.2652	2.1667	2.0976	1.9490	1.8248	1.6831
100	6.8953	4.8239	3.9837	3.5127	3.2059	2.9877	2.8233	2.6943	2.5898	2.5033	2.2230	2.0666	1.9651	1.8933	1.7353	1.5977	1.4273
∞	6.6349	4.6052	3.7816	3.3192	3.0172	2.8020	2.6393	2.5113	2.4073	2.3209	2.0385	1.8783	1.7726	1.6964	1.5231	1.3581	1.0000

Table A.3 One-tailed F distribution at the 5% level.

df	1	2	3	4	5	6	7	8	9	10	15	20	25	30	50	100	∞
1	161.45	199.50	215.71	224.58	230.16	233.99	236.77	238.88	240.54	241.88	245.95	248.02	249.26	250.10	251.77	253.04	254.32
2	18.5128	19.0000	19.1642	19.2467	19.2963	19.3295	19.3531	19.3709	19.3847	19.3959	19.4291	19.4457	19.4557	19.4625	19.4757	19.4857	19.4957
3	10.1280	9.5521	9.2766	9.1172	9.0134	8.9407	8.8867	8.8452	8.8123	8.7855	8.7028	8.6602	8.6341	8.6166	8.5810	8.5539	8.5265
4	7.7086	6.9443	6.5914	6.3882	6.2561	6.1631	6.0942	6.0410	5.9988	5.9644	5.8578	5.8025	5.7687	5.7459	5.6995	5.6640	5.6281
5	6.6079	5.7861	5.4094	5.1922	5.0503	4.9503	4.8759	4.8183	4.7725	4.7351	4.6188	4.5581	4.5209	4.4957	4.4444	4.4051	4.3650
6	5.9874	5.1432	4.7571	4.5337	4.3874	4.2839	4.2067	4.1468	4.0990	4.0600	3.9381	3.8742	3.8348	3.8082	3.7537	3.7117	3.6689
7	5.5915	4.7374	4.3468	4.1203	3.9715	3.8660	3.7871	3.7257	3.6767	3.6365	3.5107	3.4445	3.4036	3.3758	3.3189	3.2749	3.2298
8	5.3176	4.4590	4.0662	3.8379	3.6875	3.5806	3.5005	3.4381	3.3881	3.3472	3.2184	3.1503	3.1081	3.0794	3.0204	2.9747	2.9276
9	5.1174	4.2565	3.8625	3.6331	3.4817	3.3738	3.2927	3.2296	3.1789	3.1373	3.0061	2.9365	2.8932	2.8637	2.8028	2.7556	2.7067
10	4.9646	4.1028	3.7083	3.4780	3.3258	3.2172	3.1355	3.0717	3.0204	2.9782	2.8450	2.7740	2.7298	2.6996	2.6371	2.5884	2.5379
11	4.8443	3.9823	3.5874	3.3567	3.2039	3.0946	3.0123	2.9480	2.8962	2.8536	2.7186	2.6464	2.6014	2.5705	2.5066	2.4566	2.4045
12	4.7472	3.8853	3.4903	3.2592	3.1059	2.9961	2.9134	2.8486	2.7964	2.7534	2.6169	2.5436	2.4977	2.4663	2.4010	2.3498	2.2962
13	4.6672	3.8056	3.4105	3.1791	3.0254	2.9153	2.8321	2.7669	2.7144	2.6710	2.5331	2.4589	2.4123	2.3803	2.3138	2.2614	2.2064
14	4.6001	3.7389	3.3439	3.1122	2.9582	2.8477	2.7642	2.6987	2.6458	2.6022	2.4630	2.3879	2.3407	2.3082	2.2405	2.1870	2.1307
15	4.5431	3.6823	3.2874	3.0556	2.9013	2.7905	2.7066	2.6408	2.5876	2.5437	2.4034	2.3275	2.2797	2.2468	2.1780	2.1234	2.0659
16	4.4940	3.6337	3.2389	3.0069	2.8524	2.7413	2.6572	2.5911	2.5377	2.4935	2.3522	2.2756	2.2272	2.1938	2.1240	2.0685	2.0096
17	4.4513	3.5915	3.1968	2.9647	2.8100	2.6987	2.6143	2.5480	2.4943	2.4499	2.3077	2.2304	2.1815	2.1477	2.0769	2.0204	1.9604
18	4.4139	3.5546	3.1599	2.9277	2.7729	2.6613	2.5767	2.5102	2.4563	2.4117	2.2686	2.1906	2.1413	2.1071	2.0354	1.9780	1.9168
19	4.3808	3.5219	3.1274	2.8951	2.7401	2.6283	2.5435	2.4768	2.4227	2.3779	2.2341	2.1555	2.1057	2.0712	1.9986	1.9403	1.8780
20	4.3513	3.4928	3.0984	2.8661	2.7109	2.5990	2.5140	2.4471	2.3928	2.3479	2.2033	2.1242	2.0739	2.0391	1.9656	1.9066	1.8432
25	4.2417	3.3852	2.9912	2.7587	2.6030	2.4904	2.4047	2.3371	2.2821	2.2365	2.0889	2.0075	1.9554	1.9192	1.8421	1.7794	1.7110
30	4.1709	3.3158	2.9223	2.6896	2.5336	2.4205	2.3343	2.2662	2.2107	2.1646	2.0148	1.9317	1.8782	1.8409	1.7609	1.6950	1.6223
35	4.1213	3.2674	2.8742	2.6415	2.4851	2.3718	2.2852	2.2167	2.1608	2.1143	1.9629	1.8784	1.8239	1.7856	1.7032	1.6347	1.5580
40	4.0847	3.2317	2.8387	2.6060	2.4495	2.3359	2.2490	2.1802	2.1240	2.0773	1.9245	1.8389	1.7835	1.7444	1.6600	1.5892	1.5089
45	4.0566	3.2043	2.8115	2.5787	2.4221	2.3083	2.2212	2.1521	2.0958	2.0487	1.8949	1.8084	1.7522	1.7126	1.6264	1.5536	1.4700
50	4.0343	3.1826	2.7900	2.5572	2.4004	2.2864	2.1992	2.1299	2.0733	2.0261	1.8714	1.7841	1.7273	1.6872	1.5995	1.5249	1.4383
100	3.9362	3.0873	2.6955	2.4626	2.3053	2.1906	2.1025	2.0323	1.9748	1.9267	1.7675	1.6764	1.6163	1.5733	1.4772	1.3917	1.2832
∞	3.8414	2.9957	2.6049	2.3719	2.2141	2.0986	2.0096	1.9384	1.8799	1.8307	1.6664	1.5705	1.5061	1.4591	1.3501	1.2434	1.0000

different. If the F statistic exceeds this value then we have more than 99 % confidence that there is a significant difference in the two variances, for example that the lack-of-fit error really is bigger than the analytical error.

We present here the 1 and 5 % one-tailed F-test (Tables A.2 and A.3). The number of degrees of freedom belonging to the data with the highest variance is always along the top, and the degrees of freedom belonging to the data with the lowest variance down the side. It is easy to use Excel or most statistically based packages to calculate critical F values for any probability and combination of degrees of freedom, but it is still worth being able to understand the use of tables.

If one error (e.g. the lack-of-fit) is measured using eight degrees of freedom and another error (e.g. replicate) is measured using six degrees of freedom, then if the F ratio between the mean lack-of-fit and replicate errors is 7.89, is it significant? The critical F statistic at 1 % is 8.1017 and at 5 % is 4.1468. Hence the F ratio is significant at almost the 99 % ($=100 - 1$ %) level because 7.89 is almost equal to 8.1017. Hence we are 99 % certain that the lack-of-fit is really significant.

Some texts also present tables for a two-tailed F-test, but because we do not employ this in this book, we omit it. However, a two-tailed F statistic at 10 % significance is the same as a one-tailed F statistic at 5 % significance, and so on.

Table A.4 Critical values of two-tailed t distribution.

df	10 %	5 %	1 %	0.1 %
1	6.314	12.706	63.656	636.578
2	2.920	4.303	9.925	31.600
3	2.353	3.182	5.841	12.924
4	2.132	2.776	4.604	8.610
5	2.015	2.571	4.032	6.869
6	1.943	2.447	3.707	5.959
7	1.895	2.365	3.499	5.408
8	1.860	2.306	3.355	5.041
9	1.833	2.262	3.250	4.781
10	1.812	2.228	3.169	4.587
11	1.796	2.201	3.106	4.437
12	1.782	2.179	3.055	4.318
13	1.771	2.160	3.012	4.221
14	1.761	2.145	2.977	4.140
15	1.753	2.131	2.947	4.073
16	1.746	2.120	2.921	4.015
17	1.740	2.110	2.898	3.965
18	1.734	2.101	2.878	3.922
19	1.729	2.093	2.861	3.883
20	1.725	2.086	2.845	3.850
25	1.708	2.060	2.787	3.725
30	1.697	2.042	2.750	3.646
35	1.690	2.030	2.724	3.591
40	1.684	2.021	2.704	3.551
45	1.679	2.014	2.690	3.520
50	1.676	2.009	2.678	3.496
100	1.660	1.984	2.626	3.390
∞	1.645	1.960	2.576	3.291

A.3.4 t Distribution

The t distribution is somewhat similar in concept to the F distribution but only one degree of freedom is associated with this statistic. In this text the t-test is used to determine the significance of coefficients obtained from an experiment, but in other contexts it can be widely employed; for example, a common application is to determine whether the means of two datasets are significantly different.

Most tables of t distributions look at the critical value of the t statistics for different degrees of freedom. Table A.4 relates to the two-tailed t-test, which we employ in this text, and asks whether a parameter differs significantly from another. If there are 10 degrees of freedom and the t statistic equals 2.32, then the probability is fairly high, slightly above 95 % (5 % critical value), that it is significant.

The one-tailed t-test is used to see if a parameter is significantly larger than another; for example, does the mean of a series of samples significantly exceed that of a series of reference samples? In this book we are mainly concerned with using the t statistic to determine the significance of coefficients when analysing the results of designed experiments; in such cases both a negative and positive coefficient are equally significant, so a two-tailed t-test is most appropriate.

A.4   Excel for Chemometrics

There are many excellent books on Excel in general, and the package in itself is associated with an extensive system. It is not the purpose of this text to duplicate these books, which in themselves are regularly updated as new versions of Excel become available, but primarily to indicate features that the user of advanced data analysis might find useful. The examples in this book are illustrated using Office 2000, but most features are applicable to Office 97, and are also likely to be upwards compatible. It is assumed that the reader already has some experience in Excel, and the aim of this section is to indicate some features that will be useful to the scientific user of Excel, especially the chemometrician. This section should be regarded primarily as one of tips and hints, and the best way forward is by practice. There are comprehensive help facilities and Websites are also available for the specialist user of Excel. The specific chemometric Add-ins available to accompany this text are also described.

A.4.1 Names and Addresses

There are a surprisingly large number of methods for naming cells and portions of spreadsheets in Excel, and it is important to be aware of all the possibilities.

A.4.1.1 Alphanumeric Format

The default naming convention is alphanumeric, each cell's address involving one or two letters referring to the column followed by a number referring to the row. So cell *C5* is the address of the fifth row of the third column. The alphanumeric method is a historic feature of this and most other spreadsheets, but is at odds with the normal scientific convention of quoting rows before columns. After the letters A–Z are exhausted (the first 26 columns), two-letter names are used starting at AA, AB, AC up to AZ and then BA, BB, etc.

A.4.1.2 Maximum Size

The size of a worksheet is limited, to a maximum of 256 ($=2^8$) columns and 65 536 ($=2^{16}$) rows, meaning that the highest address is *IV65536*. This limitation is primarily because a very large worksheet would require a huge amount of memory and on most systems be very slow, but with modern PCs that have much larger memory the justification for this limitation is less defensible. It can be slightly irritating to the chemometrician, because of the convention that variables (e.g. spectroscopic wavelengths) are normally represented by columns and in many cases the number of variables can exceed 256. One way to solve this is to transpose the data, with variables along the rows and samples along the columns; however, it is important to make sure that macros and other software can cope with this. An alternative is to split the data on to several worksheets, but this can become unwieldy. For big files, if it is not essential to see the raw numbers, it is possible to write macros (as discussed below) that read the data in from a file and only output the desired graphs or other statistics.

A.4.1.3 Numeric Format

Some people prefer to use a numeric format for addressing cells. The columns are numbered rather than labelled with letters. To change from the default, select the 'Tools' option, choose 'General' and select R1C1 from the dialog box (Figure A.1).

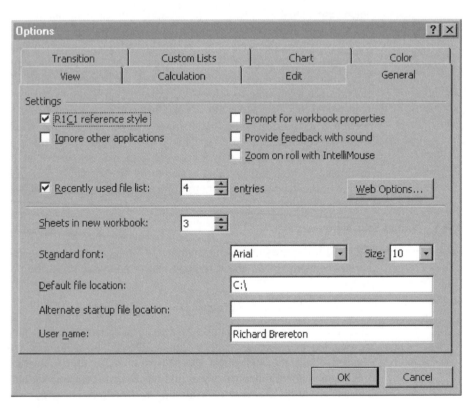

Figure A.1
Changing to numeric cell addresses

Deselecting this returns to the alphanumeric system. When using the RC notation, the rows are cited first, as is normal for matrices, and the columns second, which is the opposite to the default convention. Cell *B5* is now called cell *R5C2*. Below we employ the alphanumeric notation unless specifically indicated otherwise.

A.4.1.4 Worksheets

Each spreadsheet file can contain several worksheets, each of which either contain data or graphs and which can also be named. The default is *Sheet1*, *Sheet2*, etc., but it is possible to call the sheets by almost any name, such as *Data*, *Results*, *Spectra*, and it is also possible to contain spaces, e.g. *Sample 1A*. A cell specific to one sheet has the address of the sheet included, separated by a !, so the cell *Data!E3* is the address of the third row and fifth column on a worksheet *Data*. This address can be used in any worksheet, and so allows results of calculations to be placed in different worksheets. If the name contains a space, it is cited within quotation marks, for example *'Spectrum 1'!B12*.

A.4.1.5 Spreadsheets

In addition, it is possible to reference data in another spreadsheet file, for example *[First.xls]Result!A3* refers to the address of cell *A3* in worksheet *Result* and file *First*. This might be potentially useful, for example, if the file *First* consists of a series of spectra or a chromatogram whereas the current file consists of the results of processing the data, such as graphs or statistics. This flexibility comes with a price, as all files have to be available simultaneously, and is somewhat elaborate especially in cases where files are regularly reorganised and backed up.

A.4.1.6 Invariant Addresses

When copying an equation or a cell address within a spreadsheet, it is worth remembering another convention. The $ sign means that the row or column is invariant. For example if the equation '=A1' is placed in cell *C2*, and then this is copied to cell *D4*, the relative references to the rows and columns are also moved, so cell *D4* will actually contain the contents of the original cell *B3*. This is often useful because it allows operations on entire rows and columns, but sometimes we need to avoid this. Placing the equation '=$A1' in cell *C2* has the effect of fixing the column, so that the contents of cell *D4* will now be equal to those of cell *A3*. Placing '=A$1' in cell *C2* makes the contents of cell *D4* equal to *B1*, and placing '=$A$1' in cell *C2* makes cell *D4* equal to *A1*. The naming conventions can be combined, for example, *Data!B$3*. Experienced users of Excel often combine a variety of different tricks and it is best to learn these ways by practice rather than reading lots of books.

A.4.1.7 Ranges

A range is a set of cells, often organised as a matrix. Hence the range *A3:C4* consists of six numbers organised into two rows and three columns as illustrated in Figure A.2. It is normal to calculate the function of a range, for example = *AVERAGE(A3:C4)*, and these will be described in more detail in Section A.4.2.4. If this function is placed in cell *D4*, then if it is copied to another cell, the range will alter correspondingly;

	A	B	C	D
1				
2				
3	9	7	3	
4	0	-4	6	
5				
6				

Figure A.2
The range A3: C4

A14 = =AVERAGE(A1:B5,C8,B9:D11)

	A	B	C	D	E	F
1	7.2	9.3				
2	5.1	-4.0				
3	0.9	6.7				
4	-8.9	2.4				
5	3.8	-1.7				
6						
7						
8			6.6			
9		4.2	0.3	-8.1		
10		-2.0	0.9	7.6		
11		-3.5	1.8	4.6		
12						
13						
14	1.7					
15						

Figure A.3
The operation =AVERAGE(A1: B5,C8,B9: D11)

for example, moving to cell *E6* would change the range, automatically, to *B5*: *D6*. There are various ways to overcome this, the simplest being to use the $ convention as described above. Note that ! and [ ] can also be used in the address of a range, so that =*AVERAGE*([*first.xls*]*Data!$B$3*: *$C$10*) is an entirely legitimate statement.

It is not necessary for a range to be a single contiguous matrix. The statement =*AVERAGE*(*A1*: *B5*, *C8* , *B9*: *D11*) will consist of $10 + 1 + 9 = 20$ different cells, as indicated in Figure A.3.

There are different ways of copying cells or ranges. If the middle of a cell is 'hit' the cursor turns into an arrow. Dragging this arrow drags the cell but the original reference remains unchanged: see Figure A.4(a), where the equation =*AVERAGE*(*A1*: *B3*) has been placed originally in cell *A7*. If the bottom right-hand corner is hit, the cursor changes to a small cross, and dragging this fills all intermediate cells, changing the reference as appropriate, unless the $ symbol has been used: see Figure A.4(b).

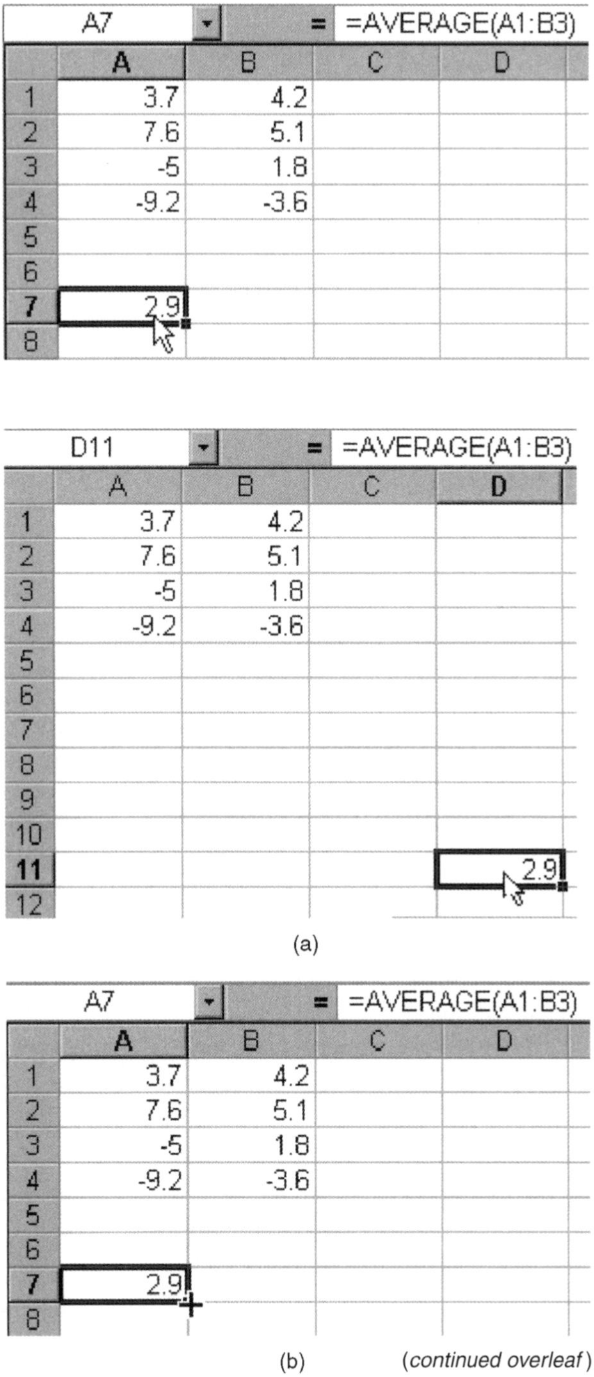

(a)

(b)   (*continued overleaf*)

Figure A.4
(a) Dragging a cell (b) Filling cells

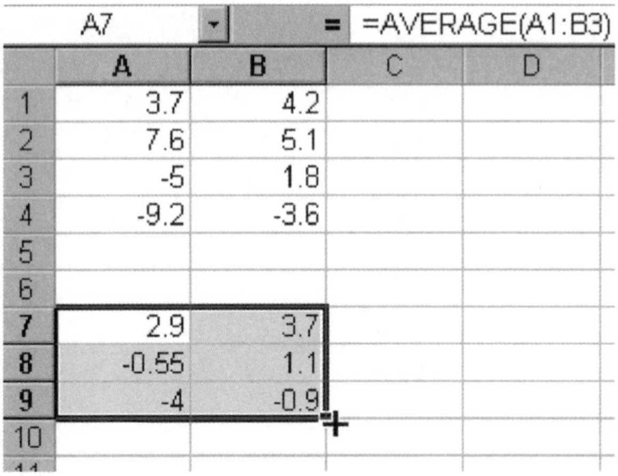

(b)

Figure A.4
(*continued*)

A.4.1.8 Naming Matrices

It is particularly useful to be able to name matrices or vectors. This is straightforward. Using the 'Insert' menu and choosing 'Name', it is possible to select a portion of a worksheet and give it a name. Figure A.5 shows how to call the range A1–B4 'X', creating a 4×2 matrix. It is then possible to perform operations on these matrices, so that $=SUMSQ(X)$ gives the sum of squares of the elements of X and is an alternative to $=SUMSQ(A1:B4)$. It is possible to perform matrix operations, for example, if X and Y are both 4×2 matrices, select a third 4×2 region and place the command $=X+Y$ in that region (as discussed in Section A.4.2.2, it is necessary to end all matrix operations by simultaneously pressing the ⟨SHIFT⟩⟨CTL⟩⟨ENTER⟩ keys). Fairly elaborate commands can then be nested, for example $=3*(X+Y)-Z$ is entirely acceptable; note that the '3*' is a scalar multiplication – more about this will be discussed below.

The name of a matrix is common to an entire spreadsheet, rather than any individual worksheet. This means that it is possible store a matrix called 'Data' on one worksheet and then perform operations on it in a separate worksheet.

A.4.2 Equations and Functions

A.4.2.1 Scalar Operations

The operations are straightforward and can be performed on cells, numbers and functions. The operations $+$, $-$, $*$ and $/$ indicate addition, subtraction, multiplication and division as in most environments; powers are indicated by ^. Brackets can be used and there is no practicable limit to the size of the expression or the amount of nesting. A destination cell must be chosen. Hence the operation $=3*(2-5)^2/(8-1)+6$ gives a value of 9.857. All the usual rules of precedence are obeyed. Operations can mix cells, functions and numbers, for example $=4*A1+SUM(E1:F3)-SUMSQ(Y)$ involves adding

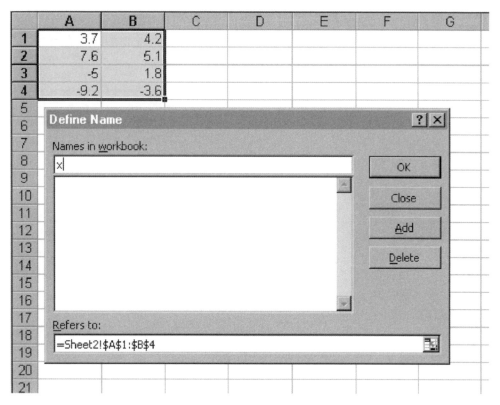

Figure A.5
Naming a range

- four times the value of cell *A1* to
- the sum of the values of the six cells in the range *E1* to *F3* and
- subtracting the sum of the squares of the matrix *Y*.

Equations can be copied around the spreadsheet, the $, ! and [ ] conventions being used if appropriate.

A.4.2.2 Matrix Operations

There are a number of special matrix operations of particular use in chemometrics. These operations must be terminated by simultaneously pressing the ⟨SHIFT⟩, ⟨CTL⟩ and ⟨ENTER⟩ keys. It is first necessary to select a portion of the spreadsheet where the destination matrix will be displayed. The most useful functions are as follows.

- *MMULT* multiplies two matrices together. The inner dimensions must be identical. If not, there is an error message. The destination (third) matrix must be selected, if its dimensions are wrong a result is still given but some numbers may be missing or duplicated. The syntax is =*MMULT(A,B)* and Figure A.6 illustrates the result of multiplying a 4 × 2 matrix with a 2 × 3 matrix to give a 4 × 3 matrix.
- *TRANSPOSE* gives the transpose of a matrix. Select the destination of the correct shape, and use the syntax =*TRANSPOSE(A)* as illustrated in Figure A.7.

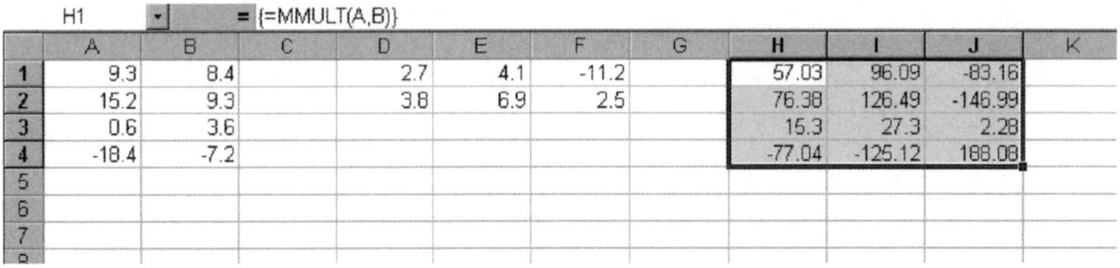

H1		▼	= {=MMULT(A,B)}							
	A	B	C	D	E	F	G	H	I	J	K
1	9.3	8.4		2.7	4.1	-11.2		57.03	96.09	-83.16	
2	15.2	9.3		3.8	6.9	2.5		76.38	126.49	-146.99	
3	0.6	3.6						15.3	27.3	2.28	
4	-18.4	-7.2						-77.04	-125.12	188.08	
5											
6											
7											

Figure A.6
Matrix multiplication in Excel

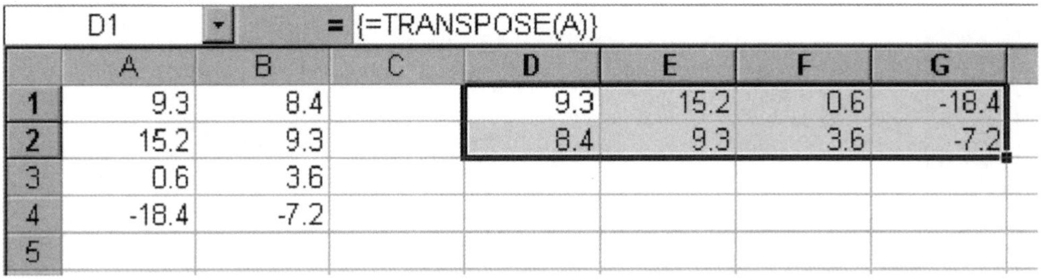

D1		▼	= {=TRANSPOSE(A)}			
	A	B	C	D	E	F	G
1	9.3	8.4		9.3	15.2	0.6	-18.4
2	15.2	9.3		8.4	9.3	3.6	-7.2
3	0.6	3.6					
4	-18.4	-7.2					
5							

Figure A.7
Matrix transpose in Excel

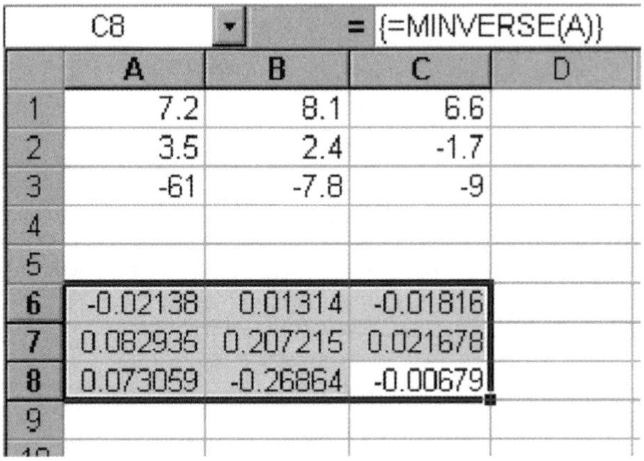

C8		▼	= {=MINVERSE(A)}	
	A	B	C	D
1	7.2	8.1	6.6	
2	3.5	2.4	-1.7	
3	-61	-7.8	-9	
4				
5				
6	-0.02138	0.01314	-0.01816	
7	0.082935	0.207215	0.021678	
8	0.073059	-0.26864	-0.00679	
9				

Figure A.8
Matrix inverse in Excel

- *MINVERSE* is an operation that can only be performed on square matrices and gives the inverse. An error message is presented if the matrix has no inverse or is not square. The syntax is =*MINVERSE(A)* as illustrated in Figure A.8.

It is not necessary to give a matrix a name, so the expression =*TRANSPOSE(C11: G17)* is acceptable, and it is entirely possible to mix terminology, for example, =*MMULT (X,B6: D9)* will work, provided that the relevant dimensions are correct.

	D2	▾		= {=MMULT(MINVERSE(MMULT(TRANSPOSE(Y),Y)),TRANSPOSE(Y))}					
	A	B	C	D	E	F	G	H	I
1	Y								
2	2.9	4.1		0.01249	-0.02003	-0.00498	0.01249	-0.02003	
3	1.8	5.0		0.035677	0.063161	0.074532	0.035677	0.063161	
4	3.6	6.9							
5	7.3	0.7							
6	1.1	2.5							
7									

Figure A.9
Pseudo-inverse of a matrix

It is a very useful facility to be able to combine matrix operations. This means that more complex expressions can be performed. A common example is to calculate the pseudoinverse $(Y'.Y)^{-1}.Y'$ which, in Excel, is

$$=MMULT(MINVERSE(MMULT(TRANSPOSE(Y),Y)),TRANSPOSE(Y)).$$

Figure A.9 illustrates this, where Y is a 5×2 matrix, and its pseudoinverse a 2×5 matrix. Of course, each intermediate step of the calculation could be displayed separately if required. In addition, via macros (as described below) we can automate this to save keying long equations each time. However, for learning the basis of chemometrics, it is useful, in the first instance, to present the equation in full.

It is possible to add and subtract matrices using $+$ and $-$, but remember to ensure that the two (or more) matrices have the same dimensions as has the destination. It is also possible to mix matrix and scalar operations, so that the syntax $=2*MMULT(X,Y)$ is acceptable. Furthermore, it is possible to add (or subtract) matrices consisting of a constant number, for example $=Y+2$ would add 2 to each element of Y. Other conventions for mixing matrix and scalar variables and operations can be determined by practice, although in most cases the result is what we would logically expect.

There are a number of other matrix functions in Excel, for example to calculate determinant and trace of a matrix, to use these select the 'Insert' and 'Function' menus or use the Help system.

A.4.2.3 Arithmetic Functions of Scalars

There are numerous arithmetic functions that can be performed on single numbers. Useful examples are $SQRT$ (square root), LOG (logarithm to the base 10), LN (natural logarithm), EXP (exponential) and ABS (absolute value), for example $=SQRT(A1+2*B1)$. A few functions have no number to operate on, such as $ROW()$ which is the row number of a cell, $COLUMN()$ the column number of a cell and $PI()$ the number π. Trigonometric functions operate on angles in radians, so be sure to convert if your original numbers are in degrees or cycles, for example $=COS(PI())$ gives a value of -1.

A.4.2.4 Arithmetic Functions of Ranges and Matrices

It is often useful to calculate the function of a range, for example $=SUM(A1:C9)$ is the sum of the 27 numbers within the range. It is possible to use matrix notation so that $=AVERAGE(X)$ is the average of all the numbers within the matrix X.

Note that since the answer is a single number, these functions are not terminated by ⟨SHIFT⟩⟨CTL⟩⟨ENTER⟩. Useful functions include *SUM, AVERAGE, SUMSQ* (sum of squares) and *MEDIAN*.

Some functions require more than one range. The *CORREL* function is useful for the chemometrician, and is used to compute the correlation coefficient between two arrays, syntax =*CORREL(A,B)*, and is illustrated in Figure A.10. Another couple of useful functions involve linear regression. The functions =*INTERCEPT(Y,X)* and =*SLOPE(Y,X)* provide the parameters b_0 and b_1 in the equation $y = b_0 + b_1x$ as illustrated in Figure A.11.

Standard deviations, variances and covariances are useful common functions. It is important to recognise that there are both population and sample functions, so that *STDEV* is the sample standard deviation and *STDEVP* the equivalent population standard deviation. Note that for standardising matrices it is a normal convention to use the population standard deviation. Similar comments apply to *VAR* and *VARP*.

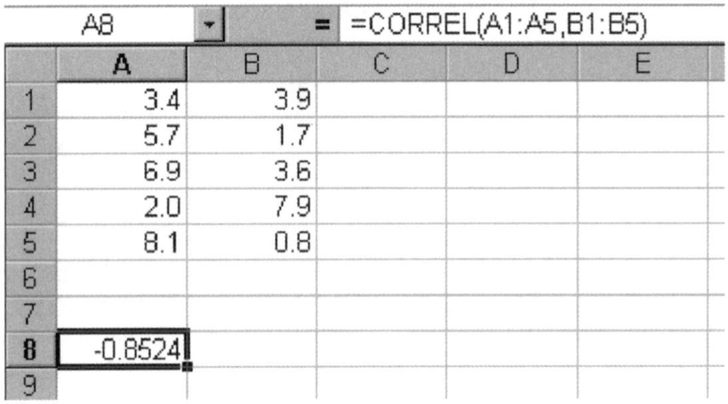

A8	▾	=	=CORREL(A1:A5,B1:B5)		
	A	B	C	D	E
1	3.4	3.9			
2	5.7	1.7			
3	6.9	3.6			
4	2.0	7.9			
5	8.1	0.8			
6					
7					
8	-0.8524				
9					

Figure A.10
Correlation between two ranges

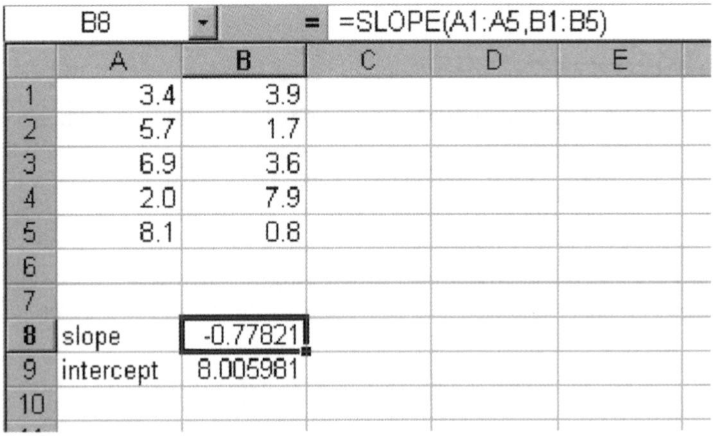

B8	▾	=	=SLOPE(A1:A5,B1:B5)		
	A	**B**	C	D	E
1	3.4	3.9			
2	5.7	1.7			
3	6.9	3.6			
4	2.0	7.9			
5	8.1	0.8			
6					
7					
8	slope	-0.77821			
9	intercept	8.005981			
10					

Figure A.11
Finding the slope and intercept when fitting a linear model to two ranges

Note that, rather eccentrically, only the population covariance is available although the function is named *COVAR* (without the *P*).

A.4.2.5 Statistical Functions

There are a surprisingly large number of common statistics available in Excel. Conventionally many such functions are presented in tabular format as in this book, for completeness, but most information can easily be obtained from Excel.

The inverse normal distribution is useful, and allows a determination of the number of standard deviations from the mean to give a defined probability; for example, $=NORMINV(0.9,0,1)$ is the value within which 90 % (0.9) of the readings will fall if the mean is 0 and standard deviation 1, and equals 1.282, which can be verified using Table A.1. The function *NORMDIST* returns the probability of lying within a particular value; for example, $=NORMDIST(1.5,0,1,TRUE)$ is the probability of a value which is less than 1.5, for a mean of 0 and standard deviation of 1, using the cumulative normal distribution ($=TRUE$), and equals 0.993 19 (see Table A.1). Similar functions *TDIST, TINV, FDIST* and *FINV* can be employed if required, eliminating the need for tables of the *F* statistic or *t* statistic, although most conventional texts and courses still employ these tables.

A.4.2.6 Logical Functions

There are several useful logical functions in Excel. *IF* is a common feature. Figure A.12 represents the function $=IF(A1<B1,A1,B1)$ and places the lower of the values of columns A and B in column D. Note that this has been copied down the column and also that there are no $ signs in the arguments in this case. *COUNTIF* can be used to determine how many times an expression is valid within a region of a worksheet. This is useful, for example, to determine how many values of a matrix are above a threshold.

A.4.2.7 Nesting and Combining Functions and Equations

It is possible to nest and combine functions and equations. The expression $=\$C6+IF(A\$7>1,10,IF(B\$3*\$C\$2>5,15,0))\hat{\ }2-2*SUM\ SQ(X+Y)$ is entirely legitimate, although it is important to ensure that each part of the expression results in an equivalent type of information (in this case the result of using the *IF* function is a number that is squared). Note that spreadsheets are not restricted to numerical information; they

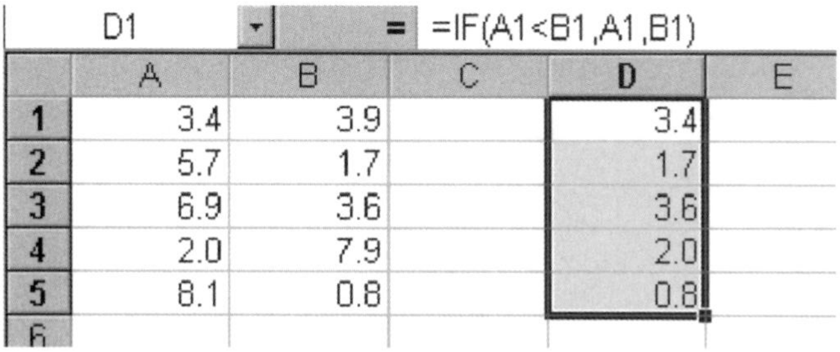

	A	B	C	D	E
1	3.4	3.9		3.4	
2	5.7	1.7		1.7	
3	6.9	3.6		3.6	
4	2.0	7.9		2.0	
5	8.1	0.8		0.8	

D1 = =IF(A1<B1,A1,B1)

Figure A.12
Use of *IF* in Excel

may, for example, also contain names (characters) or logical variables or dates. In this section we have concentrated primarily on numerical functions, as these are the most useful for the chemometrician, but it is important to recognise that nonsensical results would be obtained, for example, if one tries to add a character to a numerical expression to a date.

A.4.3   Add-ins

A very important feature of Excel consists of Add-ins. In this section we will describe only those Add-ins that are part of the standard Excel package. It is possible to write one's own Add-ins, or download a number of useful Add-ins from the Web. This book is associated with some Add-ins specifically for chemometrics as will be described in Section A.4.6.2.

If properly installed, there should be a 'Data Analysis' item in the 'Tools' menu. If this does not appear you should select the 'Add-ins' option and the tick the 'Analysis Toolpak'. Normally this is sufficient, but sometimes the original Office disk is required. One difficulty is that some institutes use Excel over a network. The problem with this is that it is not always possible to install these facilities on an individual computer, and this must be performed by the Network administrator.

Once the menu item has been selected, the dialog box shown in Figure A.13 should appear. There are several useful facilities, but probably the most important for the purpose of chemometrics is the 'Regression' feature. The default notation in Excel differs from that in this book. A multiple linear model is formed between a single response y and any number of x variables. Figure A.14 illustrates the result of performing regression on two x variables to give the best fit model $y \approx b_0 + b_1 x_1 + b_2 x_2$. There are a number of statistics produced. Note that in the dialog box one selects 'constant is zero' if one does not want to have a b_0 term; this is equivalent to forcing the intercept to be equal to 0. The answer, in the case illustrated, is $y \approx 0.0872 - 0.0971 x_1 + 0.2176 x_2$; see cells *H17–H19*. Note that this answer could also have been performed using matrix multiplication with the pseudoinverse, after first adding a column of 1s to the X matrix, as described in Section A.1.2.5 and elsewhere. In addition, squared or interaction terms can easily be introduced to the x values, simply by producing additional columns and including these in the regression calculation.

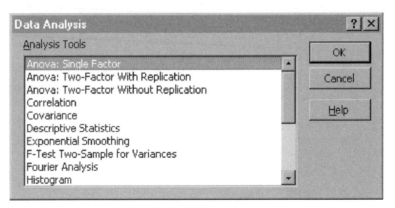

Figure A.13
Data Analysis Add-in dialog box

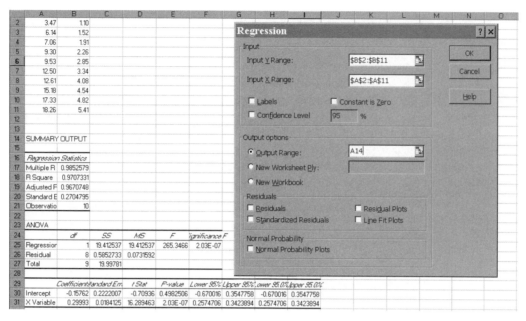

Figure A.14
Linear regression using the Excel Data Analysis Add-in

A second facility that is sometimes useful is the random number generator function. There are several possible distributions, but the most usual is the normal distribution. It is necessary to specify a mean and standard deviation. If one wants to be able to return to the distribution later, also specify a seed, which must be an integer number. Figure A.15 illustrates the generation of 10 random numbers coming from a distribution of mean 0 and standard deviation 2.5 placed in cells $A1$–$A10$ (note that the standard deviation is of the parent population and will not be exactly the same for a sample). This facility is very helpful in simulations and can be employed to study the effect of noise on a dataset.

The 'Correlation' facility that allows one to determine the correlation coefficients between either rows or columns of a matrix is also useful in chemometrics, for example, as the first step in cluster analysis. Note that for individual objects it is better to use the *CORREL* function, but for a group of objects (or variables) the Data Analysis Add-in is easier.

A.4.4 Visual Basic for Applications

Excel comes with its own programming language, VBA (or Visual Basic for Applications). This can be used to produce 'macros', which are programs that can be run in Excel.

A.4.4.1 Running Macros

There are several ways of running macros. The simplest is via the 'Tools' menu item. Select 'Macros' menu item and then the 'Macros' option (both have the same name).

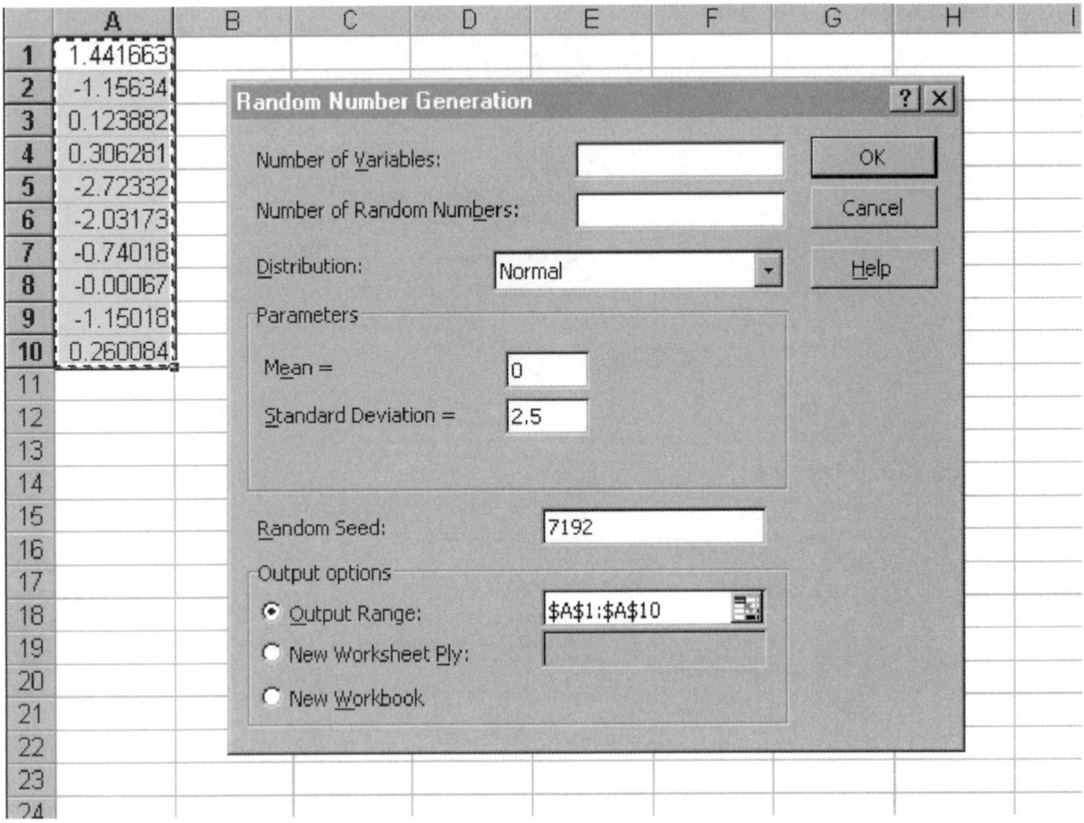

Figure A.15
Generating random numbers in Excel

A dialog box should appear; see Figure A.16. This lists all macros available associated with all open XLS workbooks. Note that there may be macros associated with XLA files (see below) that are not presented in this dialog box. However, if you are developing macros yourself rather than using existing Add-ins, it is via this route that you will first be able to run home-made or modified programs, and readers are referred to more advanced texts if they wish to produce more sophisticated packages. This text is restricted to guidance in first steps, which should be sufficient for all the data analysis in this book. To run a macro from the menu, select the option and then either double click it, or select the right-hand-side 'Run' option.

It is possible to display the code of a macro, either by selecting the right-hand-side edit option in the dialog box above, or else by selecting the 'Visual Basic Editor' option of the 'Macros' menu item of the 'Tools' menu. A screen similar to that in Figure A.17 will appear. There are various ways of arranging the windows in the VB Editor screen, and the first time you use this feature the windows may not be organised in exactly as presented in the figure; if no code is displayed, you can find this using the 'Project Explorer' window. To run a procedure, either select the 'Run' menu, or press the ▶ symbol. For experienced programmers, there are a number of other ways of running programs that allow debugging.

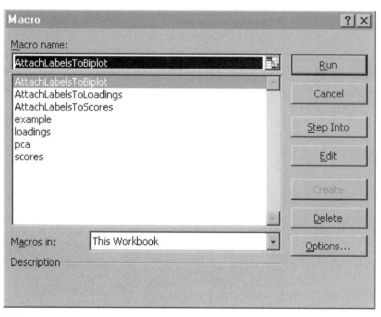

Figure A.16
Macros dialog box

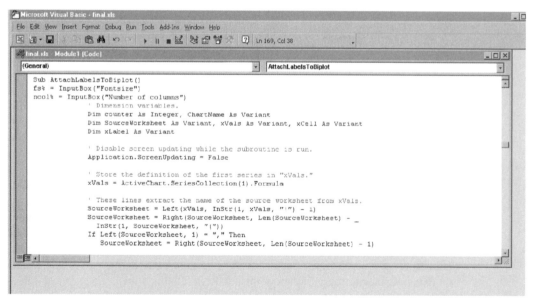

Figure A.17
VBA editor

Macros can be run using control keys, for example, 'CTL' 'f' could be the command for a macro called 'loadings'. To do this for a new macro, select 'Options' in the 'Macro' dialog box and you will be presented with a screen as depicted in Figure A.18. After this, it is only necessary to press the CTL and f keys simultaneously to run this

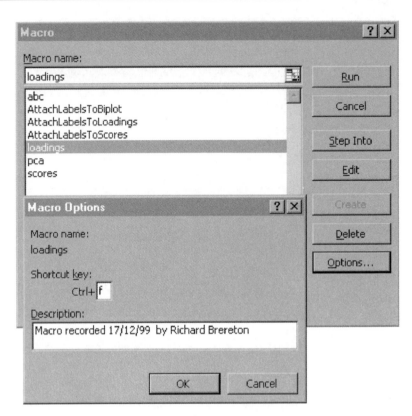

Figure A.18
Associating a Control key with a macro

macro. However, the disadvantage is that standard Excel functions are overwritten, and this facility is a left-over from previous Excel versions.

A better approach is to run the macro as a menu item or button. There are a number of ways of doing this, the simplest being via the 'Customize' dialog box. In the 'View' menu, select 'Toolbars' and then 'Customize'. Once this dialog box is obtained, select the 'Commands' and 'Macros' options as shown in Figure A.19. Select either 'Custom Button' or 'Custom Menu Item' and then drag to where you wish it. Right clicking whilst the 'Customize' menu is open allows the properties of the button or menu item to be changed; see Figure A.20. The most useful are the facilities to 'Assign Macro' or to change the name (for a menu item). Some practice is necessary to enable the user to place macros within menus or as buttons or both, and the best way to learn is to experiment with the various options.

A.4.4.2   Creating and Editing Macros

There are a number of ways of creating macros. The most straightforward is via the Visual Basic Editor, which can be accessed via the 'Macro' option of the 'Tools' menu. Depending on your set-up and the history of the spreadsheet, you may then enter a blank screen. If this is so, insert a 'Module'. This contains a series of user defined procedures (organised into Subroutines and Functions) that can be run from Excel.

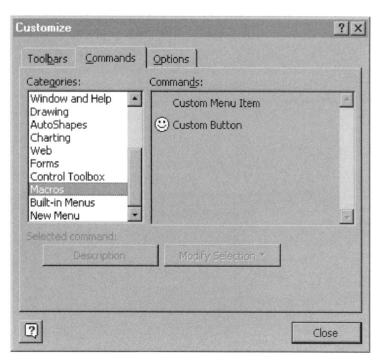

Figure A.19
Associating a macro with a menu item or a button

It is possible to have several modules, which helps organisation and housekeeping during programming, but the newcomer will probably wish to keep this aspect simple at first. Using the 'View' menu and 'Project Explorer', it is possible to obtain a window with all the open modules and workbooks, allowing reorganisation and renaming of modules; this is illustrated in Figure A.21.

Creating a new macro that is called directly from Excel is simple. In the code window, on a new line, type Sub followed by the macro name and terminated by a carriage return. VBA will automatically create a new macro as illustrated in Figure A.22 for a subroutine Calc. All subroutines without arguments may be called by Excel. Subroutines with arguments or functions are not able to be directly called by Excel. However, most programs of any sophistication are structured, so it is possible to create several procedures with arguments, and use a small routine from Excel to call these. To develop the program, simply type statements in between the Sub and End Sub statements.

There are a number of ways of modifying existing procedures. The most straightforward is in the Visual Basic Editor as discussed above, but it is also possible to do this using the 'Macro' option of the 'Tools' menu. Instead of running a macro, choose to 'Edit' the code. There are numerous tricks for debugging macros, which we will not cover in this book. In addition, the code can be hidden or compiled. This is useful when distributing software, to prevents users copying or modifying code. However, it is best to start simply and it is very easy to get going and then, once one has developed a level of expertise, to investigate more sophisticated ways of packaging the software.

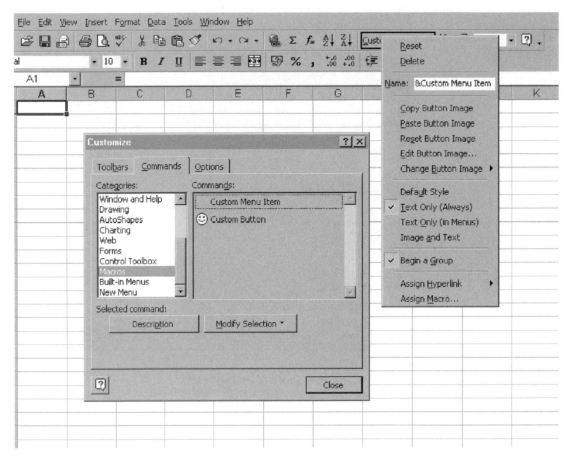

Figure A.20
Changing the properties of a menu item

Another way of creating macros is by the 'Record New Macro' facility, which can be entered via the 'Macro' option of the 'Tools' menu. This allows the user to produce a procedure which exactly corresponds to a series of operations in Excel. Every facility in Excel, whether changing the colour of a cell, multiplying matrices, changing the shape of a symbol in a graph or performing regression, has an equivalent statement in VBA. Hence this facility is very useful. However, often the code is rather cumbersome and the mechanisms for performing the operations are not very flexible. For example, although all the matrix operations in Excel, such as *MMULT*, can, indeed, be translated into VBA, the corresponding commands are tricky. The following is the result of

- multiplying the transpose of cells *B34* to *B65* (a vector of length 32) by
- cells *C34* to *C65*
- and placing the result in cell *E34*.

```
Selection.FormulaArray = _
    "=MMULT(TRANSPOSE(RC[-3]:R[31]C[-3]),RC[-2]:R[31]C[-2])"
```

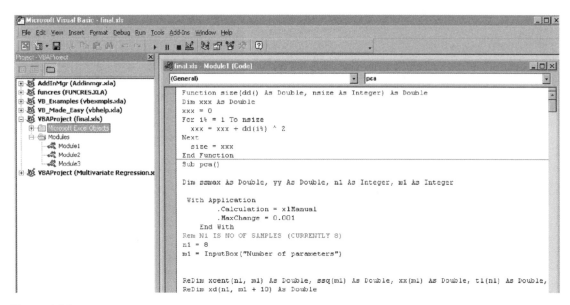

Figure A.21
Use of Project Explorer

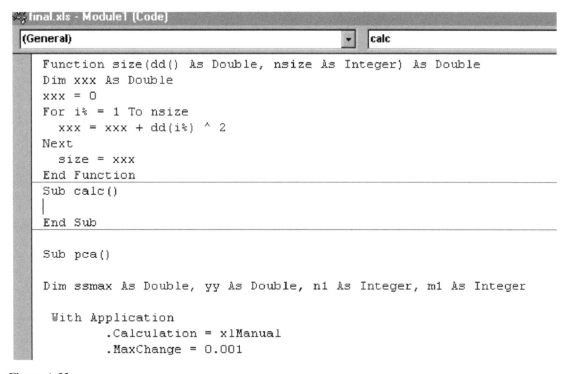

Figure A.22
Creating a new macro 'Calc'

	A	B	C	D	E
32					
33					
34		0.08523	-0.28668		4.25538854
35		1.646501	0.347964		
36		0.531818	0.663227		
37		-0.5641	-0.65751		
38		0.432639	-0.4496		
39		0.286027	-0.61291		
40		0.756513	-0.04022		
41		-0.49639	-0.42779		
42		0.633762	-0.28532		
43		-0.2727	1.388191		
44		0.657776	1.599835		
45		-1.61138	-0.7583		
46		0.793839	-0.44677		
47		0.297485	0.2496		
48		0.646408	0.301657		
49		-1.54278	-1.4E-15		
50		1.448003	-0.45049		
51		-0.13279	1.252869		
52		0.329738	1.452815		
53		-1.24223	-0.69166		
54		0.653458	-0.35539		
55		1.400607	0.233123		
56		1.284146	0.802224		
57		-1.54685	-0.36152		
58		0.709153	-0.2279		
59		-0.23936	1.03123		
60		-0.49387	1.299113		
61		0.030774	-0.75703		
62		0.757029	-0.38148		
63		-0.21051	1.243771		
64		0.379931	1.289573		
65		-1.15841	-0.40789		
66					

Figure A.23
Result of multiplying the transpose of vector *B34*: *B65* by *C34*: *C65* to give a scalar in cell *E34*

The result is illustrated in Figure A.23. The problem is that the expression is a character expression, and it is not easy to substitute the numbers, for example −3, by a variable, e.g. j. There are ways round this, but they are awkward.

Perhaps the most useful feature of the 'Record Macro' facility is to obtain portions of code, and then edit these into a full program. It is a very convenient way of learning some of the statements in VBA, particularly graphics and changing the appearance of the spreadsheet, and then incorporating these into a macro.

A.4.4.3   VBA Language

It is not necessary to have an in-depth knowledge of VBA programming to understand this text, so we will describe only a small number of features that allow a user to understand simple programs. It is assumed that the reader has some general experience of programming, and this section mainly provides tips on how to get started in VBA.

A module is organised into a number of procedures, either subroutines of functions. A subroutine without arguments is declared as `Sub new()` and is recognised as a macro that can be called directly from Excel by any of the methods described in Section A.4.4.1. A subroutine with arguments is recognised only within VBA; an example is

```
Sub new(x() as integer, d as double)
```

where `x()` is an array of integers, and `d` is a double precision variable. A typical function is as follows:

```
Function size(dd() As Double, nsize As Integer) As Double
```

In this case the function returns a double precision number. Subroutines may be called using the `Call` statement and function names are simply used in a program; for example,

```
ss0 = size(t1(), n1)
```

At first it is easiest to save all subroutines and functions in a single module, but more experienced programmers can experiment with fairly sophisticated methods for organising and declaring procedures and variables.

All common variable types are supported. It is best to declare variables, using either the `Dim` or `Redim` statement, which can occur anywhere in a procedure. Arrays must be declared. There are some leftovers from older versions of the language whereby a variable that ends in % is automatically an integer, for example i%, and a variable ending in a $ character, for example n$. However, it is not necessary for these variable types to end in these characters providing they are specifically declared, e.g.

```
Dim count as integer
```

is a legitimate statement. Because Visual Basic has developed over the years from a fairly unstructured language to a powerful programming environment, facilities have to be compatible with earlier versions.

Comments can be made in two ways, either using a `Rem` statement, in which case the entire line is considered a comment, or by using a quotation mark, in which case all text to the right of the quotation mark also constitutes a comment.

Loops are started with a `For` statement, e.g.

```
For x = -2 to 8 step 0.5
```

The default stepsize is +1, and under these circumstances it is possible to omit the `step` statement. Loops are concluded by a `Next` statement. `If` statements must be on one line ending with `Then`, and a typical syntax is as follows:

```
If (x > 10 And y <> -2) Then
```

All `If` blocks must conclude with `End if`, but it is possible to have an `Else` statement if desired. Naturally, several `If` statements can be nested.

A very important facility is to be able to pass information to and from the spreadsheet. For many purposes the `Cells` statement is adequate. Each cell is addressed with its row first and column second, so that the cell number (3,2) corresponds to *B3* in alphanumeric format. It is then easy to pass data back and forth. The statement `x = Cells(5,1)` places the value of *A5* into *x*. Equivalently, `Cells(i%,k%) = scores(i%,k%)` places the relevant value of `scores` into the cell in the *i*th row and *k*th column. Note that any type of information including character and logical information can be passed to and from the spreadsheet, so `Cells(7,3) = "Result"` places the word Result in *C7*, whereas `Cells(7,3) = Result` places the numerical value of a variable called `Result` (if it exists) in this cell.

For C programmers, it is possible to write programs in C, compile them into dynamic link libraries and use these from Excel, normally via VBA. This has advantages for numerically intensive calculations such as PCA on large datasets which are slow in VBA but which can be optimised in C. A good strategy is to employ Excel as the front end, then VBA to communicate with the user and also for control of dialog boxes, and finally a C DLL for the intensive numeric calculations.

Finally, many chemometricians use matrices. Matlab (see Section A.5) is better than Excel for developing sophisticated matrix based algorithms, but for simple applications the matrix operations of Excel can also be translated into VBA. As discussed in Section A.4.4.2, this is somewhat awkward, but for the specialist programmer there is a simple trick, which is to break down the matrix expression into a character string. Strings can be concatenated using the `&` sign. Therefore, the expression

```
n$ = "R[" & "-3]"
```

would give a character new variable *n$*, the characters representing `R[-3]`. The `CStr` function converts a number to its character representation. Hence

```
n$ = "R[" & CStr(-3) & "]"
```

would do the same trick. This allows the possibility of introducing variables into the matrix expression, so the following code is acceptable:

```
i% = -3
n1$ = CStr(i%)
n2$ = CStr(i% + 1)
m$ ="=MMULT(TRANSPOSE(RC[" & n1$ & "]:R[31]C[" & n1$ &_
"]),RC[" & n2$ & "]:R[31]C[" & n2$ & "])"
Selection.FormulaArray = m$
```

Note that the _ concatenates lines into a single statement. By generalising, it is easy to incorporate flexible matrix operations into VBA programs, and modifications of this principle could be employed to produce a matrix library.

There are a huge number of tricks that the experienced VBA programmer can use, but these are best learnt with practice.

A.4.5   Charts

All graphs in this text have been produced in Excel. The graphics facilities are fairly good except for 3D representations. This section will briefly outline some of the main features of the Chart Tool useful for applications in this text.

Graphs can be produced either by selecting the 'Insert' menu and the 'Chart' option, or by using the Chart Wizard symbol ▥ . Most graphs in this book are produced using an *xy* plot or scatterplot, allowing the value of one parameter (e.g. the score of PC2) to be plotted against another (e.g. the score of PC1).

It is often desirable to use different symbols for groups of parameters or classes of compounds. This can be done by superimposing several graphs, each being represented by a separate series. This is illustrated in Figure A.24 in which cells *AR2* to *AS9* represent Series 1 and *AR10* to *AS17* represent Series 2, each set of measurements having a different symbol. When opening the Chart Wizard, use the 'Series' rather than 'Data Range' option to achieve this.

The default graphics options are not necessarily the most appropriate, and are designed primarily for display on a screen. For printing out, or pasting into Word, it is best to remove the gridlines and the legend, and also to change the background to white. These can be done either by clicking on the appropriate parts of the graph when it has been completed or using the 'Chart Options' dialog box of the Chart Wizard. The final graph can be displayed either as an object on the current worksheet or, better, as in Figure A.25, on a separate sheet. Most aspects of the chart can be changed, such as symbol sizes and types, colours and axes, by clicking on the appropriate part of the chart and then following the relevant menus. It is possible to join lines up using the 'Format Data Series' dialog box, or by selecting an appropriate display option of the scatterplot. Using the Drawing toolbox, arrows and labelling can be added to charts.

One difficulty involves attaching a label to each point in a chart, such as the name of an object or a variable. With this text we produce a downloadable macro that can be edited to permit this facility as described in Section A.4.6.1.

A.4.6   Downloadable Macros

A.4.6.1   VBA Code

Most of the exercises in this book can be performed using simple spreadsheet functions, and this is a valuable exercise for the learner. However, it is not possible to perform PCA calculations in Excel without using VBA or an Add-in. In addition, a facility for attaching labels to points in a graph is useful.

A VBA macro to perform PCA is provided. The reader or instructor will need to edit the subroutine for appropriate applications. The code as supplied performs the following. The user is asked

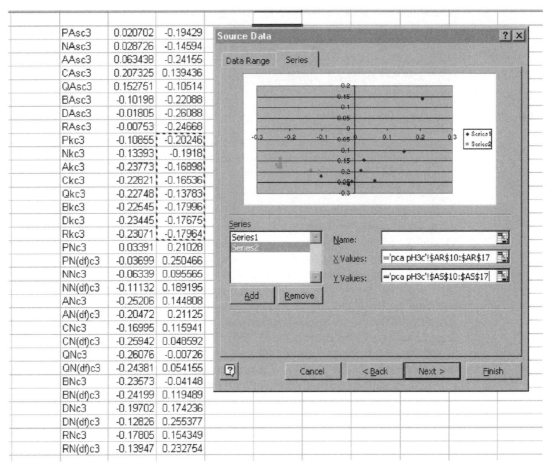

PAsc3	0.020702	-0.19429
NAsc3	0.028726	-0.14594
AAsc3	0.063438	-0.24155
CAsc3	0.207325	0.139436
QAsc3	0.152751	-0.10514
BAsc3	-0.10198	-0.22088
DAsc3	-0.01805	-0.26088
RAsc3	-0.00753	-0.24668
Pkc3	-0.10855	-0.20246
Nkc3	-0.13393	-0.1918
Akc3	-0.23773	-0.16898
Ckc3	-0.22821	-0.16536
Qkc3	-0.22748	-0.13783
Bkc3	-0.22545	-0.17996
Dkc3	-0.23445	-0.17675
Rkc3	-0.23071	-0.17964
PNc3	0.03391	0.21028
PN(df)c3	-0.03699	0.250466
NNc3	-0.06339	0.095565
NN(df)c3	-0.11132	0.189195
ANc3	-0.25206	0.144808
AN(df)c3	-0.20472	0.21125
CNc3	-0.16995	0.115941
CN(df)c3	-0.25942	0.048592
QNc3	-0.26076	-0.00726
QN(df)c3	-0.24381	0.054155
BNc3	-0.23573	-0.04148
BN(df)c3	-0.24199	0.119489
DNc3	-0.19702	0.174236
DN(df)c3	-0.12826	0.255377
RNc3	-0.17805	0.154349
RN(df)c3	-0.13947	0.232754

Figure A.24
Chart facility in Excel

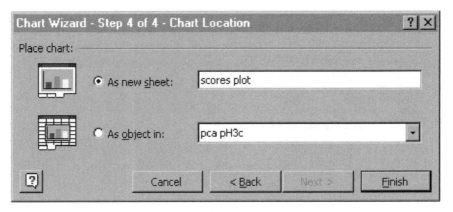

Figure A.25
Placing the chart in a new sheet

- how many samples;
- how many variables;
- how many PCs are to be calculated;
- whether the columns are to be centred or not.

The data must occupy a region of the spreadsheet starting at cell *B2* so that the first row and column can be used for names of the objects or variables if necessary, or left blank. Of course, this limitation can easily be changed by instructors or readers wishing to modify the macro. A function `size` is also required. Each row must represent one sample, and each column a variable.

NIPALS, as described in Section A.2.1, is used to extract the components sequentially. The scores of each PC are printed out to the right of the data, one column per PC, and the loadings below the data as rows. The eigenvalues are calculated in the program and may be printed out if required, by adding statements, or simply calculated as the sum of squares of the scores of each PC.

The `AddChartLabels` subroutine allows labels to be added to points in a graph. First, produce a scatterplot in Excel using the Chart Wizard. Make sure that the column to the left of the first ('*x*') variable contains the names or labels you wish to attach to the chart; see Figure A.26(a). Then simply run the macro, and each point should be labelled as in Figure A.26(b). If you want to change the font size or colour, edit the program as appropriate. If some of the labels overlap after the macro has been run, for example if there are close points in a graph, you can select each label manually and move it around the graph, or even delete selective labels. This small segment of code can, of course, be incorporated into a more elaborate graphics package, but in this text we include a sample which is sufficient for many purposes.

A.4.6.2 Multivariate Analysis Add-in

Accompanying the text is also an Add-in to perform several methods for multivariate analysis. The reader is urged first to understand the methods by using matrix commands in Excel or editing macros, and several examples in this book guide the reader to setting up these methods from scratch. However, after doing this once, it is probably unnecessary to repeat the full calculations from scratch and convenient to have available Add-ins in Excel. Although the performance has been tested on computers of a variety of configurations, we recommend a minimum of Office 2000 and Windows 98, together with at least 64 Mbyte memory. There may be problems with lower configurations. The VBA software was written by Tom Thurston based on an original implementation from Les Erskine.

You need to download the Add-ins from the publisher's Website. You will obtain a setup file, click this to obtain the screen in Figure A.27, and follow the instructions. If in doubt, contact whoever is responsible for maintaining computing facilities within your department or office. Note that sometimes there can be problems if you use networks, for example using NT, and under such circumstances you may be required to consult the systems manager. The setup program will install the Add-in (an XLA file) and support files on your computer.

Next, start Excel, and select 'Tools' then 'Add-ins …' from the menu. The 'Add-Ins' dialog box should now appear. If 'Multivariate Analysis' is not listed, then click

	A	B	C
1	**Element**	**Melting P. (K)**	**Boiling P. (K)**
2	Li	453.69	1615
3	Na	371	1156
4	K	336.5	1032
5	Rb	312.5	961
6	Cs	301.6	944
7	Be	1550	3243
8	Mg	924	1380
9	Ca	1120	1760
10	Sr	1042	1657
11	F	53.5	85
12	Cl	172.1	238.5
13	Br	265.9	331.9
14	I	386.6	457.4
15	He	0.9	4.2
16	Ne	24.5	27.2
17	Ar	83.7	87.4
18	Kr	116.5	120.8
19	Xe	161.2	166
20	Zn	692.6	1180
21	Co	1765	3170
22	Cu	1356	2868
23	Fe	1808	3300
24	Mn	1517	2370
25	Ni	1726	3005
26	Bi	544.4	1837
27	Pb	600.61	2022
28	Tl	577	1746
29			

(a)

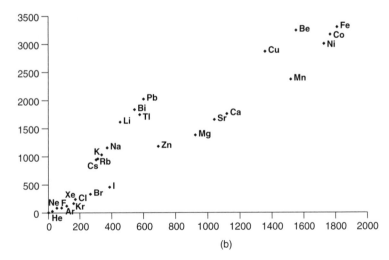

(b)

Figure A.26
Labelling a graph in Excel

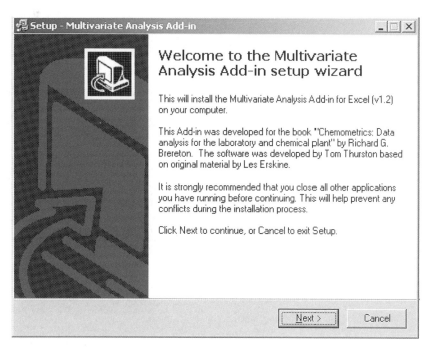

Figure A.27
Setup screen for the Excel chemometrics Add-in

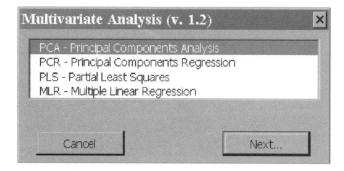

Figure A.28
Multivariate Analysis dialog box

the 'Browse ...' button and find an XLA file which would have been created during the setup procedure. Make sure the box next to 'Multivariate Analysis' is ticked, then click 'OK' to close the Add-Ins dialog box. You should now have a new item in the 'Tools' menu in Excel, titled 'Multivariate Analysis', which gives access to the various chemometric methods. Once selected, the dialog box in Figure A.28 should appear, allowing four options which will be described below.

The PCA dialog box is illustrated in Figure A.29. It is first necessary to select the data range and the number of PCs to be calculated. By default the objects are along

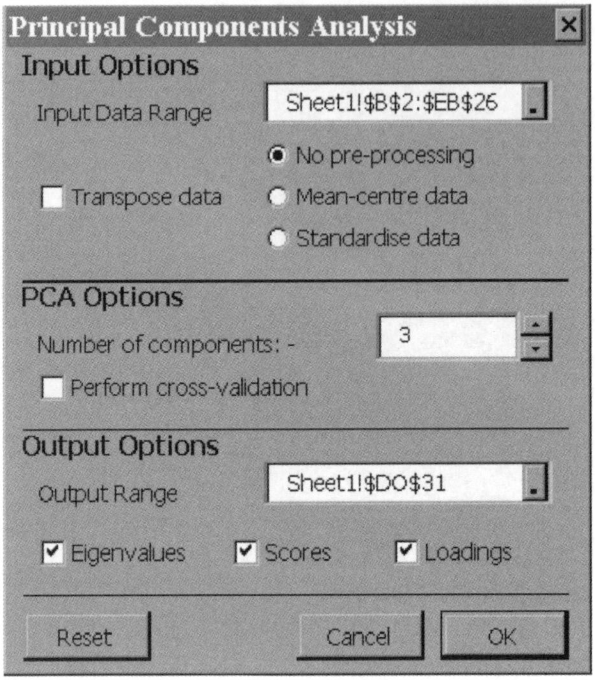

Figure A.29
PCA dialog box

the rows and the variables down the columns, but it is possible to transpose the data matrix, in PCA and all other options, which is useful, for example, when handling large spectra with more the 256 wavelengths, remembering that there is a limit to the number of columns in an Excel worksheet; use the 'Transpose data' option in this case. The data may be mean centred in the direction of variables, or standardised (this uses the population rather than sample standard deviation, as recommended in this book).

It is possible to cross-validate the PCs using a 'leave one sample out at a time' approach (see Chapter 4, Section 4.3.3.2); this option is useful if one wants guidance as to how many PCs are relevant to the model. You are also asked to select the number of PCs required.

An output range must be chosen; it is only necessary to select the top left-hand cell of this range, but be careful that it does not overwrite existing data. For normal PCA, choose which of eigenvalues, scores and loadings you wish to display. If you select eigenvalues you will also be given the total sum of squares of the preprocessed (rather than raw) data together with the percentage variance of each eigenvalue.

Although cross-validation is always performed on the preprocessed data, the RSS and PRESS values are always calculated on the 'x' block in the original units, as discussed in Chapter 4, Section 4.3.3.2. The reason for this relates to rather complex problems that occur when standardising a column after one sample has been removed. There are, of course, many other possible approaches. When performing cross-validation, the only output available involves error analysis.

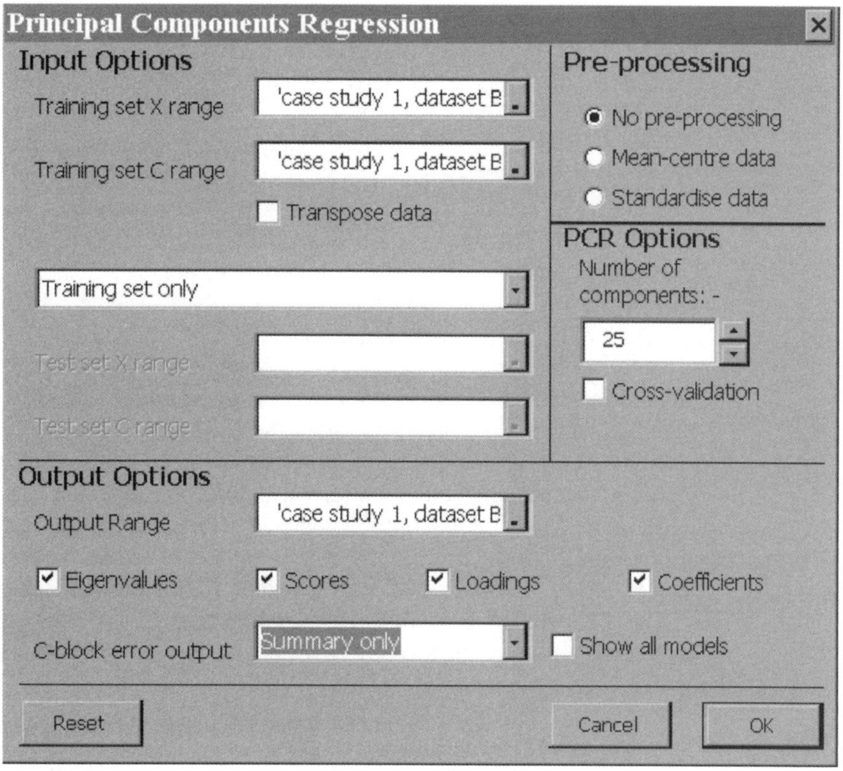

Figure A.30
PCR dialog box

The PCR dialog box, illustrated in Figure A.30, is considerably more complicated. It is always necessary to have a training set consisting of an 'x' block and a 'c' block. The latter may consist of more than one column. For PCR, unlike PLS, all columns are treated independently so there is no analogy to PLS2. You can choose three options. (1) 'Training set only' is primarily for building and validating models. It only uses the training set. You need only to specify an 'x' and 'c' block training set. The number of objects in both sets must be identical. (2) 'Prediction of unknowns' is used to predict concentrations from an unknown series of samples. It is necessary to have an 'x' and 'c' block training set as well as an 'x' block for the unknowns. A model will be built from the training set and applied to the unknowns. There can be any number of unknowns, but the number of variables in the two 'x' blocks must be identical. (3) 'Use test set (predict and compare)' allows two sets of blocks where concentrations are known, a training set and a test set. The number of objects in the training and test set will normally differ, but the number of variables in both datasets must be identical.

There are three methods for data scaling, as in PCA, but the relevant column means and standard deviations are always obtained from the training set. If there is a test set, then the training set parameters will be used to scale the test set, so that the test set is unlikely to be mean centred or standardised. Similar scaling is performed on both the 'c' and 'x' block simultaneously. If you want to apply other forms of scaling (such as summing rows to a constant total), this can be performed manually in Excel and PCA can be performed without further preprocessing. Cross-validation is performed only on the 'c' block or concentration predictions; if you choose cross-validation you can only do this on the training set. If you want to perform cross-validation on the 'x' block, use the PCA facility.

There are a number of types of output. Eigenvalues, scores and loadings (of the training set) are the same as in PCA, whereas the coefficients relate the PCs to the concentration estimates, and correspond to the matrix R as described in Chapter 5, Section 5.4.1. This information is available if requested in all cases except for cross-validation. Separate statistics can be obtained for the 'c' block predictions. There are three levels of output. 'Summary only' involves just the errors including the training set error (adjusted by the number of degrees of freedom to give $^1E_{cal}$ as described in Chapter 5, Section 5.6.1), the cross-validated error (divided by the number of objects in the training set) and the test set error, as appropriate to the relevant calculation. If the 'Predictions' option is selected, then the predicted concentrations are also displayed, and 'Predictions and Residuals' provides the residuals as well (if appropriate for the training and test sets), although these can also be calculated manually. If the 'Show all models' option is selected, then predicted 'c' values and the relevant errors (according to the information required) for 1, 2, 3, up to the chosen number of PCs is displayed. If this option is not selected, only information for the full model is provided.

The PLS dialog box, illustrated in Figure A.31, is very similar to PCR, except that there is an option to perform PLS1 ('one c variable at a time') (see Chapter 5,

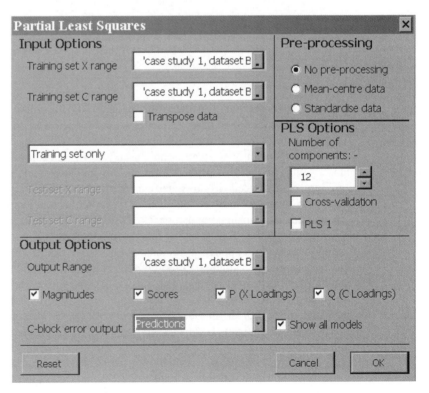

Figure A.31
PLS dialog box

Section 5.5.1) as well as PLS2 (Chapter 5, Section 5.5.2). However, even when performing PLS1, it is possible to use several variables in the 'c' block; each variable, however, is modelled independently. Instead of coefficients (in PCR) we have 'C-loadings' (Q) for PLS, as well as the 'X-loadings' (P), although there is only one scores matrix. Strictly, there are no eigenvalues for PLS, but the size of each component is given by the magnitude, which is the product of the sum of squares of the scores and X-loadings for each PLS component. Note that the loadings in the method described in this text are neither normalised nor orthogonal. If one selects PLS2, there will be a single set of 'Scores' and 'X-loadings' matrices, however many columns there are in the 'c' block, but 'C-loadings' will be in the form of a matrix. If PLS1 is selected and there is more than one column in the 'c' block, separate 'Scores' and 'X-loadings' matrices are generated for each compound, as well as an associated 'C-loadings' vector, so the output can become extensive unless one is careful to select the appropriate options.

For both PCR and PLS it is, of course, possible to transpose data, and this can be useful if there are a large number of wavelengths, but *both* the 'x' block and the 'c' block must be transposed. These facilities are not restricted to predicting concentrations in spectra of mixtures and can be used for any purpose, such as QSAR or sensory statistics.

The MLR dialog box, illustrated in Figure A.32, is somewhat simpler than the others and is mainly used if two out of X, C and S are known. The type of unknown matrix is chosen and then regions of the spreadsheet of the correct size must be selected. For small datasets MLR can be performed using standard matrix operations in Excel as described in Section A.4.2.2, but for larger matrices it is necessary to have a separate

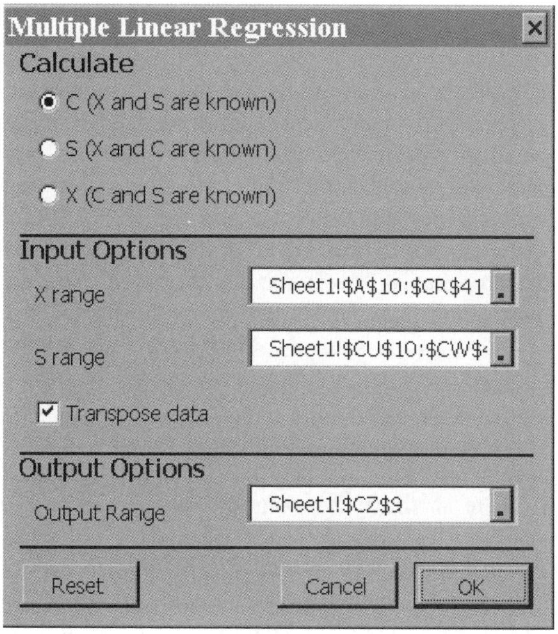

Figure A.32
MLR dialog box

tool, as there is a limitation in the Excel functions. This facility also performs regression using the pseudoinverse, and is mainly provided for completion. Note that it is not necessary to restrict the data to spectra or concentrations.

This Add-in provides a basic functionality for many of the multivariate methods described in Chapters 4–6 and can be used when solving the problems.

A.5   Matlab for Chemometrics

Many chemometricians use Matlab. In order to appreciate the popularity of this approach, it is important to understand the vintage of chemometrics. The first applications of quantum chemistry, another type of computational chemistry, were developed in the 1960s and 1970s when Fortran was the main numerical programming environment. Hence large libraries of routines were established over this period and to this day most quantum chemists still program in Fortran. Were the discipline of quantum chemistry to start over again, probably Fortran would not be the main programming environment of choice, but tens of thousands (or more) man-years would need to be invested to rewrite entire historical databases of programs. If we were developing an operating system that would be used by tens or hundreds of millions of people, that investment might be worthwhile, but the scientific market is much smaller, so once the environment is established, new researchers tend to stick to it as they can then exchange code and access libraries.

Although some early chemometrics code was developed in Fortran (the Arthur package of Kowalski) and Basic (Wold's early version of SIMCA) and commercial packages are mainly written in C, most public domain code first became available in the 1980s when Matlab was an up and coming new environment. An advantage of Matlab is that it is very much oriented towards matrix operations and most chemometrics algorithms are best expressed in this way. It can be awkward to write matrix based programs in C, Basic or Fortran unless one has access to or develops specialised libraries. Matlab was originally a technical programming environment mainly for engineers and physical scientists, but over the years the user base has expanded strongly and Matlab has kept pace with new technology including extensive graphics, interfaces to Excel, numerous toolboxes for specialist use and the ability to compile software. In this section we will concentrate primarily on the basics required for chemometrics and also to solve the problems in this book; for the more experienced user there are numerous other outstanding texts on Matlab, including the extensive documentation produced by the developer of the software, MathWorks, which maintains an excellent Website. In this book you will be introduced to a number of main features, to help you solve the problems, but as you gain experience you will undoubtedly develop your own personal favourite approaches. Matlab can be used at many levels, and it is now possible to develop sophisticated packages with good graphics in this environment.

There are many versions of Matlab and of Windows and for the more elaborate interfaces between the two packages it is necessary to refer to technical manuals. We will illustrate this section with Matlab version 5.3, although many readers may have access to more up-to-date editions. All are forward compatible. There is a good on-line help facility in Matlab: type `help` followed by the command, or follow the appropriate menu item. However, it is useful first to have a grasp of the basics which will be described below.

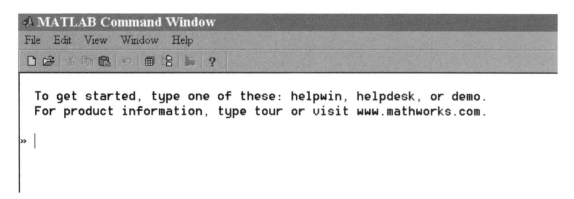

Figure A.33
Matlab window

A.5.1   Getting Started

To start Matlab it is easiest to simply click the icon which should be available if properly installed, and a blank screen as in Figure A.33 will appear. Each Matlab command is typed on a separate line, terminated by the ⟨ENTER⟩ key. If the ⟨ENTER⟩ key is preceded by a semi-colon (;) there is no feedback from Matlab (unless you have made an error) and on the next line you type the next command and so on. Otherwise, you are given a response, for example the result of multiplying matrices together, which can be useful but if the information contains several lines of numbers which fill up a screen and which may not be very interesting, it is best to suppress this.

Matlab is case sensitive (unlike VBA), so the variable x is different to X. Commands are all lower case.

A.5.2   Directories

By default Matlab will be installed in the directory C:\matlabr$xx$\ on your PC, where xx relates to the edition of the package. You can choose to install elsewhere but at first it is best to stick to the standard directories, which we will assume below. You need some knowledge of DOS directory structure to use the directory commands within Matlab. According to particular combinations of versions of Windows and Matlab there is some flexibility, but keeping to the commands below is safe for the first time user.

A directory `c:\matlabrxx\work` will be created where the results of your session will be stored unless you specify differently. There are several commands to manage directories. The `cd` command changes directory so that `cd c:\` changes the directory to `c:\`. If the new directory does not exist you must first create it with the `mkdir` command. It is best not to include a space in the name of the directory. The following code creates a directory called `results` on the c drive and makes this the current Matlab directory:

```
cd c:\
mkdir results
cd results
```

To return to the default directory simply key

```
cd c:\matlabrxx\work
```

where *xx* relates to the edition number, if this is where the program is stored.

If you get in a muddle, you can check the working directory by typing `pwd` and find out its contents using `dir`.

You can also use the pull down `Set Path` item on the `File` menu, but be careful about the compatibility between Matlab and various versions of Windows; it is safest to employ the line commands.

A.5.3 File Types

There are several types of files that one may wish to create and use, but there are three main kinds that are useful for the beginner.

A.5.3.1 mat Files

These files store the 'workspace' or variables created during a session. All matrices, vectors and scalars with unique names are saved. Many chemometricians exchange data in this format. The command `save` places all this information into a file called `matlab.mat` in the current working directory. Alternatively, you can use the `Save Workspace` item on the `File` menu. Normally you wish to save the information as a named file, in which case you enter the filename after the `save` command. The following code saves the results of a session as a file called `mydata` in the directory `c:\results`, the first line being dependent on the current working directory and requires you to have created this first:

```
cd c:\results
save mydata
```

If you want a space in the filename, enclose in single quotes, e.g. `'Tuesday file'`.

In order to access these data in Matlab from an existing file, simply use the `load` command, remembering what directory you are in, or else the `Load Workspace` item on the `File` menu. This can be done several times to bring in different variables, but if two or more variables have the same names, the most recent overwrite the old ones.

A.5.3.2 m Files

Often it is useful to create programs which can be run again. This is done via m files. The same rules about directories apply as discussed above.

These files are simple text files and may be created in a variety of ways. One way is via the normal Notepad text editor. Simply type in a series of statements, and store them as a file with extension `.m`. There are five ways in which this file can be run from the Matlab command window.

1. Open the m file, cut and paste the text, place into the Matlab command window and press the return key. The program should run provided that there are no errors.

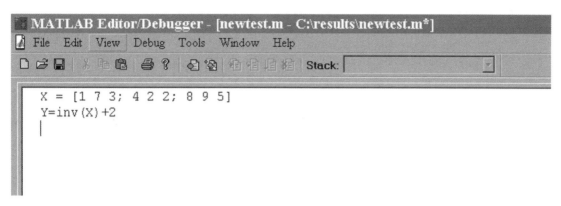

```
X = [1 7 3; 4 2 2; 8 9 5]
Y=inv(X)+2
```

Figure A.34
The m file window

2. Start Matlab, type `open` together with the name of the .m file, e.g. `open myprog.m`, and a separate window should open; see Figure A.34. In the `Tools` menu select `Run` and then return to the main Matlab screen, where the results of the program should be displayed.
3. Similarly to method 2, you can use the `Open` option in the `File` menu.
4. Provided that you are in the correct directory, you can simply type the name of the m file, and it will run; for example, if a file called `prog.m` exists in the current directory, just type `prog` (followed by the ⟨ENTER⟩ key).
5. Finally, the program can be run via the `Run Script` facility in the `File` menu.

Another way of creating an. m file is in the Matlab command window. In the `File` menu, select `New` and then `M-file`. You should be presented with a new Matlab Editor/Debugger window (see Figure A.34) where you can type commands. When you have finished, save the file, best done using the `Save As` command. Then you can either return to the Matlab window (an icon should be displayed) and run the file as in option 4 above, or run it in the editing window as in option 2 above, but the results will be displayed in the Matlab window. Note that if you make changes you must save this file to run it. If there are mistakes in the program an error message will be displayed in the Matlab window and you need to edit the commands until the program is correct.

A.5.3.3 Diary Files

These files keep a record of a session. The simplest approach is not to use diary files but just to copy and paste the text of a Matlab session, but diary files can be useful because one can selectively save just certain commands. In order to start a diary file type `diary` (a default file called `diary` will be created in the current directory) or `diary filename` where `filename` is the name of the file. This automatically opens a file into which all subsequent commands used in a session, together with their results, are stored. To stop recording simply type `diary off` and to start again (in the same file) type `diary on`.

The file can be viewed as a text file, in the Text Editor. Note that you must close the diary session before the information is saved.

A.5.4   Matrices

The key to Matlab is matrices. Understanding how Matlab copes with matrices is essential for the user of this environment.

A.5.4.1   Scalars, Vectors and Matrices

It is possible to handle scalars, vectors and matrices in Matlab. The package automatically determines the nature of a variable when first introduced. A scalar is simply a number, so

```
P = 2
```

sets up a scalar *P* equal to 2. Notice that there is a distinction between upper and lower case, and it is entirely possible that another scalar *p* (lower case) co-exists:

```
p = 7
```

It is not necessary to restrict a name to a single letter, but all matrix names must start with an alphabetic rather than numeric character and not contain spaces.

For one- and two-dimensional arrays, it is important to enclose the information within square brackets. A row vector can be defined by

```
Y = [2 8 7]
```

resulting in a 1×3 row vector. A column vector is treated rather differently as a matrix of three rows and one column. If a matrix or vector is typed on a single line, each new row starts a semicolon, so a 3×1 column vector may be defined by

```
Z = [1; 4; 7]
```

Alternatively, it is possible to place each row on a separate line, so

```
Z = [1
       4
       7]
```

has the same effect. Another trick is to enter as a row vector and then take the transpose (see Section A.5.4.3).

Matrices can be similarly defined, e.g.

```
W = [2 7 8; 0 1 6]
```

and

```
W = [2 7 8
      0 1 6]
```

are alternative ways, in the Matlab window, of setting up a 2×3 matrix.

One can specifically obtain the value of any element of a matrix, for example $W(2,1)$ gives the element on the second row and first column of W which equals

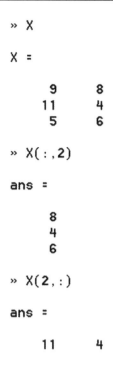

Figure A.35
Obtaining vectors from matrices

0 in this case. For vectors, only one dimension is needed, so Z(2) equals 4 and Y(3) equals 7.

It is also possible to extract single rows or columns from a matrix, by using a colon operator. The second row of matrix *X* is denoted by X(2,:). This is exemplified in Figure A.35. It is possible to define any rectangular region of a matrix, using the colon operator. For example, if *S* is a matrix having dimensions 12 × 8 we may want a sub-matrix between rows 7 to 9 and columns 5 to 12, and it is simply necessary to define S(7:9, 5:12).

If you want to find out how many matrices are in memory, use the function who, which lists all current matrices available to the program, or whos, which contains details about their size. This is sometimes useful if you have had a long Matlab session or have imported a number of datasets; see Figure A.36.

There is a special notation for the identity matrix. The command eye(3) sets up a 3 × 3 identity matrix, the number enclosed in the brackets referring to the dimensions.

A.5.4.2 Basic Arithmetic Matrix Operations

The basic matrix operations +, − and * correspond to the normal matrix addition, subtraction and multiplication (using the dot product); for scalars these are also defined in the usual way. For the first two operations the two matrices should generally have the same dimensions, and for multiplication the number of columns of the first matrix should equal the number of rows of the second matrix. It is possible to place the results in a target or else simply display them on the screen as a default variable called ans.

```
» whos
  Name          Size              Bytes  Class

  X             3x2                  48  double array
  Y             2x3                  48  double array
  Z             3x1                  24  double array
  Z1            1x1                   8  double array
  a             3x1                  24  double array
  ans           1x1                   8  double array

Grand total is 20 elements using 160 bytes
```

Figure A.36
Use of whos command to determine how many matrices are available

```
» X = [9 8; 11 4; 5 6];
» Y = [2 7 1; 5 3 8];
» Z=[2 3 5; 6 0 1; 11 4 8];
» X*Y+Z

ans =

    60    90    78
    48    89    44
    51    57    61
```

Figure A.37
Simple matrix operations in Matlab

Figure A.37 exemplifies setting up three matrices, a 3×2 matrix X, a 2×3 matrix Y and a 3×3 matrix Z, and calculating $X.Y + Z$.

There are a number of elaborations based on these basic operations, but the first time user is recommended to keep things simple. However, it is worth noting that it is possible to add scalars to matrices. An example involves adding the number 2 to each element of W as defined above: either type W + 2 or first define a scalar, e.g. P = 2, and then add this using the command W + P. Similarly, one can multiply, subtract or divide all elements of a matrix by a scalar. Note that it is not possible to add a vector to a matrix even if the vector has one dimension identical with that of the matrix.

A.5.4.3 Matrix Functions

A significant advantage of Matlab is that there are several further very useful matrix operations. Most are in the form of functions; the arguments are enclosed in brackets. Three that are important in chemometrics are as follows:

- transpose is denoted by ', e.g. W' is the transpose of W;
- inverse is a function inv so that inv(Q) is the inverse of a square matrix Q;

```
» W

W =

        2     7     8
        0     1     6

» pinv(W)

ans =

     0.0567   -0.0844
     0.1564   -0.2055
    -0.0261    0.2009
```

Figure A.38
Obtaining the pseudoinverse in Matlab

- the pseudoinverse can simply be obtained by the function pinv, without any further commands; see Figure A.38.

For a comprehensive list of facilities, see the manuals that come with Matlab, or the help files; however, a few that are useful to the reader of this book are as follows. The size function gives the dimensions of a matrix, so size(W) will return a 2 × 1 vector with elements, in our example, of 2 and 3. It is possible to create a new vector, for example, s = size(W); in such a situation s(1) will equal 2, or the number of rows. The element W(s(1), s(2)) represents the last element in the matrix W. In addition, it is possible to use the functions size(W,1) and size(W,2) which provide the number of rows and columns directly. These functions are very useful when writing simple programs as discussed below.

The mean function can be used in various ways. By default this function produces the mean of each column in a matrix, so that mean(W) results in a 1 × 3 row vector containing the means. It is possible to specify which dimension one wishes to take the mean over, the default being the first one. The overall mean of an entire matrix is obtained using the mean function twice, i.e. mean(mean(W)). Note that the mean of a vector is always a single number whether the vector is a column or row vector. This function is illustrated in Figure A.39. Similar syntax applies to functions such as min, max and std, but note that the last function calculates the sample rather than population standard deviation and if employed for scaling in chemometrics, you must convert back to the sample standard deviation, in the current case by typing std(W)/sqrt(((s(1))/(s(1)-1))), where sqrt is a function that calculates the square root and s contains the number of rows in the matrix. Similar remarks apply to the var function, but it is not necessary use a square root in the calculation.

The norm function of a matrix is often useful and consists of the square root of the sum of squares, so in our example norm(W) equals 12.0419. This can be useful when scaling data, especially for vectors. Note that if Y is a row vector, then sqrt(Y*Y') is the same as norm(Y).

It is useful to combine some of these functions, for example min(s) would be the minimum dimension of matrix **W**. Enthusiasts can increase the number of variables

```
» mean(W)

ans =

        1       4       7

» mean (W,1)

ans =

        1       4       7

» mean (W,2)

ans =

      5.6667
      2.3333

» mean(mean(W))

ans =

      4
```

Figure A.39
Mean function in Matlab

within a function, an example being min([s 2 4]), which finds the minimum of all the numbers in vector s together with 2 and 4. This facility can be useful if it is desired to limit to number of principal components or eigenvalues displayed. If Spec is a spectral matrix of variable dimensions, and we know that we will never have more than 10 significant components, then min([size(Spec)] 10) will choose a number that is the minimum of the two dimensions of Spec or equals 10 if this value is larger.

Some functions operate on individual elements rather than rows or columns. For example, sqrt(W) results in a new matrix of dimensions identical with W containing the square root of all the elements. In most cases whether a function returns a matrix, vector or scalar is commonsense, but there are certain linguistic features, a few rather historical, so if in doubt test out the function first.

A.5.4.4 Preprocessing

Preprocessing is slightly awkward in Matlab. One way is to write a small program with loops as described in Section A.5.6. If you think in terms of vectors and matrices, however, it is fairly easy to come up with a simple approach. If W is our original 2×3 matrix and we want to mean centre the columns, we can easily obtain a 1×3 vector $\overline{w}$

```
» W

W =

         2     7     8
         0     1     6

» one = [1;1]

one =

         1
         1

» V=W-one×mean(W)

V =

         1     3     1
        -1    -3    -1
```

Figure A.40
Mean centring a matrix in Matlab

which corresponds to the means of each column, multiply this by a 2×1 vector $\mathbf{1}$ giving a 2×3 vector consisting of the means, and so our new mean centred matrix $\mathbf{V}$ can be calculated as $\mathbf{V} = \mathbf{W} - \mathbf{1}.\overline{\mathbf{w}}$ as illustrated in Figure A.40. There is a special function in Matlab called ones that also creates vectors or matrices that just consist of the number 1, there being several ways of using this, but an array ones (5,3) would create a matrix of dimensions 5×3 solely of 1s, so a 2×1 vector could be specified using the function ones(2,1) as an alternative to the approach illustrated in the figure.

The experienced user of Matlab can build on this to perform other common methods for preprocessing, such as standardisation.

A.5.4.5 Principal Components Analysis

PCA is simple in Matlab. The singular value decomposition (SVD) algorithm is employed, but this should normally give equivalent results to NIPALS except that all the PCs are calculated at once. One difference is that the scores and loadings are both normalised, so that for SVD

$$X = U.S.V$$

where, using the notation elsewhere in the text,

$$T = U.S$$

and

$$V = P$$

The matrix V is equivalent to T but the sum of squares of the elements of each row equals 1, and S is a matrix, whose dimensions equal the number of PCs whose diagonal elements equal the square root of the eigenvalues and the remaining elements equal 0. The command svd(X) will display the nonzero values of $\sqrt{g}$ or the square roots of the eigenvalues. To obtain the scores, loadings and eigenvalue matrices, use the command [U,S,V] = svd(X). Note that the dimensions of these three matrices differ slightly from those above in that S is not a square matrix, and U and V are square matrices with their respective dimensions equal to the number of rows and columns in the original data. If X is an $I \times J$ matrix then U will be a matrix of dimensions $I \times I$, S of dimensions $I \times J$ (the same as the original matrix) and V of dimensions $J \times J$. To obtain a scores matrix equivalent to that using the NIPALS algorithm, simply calculate T = U * S. The sum of squares each column of T will equal the corresponding eigenvalue (as defined in this text). Note that if $J > I$ columns $I + 1$ to J of matrix V will have no meaning, and equivalently if $I > J$ the last columns of matrix U will have no meaning. The Matlab SVD scores and loadings matrices are square matrices.

One problem about the default method for SVD is that the matrices can become rather large if there are many variables, as often happens in spectroscopy or chromatography. There are a number of ways of reducing the size of the matrices if we want to calculate only a few PCs, the simplest being to use the svds function; the second argument restricts the number of PCs. Thus svds(X,5) calculates the first five PCs, so if the original data matrix was of dimensions 25×100, U becomes 25×5, S becomes 5×5 (containing only five nonzero values down the diagonals) and V becomes 5×100.

A.5.5 Numerical Data

In chemometrics we want to perform operations on numerical data. There are many ways of getting information into Matlab generally straight into matrix format. Some of the simplest are as follows.

1. Type the numerical information in as described above.
2. If the information is available in a space delimited form with each row on a separate line, for example as a text file, copy the data, type a command such as

    ```
    X = [
    ```

    but do NOT terminate this by the enter key, then paste the data into the Matlab window and finally terminate with

    ```
    ]
    ```

    using a semicolon if you do not want to see the data displayed again (useful if the original dataset is large such as a series of spectra).
3. Information can be saved as mat files (see Section A.5.3.1) and these can be imported into Matlab. Many public domain chemometrics datasets are stored in this format.

In addition, there are a huge number of tools for translating from a variety of common formats, such as Excel, and the interested reader should refer to the relevant source manuals where appropriate.

A.5.6   Introduction to Programming and Structure

For the enthusiasts it is possible to write quite elaborate programs and develop very professional looking m files. The beginner is advised to have a basic idea of a few of the main features of Matlab as a programming environment.

First and foremost is the ability to make comments (statements that are not executed), by starting a line with the % sign. Anything after this is simply ignored by Matlab but helps make large m files comprehensible.

Loops commence with the `for` statement, which has a variety of different syntaxes, the simplest being `for i = begin : end` which increments the variable i from the number `begin` (which must be a scalar) to `end`. An increment (which can be negative and does not need to be an integer) can be specified using the syntax `for i = begin : inc : end`; notice how, unlike many programming languages, this is the middle value of the three variables. Loops finish with the `end` statement. As an example, the operation of mean centring (Section A.5.4.4) is written in the form of a loop; see Figure A.41. The interested reader should be able to interpret the commands using the information given above. Obviously for this small operation a loop is not strictly necessary, but for more elaborate programs it is important to be able to use loops, and there is a lot of flexibility about addressing matrices which make this facility very useful.

`If` and `while` facilities are also useful to the programmer.

```
» W = [2 7 8; 0 1 6]

W =

        2       7       8
        0       1       6

» Y=mean(W)

Y =

        1       4       7

» for i=1:size(W,2)
U(:,i)=W(:,i)-Y(i);
end
» U

U =

        1       3       1
       -1      -3      -1
```

Figure A.41
A simple loop used for mean centring

Many programmers like to organise their work into functions. In this introductory text we will not delve too far into this, but a library of m files that consist of different functions can be easily set up. In order to illustrate this, we demonstrate a simple function called `twoav` that takes a matrix, calculates the average of each column and produces a vector consisting of two times the column averages. The function is stored in an m file called `twoav` in the current working directory. This is illustrated in Figure A.42. Note that the m file must start with the `function` statement, and the name of the function should correspond to the name of the m file. The arguments (in this case a matrix which is called p within the function and can be called anything in

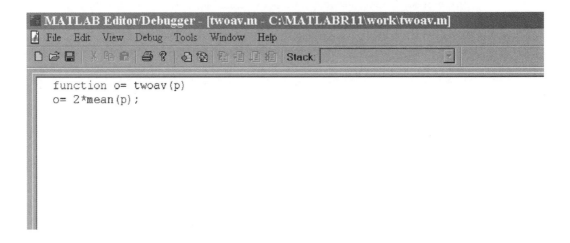

```
function o= twoav(p)
o= 2*mean(p);
```

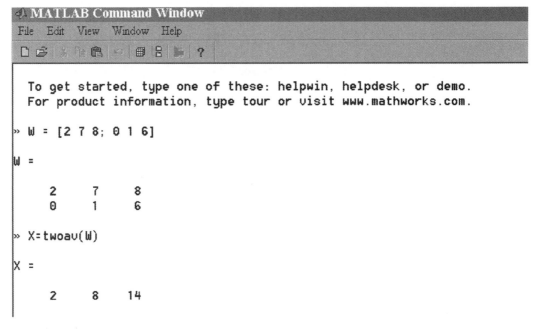

```
To get started, type one of these: helpwin, helpdesk, or demo.
For product information, type tour or visit www.mathworks.com.

» W = [2 7 8; 0 1 6]

W =

    2    7    8
    0    1    6

» X=twoav(W)

X =

    2    8    14
```

Figure A.42
A simple function and its result

the main program, $-$ W in this example) are place in brackets after the function name. The array o contains the result of the expression that is passed back.

A.5.7   Graphics

There are a large number of different types of graph available in Matlab. Below we discuss a few methods that can be used to produce diagrams of the type employed in this text. The enthusiast will soon discover further approaches. Matlab is a very powerful tool for data visualisation.

A.5.7.1   Creating Figures

There are several ways to create new graphs. The simplest is by a plotting command as discussed in the next sections. A new window consisting of a figure is created. Unless indicated otherwise, each time a graphics command is executed, the graph in the figure window is overwritten.

In order to organise the figures better, it is preferable to use the figure command. Each time this is typed in the Matlab command window, a new blank figure as illustrated in Figure A.43 is produced, so typing this three times in succession results in three blank figures, each of which is able to contain a graph. The figures are automatically numbered from 1 onwards. In order to return the second figure (number 2), simply type figure(2). All plotting commands apply to the currently open figure. If you wish to produce a graph in the most recently opened window, it is not necessary to specify a number. Therefore, if you were to type the command three times, unless specified otherwise the current graph will be displayed in Figure 3. The figures can be accessed either as small icons or through the Window menu item. It is possible to skip figure numbers, so the command figure(10) will create a figure number 10, even if no other figures have been created.

If you want to produce several small graphs on one figure, use the subplot command. This has the syntax subplot(n,m,i). It divides the figure into $n \times m$ small graphs and puts the current plot into the ith position, where the first row is numbered from 1 to m, the second from $m + 1$ to $2m$, and so on. Figure A.44 illustrates the case where the commands subplot(2,2,1) and subplot(2,2,3) have been used to divide the window into a 2×2 grid, capable of holding up to four graphs, and figures have been inserted into positions 1 (top left) and 3 (bottom left). Further figures can be inserted into the grid in the vacant positions, or the current figures can be replaced and overwritten.

New figures can also be created using the File menu, and the New option, but it is not so easy to control the names and so probably best to use the figure command.

Once the figure is complete you can copy it using the Copy Figure menu item and then place it in documents. In this section we will illustrate the figures by screen snapshots showing the grey background of the Matlab screen. Alternatively, the figures can be saved in Matlab format, using the menu item under the current directory, as a fig file, which can then be opened and edited in Matlab in the future.

A.5.7.2   Line Graphs

The simplest type of graph is a line graph. If Y is a vector then plot(Y) will simply produce a graph of each element against row number. Often we want to plot a row

Figure A.43
Blank figure window

or column of a matrix against element number, for example if each successive point corresponds to a point in time or a spectral wavelength. This is easy to do: the command plot(X(:,2)) plots the second column of *X*. Plotting a subset is also possible, for example plot(X(11:20,2)) produces a graph of rows 11–20, in practice allowing an expansion of the region of interest.

Once you have produced a line graph it is possible to change its appearance. This is easily done by first clicking the arrow tool in the graphics window, which allows editing of the properties, and then clicking on either the line to change the appearance of the data, or the axes. One useful facility is to make the lines thicker: the default line width of 0.6 is often thin when intended for publication (although it is a good size for displaying on a screen), and it is recommended to increase this to around 2. In addition, one sometimes wishes to mark the points, using the marker facility. The result is presented in Figure A.45. If you do not wish to join up the points with a line you can select a line style 'none'. The appearance of the axes can also be altered. There are various commands to change the nature of these plots, and you are recommended to use the Matlab help facility for further information.

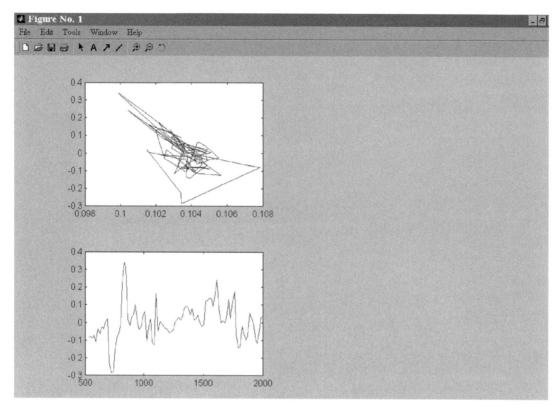

Figure A.44
Use of multiple plot facility

It is possible to superimpose several line graphs, for example if *X* is a matrix with five columns, then the command plot(X) will superimpose five graphs in one picture.

Note that you can further refine the appearance of the plot using the tools to create labels, extra lines and arrows.

A.5.7.3 *Two Variables Against Each Other*

The plot command can also be used to plot two variables against each other. It is common to plot columns of matrices against each other, for example when producing a PC plot of the scores of one PC against another. The command plot(X(:,2), X(:,3)) produces a graph of the third column of *X* against the second column. If you do not want to join the points up with a line you can either use the graphics editor as in Section A.5.7.2, or else the scatter command, which has a similar syntax but by default simply presents each point as a symbol. This is illustrated in Figure A.46.

A.5.7.4 *Labelling Points*

Points in a graph can be labelled using the text command. The basic syntax is text (A,B,name), where the A and B are arrays with the same number of elements, and it is recommended that name is an array of names or characters likewise with the

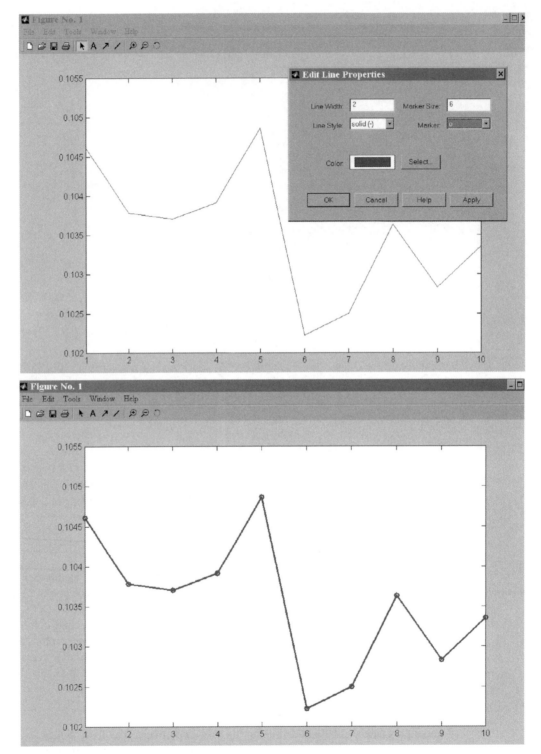

Figure A.45
Changing the properties of a graph in Matlab

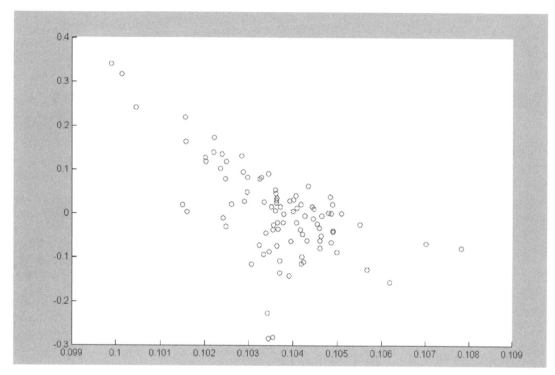

Figure A.46
Scatterplot in Matlab

identical number of elements. There are various ways of telling Matlab that a variable is a string (or character) rather than numeric variable. Any data surrounding by single quotes is treated as a string, so the array c = ['a'; 'b'; 'c'] will be treated by Matlab as a 3 × 1 character array. Figure A.47 illustrates the use of this method. Note that in order to prevent the labels from overlapping with the points in the graph, leaving one or two spaces before the actual text helps. It is possible to move the labels later in the graph editor if there is still some overlap.

Sometimes the labels are originally in a numerical format, for example they may consist of points in time or wavelengths. For Matlab to recognise this, the numbers can be converted to strings using the num2str function. An example is given in Figure A.48, where the first column of the matrix consists of the numbers 10, 15 and 20 which may represent times, the aim being to plot the second against the third column and use the first for labelling. Of course, any array can contain the labels.

A.5.7.5 *Three-dimensional Graphics*

Matlab can be very useful for the representation of data in three dimensions, in contrast to Excel where there are no straightforward 3D functions. In Chapter 6 we used 3D scores and loadings plots.

Consider a scores matrix of dimensions 36 × 3 (**T**) and a loadings matrix of dimensions 3 × 25 (**P**). The command plot3(T(:,1),T(:,2),T(:,3)) produces a graph of all three columns against one another; see Figure A.49. Often the default

```
» x

x =

     1     2     7
     5     8     9

» plot(x(1,:),x(2,:))
» c=['  a';'  b';'  c'];
» text(x(1,:),x(2,:),c)
```

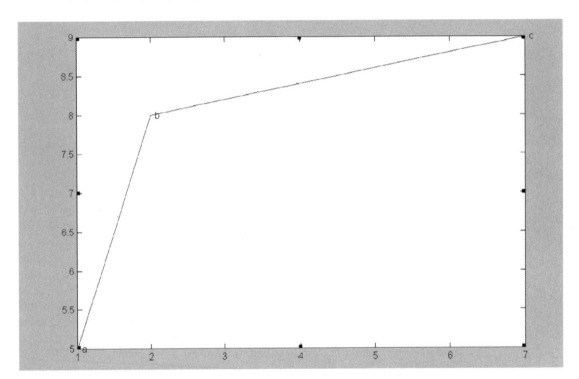

Figure A.47
Use of text command in Matlab

orientation is not the most informative for our purposes, and we may wish to change this. There are a huge number of commands in Matlab to do this, which is a big bonus for the enthusiast, but for the first time user the easiest is to select the right-hand rotation icon, and interactively change the view; see Figure 4.50. If that is the desired view, leave go of the icon.

Often we want to return to the view, and a way of keeping the same perspective is via the `view` command. Typing A = `view` will keep this information in a 4×4 matrix A. Enthusiasts will be able to interpret these in fundamental terms, but it is

```
> X= [10, 0.75 0.19; 15, 0.35, 0.62; 20, 0.81 1.20];
> name=num2str(X(:,1));
> plot(X(:,2),X(:,3))
> text(X(:,2),X(:,3),name)
```

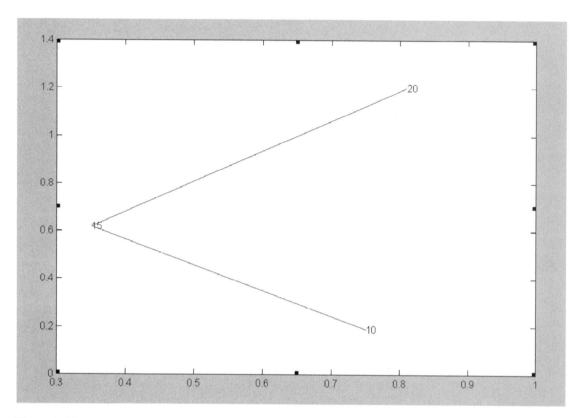

Figure A.48
Using numerical to character conversion for labelling of graphs

not necessary to understand this when first using 3D graphics in Matlab. However, in chemometrics we often wish to look simultaneously at 3D scores and loadings plots and it is important that both have identical orientations. The way to do this is to ensure that the loadings have the same orientation as the scores. The commands

```
figure(2)
plot3(P(:,1),P(:,2),P(:,3))
view(A)
```

should place a loadings plot with the same orientation in Figure 2. Sometimes this does not always work the first time; the reasons are rather complicated and depend on

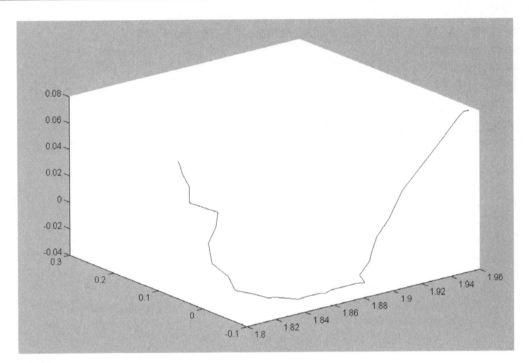

Figure A.49
A 3D scores plot

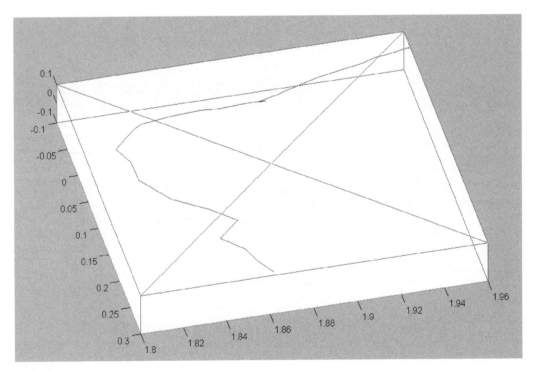

Figure A.50
Using the rotation icon

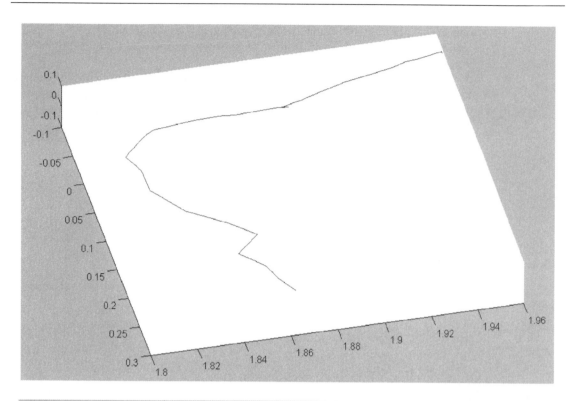

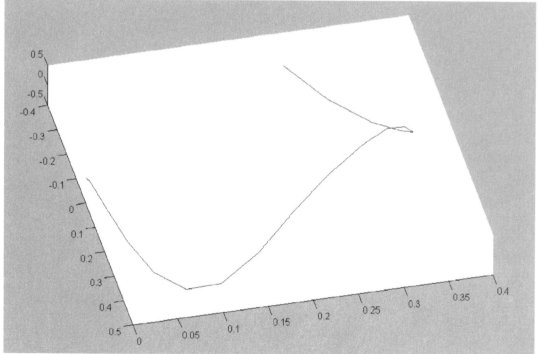

Figure A.51
Scores and loadings plots with identical orientations

the overall starting orientation, but it is usually easy to see when it has succeeded. If you are in a mess, start again from scratch. Scores and loadings plots with the same orientation are presented in Figure A.51.

The experienced user can improve these graphs just as the 2D graphs, for example by labelling axes or individual points, using symbols in addition to or as an alternative to joining using a line. The `scatter3` statement has similar properties to `plot3`.

Index

Note: Figures and tables are indicated by *italic page numbers*

Index compiled by Paul Nash